Aerodynamics for Engineers

Aerodynamics for Engineers

Edited by **Russell Mikel**

Willford Press
New York

Published by Willford Press,
118-35 Queens Blvd., Suite 400,
Forest Hills, NY 11375, USA
www.willfordpress.com

Aerodynamics for Engineers
Edited by Russell Mikel

International Standard Book Number: 978-1-68285-230-9 (Hardback)

Printed in China.

Contents

Preface

Aerodynamics has emerged as a significant field of study and research in the recent years. This book brings forth advanced details of the branches of aerodynamics. It also covers aerodynamic methods that describe forces and momentums in context of aircrafts. It aims to elucidate computational methods that have expanded and transformed the modern approaches in aerodynamics design. The text outlines various researches in aeronautical industry that present a practical approach to interested readers. It will also serve as a valuable reference to professionals in the field.

The researches compiled throughout the book are authentic and of high quality, combining several disciplines and from very diverse regions from around the world. Drawing on the contributions of many researchers from diverse countries, the book's objective is to provide the readers with the latest achievements in the area of research. This book will surely be a source of knowledge to all interested and researching the field.

In the end, I would like to express my deep sense of gratitude to all the authors for meeting the set deadlines in completing and submitting their research chapters. I would also like to thank the publisher for the support offered to us throughout the course of the book. Finally, I extend my sincere thanks to my family for being a constant source of inspiration and encouragement.

Editor

1

Simulations of the Atmospheric Boundary Layer in a Wind Tunnel with Short Test Section

Luciana Bassi Marinho Pires[1], Igor Braga de Paula[2], Gilberto Fisch[3], Ralf Gielow[2], Roberto da Mota Girardi[4]

ABSTRACT: This article presents a study of three different passive devices (spires, screens, and a carpet) separately and in various combinations, to simulate the atmospheric boundary layer (ABL) in a wind tunnel with a short test chamber (465x465x1200mm). The influence of distances between these devices on the formation of the ABL is established, and optimization of variation of thicknesses of the screens (thin, medium, and coarse) on pressure loss is explored. The results obtained in this work gave support for the analysis of the atmospheric flow and turbulence at Alcantara Launching Center (ALC) in order to launch Brazilian space vehicles under safe conditions. The results show that the "spires" and the thin screen are the devices that require the least area to form an ABL in a test chamber. The physical proximity of two devices (the spires and the medium screen) also influences the size of the ABL formed, which varies from 180 to 200 mm. The power law exponent ranged from 0.12 up to 0.14 after the insertion of a carpet.

KEYWORDS: Passive methods, Spires, Carpet, Screens, Wind power law.

INTRODUCTION

The first boundary layer (BL) wind tunnel was built in 1965 by Davenport (1967) at Western University (formerly University of Western Ontario), Ontario, Canada. The studies of Jensen and Franck (1963, 1965) and Franck (1963) in wind tunnels concluded that it would be possible to obtain an appropriate scale of the natural wind structure by covering a considerable length of the wind tunnel's floor with a material of suitable roughness. However, the disadvantage of this process is that it requires a length of about 25 m to form a BL with 60 to 120 cm height, which is possible only in tunnels with a long test chamber (Blessmann, 1973). Thus, improved techniques for reproduction of the main characteristics of natural winds, as well as the formation of the atmospheric boundary layer (ABL), are needed. These techniques will permit shorter test chambers, so that existing aeronautical tunnels could be used for atmospheric simulations of meteorological interest, with the advantage of flow control and improved data collection. The objective of the current article is to provide directions to fulfill the challenges of adaptation of aeronautic wind tunnels for meteorology purposes, cataloging the steps of modification and their expected results. This work was done at Departamento de Ciência e Tecnologia Aeroespacial (DCTA) wind tunnel to give support for the analysis of the atmospheric flow and turbulence at Alcantara Launching

1.Instituto Nacional de Pesquisas Espaciais – São José dos Campos/SP – Brazil 2.Pontifícia Universidade Católica do Rio de Janeiro – Rio de Janeiro/RJ – Brazil 3.Instituto de Aeronáutica e Espaço – São José dos Campos/SP – Brazil 4.Instituto Tecnológico de Aeronáutica – São José dos Campos/SP – Brazil

Author for correspondence: Luciana Bassi Marinho Pires | Instituto Nacional de Pesquisas Espaciais – INPE/CPTEC | Avenida dos Astronautas, 1.758 – Jardim da Granja | CEP 12.227-010 São José dos Campos/SP – Brazil | Email: lubassimp@gmail.com

Center (ALC). This is an ongoing project that aims to develop technical tools for Aerospace Meteorology associated with the launching of the Satellite Vehicle Launcher (VLS) within the Brazilian Space Program (Pires *et al.*, 2010).

There are several studies using the wind tunnel for simulating the characteristics and behavior of the atmosphere. Some examples are the works of Novak *et al.*, (2000) analyzing the turbulent structure of the atmosphere within and above canopies; Kwon *et al.*, (2003) simulating the atmospheric wind field over a complex topography in order to plan the Naro Space Center in South Korea; Mavroidis *et al.*, (2003), studying pollutant dispersion fields immersed in obstacles; Cao and Tamura (2006), simulating the turbulent BL flow over 2D steep, smooth, and rough hills in Tokyo. Kozmar (2009) reproduced the BL at different scales inside the wind tunnel to address the proper choice of simulation length scale and concluded that the length-scale factor does not influence the generated ABL models when using similarity criteria; and recently, Carpentieri *et al.*, (2012) worked to quantify the mean and turbulent flow in geometries of real street canyon intersections, focusing on the area surrounding the intersection between Marylebone Road and Gloucester Place in Central London, UK. Recently, Avelar *et al.*, (2012) did a detailed analysis of the atmospheric flow at ALC using a large wind tunnel. They have constructed the ABL profile using a combination of barrier and small blocks of wood. As all of them used wind tunnels for their experiments, new techniques and methods to develop the ABL are necessary in order to use different sizes of wind tunnel section tests.

The methods of simulating the ABL in wind tunnels are divided into passive and active types. The passive methods use barriers such as grilles, flat plates, triangular plates, writing pads, spires, carpets, etc. The active methods are those that use air jets to form a wall of fluid. The objective of this work is to present some of the passive devices (screens, spires, and carpet) used for the formation of the BL in wind tunnels. It will be done analyzing the optimizations of the formation of the ABL with the use of such devices as a function of both the distance among them and their characteristics (grids with larger or smaller spacing) based primarily on the mean velocity profile and turbulence intensity profiles. The main focus was on rural terrain. However, additional studies could investigate whether this approach works well when applied to other types of terrains (e.g., suburban, urban) with different geometries.

MATERIAL AND METHODS

This work was carried out in the commercially available open subsonic wind tunnel *Plinth & Partners LDD Wokingham Berkshire England* (Serial No. 44/5065 TE) at the Kwein Lien Feng Laboratory at Instituto Tecnológico de Aeronáutica (ITA)/DCTA, São José dos Campos, SP, Brazil (Fig. 1a). Its test section is square with cut corners (465x465 mm) with a length of 1200 mm. For this study, a channel apparatus with a width of 410 mm and length of 1200 mm, consisting of an uncovered wooden frame, with free edges and side walls parallel to each other and perpendicular to the floor of the wind tunnel, was used to extend the test section for the formation of the ABL. The walls of the tunnel and of the "channel" (Fig. 1b) do not coincide, to minimize the lateral BL of the tunnel. An automatic positioner device with an arm capable of moving in three directions perpendicular to each other and with an accuracy of tenths of a millimeter was used, coupled to a computer, with the data collection points inserted by automatic handling equipment or manually (Fig. 1b). The positioner consists of the following elements: (i) positioner, ("traversing") from Dantec Dynamics serial code 9057 h 0123, Denmark, (ii) controls for the positioner, Dantec type 57 b 100; and (iii) command manual positioner, serial code 9055X530. The same configuration was utilized by Roballo (2007). The atmospheric flow is simulated by 22 kW (30 hp) electric fans for the range of the windspeeds. The maximum windspeed was 33 m/s (120 km/h).

For initial adjustments, passive devices — screen and spires — were used and for fine tuning a wrinkled carpet was added. It is important to note that the insertion of screens within wind tunnels is usually done with a primary objective of reduction of turbulence of flow, because they have a tendency to uniform the flow and to break the big vortices into smaller ones that decay quickly (energy cascade process). However, in this study, the screens were used to accelerate the formation of the ABL in the test section.

Three types of screens with different meshes were used: (i) nylon square mesh with spacing of 2×2 mm^2 and 0.4 mm in diameter, called "thin screen," (ii) metal square mesh with spacing of 5.5×5.5 mm^2 and 1.0 mm diameter, called "medium screen"; and (iii) metal mesh type "beehive" with a spacing of 19.0×17.0 mm^2 and 0.5 mm in diameter, called "coarse screen" (Fig. 2). All these screens were available in local shops. The screens were assembled in a rigid frame made of steel stripes

(a) (b)

Figure 1. Wind tunnel of Prof. Kwein Lieng Feng Laboratory (ITA) (a) and automatic positioner and extension attached (b).

of 2 mm thick. Their tension was adjusted with small plates used to tighten the mesh to the frame. This was necessary in order to avoid the screen bending from aerodynamic loads during measurements. The mechanism created a small step of approximately 3 mm (approximately 1% of the BL thickness) at the basis of the BL. This height is considered small for the current experiments. For instance, the carpet used to form the ABL had nearly the same height as the step, but a much longer length. Thus, the influence of such a small step on the ABL was assumed to be negligible.

The spires consist of triangular plates arranged in the test chamber entrance which, combined with the surface roughness, are used to generate the windspeed profile of the BL. The dimensions of the spires depend on the desired BL and the dimensions of the tunnel. It is recommended that the distance between two spires should be half of the height of one spire, and their heights should be smaller than the height of the wind tunnel (Pires, 2009). It is assumed, according with Blessman (1973), that the BL real δ is around 280 m. The tunnel test chamber used has a height of 460 mm and the reduction of scale with respect to the actual size was 1:1000. Therefore it should produce a ABL as close as possible to 280 mm, which was also checked by numerical simulations

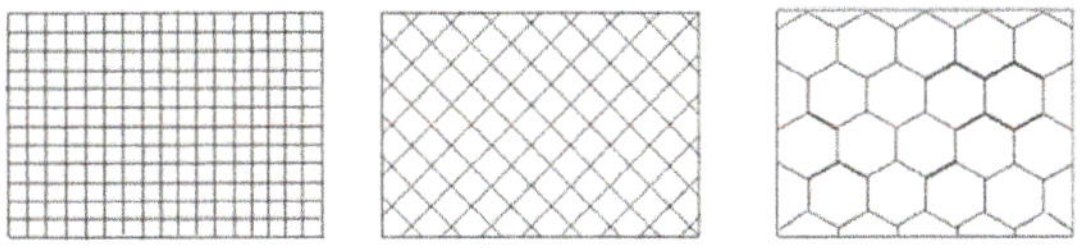

Figure 2. Screens used for the formation of atmospheric boundary layer.

(Pires, 2009). Details regarding the calculation of the quantity and characteristics of spires that need to be used for different wind tunnel dimensions can be found in Blessmann (1973) and Pires (2009). For this work, three spires, each with heights of 307.7 mm and base width of 32.6 mm, were used. The spires (Irwin, 1981) were produced using steel plates, with each spire consisting of two plates cut and folded together by spot welding (Fig. 3).

Hot-wire measurements were performed with a single-wire BL probe ("Dantec 55P15") of the constant-temperature hot-wire anemometer ("DANTEC Streamline bridge"). The calibration of the probe was performed using a DANTEC calibration unit. The equipment enables a precise adjustment of flow velocity at the exit of a nozzle. The dynamic pressure of the jet was measured with a standard Betz type manometer, and the respective velocities were correlated with

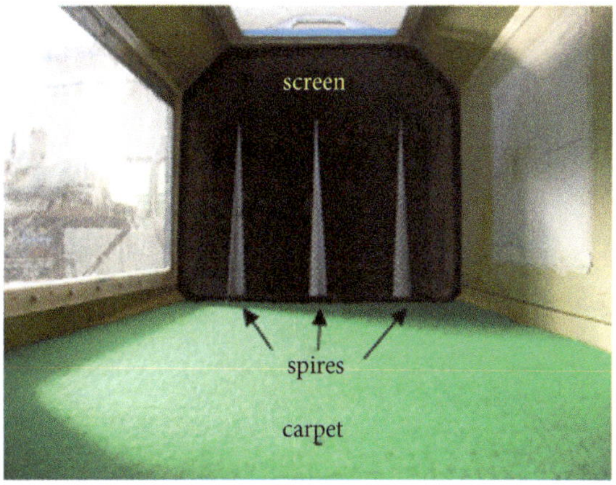

Figure 3. Devices used: carpet, screen, and spires.

the hot-wire signal by the King's law. The procedure resulted in calibration errors below 0.5%, which are typical of hot-wire applications. During the measurements, the hot-wire signal was acquired with an AD card (NI-PCI-MIO-16) with a resolution of 16 bits. To avoid clipping effects, the gain applied to analog signal was fixed; this kept the maximum voltage always below 80% of the range of the AD card. The DC and AC parts of the signal were not split prior to the acquisition; therefore no high-pass filtering was applied. Only low-pass filters were used to avoid aliasing effects. The positioning of the probe was done with a DANTEC traverse mechanism, which provides a resolution of 6.25 μm in all three axes. One important issue of BL measurements is the definition of the surface location. In the current experiments, the initial position was set by moving slowly the probe toward the wall until it visually touched the surface. Afterward, a correction based on the extrapolation of the mean flow data was applied. The procedure resulted in a nearly constant shift of approximately 1 mm. The experimental apparatus is shown in Figs. 4 and 5. It uses an x (longitudinal), y (lateral), and z (vertical) coordinate system to describe the positions of measurements. Hot-wire anemometers of constant temperature were used to measure the flow. These anemometers utilize a several millimeter long tungsten

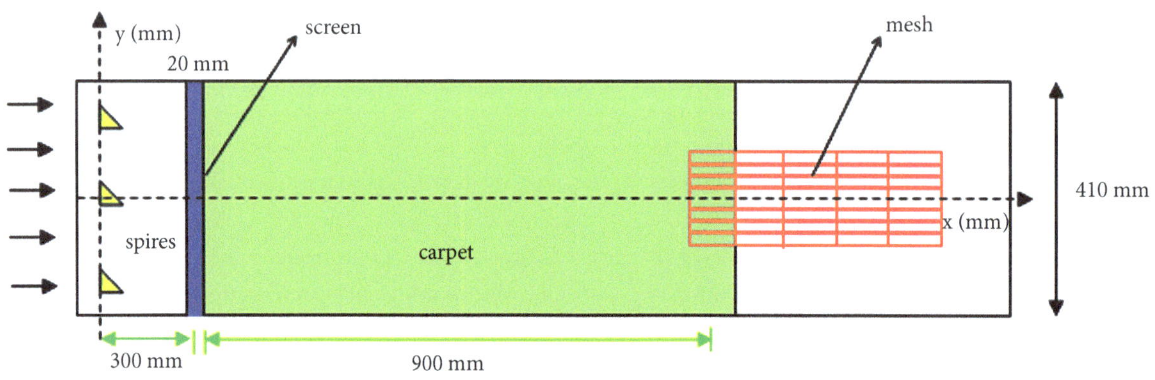

Figure 4. View of the arrangement of the experimental apparatus with the x (longitudinal) and y (lateral) coordinates.

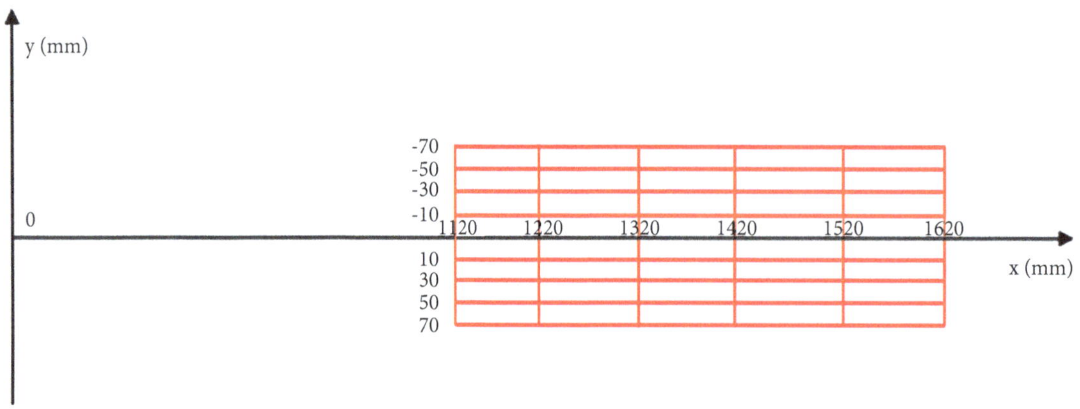

Figure 5. View of the mesh where measurements were taken.

wire with a diameter of around 4 μm. The system functions as one of the resistors in a Wheatstone bridge, allowing measurements with high spatial and temporal resolutions. Further details about this technique can be seen in Roballo (2007). For data acquisition, we used the Labview computer program. The positions on the longitudinal and lateral axes are shown in Fig. 5. The vertical profile was performed at the heights 1, 3, 6, 10, 20, 30, 40, 50, 60, 80, 100, 130, 160, 190, 220, and 260 mm, using a scale factor of 1:1000.

The experimental uncertainty in the measurement of pressure and temperature is 0.1 and 0.4% respectively. This result is an uncertainty of 3.1% in the estimation of the Reynolds number, with the confidence interval of 95%. More details can be found in Roballo (2007).

Initially, the spires and coarse, medium, and fine screen devices were tested separately.

Afterward, combinations of these devices were tested. Then, an optimization of the positions of spires and screen was done with distances of 150 and 300 mm from the screen. Finally, the carpet was inserted to perform the fine tuning of the desired roughness. The metric used to optimize the simulation of rural land was by the coefficient α of the wind power law, which was assumed equal to 0.15.

RESULTS AND DISCUSSION

Table 1 presents the values of the coefficient α of the power law versus the Reynolds number (Re) for the four devices (spires and fine, medium, and coarse screens) individually deployed, corresponding to the positions x=1120 mm and x=1620 mm grid points. Re was calculated for an ABL height of 280 mm as suggested by Pires *et al.*, (2009), and the windspeed used was u_∞. The largest pressure loss occurred for the fine screen, with a dynamic pressure of about 28.8 mm H_2O and a Re of 3.81×10^5. The pressure losses measured by the Pitot tube were smaller for the coarse screen (v=27.6 m/s). The spires presented the second highest pressure loss with Re equal to 4.16×10^5, which is not very relevant, since it will be coupled to another device for the generation of the ABL. The ideal configuration causes the smallest pressure loss and generates the ABL in the shortest possible distance of the test section. Note that the coefficient α is independent of Re or the longitudinal position (x) and is always smaller than 0.1 (ranging from 0.05 up to 0.09). This small value of α is due to absence of other passive devices to fine tune the generation of ABL.

The ABL was fully developed at x=1120 mm, showing nonsignificant differences with the combination of spires and thin screen, due to different gaps between the spires and the screen (150 mm or 300 mm). The combination of the spires with the medium screen moved the ABL to the position x=1420 mm, with the highest mean square deviation for spires 300 mm distant from the screen reaching 3.0 m/s near the surface and 0.5 m/s at the height of 260 mm. For the combination of the spires with coarse screen, the ABL was not developed inside the test chamber section, presenting an irregular windspeed profile, requiring a longer test section for its formation. Table 2 summarizes the values of α and Re for spires combined with the fine and medium screens. The higher values of Re, of about 4.8×10^5, were obtained for the medium screen placed further away (300 mm) from the spires. Spires with the thin screen display lower values of Re (3.9×10^5) for both small and greater separations. The largest coefficient α occurs for spires close (150 mm) to medium screens, with values around 0.10.

Table 1. Re and α for isolated spires or screens.

Devices	Positions (mm)	Windspeed (m/s)	Dynamic pressure (mm H_2O)	α	Re
Spires	x=1220	24	32.8	0.06	4.16×10^5
	x=1620	24	32.8	0.09	4.16×10^5
Thin screen	x=1220	22	28.8	0.02	3.81×10^5
	x=1620	22	28.8	0.07	3.81×10^5
Medium screen	x=1220	27	41	0.07	4.68×10^5
	x=1620	27	41	0.04	4.68×10^5
Coarse screen	x=1220	27.6	44	0.06	4.78×10^5
	x=1620	27.6	44	0.05	4.78×10^5

Table 2. Values of Re and α for spires combined with fine and medium screens.

Devices	Positions (mm)	Windspeed (m/s)	α	Re
Spires+thin screen	x=1120	22.5	0.03	3.9×10^5
	x=1220	22.5	0.03	3.9×10^5
Spires+medium screen	x=1320	28	0.06	4.8×10^5
	x=1420	28	0.05	4.8×10^5
	x=1520	28	0.04	4.8×10^5
Spires close+thin screen	x=1120	23	0.08	3.9×10^5
	x=1220	23	0.07	3.9×10^5
Spires close+medium screen	x=1320	24	0.11	4.2×10^5
	x=1420	24	0.11	4.2×10^5
	x=1520	24	0.10	4.2×10^5

Gray: position where the ABL is formed.

Measurements were carried out at several locations along the stream and spanwise directions, as shown schematically in Fig. 5. The variations of mean and turbulent flow profiles along the spanwise direction were used as a criterion to ensure that the ABL was developed. Figure 6 shows an example of developed BLs, measured with medium screens. No significant variation of the profiles is present in the results, suggesting the presence of a well-developed flow at position x=1420.

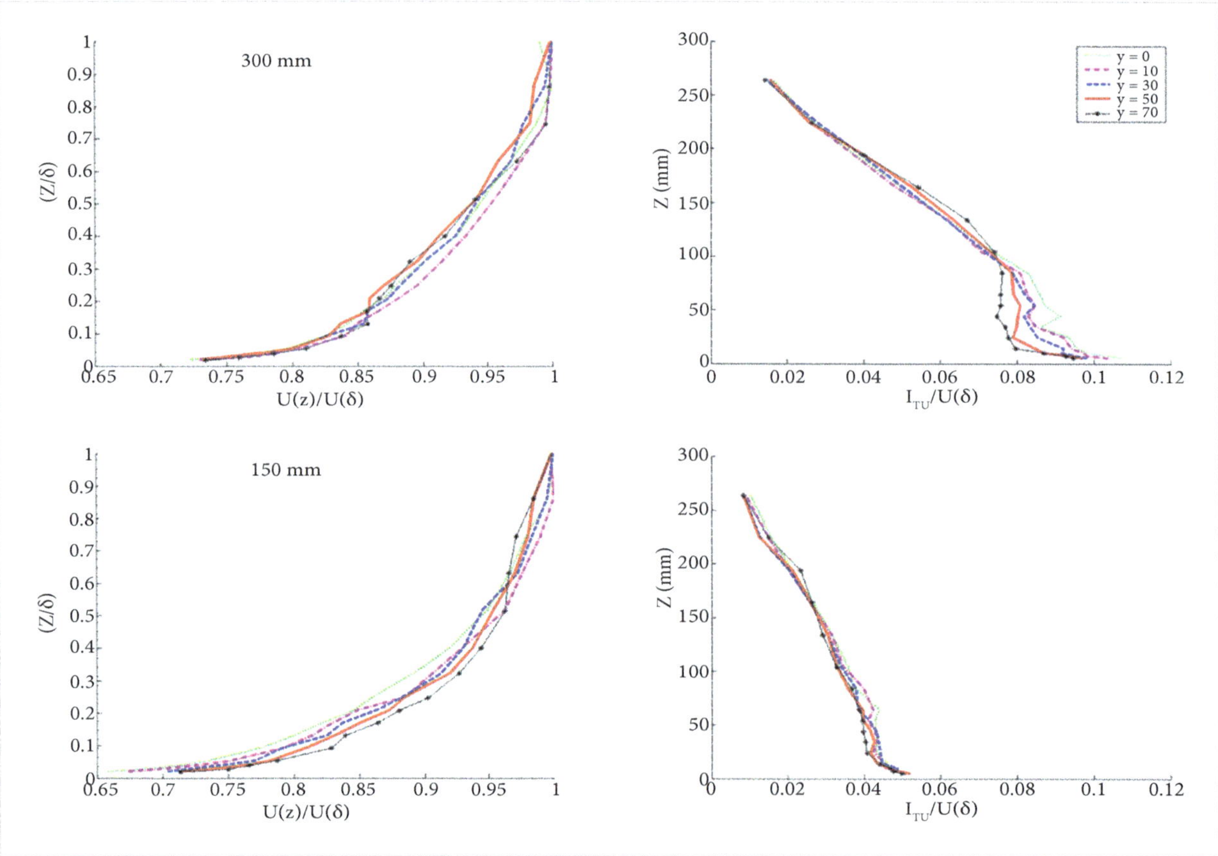

Figure 6. Nondimensional windspeed and turbulence profile at x=1420 mm. Cases with medium screen (see Table 2). The turbulence level is normalized with respect to the velocity at top of the boundary layer.

Figure 7 shows the windspeed profile for a medium screen positioned at 150 mm (named as near) and 300 mm (far) from the spires, with and without the carpet for fine adjustment. Note that the drag is higher with the carpet, and the ABL is better formed with the carpet and a medium screen positioned at a 300 mm distance from the spires. Also, for spires further away from the screen (Fig. 7a), the velocities are greater for low heights, 15 and 25 m/s, respectively.

Figure 8 shows that when the carpet is used in conjunction with screens and spires, the windspeed profiles become more convex, with the greatest convexity noted for the smallest separation between the spires and screens (150 mm).

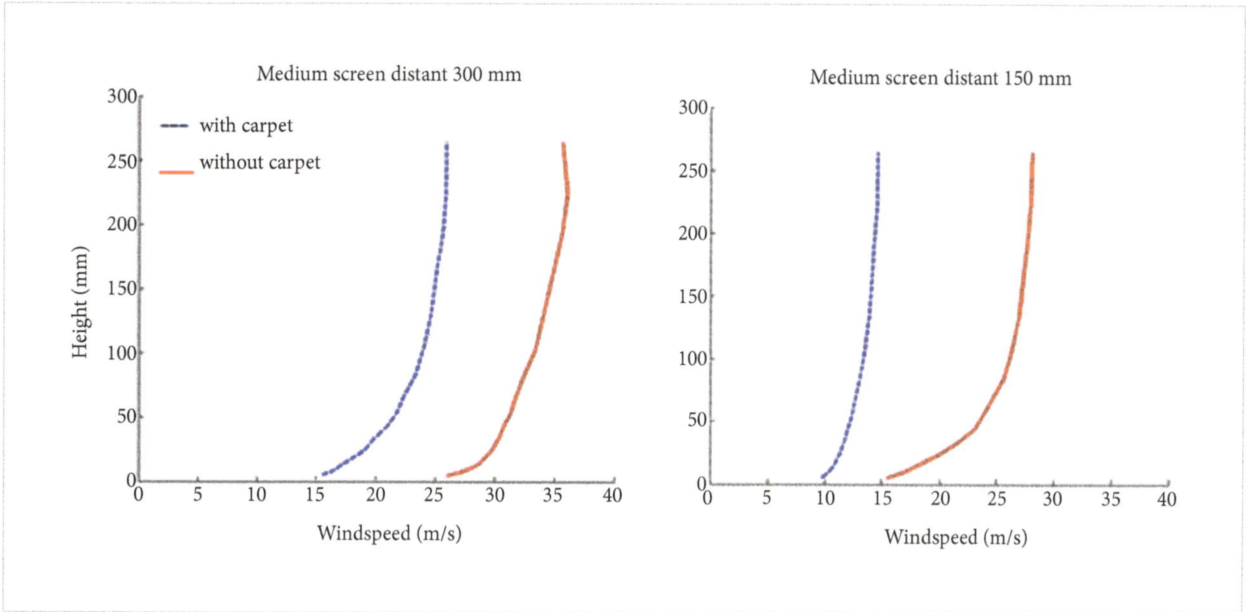

Figure 7. Windspeed profile at x=1420 mm.

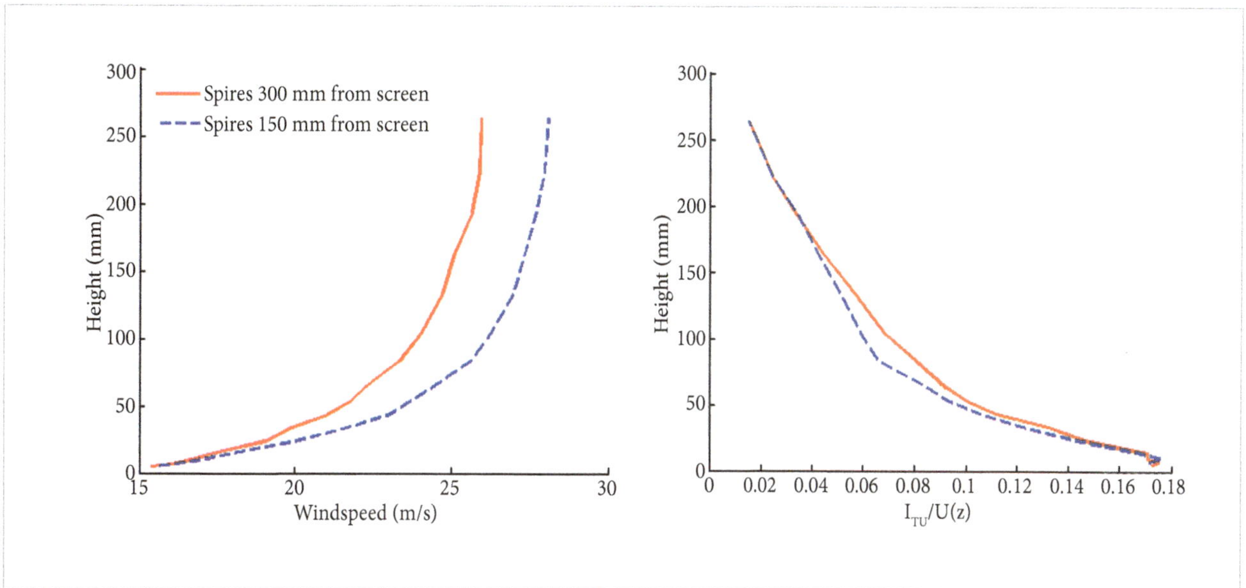

Figure 8. Windspeed profile and turbulent intensity at the point x=1420 mm with medium screen and carpet.

Table 3. Values of Re and α for tests with the carpet.

Devices	Positions (mm)	Windspeed (m/s)	Alpha	Re	Height of ABL (mm)
Without carpet	x=1420	28	0.05	4.9×10^5	–
Without carpet, with spires close to the screen	x=1420	24	0.11	4.2×10^5	–
With carpet	x=1420	25.5	0.14	4.4×10^5	200
With carpet, with spires close to the screen	x=1420	25.5	0.12	4.4×10^5	180

Thus, the height of the ABL depends on the distance separating the devices. Adjusting the data to the theoretical power law profile, with α=0.15, with the spires 300 mm away from the screen, an ABL around 200 mm was obtained, while for the spires closer to the screen (distance of 150 mm), the result was an ABL of 180 mm. The intensity of turbulence showed nonsignificant differences between the two cases.

Table 3 summarizes that Re is not sensitive to the inclusion (or not) of the carpet, varying from 4.2×10^5 to 4.9×10^5. However, the inclusion of the carpet resulted in values of α closer to the ideal 0.15. The height of the ABL increased from 180 to 200 mm as the distance between the spires and the medium screen increased (from 150 to 300 mm). Thus, the mean flow characteristics were apparently met with the combination of spires, medium screen, and carpet. In order to evaluate the results with respect to real ABLs, the mean flow profile was compared to theoretical power law and log-law profiles (Fig. 9). This last one is given by the relation $U(z)/u_*=1/0.4\ log[(z-z_d)/z_0]$ (Blessman, 1973), where u_* is the friction velocity, z_0 is the roughness factor, and z_d is the zero-plane displacement for very rough surface. Here z_d is assumed to be zero due to the flat terrain conditions and u_* and z_0 were obtained using the Clauser method. The results of Fig. 9 show a good agreement of the experimental data with the theory. This confirms a good modeling of the mean flow from the ABL at the wind tunnel.

The turbulence intensity is also an important feature of the ABL; therefore its intensities measured in the current experiments were compared to a benchmark suggested by ESDU (Engineering Science Data Unit). Such a benchmark consist of an empirical correlation based on a large experimental database of ABLs measured at height up to 100 m. The expression for this correlation can be found in ESDU (1985) and Liu *et al.*, (2003). Therefore, the current data had to be scaled to the real ABL in order to allow a proper comparison. The results in Fig. 10 show a good agreement between the correlation and the experiments, especially for the streamwise stations where the BL was fully developed (x=1420 and 1520). At these stations, all experimental results lie between the limit of ±15%, represented in the figure

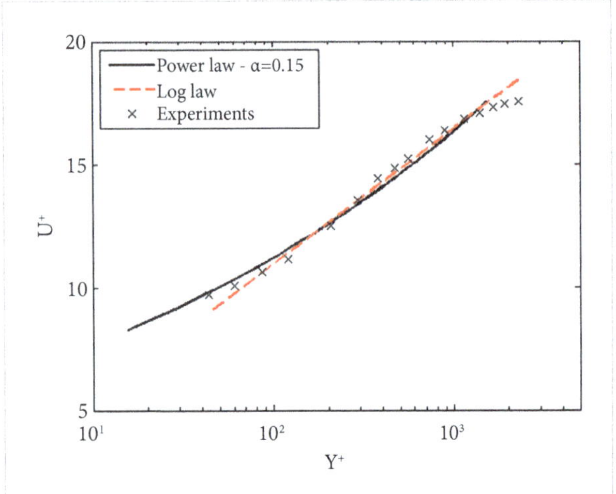

Figure 9. Windspeed profile in comparison to theoretical power law (α=0.15) and log law profiles.

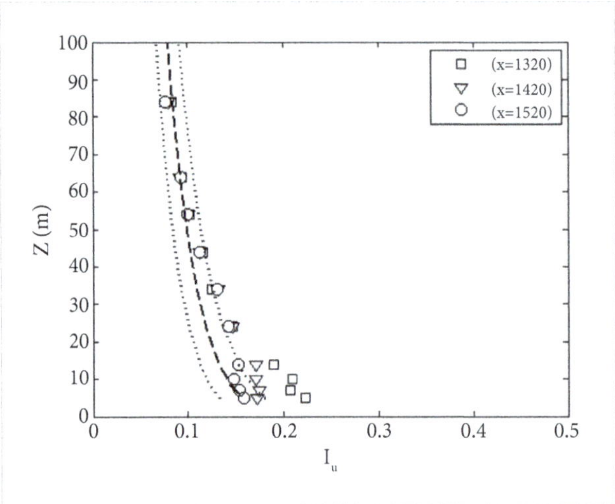

Figure 10. Turbulent intensity at the first 100 m (equivalent to the boundary layer scaling) and comparison with the correlation suggested by ESDU 7430 and 7431 (Engineering Science Data Unit). The dotted lines represent the ±15% deviation from the correlation.

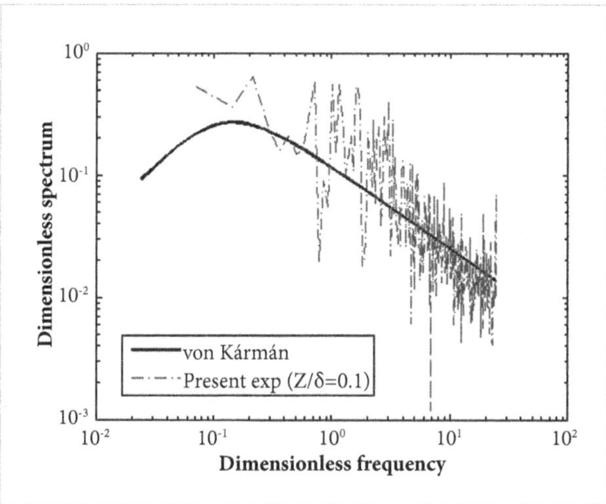

Figure 11. Comparison of the dimensionless spectrum obtained at z/δ=0.1 and the von Kármán spectrum.

by dotted lines. This suggests that the characteristics of the BL turbulence were also well modeled at the wind tunnel.

In order to complete the assessment of wind tunnel data, the spectrum of flow fluctuations was compared with the von-Kármán spectrum (ESDU, 1985), and the results are displayed in Fig. 11. According to ESDU, von-Kármán spectrum is a good model to represent the main features of the longitudinal component of turbulence in ABL. The formula of the dimensionless spectrum is $fS_u/\sigma^2=4X_u(z)/[1+70{,}78X_u(z)^2]^{5/6}$, where S_u is the spectral density function, f the dimensional frequency, σ^2 is the variance, and X_u is the dimensionless frequency. The last term is given by $fL(z)/U(z)$, $L(z)$ being the integral scale calculated using the autocorrelation spectra.

A qualitatively good agreement is found between the experiments and the von Kármán spectrum. The region of constant slope that is equivalent to the -5/3 exponential decay in the inertial subrange of Kolmogorov's cascade is nearly the same in both cases. Unfortunately, a more quantitative comparison could not be made with the current data because the period set for data acquisition was not long enough to enable the use of averaging procedures. Nevertheless, it is clearly observable that the main features of both spectra are the same. Therefore, it is reasonable to assume that mean flow and the turbulence characteristics of the ABL, expected to occur at the coast close to ALC, could be well simulated in a short wind tunnel with the use of passive devices. It is important to point out that for other terrain conditions, especially those at urban regions, this might not be the case.

CONCLUSION

The ideal situation is the formation of the ABL in the shortest possible extension of the wind tunnel, with the lowest pressure loss possible, to obtain a large Re, similar to the values between 10^6 and 10^7 observed in nature. An ABL can be formed by combinations of passive devices, which do not require a long test section. So, in some situations, existing aeronautical wind tunnels can be used for atmospheric simulations, in particular those with small values of the coefficient α in the power law.

The lowest pressure loss was obtained with the coarse screen alone, while spires followed by a thin screen resulted in the highest pressure loss. The medium screen generated a well-developed ABL at the mesh point x=1420 mm, while the use of thin screen moved the ABL to the position x=1120 mm. In view of these findings, if the length of the wind tunnel is critical (being short) and the pressure loss is not so important; the best choice would be the thin screen.

The distance between spires and screen was the other factor examined. A lower pressure loss, with a larger Re equal to 4.8×10^5, was observed with the medium screen more distant from the spires, while for the thin screen Re was equal to 3.9×10^5. However, higher values of α (equal to or greater than 0.10) were found for all cases of spires closer to the screens.

The insertion of a carpet caused a sharp increase of α, which became closer to the desired value of 0.15. The ratio of pressure loss between the distances of the spires and the screen remained the same, but demonstrated a smaller drop with the increase of the distance between the devices; however, for the same screen, the height of the ABL was increased with a smaller distance between the screens and spires, resulting in 180 mm for the distance of 150 mm, and 200 mm for 300 mm.

These findings indicate that this technique is a suitable tool for generation of an ABL in short-chamber wind tunnel experiments. The main characteristics of wind in an ABL seem to be well reproduced in the current experiments. This is confirmed by comparisons with established benchmarks from the literature. However, an important caveat is that the technique employed here seems to be limited to the reproduction of ABLs of nearly flat terrains, which have very small values of power law exponent. For regions with higher terrain roughness, such as forests and urban areas, the ABL wind tunnels clearly are the most suitable tools.

ACKNOWLEDGMENTS

Authors acknowledge a doctoral fellowship (CNPq 141861/2006-1) and a Research Scholarship 302117/2004-0, both granted by CNPq (Conselho Nacional de Desenvolvimento Científico e Tecnológico, Brazil).

REFERENCES

Avelar, A.C., Brasileiro, F., Marto, A.G., Marciotto, E. and Fisch, G., 2012, "Wind Tunnel Simulation of the Atmospheric Boundary Layer for Study of the Wind Pattern at the Alcantara Space Center", Journal of Aerospace and Technology Management, Vol. 4, No. 4, doi: 10.5028/jatm.2012.04044912.

Blessmann, J., 1973, "Simulation of the Natural Wind Structure in an Aerodynamic Wind Tunnel" (in Portuguese), Ph.D. thesis, Instituto Tecnológico de Aeronáutica (ITA), São José dos Campos, S.P., Brazil, 169 p.

Cao, S. and Tamura, T., 2006, "Experimental Study on Roughness Effects on Turbulent Boundary Layer Flow over a Two-Dimensional Steep Hill", Journal of Wind Engineering and Industrial Aerodynamics, Vol. 94, pp. 1-19. doi: 10.1016/j.jweia.2005.10.001.

Carpentieri, M., Hayden, P. and Robins, A.G., 2012, "Wind Tunnel Measurements of Pollutant Turbulent Fluxes in Urban Intersections", Atmospheric Environment, Vol. 46, pp. 669-674. doi: 10.1016/j.atmosenv.2011.09.083.

Davenport, A.G., 1967, "The Dependence of Wind Loads on Meteorological Parameters", Proceedings of the International Seminar on Wind Effects on Buildings and Structures, Ottawa, pp. 19-82.

ESDU, 1985, "Characteristics of atmospheric turbulence near the ground. Part II: Single point data for strong winds (neutral atmosphere)", ESDU International, Item No. 85020, London.

Franck, N., 1963, "Model law and Experimental Technique for Determination of Wind Loads on Buildings", in Symposium on Wind Effects on Buildings and Structures, Vol. 16, National Physical Laboratory, Teddington, UK, pp. 181-196.

Irwin, H.P.A.H., 1981, "The Design of Spires for Wind Simulation", Journal of Wind Engineering and Industrial Aerodynamics, Vol. 7, pp. 361–366, doi: 10.1016/0167-6105(81)90058-1.

Jensen, M. and Franck, N., 1963, "Model Scale Tests in Turbulent Wind", Part I, Danish Technical Press, Copenhagen.

Jensen, M. and Franck, N., 1965, "Model Scale Tests in Turbulent Wind", Part II, Danish Technical Press, Copenhagen.

Kozmar, H., 2009, "Scale Effects in Wind Tunnel Modeling of an Urban Atmospheric Boundary Layer", Theoretical and Applied Climatology, Vol. 100, pp. 153–162, doi: 10.1007/s00704-009-0156-3.

Kwon, K.J., Lee, J.Y. and Sung, B., 2003, "PIV Measurements on the Boundary Layer Flow around Naro Space Center", Proceedings of the 5th International Symposium on Particle Image Velocimetry, pp. 22-24.

Liu, G., Xuan, J. and Park, S., 2003, "A new method to calculate wind profile parameters of the wind tunnel boundary layer", Journal of Wind Engineering and Industrial Aerodynamics, Vol. 91, pp. 1155-1162.

Mavroidis, I., Griffiths, R.F. and Hall, D.J.H., 2003, "Field and Wind Tunnel Investigations of Plume Dispersion around Single Surface Obstacles". Atmospheric Environment, Vol. 37, pp. 2903-2918, doi: 10.1016/S1352-2310(03)00300-5.

Novak, M.D., Warland, J.S., Orchansky, A.L., Ketler, R. and Green, S., 2000, "Wind Tunnel and Field Measurements of Turbulent Flow in Forests. Part I: Uniformly Thinned Stands", Boundary Layer Meteorology, Vol. 95, pp. 457-495, doi: 10.1023/A:1002693625637.

Pires, L.B.M., 2009, "Study of the Internal Boundary Layer Developed on Coastal Cliffs with Application to the Alcântara Launching Center, Brazil" (in Portuguese), Ph.D. thesis, Instituto Nacional de Pesquisas Espaciais (INPE), São José dos Campos, SP, Brazil. 165p, (INPE-16566-TDI/1562), http://pct.capes.gov.br/teses/2009/33010013003P8/TES.PDF.

Pires, L.B.M., Souza, L.F., Fisch, G. and Gielow, R., 2009, "La Influencia de la Altura de la Capa Límite Oceánica en la Región del Centro de Lanzamientos de Alcántara en Brasil", Información Tecnológica, Vol. 20, pp. 119-128, doi: 10.1612/inf.tecnol.4071it.08.

Pires, L.B.M., Roballo, S.T., Fisch, G., Avelar, A.C., Girardi, R.M. and Gielow, R., 2010, "Atmospheric Flow Measurements Using the PIV and HWA Techniques", Journal of Aerospace Technology and Management, v.2, pp.127-136, doi:10.5028/jatm.2010.02027410.

Roballo, S.T., 2007, "Estudo do Escoamento Atmosférico na CLA, através de Medidas em Torre Anemométrica e em Túnel de Vento" (in Portuguese), Master thesis, Instituto Nacional de Pesquisas Espaciais (INPE), São José dos Campos, SP, Brazil, 137 p, (INPE-14824-TDI/1264).http://mtc-m17.sid.inpe.br/col/sid.inpe.br/mtc-m17@80/2007/06.12.18.07/doc/ publicacao.pdf.

2

Simulation of Rocket Exhaust Clouds at the *Centro de Lançamento de Alcântara* Using the WRF-CMAQ Modeling System

Erick Giovani Sperandio Nascimento[1], Davidson Martins Moreira[2,3], Gilberto Fisch[2], Taciana Toledo de Almeida Albuquerque[1]

ABSTRACT: In this work we report numerical simulations of the contaminant dispersion and photochemical reactions of rocket exhaust clouds at the *Centro de Lançamento de Alcântara* (CLA) using the CMAQ modeling system. The simulations of carbon monoxide (CO), hydrogen chloride (HCl) and alumina (solid Al_2O_3) pollutants emission represent an effort in the construction of a computational tool in order to simulate normal and/or accidental events during rocket launches, making possible to predict the contaminant concentrations in accordance with emergency plans and pre and post-launchings for environmental management. The carbon monoxide and the alumina concentrations showed the formation of the ground and contrail cloud. The results also showed that hydrogen chloride concentrations would be harmful to human health, demonstrating the importance of assessing the impact of rocket launches in the environment and human health.

KEYWORDS: *Centro de Lançamento de Alcântara*, Rocket launch, CMAQ, Dispersion modeling.

INTRODUCTION

An important and singular air pollution problem is related to rocket launches. The burning of rocket engines during the first few seconds prior to and immediately after vehicle launchings results in the formation of a large cloud of hot, buoyant exhaust products near ground level, which subsequently rises and entrains ambient air until the temperature and density of the cloud reach an approximate equilibrium with ambient conditions. The United States' space activities are conducted by military personel (Air Force) and civilians (NASA). The US Air Force has 2 rocket launching centers, being one at the east coast at Cape Canaveral Air Force Base (AFB), in Florida, and another one at the west part, at Vandenberg AFB, California. These launching centers are close to big and populated region/cities which may be affected by the gases released during the launchings. In order to estimate the risks and environmental impacts associated to the launchings (either normal or failed ones), a special model named Rocket Exhaust Effluent Diffusion Model (REEDM) was developed by Bjorklund *et al.* (1982). This model assumes a constant wind profile and Gaussian plume turbulence to access the movement of the clouds derived from the exhausted gases. According to Nappo and Essa (2001), despite the fact that the REEDM model was operationally used for a long time (since the 1980s), it may not be fully checked using field data collected over the US launching centers. During the period from 1995 up to 1997, a Model Validation Program (MVP) was established to address these concerns and several field experiments with special tracers made at Cape Canaveral AFB and Vandenberg AFB. Furthermore, regarding the numerical simulations for the prediction of rocket exhaust flow fields, we can cite the works of Reis *et al.* (1970) and Rao *et al.* (2001).

1. Universidade Federal do Espírito Santo – Vitoria/ES – Brazil 2. Instituto de Aeronáutica e Espaço – São José dos Campos/SP – Brazil 3. Centro Integrado de Manufatura e Tecnologia – Salvador/BA – Brazil

Author for correspondence: Davidson Martins Moreira | Instituto de Aeronáutica e Espaço | Praça Marechal Eduardo Gomes, 50, São José dos Campos/SP | CEP: 12.228-901 | Brazil | Email: davidson@pq.cnpq.br

Unfortunately, there is no model fully ready to meet these demands in Brazil, or some experimental data dispersion of contaminants related to the Centro de Lançamento de Alcântara (CLA). Therefore, researches are very important as well as the developing of a modeling system designed to calculate peak concentrations, dosage and deposition (resulting from both gravitational settling and precipitation scavenging), downwind from normal and aborted launchings, to be used in mission planning activities and environmental assessments, pre-launch forecasts of the environmental effects of launch operations and post-launch environmental analysis in the Brazilian site. To this end, this paper aims at providing a framework which will allow the development of a model that considers the Brazilian site's characteristics. Therefore, the main purpose of this study is to report a numerical study (numerical experiment) of the contaminant dispersion and photochemical reactions for the pollutants HCl, Al_2SO_3 and CO in CLA using the Community Multi-scale Air Quality (CMAQ) modeling system. The paper describes the main CMAQ changes needed to incorporate rocket launches in this modeling system. In fact, CMAQ is a multi-pollutant, multiscale air quality model that contains state-of-science techniques for simulating all atmospheric and land processes which affect the transport, transformation, and deposition of atmospheric pollutants and/or their precursors on both regional and urban scales (Byun e Ching, 1999). These simulations represent an effort in the construction of a computational tool for normal and/or accidental events during rocket launches, making it possible to both predict and simulate the concentration needed, in accordance to emergency plans and pre and post-launchings for environmental management.

THE MODELING SYSTEM

The CMAQ model is a fully 3-dimensional modeling system consisting of three main components (meteorology, emissions and a chemical model) and various interfaces. Therefore, with this modular structure, the CMAQ has enough flexibility to change other meteorological models and emissions modeling systems (Byun and Ching, 1999). It is important to mention that the CMAQ model is a state-of-the-art modeling tool in air quality problems.

SITE DESCRIPTION

The CLA is the space gate of Brazil and it is located at the North part of Brazil (2 degrees south), close to Equator (Fig. 1). More information about the CLA, its topography and climate can be found at Pires *et al.* (2009).

Climatologically, there is a very distinctive rainfall characteristic: a wet season from January up to June, with highest rainfall in March and April (monthly rainfall about 300 mm/mo) and a dry season from July up to December (rainfall almost null or very low – less than 30 mm/mo). The winds follow this climatic behavior presenting high wind speed during the dry season (winds around 7-8 m/s at 70 m height) and a moderate winds (winds around 5-6 m/s) during the wet season. To represent this behavior, it was released a radiosonde in the CLA on April 2nd, 2010. The radiosonde uses a Vaisala Oy system (sonde RSV92G), which is the state-of-art in terms of upper air measurements. The winds (wind speed and direction up to 5,000 m) are presented at Fig. 2 and they are quite characteristic for the wet season: there is a linear increase of the wind speed from the surface up to a maximum value of approximately 16 m/s at 3,000 m. Between 2,000 and 4,000 m, the winds are very persistent, ranging from 14-16 m/s. The direction of the wind is NE/E (around 60° at 500 m) and then rotating southward (around 100-110°). The synoptic pattern is typical from the wet season, with the intertropical convergence zone (ITCZ) reaching the north part of Brazil, with a large cloud cluster over the ocean and inland at the CLA (Fig. 3).

Figure 4 presents the modeling domains. These domains were modeled using the Weather Research and Forecasting (WRF) Model (version 3.3) (Skamarock and Klemp, 2008) to generate meteorological fields in a one-day simulation. The horizontal

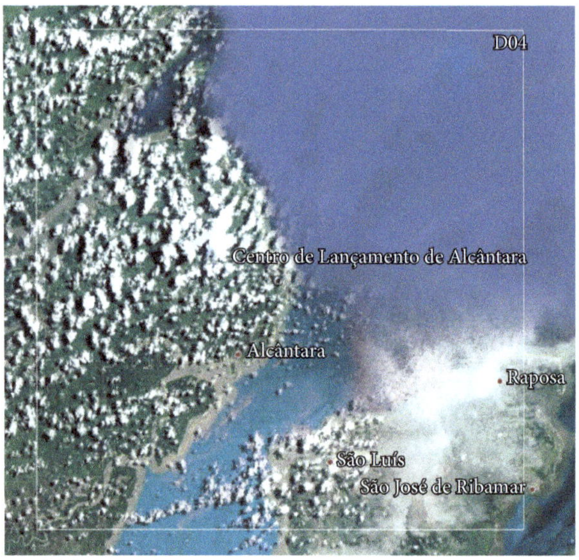

Figure 1. The location of the domain used in CMAQ to model the rocket launch case.

resolutions of the nested grid were 27 km, 9 km, 3 km and 1 km, and the horizontal dimensions, in grid cells, were 36 x 36, 54 x 54, 84 x 84 and 66 x 66, for domain 1 to 4, respectively. Recently, Silva and Fisch (2014) did a detailed analysis about the use of WRF for the CLA using radiosonde data collected during dry (2008) and wet season (2010) as a comparison. They found that the WRF, using the default parameterizations, can represent the wind speed at the site reasonably well (difference between the model and the observed is less than 2.0 m/s). The innermost domain, highlighted in Fig. 1, was used in order to model a hypothetical average rocket launch event in CLA using the CMAQ model, version 5.0.1 (Byun and Schere, 2006).

In order to run the WRF model to properly generate the meteorological fields, we used final analysis data from the Global Forecast System (GFS) with a resolution of 1 arc degree. Figure 5 presents the surface wind field simulated by the WRF model at the time of the launch (April 2nd, 2010 – 2-5PM). It is possible to note that the wind is predominantly blowing in the opposite direction of the large city of São Luís and in the direction of an inhabited area, and it keeps on until later in the same day, at 9PM local time, when the wind field entirely changes and it starts blowing in the direction of the city of São Luís. This situation, which is presented in Fig. 6, shows that a major concern is to determine when launching operations should be released, in order to prevent the environmental impact and air contamination of a metropolitan area, due to the large cloud of hazardous pollutants generated by the launch. In this simulation, the pollutants were already been dispersed to out of the domain at the time when the wind field changed, meaning that there were no more environmental risk of the cloud reaching the city of São Luís, as observed in the "Results and Discussion" section.

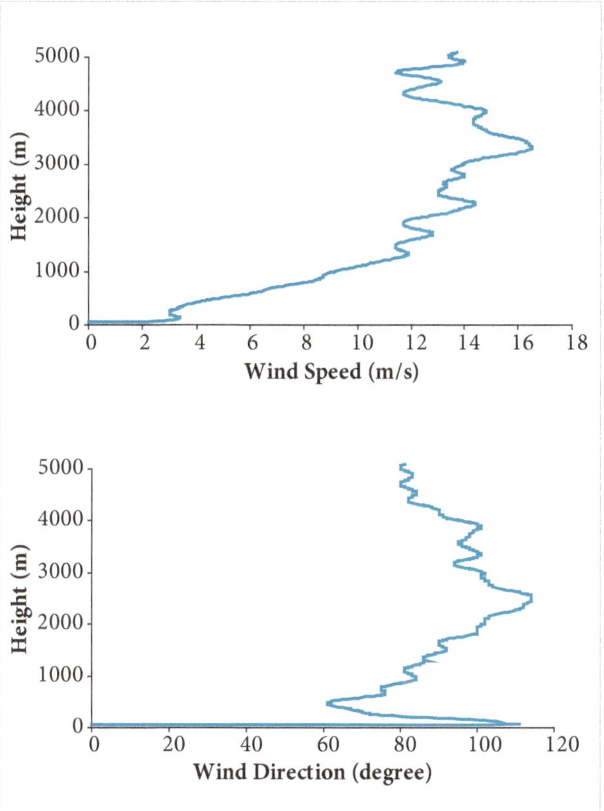

Figure 2. The wind profile at CLA for April 2nd, 2010.

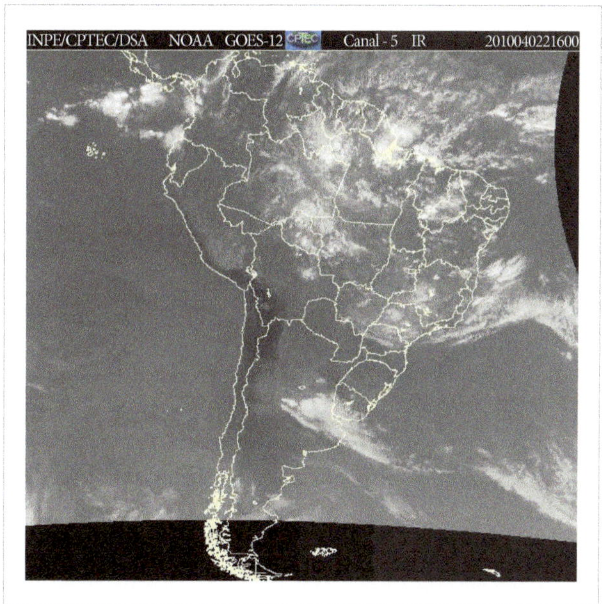

Figure 3. The synoptic pattern for April 2nd, 2010.

Figure 4. The location and distribution of the domains modeled in WRF Model.

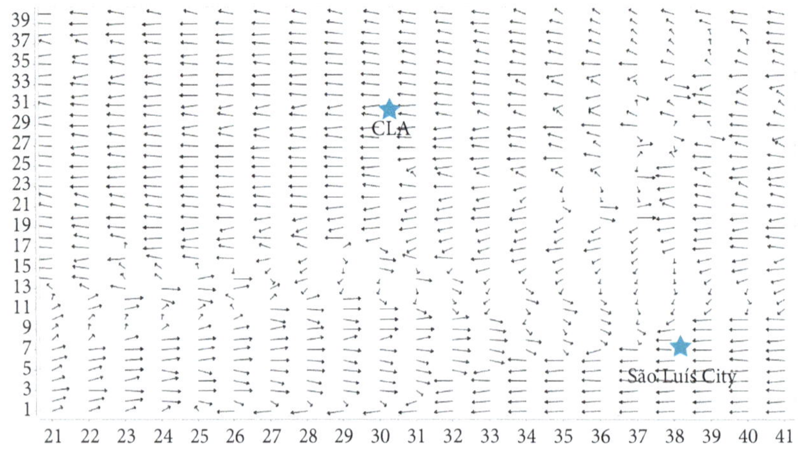

Figure 5. The surface wind field on April 2nd, 2010, at the time of the launching.

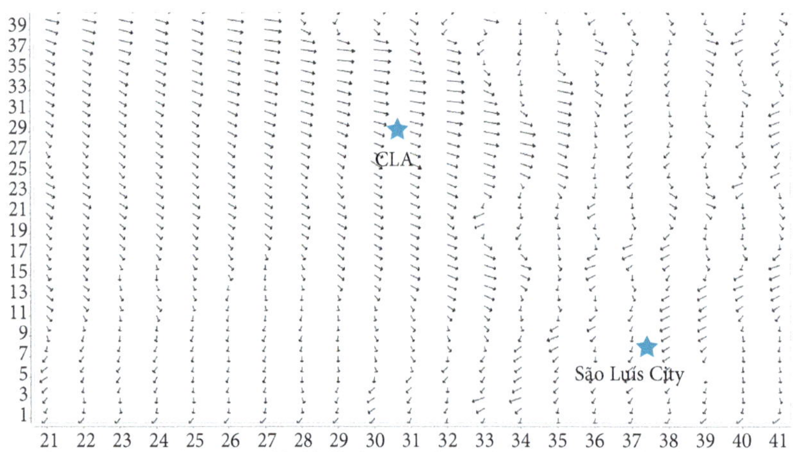

Figure 6. The surface wind field in domain 4 at 9PM, when it changed to the direction of São Luís.

BUILDING THE EMISSIONS FILE

The emission file is the critical point in the air quality context. To this end, it was developed the Sparce Matrix Operator Kernel System (SMOKE), which is used in order to generate the input information from emissions compatible with the model (Borge *et al.*, 2008). In this study we did some adaptations in the SMOKE system to work with rocket launching emissions, which has some particularities that we had to address to in the system. The system has been applied to the CLA to process a hypothetical emission inventory for a test.

Figure 7 shows the launching of Titan IV, in Cape Canaveral, USA, and the formation of the so called ground-cloud and contrail-cloud (Nyman, 2009). It has been a practice in USA to consider only the emissions from ground to 3000 m in simulations, and the emission rate of a normal rocket launch is $5{,}2 \times 10^5$ g/s (Bjorklund *et al.*, 1982).

Since SMOKE was not designed to deal with such kind of source, for each vertical level below 3,000 m, a virtual stack source emitting a certain hypothetical amount of the total emissions was configured, following the distribution presented in Table 1. These constituents are supposed to be released by the Brazilian Satellite Vehicle Launcher, for instance. Each virtual stack was positioned in the middle of its respective level. The virtual stacks located nearest to the ground have the main contribution in the emissions of simulating the ground-cloud effect. The remaining emissions were distributed along the highest levels. Input parameters, which entail significant uncertainty, were treated in a conservative fashion, in the sense that choices were made in order to favor overestimating rather than underestimating the toxic chemical concentrations being evaluated for the environmental assessment study. For example, we assume that all the material released by the rocket remains in the area below 3,000 m.

The emissions inventory was composed of only one pollutant, representing the combustion gases of the propellant. Table 2 shows the distribution of compounds that are emitted during the combustion, and their corresponding species in the chosen chemical mechanism (Bjorklund *et al.*, 1982). For this simulation, we considered the three major emitted compounds: carbon monoxide (CO), hydrogen chloride (HCl) and alumina (solid Al_2O_3). It is important to mention that only the emissions from the rocket launch were considered in this work. Regarding temporal allocation the temporal profiles were setup in SMOKE, in order to have the entire emissions only at noon (time of the hypothetical launch).

Figure 7. Illustration of the formation of the ground and contrail clouds, during a rocket launch (Nyman, 2009).

Table 1. Hypothetical distribution of the total emissions along the levels represented by their corresponding virtual stack.

Virtual Stack Id	Stack Height (m)	Distribution of Total Emissions (%)
0	11	20
1	53	30
2	81	10
3	120	7
4	180	6
5	233	5
6	285	5
7	524	4
8	704	4
9	1446	3
10	2195	3
11	2788	3

CONFIGURING THE COMMUNITY MULTI-SCALE AIR QUALITY (CMAQ) MODEL

For this study, we used the CMAQ model, version 5.0.1. Though our work has no intention to evaluate the impact of the emissions in the region, but only to study how we can apply CMAQ to such case, we chose to evaluate how it would impact the environment in this situation. Table 3 shows the options used in order to build the CMAQ. No initial or boundary conditions were applied, and the photolysis rates were calculated by the photolysis rate preprocessor (JPROC) using a table of clear-sky photolysis rates (or J-values) for a specific date according to the selected options used in order to build the CMAQ.

Table 2. Basic composition of the combustion gases of the propellant.

Product	% m/m
Carbon Dioxide (CO_2)	4.4
Carbon Monoxide (CO)	27.6
Hydrogen Chloride (HCl)	21.6
Alumina (Al_2O_3 sólido)	28.2
Water Vapor (H_2O)	7.0
Nitrogen (N_2)	8.4
Hydrogen (H_2)	2.8

Table 3. List of the options used in Community Multi-scale Air Quality, for this work.

Property	Selected Option
Gas Chemistry	cb05tucl_ae6_aq
Aerosol Chemistry	aero6
Advection	vwrf
Vertical Diffusion	acm2
Solver	ebi_cb05tucl
Cloud Module	cloud_acm_ae6

RESULTS AND DISCUSSIONS

Figure 8 shows a sequential tile plot for alumina, which was modeled as fine particulate matter (PMFINE), and Fig. 9 shows a sequential tile plot for carbon monoxide, both at the first level, at an approximate 20 m height. In these figures, we can see how the ground cloud behaviors after the rocket launch. Initially, it presents high concentrations at the location where the launching occurred, and after the first hour it presents a sensible decrease in the concentrations. Six hours later the ground cloud was totally dispersed from the domain.

Figures 10 and 11 show a sequential 3D plot for both alumina and carbon monoxide, respectively. As it can be observed, these figures shows the same behavior, where we can easily identify the ground cloud and contrail cloud formed after the launching. In the subsequent hours, most of the concentrations remain near the ground.

Hydrogen chloride emissions were present in a considerable amount (Fig. 12), which led to harmful

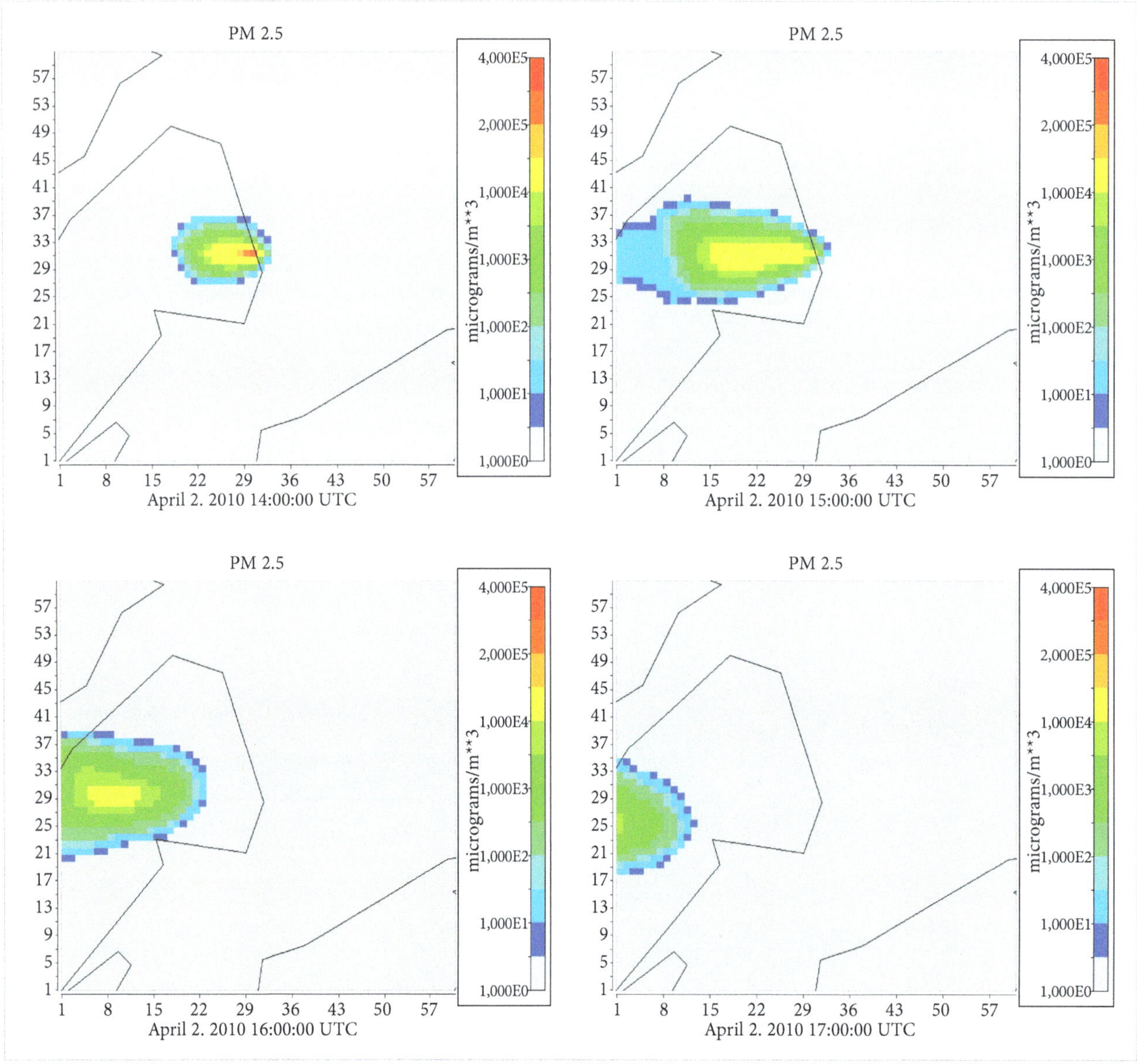

Figure 8. Sequential tile plot for PM $_{2.5}$, representing the concentrations of alumina, for the first layer.

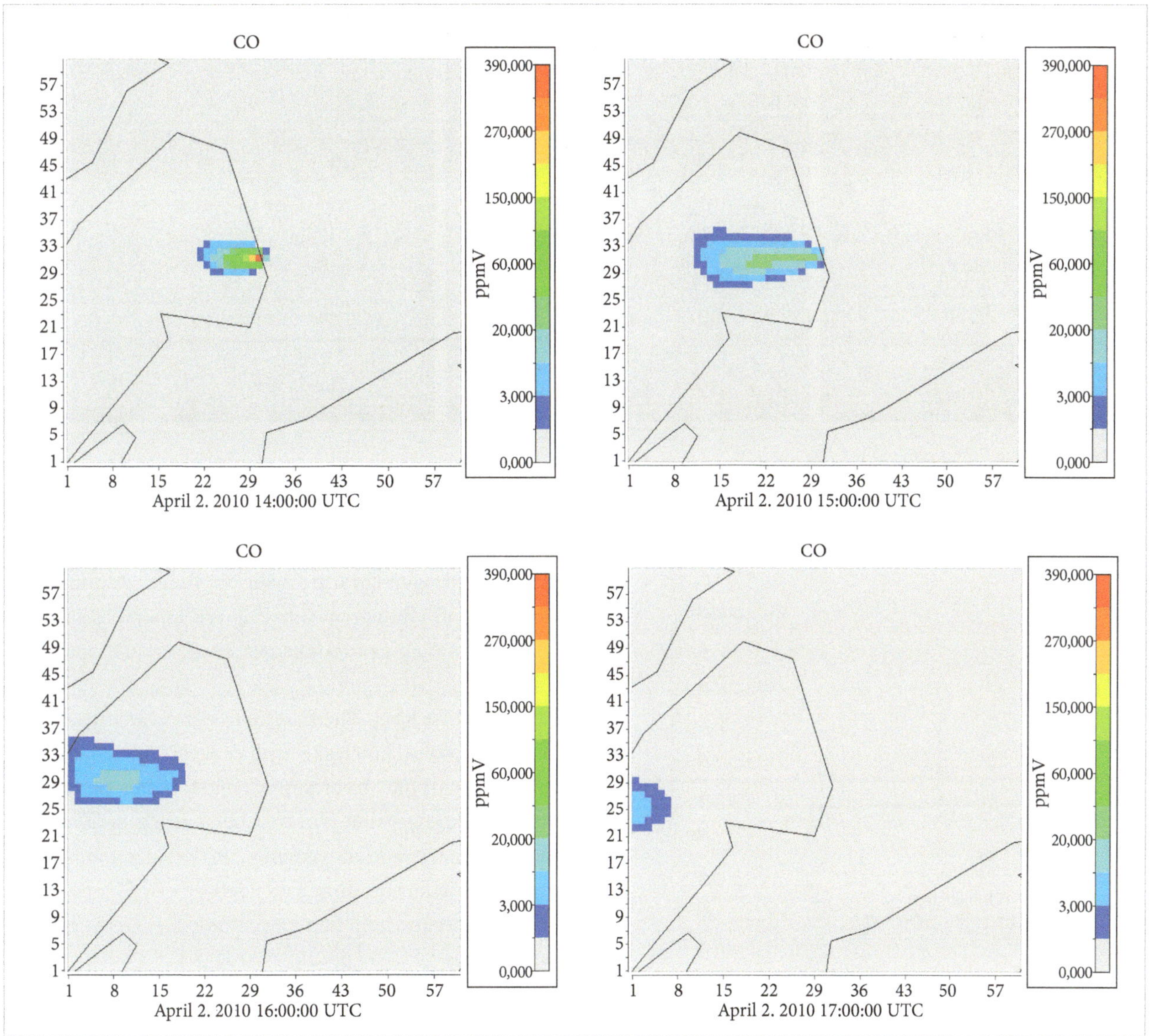

Figure 9. Sequential tile plot for CO for the first layer.

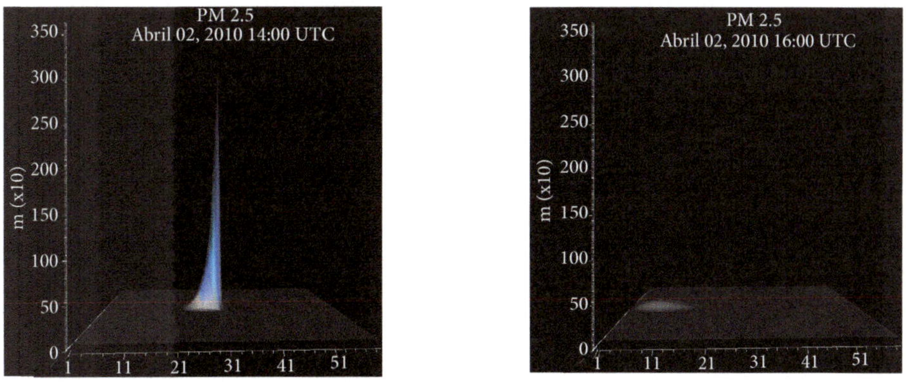

Figure 10. Sequential 3D plot for PM 2.5.

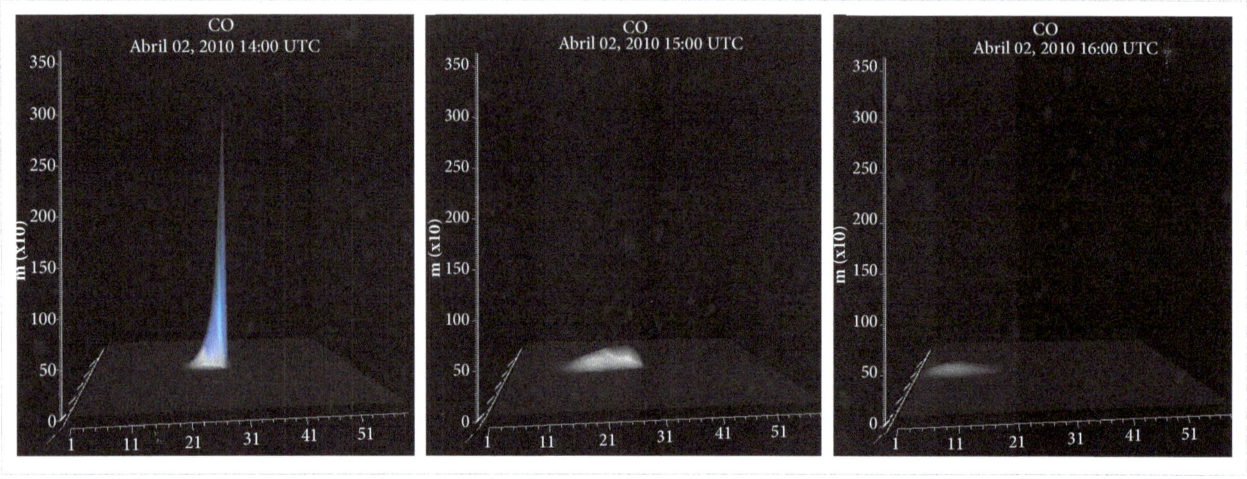

Figure 11. Sequential 3D plot for CO.

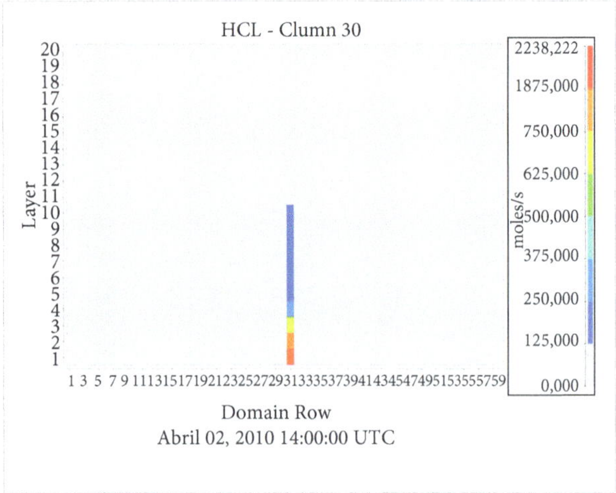

Figure 12. Vertical cross section plot of the emissions of hydrogen chloride. The emissions near ground are in the order of 2,000 moles/s.

Table 4. Health effects of respiratory exposure to hydrogen chloride (HCl).

Exposure Limits (ppm)	Health Effects
5 <	Coughing
35	Throat irritation occurs after only a short time
35 <	Severe breathing difficulties and skin inflammation or burns
10-50	Maximum level that can be sustained for several hours
100 <	Swelling of the lungs and often throat spasm
50-1,000	Maximum possible exposure: 1 hour
1,000-2,000	Very dangerous even for a very short exposure

concentration levels in CMAQ. Table 4 (Baxter *et al.*, 2000) presents the hydrogen chloride exposure limits, in ppm, and their impact in human health. HCl is a colourless gas with an irritating pungent odour perceivable at 0.8 ppm (Lide, 2003). Figure 13 presents the sequential tile plot for HCl at the first level. The first hour shows the higher concentrations at the launch site, with concentration levels between 100-1,000 ppm. According to Table 4, at this region, the maximum exposure time is only 1 hour, causing swelling of the lungs, throat spasm and irritation. In the surrounding area, it presents concentration levels between 35-50 ppm, and even at the second hour, the concentration still remains in a dangerous level (5-35 ppm).

Based on the results, it is considered that the use of CMAQ to model rocket launch cases is very interesting and promising. Some difficulties were faced, which had to be addressed in order to run the model. One of the biggest challenges here was on how to properly build the emissions file. Although the configuration and the running of the SMOKE, in this case, were manageable, it was considered that a specific tool should be created in order to achieve this purpose. The results of the concentrations of HCl, CO and PM 2.5 (Al_2SO_3) showed to be very interesting: the formation of the ground and contrail cloud and the dispersion of the ground cloud were very clear in the simulations, and the concentration levels of HCl, reported by the model, showed the importance to assess the impact of rocket launch events in the environment and human health, since chlorine has a major impact in the emissions of the combustion gases of the propellant during rocket launch events.

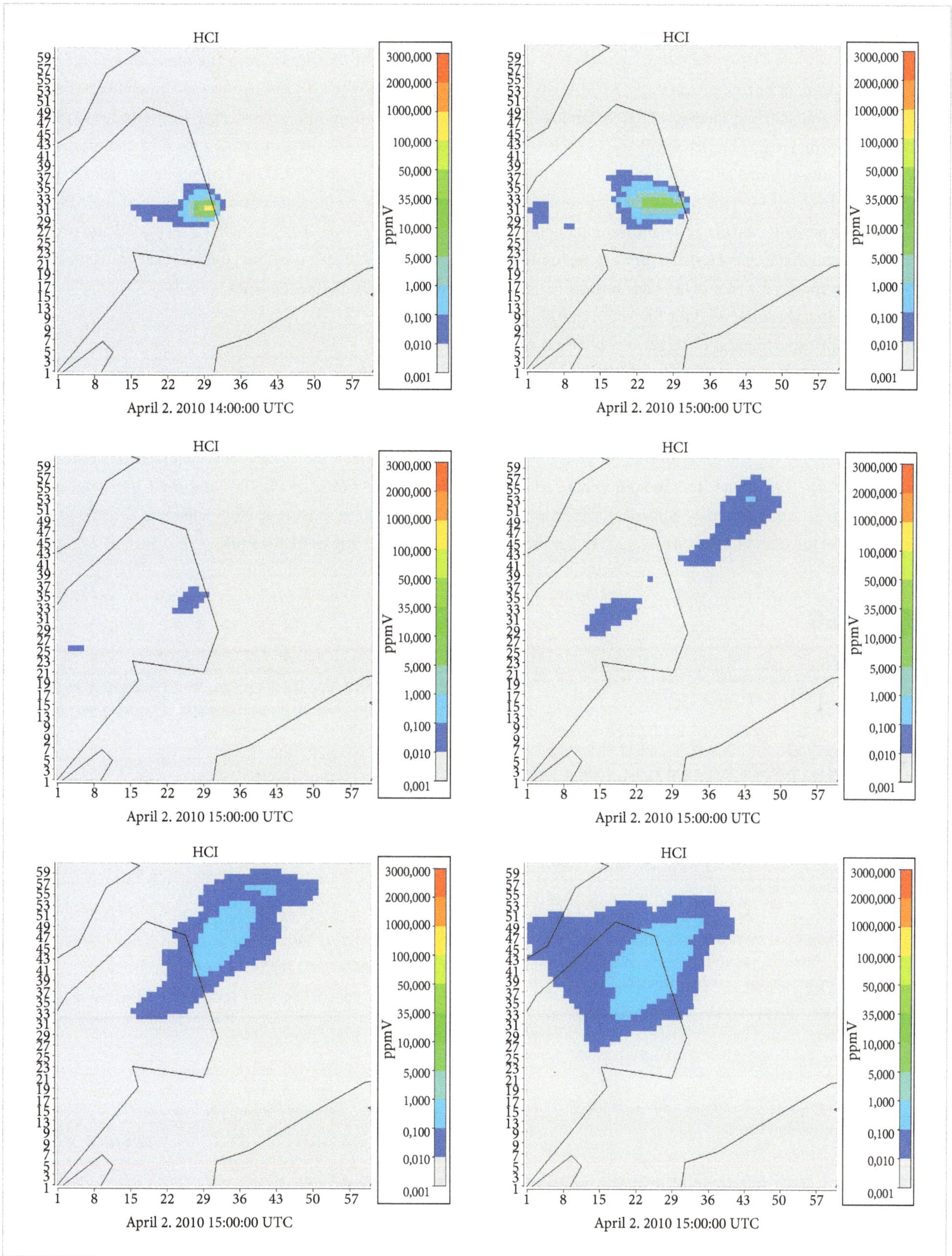

Figure 13. Sequential tile plot for HCl for the first layer.

SUMMARY AND PERSPECTIVES

In this work, our main purpose was to apply CMAQ as a numerical model to simulate the dispersion of the contaminants emitted during a normal rocket launch event and to evaluate the entire CMAQ suite - model and its accessories - for this case. Since it is a first step towards providing a framework which will allow the development of a more meaningful model considering the Brazilian site characteristics, in CLA, we found that the use of the model for this case is quite promising. These results show that there is a need for thorough checking of the schemes with experimental observations planned in the region. We focus our attention in the future to evaluate the construction of the cloud released by the rocket, which can be initially defined as a single cloud that grows and moves, but remains the same during the formation of its ascending phase (ground cloud). This will address the different spatial and temporal scales of the problem more consistently. For more details about this, see the work of Moreira *et al.* (2011). For this purpose, we plan to develop a tool that is capable of calculating the formation of the cloud, from the moment it is released to the moment when it stabilizes in the atmosphere, then to build the emissions file accordingly to the input into the CMAQ. Furthermore, it is important to construct numerical experiments from the CMAQ system model in order to have them compared to the concentration results of the MSDEF model (Moreira *et al.*, 2011), which is a very good and proven tool for a fast screening of the contaminant concentration field during rocket launch events.

ACKNOWLEDGEMENTS

The authors thank the Conselho Nacional de Desenvolvimento Científico e Tecnológico (CNPq) and the Coordenação de Aperfeiçoamento de Pessoal de Nível Superior (CAPES) for the partial financial support of this work.

REFERENCES

Baxter, P.J., Adams P.H., Aw, T.-C., Cockcroft, A. and Harrington, J.M., 2000, "Hunter's Diseases of Occupations", Arnold, London, Vol. 45, No. 2, pp. 123-178. doi: 10.1016/S0003-4878(00)00067-3.

Borge, R., Lumbreras, J. and Rodrıguez, E., 2008, "Development of a High-Resolution Emission Inventory for Spain Using the SMOKE Modelling System: A Case Study for the Years 2000 and 2010", Environmental Modelling & Software, Vol. 23, pp. 1026-1044.

Bjorklund, J.R., Dumbauld, J.K, Cheney, C.S. and Geary, H.V., 1982, "User's Manual for the REEDM (Rocket Exhaust Effluent Diffusion Model) Computer Program", NASA contractor report 3646. NASA George C. Marshall Space Flight Center, Huntsville, AL.

Byun, D.W. and Ching, J.K.S., 1999, "Science algorithms of the EPA Models-3 Community Multiscale Air Quality (CMAQ) Modeling System", EPA/600/R-99/030, Office of Research and Development, United States Environmental Protection Agency, Washington, DC.

Byun, D. and Schere, K.L., 2006, "Review of the Governing Equations, Computational Algorithms, and Other Components of the Models-3community Multiscale Air Quality (CMAQ) Modeling System", Appl. Mech. Rev., Vol. 59, pp. 51-77.

Lide, D.R., 2003, "CRC Handbook of Chemistry and Physics", 84th edition. CRC Press, Boca Raton, Florida.

Moreira, D.M., Trindade, L.B., Fisch, G., Moraes, M.R., Dorado, R.M. and Guedes, R.L., 2011 "A Multilayer Model to Simulate Rocket Exhaust Clouds". Journal of Aerospace Technology and Management, Vol. 3, pp. 41-52. doi: 10.5028/jatm.2011.03010311.

Nappo, C.J. and Essa, K.S.M., 2001, "Modeling Dispersion from Near-Surface Tracer Releases at Cape Canaveral, Florida". Atmospheric Environment, Vol. 35, pp. 3999–4010.

Nyman, R.L., 2009, "NASA Report: Evaluation of Taurus II Static Test Firing and Normal Launch Rocket Plume Emissions".

Pires, L.B.M., Avelar, A.C., Fisch, G., Roballo, S.T., Souza, L.F., Gielow, R. and Girardi, R., 2009, "Studies Using Wind Tunnel to Simulate the Atmospheric Boundary Layer at the Alcântara Space Center". Journal of the Aerospace and Technology Management, Vol. 1, No. 1, pp. 91-98. doi: 10.5028/jatm. 2009.01019198.

Rao, R.M., Sinha, K., Candler, G.V., Wright, M.J. and Levin, D.A., 2001, "Numerical Simulation of Atlas II Rocket Motor Plumes", 39th AIAA Aerospace Sciences Meeting and Exhibit, Reno, NV.

Reis, R.J., Aucoin, P.J. and Stechman, R.C., 1970, "Prediction of Rocket Exhaust Flow Fields", Journal of Spacecraft and Rockets, Vol. 7, No. 2, pp. 155-159. doi: 10.2514/3.29891.

Silva, A.F.G. and Fisch, G., 2014, "Avaliação do Modelo WRF para a Previsão do Perfil do Vento no Centro de Lançamento de Alcântara", Brazilian Journal of Meteorology, In press.

Skamarock, W.C. and Klemp, J.B., 2008, "A Time-Split Nonhydrostatic Atmospheric Model for Weather Research and Forecasting Applications", Journal of Computational Physics, Vol. 227, pp. 3465-3485. doi:10.1016/j.jcp.2007.01.037.

3

Relative Motion Guidance, Navigation and Control for Autonomous Orbital Rendezvous

Mohamed Okasha[1,2], Brett Newman[2]

ABSTRACT: In this paper, the dynamics of the relative motion problem in a perturbed orbital environment are exploited based on Gauss' variational equations. The relative coordinate frame (Hill frame) is studied to describe the relative motion. A linear high fidelity model is developed to describe the relative motion. This model takes into account primary gravitational and atmospheric drag perturbations. In addition, this model is used in the design of a control, guidance, and navigation system of a chaser vehicle to approach towards and to depart from a target vehicle in proximity operations. Relative navigation uses an extended Kalman filter based on this relative model to estimate the relative position and velocity of the chaser vehicle with respect to the target vehicle and the chaser attitude and gyros biases. This filter uses the range and angle measurements of the target relative to the chaser from a simulated Light Detection and Ranging (LIDAR) system, along with the star tracker and gyro measurements of the chaser. The corresponding measurement models, process noise matrix and other filter parameters are provided. Numerical simulations are performed to assess the precision of this model with respect to the full nonlinear model.The analyses include the navigations errors, trajectory dispersions, and attitude dispersions.

KEYWORDS: Satellite relative motion, Orbital rendezvous.

INTRODUCTION

Although significant progress and technical development have been achieved with regards to orbital rendezvous such as International Space Station supply and repair and automated inspection, servicing, and assembly of space systems, there are limitations with the traditional methods that struggle to meet the new demands for orbital rendezvous. Presently, in order to perform such close proximity operations, mission controllers generally require significant cooperation between vehicles and utilize man-in-the-loop to ensure successful maneuvering of both spacecraft. The interest in autonomous rendezvous and proximity operations has increased with the recent demonstration of XSS-11, Demonstration of Autonomous Rendezvous Technology (DART), and Orbital Express. Autonomous rendezvous and proximity operations have also been demonstrated by Japanese EST-VII, and the Russian Progress vehicles. In addition future missions to the ISS will require autonomous rendezvous and proximity operations (Fehse, 2003; Woffinden and Geller, 2007).

Many relative motion modeling and control strategies have been designed using the linearized Clohessy-Wiltshire (CW) equations to describe the relative motion between satellites. The CW equations are valid if two conditions are satisfied:

- The distance between the chaser and the target is small compared tothe distance between the target and the center of the attracting planet; and
- The target orbit is near circular (Clohessy and Wiltshire, 1960).

The CW equations do not include any disturbance forces, for example, gravitational perturbations and environmental forces

1.International Islamic University Malaysia – Kuala Lumpur – Malaysia 2.Old Dominion University – Norfolk/VA – United States

Author for correspondence: Mohamed Okasha – Assistant Professor – Department of Mechanical Engineering – Faculty of Engineering–International Islamic University Malaysia | Jalan Gombak | P.O. Box 10 50728 Kuala Lumpur – Malaysia | Email: mokasha@iium.edu.my or Ph.D. Alumni from Old Dominion University | Email: mokas001@odu.edu.my (solar radiation pressure and atmospheric drag). Alternative linear equations that have been used in the literature to model the relative motion are the Tschauner-Hempel (TH) equations (Tschauner and Hempel, 1965).These expressions generalize the CW equations and are similar to them in their derivation and types of applications. Tschauner and Hempel derived theses equations from the viewpoint of rendezvous of a spacecraft with an object in an elliptical orbit. They found complete solutions for elliptical orbits in terms of the eccentric anomaly. This advancement was followed by additional papers which present the complete analytical solution explicit in time, expanding the state transition matrix in terms of eccentricity (Yamanaka and Ankersen, 2002; Carter, 1998; Melton, 2000; Broucke, 2003; Inalhan *et al.*, 2002; Sengupta and Vadali, 2007; Cho and Park, 2009). This form of solution is used to analyze the relative motion between the chaser and the target vehicles in the relative frame of motion more efficiently and rapidly than solving the exact nonlinear differential equations in the inertial coordinate system. The TH equations do not take into account any perturbation forces. These perturbations have a significant effect on the satellite relative motion.

Due to the previous limitations of the CW and TH models, this paper proposes an innovative linear model which includes both the perturbation that reflects the Earth's oblateness effect and atmospheric drag perturbation in the Cartesian coordinates orbital frame with little complication. Especially in low Earth orbits (LEOs), these perturbations have a deep influence on the relative dynamics, and their inclusion in the linear model can sensibly increase the performance of the linear filters, allowing greater insight of satellite relative motion, and providing an opportunity to investigate alternative feedback control strategies for the proximity operations.

Unlike the relative translation motion control, the relative rotational control is a traditional feedback control system. During the mission scenarios, the chaser vehicle may need to track the target vehicle to achieve proper docking maneuvers and or visual inspection tasks. The paper uses an extended Kalman filter formulation to estimate the relative motion and chaser attitude using range and angle measurements from a LIDAR system coupled with gyro and star tracker measurements of the chaser (Woffinder and Geller, 2007; Jenkins and Geller, 2007; Junkins *et al.*, 2005; Woffinden, 2004). The Kalman filter basically consists of two main stages. The first stage is the propagation stage, where the states are propagated numerically and it is based on the proposed linear model. The second stage comes when the measurements from the sensors are available and it is used to update the states of the first stage. The corresponding measurement models, process noise matrix, and other filter parameters are provided. Momentum wheels are assumed for attitude control and thrusters are assumed for translation control. The effects of the navigation filter, pointing algorithms, and control algorithms are included in the analysis.

The objective of this paper is as follows:

- To develop linearized high fidelity models for relative motion in a perturbed orbit;
- To design a navigation filter that can determine the relative position and velocity between target and chaser vehicles as well as orientations and angular rates of the chaserthat support closed-loop proximity attitude control operations and maneuvers; and
- To design a control system for the chaser vehicle to either approach or depart fromthe target vehicle in proximity operations in a general perturbed orbit for coupled translation and rotation relative motion.

The analysis in the current paper is summarized as follows. First, we present the relative dynamics equation of motion for the chaser with respect to the target in a general perturbed orbit, along with attitude dynamics models.Next, a linear high fidelity relative motion model is developed to describe relative motion in proximity operations based on Gauss' variational method. Then, the relative navigation and an extended Kalman filter are presented for the relative motion and attitude estimations, along with the relative translational and rotational controller. In the simulation section, the accuracy and performanceof the relative navigation and controller, based on the high fidelity model, are illustrated through different numerical examples and comparisons are made with the truth nonlinear model. Finally, conclusion of the work is presented and suggestions are made for future work.

RELATIVE MOTION MODELS

Consider an Earth-centered inertia (ECI) frame, with orthonormal basis $\{i_X, i_Y, i_Z\}$. The vectors i_X and i_Y lie in the equatorial plane, with i_X coinciding with the line of equinoxes, and i_Z passing through the North Pole. Relative motion is conveniently described in a Local-Vertical-Local-Horizontal

(LVLH) frame,which is attached to the target spacecraft, as shown in Fig.1. This frame has basis $\{i_X, i_Y, i_Z\}$ with i_X lying along the radius vector form the Earth's center to the spacecraft, i_Z coinciding with the normal to the plane defined by the position and velocity vectors of the spacecraft, and $i_Y = i_Z \times i_X$. The LVLH frame rotates with angular velocityvector ω, and its current orientation with respect to the ECI frame is given by the 3-1-3 direction cosine matrix, comprising right ascension of ascending node Ω, inclination i, perigee argument ω plus true anomaly f, respectively (Fig.2). The angular velocity can also be expressed in terms of orbital elements and their rates.

Let the position of the chaser vehicle in the target's LVLH frame be denoted by $\rho = xi_x + yi_y + zi_z$, where x, y and z denote the components of the position vector along the radial, transverse,

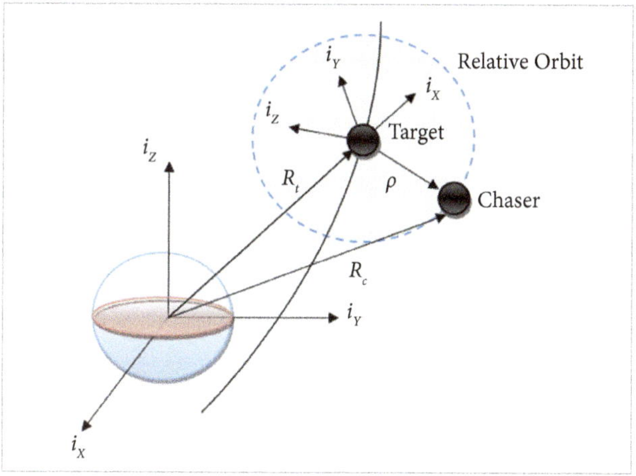

Figure 1. Relative Motion Coordinates.

and out-of-plane directions, respectively. $\rho$is determined from $\rho = R_c - R_t$, where R_c and $R_t$are the chaser and target absolute position vectors. Then, the most general equations modeling relative motion are given by the following:

$$\ddot{\rho} = [\mathbf{f}_c]^{\mathrm{LVLH}} - [\mathbf{f}_t]^{\mathrm{LVLH}} - 2\boldsymbol{\omega} \times \dot{\boldsymbol{\rho}} - \boldsymbol{\omega} \times \boldsymbol{\omega} \times \boldsymbol{\rho} - \dot{\boldsymbol{\omega}} \times \boldsymbol{\rho} \quad (1)$$

where $[f_c]^{LVLH}$ and $[f_t]^{LVLH}$ are the external accelerations acting on the chaser and the target, respectively in the LVLH frame of the target vehicle. In Eq. (1), (..) and (.) denote the first and second derivatives with respect to time.

It is assumed, in this paper, that the externalaccelerations arise due to two basic groups of accelerations, defined by the following equation:

$$f = f_g + f_a + f_c + f_w \quad (2)$$

The first group of accelerationsis due to gravitational effects, f_g, atmospheric drag, f_a, and control, f_c. Since Earth isn't perfectly spherical, more accurate gravity models exist, taking into account Earth's irregular shape. One irregularity that has a significant influence on space missions is the Earth's bulge at the equator. This phenomenon is captured in the J_2 gravity model (Vallado, 2001; Schaub and Junkins, 2003). The second group of accelerations, f_w, is considered to be small accelerations, due to the gravity fields of other planets, solar pressure, or venting, which also perturbs the spacecraft's motion. These small accelerationsare grouped together and modeled as zero mean normally distributed random variables (Woffinden and Geller, 2007).

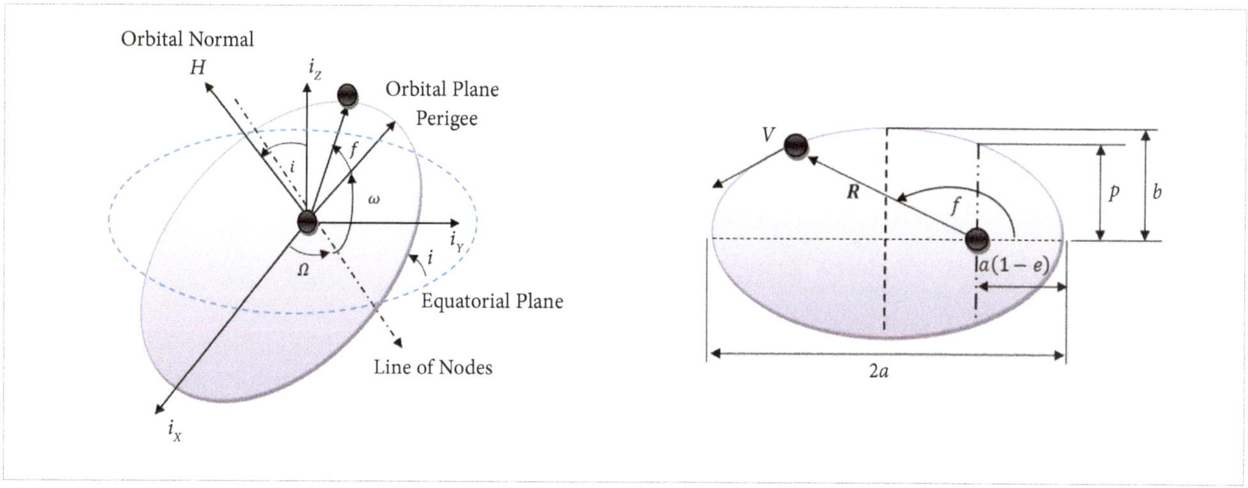

Figure 2. Orbital Elements.

In the literature, the most popular methods to model the spacecraft's orbit are known as Cowell's method and Gauss's method (Vallado, 2001; Schaub and Junkins, 2003). The Cowell's method is basically defined by specifying the position (R) and velocity (V) vectors of the spacecraft in the inertial coordinate frame, while Gauss' method is defined by an equivalent set of elements called orbital elements (a,e,i,Ω,ω,f) which correspond to the semi-major axis, eccentricity, inclination, right ascension of the ascending node, argument of periapsis, and true anomaly, respectively, as shown in Fig. 2.

Table 1 summarizes the dynamic equations that are used in order to describe all of these methods. In this table, $[.]^{I}$ and $[.]^{LVLH}$ denote that the forces are defined in inertial and LVLH coordinate frames, respectively; a_x, a_y and a_z are the components of disturbance accelerations acting on the target in the LVLH reference coordinate frame; $s_{(\cdot)}=sin(\cdot)$ and $c_{(\cdot)}=cos(\cdot)$; and $R_\oplus$ are the Earth gravitational constant and the radius of the Earth; the terms R and V refer to the magnitude of the position and velocity vectors, respectively; the quantity H denotes the magnitude of the specific angular momentum vector defined by $H=R\times V$; X, Y and Z are the components of the spacecraft position vector; C_D

Table 1. Orbit Model Methods Summary.

Method	Dynamic equations
Cowell's	$\ddot{\boldsymbol{R}} = \mathbf{f}$ $\mathbf{f} = \mathbf{f}_g + \mathbf{f}_a + \mathbf{f}_c + \mathbf{f}_w$ $\mathbf{f}_g = \boldsymbol{a}_{two-body} + \boldsymbol{a}_{J_2}$ $\boldsymbol{a}_{two-body} = -\left(\frac{\mu}{R^3}\right)\boldsymbol{R}$ $\left[\boldsymbol{a}_{J_2}\right]^{\mathrm{I}} = -\frac{3}{2}J_2\frac{\mu R_\oplus^2}{R^4}\left[(1-5(Z/R)^2)\frac{X}{R} \quad (1-5(Z/R)^2)\frac{Y}{R} \quad (3-5(Z/R)^2)\frac{Z}{R}\right]^{\mathrm{T}}$ $[\mathbf{f}_a]^{\mathrm{I}} = -\frac{1}{2}\frac{C_D A}{m}\rho V\boldsymbol{V}$
Gauss's	$\frac{da}{dt} = \frac{2a^2}{H}\left[es_f a_x + \frac{p}{R}a_y\right]$ $\frac{de}{dt} = \frac{1}{H}\{ps_f a_x + [(p+R)c_f + Re]a_y\}$ $\frac{de}{dt} = \frac{1}{H}\{ps_f a_x + [(p+R)c_f + Re]a_y\}$ $\frac{di}{dt} = \frac{Rc_{\omega+f}}{H}a_z$ $\frac{d\omega}{dt} = \frac{1}{He}\{-pc_f a_x + (p+R)s_f a_y\} - \frac{Rs_{\omega+f}c_i}{Hs_i}a_z$ $\frac{df}{dt} = \frac{H}{R^2} + \frac{1}{He}[pc_f a_x - (p+R)s_f a_y]$ $[\boldsymbol{a}]^{\mathrm{LVLH}} = \left[\boldsymbol{a}_{J_2} + \mathbf{f}_a + \mathbf{f}_c + \mathbf{f}_w\right]^{\mathrm{LVLH}}$ $\left[\boldsymbol{a}_{J_2}\right]^{LVLH} = -\frac{3}{2}J_2\frac{\mu R_\oplus^2}{R^4}\left[1-3s_i^2 s_{\omega+f}^2 \quad s_i^2 s_{2(\omega+f)} \quad s_{2i}s_{\omega+f}\right]^{\mathrm{T}}$ $[\mathbf{f}_a]^{\mathrm{LVLH}} = -\frac{1}{2}\frac{C_D A}{m}\rho V\boldsymbol{V}^{\mathrm{LVLH}}$ $\boldsymbol{V}^{\mathrm{LVLH}} = \frac{H}{p}[es_f \quad p/R \quad 0]^{\mathrm{T}}$

is the atmospheric drag coefficient; A denotes the cross sectional area; m is the spacecraft mass; and finally, ρ is the atmospheric density. Exponential atmospheric behavior is used to model Earth's atmospheric density. This model and its corresponding parameters are defined in Vallado (2001).

In order to use the generalized relative dynamic model defined by Eq. (1), the angular velocity vector, ω, and the angular acceleration vector, $\dot{\omega}$, of the LVLH frame with respect to the ECI frame, needs to be determined. Table 2 summarizes the equations that can be used to compute these vectors. These equations are derived based on using either Cowell's method (position and velocity vectors) or Gauss's method (orbital elements). In this table, the matrix $\mathrm{T}_{\mathrm{I}}^{\mathrm{LVLH}}$ denotes the direction cosine matrix of the LVLH coordinate frame with respect to the ECI coordinate frame.

The Euler's equation of motion is used to describe the attitude dynamics for both target and chaser vehicles, and a quaternion formulation is used for attitude kinematics. The dynamics for both vehicles are given below as (Woffinden and Geller, 2007):

$$\dot{\boldsymbol{q}}_I^t = \frac{1}{2}\boldsymbol{\omega}^t \otimes \boldsymbol{q}_I^t \tag{3a}$$

$$\dot{\boldsymbol{\omega}}^t = \boldsymbol{I}_t^{-1}[\boldsymbol{\tau}_t - \boldsymbol{\omega}^t \times \boldsymbol{I}_t\boldsymbol{\omega}^t] \tag{3b}$$

$$\boldsymbol{\tau}_t = \boldsymbol{\tau}_{t_g} + \boldsymbol{\tau}_{t_d} \tag{3c}$$

$$\dot{\boldsymbol{q}}_I^c = \frac{1}{2}\boldsymbol{\omega}^c \otimes \boldsymbol{q}_I^c \tag{4a}$$

$$\dot{\boldsymbol{\omega}}^c = \boldsymbol{I}_c^{-1}[\boldsymbol{\tau}_c - \boldsymbol{\omega}^c \times \boldsymbol{I}_c\boldsymbol{\omega}^c] \tag{4b}$$

$$\boldsymbol{\tau}_c = \boldsymbol{\tau}_{c_c} + \boldsymbol{\tau}_{c_g} + \boldsymbol{\tau}_{c_d} \tag{4c}$$

where $\otimes$ is the quaternion multiplication operator defined by Lear (1985).

$$\boldsymbol{p} \otimes \boldsymbol{q} = \begin{bmatrix} p_4 & p_3 & -p_2 & p_1 \\ -p_3 & p_4 & p_1 & p_2 \\ p_2 & -p_1 & p_4 & p_3 \\ -p_1 & -p_2 & -p_3 & p_4 \end{bmatrix} \begin{bmatrix} q_1 \\ q_2 \\ q_3 \\ q_4 \end{bmatrix} \tag{5}$$

Table 2. LVLH Coordinate Frame Orientation.

	Dynamic equations
Given Inertial Position and Velocity	$\mathrm{T}_{\mathrm{I}}^{\mathrm{LVLH}} = [\boldsymbol{i}_x \;\; \boldsymbol{i}_y \;\; \boldsymbol{i}_z]^{\mathrm{T}}$ $\boldsymbol{i}_x = \frac{\boldsymbol{R}}{R}, \quad \boldsymbol{i}_z = \frac{\boldsymbol{H}}{H}, \quad \boldsymbol{i}_y = \boldsymbol{i}_z \times \boldsymbol{i}_x$ $\boldsymbol{\omega} = \begin{pmatrix} R[\mathrm{f}_z]^{\mathrm{LVLH}}/H \\ 0 \\ H/R^2 \end{pmatrix}, \quad \dot{\boldsymbol{\omega}} = \begin{pmatrix} \frac{\left(\dot{R}[\mathrm{f}_z]^{\mathrm{LVLH}} + R[\dot{\mathrm{f}}_z]^{\mathrm{LVLH}}\right)H - R[\mathrm{f}_z]^{\mathrm{LVLH}}\dot{H}}{H^2} \\ 0 \\ \frac{\dot{H}R - 2H\dot{R}}{R^3} \end{pmatrix}$ $[\dot{\mathrm{f}}]^{\mathrm{LVLH}} = \mathrm{T}_{\mathrm{I}}^{\mathrm{LVLH}}[\dot{\mathrm{f}}]^{\mathrm{I}} - \boldsymbol{\omega} \times \mathrm{T}_{\mathrm{I}}^{\mathrm{LVLH}}[\mathrm{f}]^{\mathrm{I}}$ $\dot{H} = R\mathrm{f}_y, \quad \dot{R} = \frac{\boldsymbol{R}\cdot\boldsymbol{V}}{R}, \quad \boldsymbol{H} = \boldsymbol{R} \times \boldsymbol{V}$
Given Orbital Elements	$\mathrm{T}_{\mathrm{I}}^{\mathrm{LVLH}} = \mathbf{T}^3(\omega + f)\mathbf{T}^1(i)\mathbf{T}^3(\Omega) = \begin{bmatrix} c_\Omega c_{\omega+f} - s_\Omega s_{\omega+f} c_i & s_\Omega c_{\omega+f} + c_\Omega s_{\omega+f} c_i & s_{\omega+f} s_i \\ -c_\Omega s_{\omega+f} - s_\Omega c_{\omega+f} c_i & -s_\Omega s_{\omega+f} + c_\Omega c_{\omega+f} c_i & c_{\omega+f} s_i \\ s_\Omega s_i & -c_\Omega s_i & c_i \end{bmatrix}$ $\boldsymbol{\omega} = \mathbf{T}^3(\omega + f)\mathbf{T}^1(i)\begin{bmatrix} 0 \\ 0 \\ \dot{\Omega} \end{bmatrix} + \mathbf{T}^3(\omega + f)\begin{bmatrix} di/dt \\ 0 \\ 0 \end{bmatrix} + \begin{bmatrix} 0 \\ 0 \\ \dot{\omega} + \dot{f} \end{bmatrix} = \begin{pmatrix} s_{\omega+f} s_i \dot{\Omega} + c_{\omega+f}\, di/dt \\ 0 \\ c_i\dot{\Omega} + (\dot{\omega} + \dot{f}) \end{pmatrix}$ $\mathbf{T}^1(\theta) = \begin{bmatrix} 1 & 0 & 0 \\ 0 & c_\theta & s_\theta \\ 0 & -s_\theta & c_\theta \end{bmatrix}, \quad \mathbf{T}^2(\theta) = \begin{bmatrix} c_\theta & 0 & -s_\theta \\ 0 & 1 & 0 \\ s_\theta & 0 & c_\theta \end{bmatrix}, \quad \mathbf{T}^3(\theta) = \begin{bmatrix} c_\theta & s_\theta & 0 \\ -s_\theta & c_\theta & 0 \\ 0 & 0 & 1 \end{bmatrix}$

and the i vehicle gravity gradient torque is defined by

$$\boldsymbol{\tau}_{i_g} = 3\frac{\mu}{|\boldsymbol{R}_i|^5}(\boldsymbol{R}_i \times \boldsymbol{I}_i\boldsymbol{R}_i) \quad (6)$$

In Eqs. 3 and 4, the target states include the quaternion, q_I^t, that defines the orientation of the target with respect to the inertial frame, and the target's angular rate, ω^t. Similarly the chaser states are q_I^c and I_c. andare the target and chaser inertia matrices, respectively. The gravity gradient torque, $\tau_{(ig)}$, for both vehicles ($\tau_{(cg)}$ for the chaser and $\tau_{(tg)}$ for the target) is derived from the Earth'spoint mass gravity models. The random disturbances, $\tau_{(td)}$ and, $\tau_{(tg)}$ are included in the models to account for disturbance torques acting on each vehicle such as drag, solar radiation and other unmodeled disturbances. These unmodeled disturbances are represented as uncorrelated white noise, with mean and variance defined by a trial and error technique outlined by Lear (1985). The control input, $\tau_{(cc)}$, is the torque executed by the actuators (momentum wheels) on the chaser spacecraft.

It is assumed that the available sensors are the LIDAR for tracking the target and an assembly of a star tracker and gyros for attitude determination. The parameter states for these sensors are modeled as first-order Markov processes with large time constants, causing them to behave like biases. The parameter states include the gyros bias b_ω^c, star camera misalignments ϵ_s^s, and LIDAR misalignments ϵ_l^l. The dynamics model associated with these states is given by:

$$\dot{\boldsymbol{b}}_\omega^c = -\frac{\boldsymbol{b}_\omega^c}{\tau_b^\omega} + \boldsymbol{w}_b^\omega \quad (7a)$$

$$\dot{\boldsymbol{\epsilon}}_s^s = -\frac{\boldsymbol{\epsilon}_s^s}{\tau^s} + \boldsymbol{w}^s \quad (7b)$$

$$\dot{\boldsymbol{\epsilon}}_l^l = -\frac{\boldsymbol{\epsilon}_l^l}{\tau^l} + \boldsymbol{w}^l \quad (7c)$$

where, w_b^ω, w^s and w^l are white noise terms, driving the first-order Markov processes and τ_b^ω, τ^s and τ^l are the corresponding time constants.

The actuator models used in the simulation include momentum wheels for orientation control and thrusters for translational control. The mathematical model for the actual control torque, generated by the wheels, and the impulsive thrust, by the thrusters, are:

$$\boldsymbol{\tau}_{cc} = \boldsymbol{\delta T}(\boldsymbol{\epsilon}_\tau^c)[\{\boldsymbol{I}_{3\times3} + Diag(\boldsymbol{f}_\tau^c)\}\hat{\boldsymbol{\tau}}_{cc} + \boldsymbol{b}_\tau^c + \boldsymbol{v}_\tau^c] \quad (8)$$

$$\boldsymbol{u}_{C_{\Delta V}} = \boldsymbol{\delta T}(\boldsymbol{\epsilon}_{\Delta V}^c)[\{\boldsymbol{I}_{3\times3} + Diag(\boldsymbol{f}_{\Delta V}^c)\}\Delta\boldsymbol{V}_c + \boldsymbol{b}_{\Delta V}^c + \boldsymbol{v}_{\Delta V}^c] \quad (9)$$

The generated torque and impulsive include errors such as noises v^c, biases b^c, scale factor biases f^c, and misalignments ϵ^c. These errors can be modeled also as white noises.

The simulation contains gyros, star tracker, and LIDAR sensor models. The models for these measurements are given by:

Gyro Model:

$$\tilde{\boldsymbol{\omega}}^c = \boldsymbol{\delta T}(\boldsymbol{\epsilon}_\omega^c)[\{\boldsymbol{I}_{3\times3} + Diag(\boldsymbol{f}_\omega^c)\}\boldsymbol{\omega}^c + \boldsymbol{b}_\omega^c + \boldsymbol{v}_\omega^c] \quad (10)$$

Star Tracker Model:

$$\tilde{\boldsymbol{q}}_I^s = \boldsymbol{\delta q}(\boldsymbol{v}_s^s)\otimes\boldsymbol{\delta q}(\boldsymbol{\epsilon}_s^s)\otimes\boldsymbol{q}_c^s\otimes\boldsymbol{q}_I^c \quad (11)$$

LIDAR Model:

$$\begin{pmatrix}\tilde{\alpha}\\ \tilde{\beta}\\ \tilde{\rho}\end{pmatrix} = \begin{pmatrix} tan^{-1}\left(\frac{i_y}{i_x}\right)\\ sin^{-1}(i_z)\\ \rho\end{pmatrix} + \begin{pmatrix}v_\alpha\\ v_\beta\\ v_\rho\end{pmatrix} \quad (12)$$

where:

$$\boldsymbol{i}_{los}^l = \begin{pmatrix}i_x\\ i_y\\ i_z\end{pmatrix} = \begin{pmatrix}c_\alpha c_\beta\\ s_\alpha c_\beta\\ s_\beta\end{pmatrix} = \boldsymbol{T}^{l\hat{l}}\boldsymbol{T}^{\hat{l}s}\boldsymbol{T}^{s\hat{s}}\boldsymbol{T}^{\hat{s}I}\boldsymbol{T}^{IT}\left(-\frac{\boldsymbol{\rho}}{\rho}\right) \quad (13)$$

The gyro models include bias b_ω^c, scale factor bias f_ω^c, and angular random walk noise v_ω^c. The starcamera model accounts for the uncertainty in the alignment of the star camera frame ϵ_s^s with respect to the chaser frame and sensor noise v_s^s. The q_c^s refers to the fixed orientation of the star camera coordinate frame with respect to the chaser body coordinate frame. The LIDAR model includes angle measurements (azimuth, α, and elevation, β) noises v_α, v_β and range (ρ) noise, v_ρ. The transformation matrix denoted by T^{ab} is the transformation matrix used to transform any vector from coordinate b to coordinate a. The term i_{los}^l *is* the line of sight vector in the LIDAR coordinate frame (Fig. 3). The transformations T^{ll}, T^{IS}, $T^{\$S}$, $T^{\$I}$, and T^{IT} are a series of transformation matrices to transform the line of sight vector from target LVLH coordinate frame to the LIDAR coordinate frame. These transformations include errors from sensor misalignments, noises, and attitude determination errors.

The small angle rotations can be written in terms of quaternions as

$$\boldsymbol{\delta q}(\boldsymbol{\theta}) \approx \begin{bmatrix}\boldsymbol{\theta}/2\\ 1\end{bmatrix} \quad (14)$$

or attitude matrices as

$$\delta\boldsymbol{T}(\boldsymbol{\theta}) \approx \boldsymbol{I} - \boldsymbol{\theta}^{\times} \tag{15}$$

where $\theta = \theta u$ is a small rotation vector, and $\theta^{\times}$ operating on vector ω is a cross product matrix defined by the ordinary cross product $\theta^{\times}\,\omega = \omega \times \theta$.

LINEAR HIGH FIDELITY RELATIVE MODEL

In this section, a linear time varying high fidelity model is obtained to describe the relative motion dynamics. This model is derived based on two main assumptions. The first assumption is that the relative distant between the chaser and the target vehicles is much less than the target orbital radius. The second one assumes that the main disturbance accelerations, that affect both vehicles are the gravitational acceleration and the atmospheric drag acceleration. Based on these assumptions, all terms mentioned in the general relative dynamic expression, Eq. (1), are expanded considering only first order terms to obtain the new proposed model. Table 3 summarizes the procedures that have been followed to obtain this model. In this table the linear time varying model reduces to the following form

$$\dot{\boldsymbol{x}} = \boldsymbol{A}\boldsymbol{x} \tag{16}$$

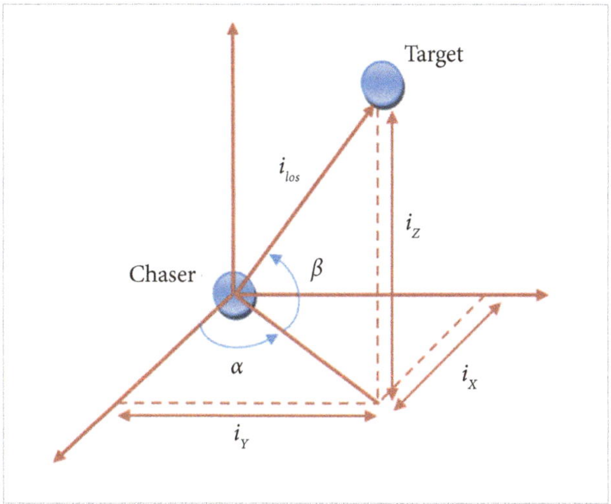

Figure 3. Line of Sight Vector.

where x is the state vector. This model can be used to approximate the time varying state transition matrix by expanding the time invariant exponential matrix solution in a Taylor series to fourth order, as follows:

$$\phi_{\mathrm{LTV}} = e^{A\Delta t} \approx \boldsymbol{I} + \boldsymbol{A}\Delta t + \frac{\boldsymbol{A}^2\Delta t^2}{2!} + \frac{\boldsymbol{A}^3\Delta t^3}{3!} + \frac{\boldsymbol{A}^4\Delta t^4}{4!} \tag{17}$$

This matrix is used in the next section as a part of the extended Kaman filter, to propagate the states forward in time and to compute the filter parameters.

NAVIGATION CONTROL MODEL ALGORITHMS

The main objective of the navigation system is to estimate the target's relative position, relative velocity and orientation given noisy sensor measurements, imperfect dynamic models, and uncertain initial conditions. The logic behind the navigation filter is to process information collected from sensors and various mathematical models to generate the best possible estimation of the states. Space navigation application of the Kalman filter is presented in this section. The dynamic models for a closed loop GN&C system are shown in Fig. 4.

The navigation model uses an extended Kalman filter to estimate the relative position and velocity of the chaser vehicle with respect to the target vehicle, and the approximated analytical state transition matrix solution. Orbital elements of the target are numerically propagated with respect to time using Gauss's variational equations, with J_2 and drag perturbations. These orbital elements are used to compute the transformation matrix of the target vehicle with respect to the inertial frame, as well as to assist in estimating LIDAR measurements. The dynamic models used to propagate the navigation states are:

$$\frac{d\hat{a}}{dt} = \frac{2\hat{a}^2}{\hat{H}}\left[\hat{e}s_{\hat{f}}\hat{a}_x + \frac{\hat{p}}{\hat{R}}\hat{a}_y\right] \tag{18a}$$

$$\frac{d\hat{e}}{dt} = \frac{1}{\hat{H}}\left\{\hat{p}s_{\hat{f}}\hat{a}_x + \left[(\hat{p} + \hat{R})c_{\hat{f}} + \hat{R}\hat{e}\right]\hat{a}_y\right\} \tag{18b}$$

$$\frac{d\hat{\imath}}{dt} = \frac{\hat{R}c_{\hat{\omega}+\hat{f}}}{\hat{H}}\hat{a}_z \tag{18c}$$

$$\frac{d\hat{\Omega}}{dt} = \frac{\hat{R}s_{\hat{\omega}+\hat{f}}}{\hat{H}s_{\hat{\imath}}}\hat{a}_z \tag{18d}$$

Table 3. Relative Orbit Model Summary.

Model	Equations
Nonlinear	$\ddot{\boldsymbol{\rho}} = [\mathbf{f}_c]^{\mathrm{LVLH}} - [\mathbf{f}_t]^{\mathrm{LVLH}} - 2\boldsymbol{\omega} \times \dot{\boldsymbol{\rho}} - \boldsymbol{\omega} \times \boldsymbol{\omega} \times \boldsymbol{\rho} - \dot{\boldsymbol{\omega}} \times \boldsymbol{\rho}$ $\mathbf{T}_{\mathrm{I}}^{\mathrm{LVLH}} = \mathbf{T}^3(\omega+f)\mathbf{T}^1(i)\mathbf{T}^3(\Omega) = \begin{bmatrix} c_\Omega c_{\omega+f} - s_\Omega s_{\omega+f} c_i & s_\Omega c_{\omega+f} + c_\Omega s_{\omega+f} c_i & s_{\omega+f} s_i \\ -c_\Omega s_{\omega+f} - s_\Omega c_{\omega+f} c_i & -s_\Omega s_{\omega+f} + c_\Omega c_{\omega+f} c_i & c_{\omega+f} s_i \\ s_\Omega s_i & -c_\Omega s_i & c_i \end{bmatrix}$ $\boldsymbol{\omega} = \mathbf{T}^3(\omega+f)\mathbf{T}^1(i)\begin{bmatrix} 0 \\ 0 \\ \dot{\Omega} \end{bmatrix} + \mathbf{T}^3(\omega+f)\begin{bmatrix} di/dt \\ 0 \\ 0 \end{bmatrix} + \begin{bmatrix} 0 \\ 0 \\ \dot{\omega}+\dot{f} \end{bmatrix} = \begin{pmatrix} s_{\omega+f} s_i \dot{\Omega} + c_{\omega+f}\, di/dt \\ 0 \\ c_i\dot{\Omega} + (\dot{\omega}+\dot{f}) \end{pmatrix}$ $\mathbf{T}^1(\theta) = \begin{bmatrix} 1 & 0 & 0 \\ 0 & c_\theta & s_\theta \\ 0 & -s_\theta & c_\theta \end{bmatrix},\ \mathbf{T}^2(\theta) = \begin{bmatrix} c_\theta & 0 & -s_\theta \\ 0 & 1 & 0 \\ s_\theta & 0 & c_\theta \end{bmatrix},\ \mathbf{T}^3(\theta) = \begin{bmatrix} c_\theta & s_\theta & 0 \\ -s_\theta & c_\theta & 0 \\ 0 & 0 & 1 \end{bmatrix}$
Linear Time Varying	$\ddot{\boldsymbol{\rho}} = [\mathbf{f}_c]^{\mathrm{LVLH}} - [\mathbf{f}_t]^{\mathrm{LVLH}} - 2\boldsymbol{\omega} \times \dot{\boldsymbol{\rho}} - \boldsymbol{\omega} \times \boldsymbol{\omega} \times \boldsymbol{\rho} - \dot{\boldsymbol{\omega}} \times \boldsymbol{\rho}$ $-2\boldsymbol{\omega} \times \dot{\boldsymbol{\rho}} - \boldsymbol{\omega} \times \boldsymbol{\omega} \times \boldsymbol{\rho} - \dot{\boldsymbol{\omega}} \times \boldsymbol{\rho} = \begin{bmatrix} \omega_z^2 & \dot{\omega}_z & -\omega_x\omega_z & 0 & 2\omega_z & 0 \\ -\dot{\omega}_z & \omega_x^2+\omega_z^2 & \dot{\omega}_x & -2\omega_z & 0 & 2\omega_x \\ -\omega_x\omega_z & -\dot{\omega}_x & \omega_x^2 & 0 & -2\omega_x & 0 \end{bmatrix}\begin{pmatrix} x \\ y \\ z \\ \dot{x} \\ \dot{y} \\ \dot{z} \end{pmatrix}$ $[\mathbf{f}_c]^{\mathrm{LVLH}} - [\mathbf{f}_t]^{\mathrm{LVLH}} \approx [\Delta\mathbf{f}_g + \Delta\mathbf{f}_a]^{\mathrm{LVLH}}$ $[\Delta\boldsymbol{a}_{two-body}]^{\mathrm{LVLH}} \approx \frac{\mu}{R^3}\begin{pmatrix} 2 & 0 & 0 \\ 0 & -1 & 0 \\ 0 & 0 & -1 \end{pmatrix}\begin{pmatrix} x \\ y \\ z \end{pmatrix}$ $[\Delta\boldsymbol{a}_{J_2}]^{\mathrm{LVLH}} \approx -\frac{3}{2}J_2\frac{\mu R_\oplus^2}{R^5}\begin{bmatrix} -4(1-3i_{x_3}^2) & -8i_{x_3}i_{y_3} & -8i_{x_3}i_{z_3} \\ -8i_{x_3}i_{y_3} & 1-7i_{x_3}^2+2(1-i_{z_3}^2) & 2i_{y_3}i_{z_3} \\ -8i_{x_3}i_{z_3} & 2i_{y_3}i_{z_3} & 3-5i_{x_3}^2-2(1-i_{z_3}^2) \end{bmatrix}\begin{pmatrix} x \\ y \\ z \end{pmatrix}$ $[\Delta\mathbf{f}_a]^{\mathrm{LVLH}} \approx \begin{bmatrix} C_c\omega_z R & 0 & 0 \\ 0 & C_c\omega_z R & 0 \\ 0 & 0 & C_c\omega_z R \end{bmatrix}\begin{pmatrix} \dot{x} \\ \dot{y} \\ \dot{z} \end{pmatrix}$ $\dot{\boldsymbol{x}} = \boldsymbol{A}\boldsymbol{x},\quad \boldsymbol{x} = [x\ \ y\ \ z\ \ \dot{x}\ \ \dot{y}\ \ \dot{z}]^{\mathrm{T}}$ $\boldsymbol{A} = \begin{bmatrix} \boldsymbol{A}_{11} & \boldsymbol{A}_{12} \\ \boldsymbol{A}_{21} & \boldsymbol{A}_{22} \end{bmatrix} \boldsymbol{A}_{11} = \boldsymbol{0}_{3\times3},\quad \boldsymbol{A}_{12} = \boldsymbol{I}_{3\times3}$ $\boldsymbol{A}_{21} = \begin{bmatrix} \omega_z^2 + 2C_1 - 4C_2(1-3i_{x_3}^2) & \dot{\omega}_z - 8C_2 i_{x_3}i_{y_3} - C_c\omega_z^2 R & -\omega_x\omega_z - 8C_2 i_{x_3}i_{z_3} \\ -\dot{\omega}_z - 8C_2 i_{x_3}i_{y_3} + C_c\omega_z^2 R & (\omega_x^2+\omega_z^2) - C_1 + C_2\left(1-7i_{x_3}^2+2(1-i_{z_3}^2)\right) & \dot{\omega}_x + 2C_2 i_{y_3}i_{z_3} - C_c\omega_x\omega_z R \\ -\omega_x\omega_z - 8C_2 i_{x_3}i_{z_3} & -\dot{\omega}_x + 2C_2 i_{y_3}i_{z_3} + C_c\omega_x\omega_z R & \omega_x^2 - C_1 + C_2\left(3-5i_{x_3}^2-2(1-i_{z_3}^2)\right) \end{bmatrix}$ $\boldsymbol{A}_{22} = \begin{bmatrix} C_c\omega_z R & 2\omega_z & 0 \\ -2\omega_z & C_c\omega_z R & 2\omega_x \\ 0 & -2\omega_x & C_c\omega_z R \end{bmatrix}$ $C_1 = \frac{\mu}{R^3},\quad C_2 = -\frac{3}{2}J_2\frac{\mu R_\oplus^2}{R^5},\quad C_c = -\frac{1}{2}\frac{C_D A}{m}\rho$

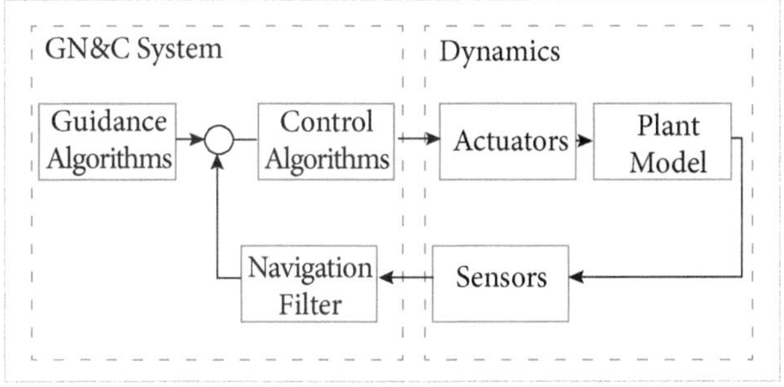

Figure 4. Closed Loop GN&C System.

$$\frac{d\hat{\omega}}{dt} = \frac{1}{\hat{H}\hat{e}}\{-\hat{p}c_{\hat{f}}\hat{a}_x + (\hat{p} + \hat{R})s_{\hat{f}}\hat{a}_y\} - \frac{\hat{R}s_{\hat{\omega}+\hat{f}}c_{\hat{\imath}}}{\hat{H}s_{\hat{\imath}}}\hat{a}_z \quad (18e)$$

$$\frac{d\hat{f}}{dt} = \frac{\hat{H}}{\hat{R}^2} + \frac{1}{\hat{H}\hat{e}}\left[\hat{p}c_{\hat{f}}\hat{a}_x - (\hat{p} + \hat{R})s_{\hat{f}}\hat{a}_y\right] \quad (18f)$$

where

$$[\hat{\boldsymbol{a}}]^{\mathrm{LVLH}} = \left[\hat{\boldsymbol{a}}_{J_2} + \hat{\mathbf{f}}_{\mathrm{a}}\right]^{\mathrm{LVLH}} \quad (19)$$

The orbit perturbed acceleration term, $\hat{a}$, is different form the term used in the truth model in which it does not contain the unmodeled disturbance acceleration term f_w. This navigation target model is used only to assist in the process of estimation. The dynamic modelfor the relative navigation states are:

$$\begin{bmatrix}\hat{\boldsymbol{\rho}}(t)\\ \dot{\hat{\boldsymbol{\rho}}}(t)\end{bmatrix} = \phi_{\mathrm{LTV}}(t, t_0)\begin{bmatrix}\hat{\boldsymbol{\rho}}(t_0)\\ \dot{\hat{\boldsymbol{\rho}}}(t_0)\end{bmatrix} \quad (20)$$

where ϕ_{LTV} is the state transition matrix, and it is defined by Eq. (17) for the relative linear time varying model.

The navigation model for the target angular motion is used only to produce a reference attitude trajectory. This trajectory will be tracked by the chaser attitude control system.

$$\dot{\hat{\boldsymbol{q}}}_I^t = \frac{1}{2}\hat{\omega}^t \otimes \hat{\boldsymbol{q}}_I^t \quad (21a)$$

$$\dot{\hat{\boldsymbol{\omega}}}^t = \hat{\boldsymbol{I}}_t^{-1}\left[\hat{\boldsymbol{\tau}}_t - \hat{\omega}^t \times \hat{\boldsymbol{I}}_t\hat{\omega}^t\right] \quad (21b)$$

$$\hat{\boldsymbol{\tau}}_t = \hat{\boldsymbol{\tau}}_{tg} \quad (21c)$$

For the chaser vehicle, the propagation of the state can be accomplished by using numerical integration techniques. However, in general, the gyros observations are sampled at a high rate (usually higher than or at least equal to the same rate as the vector attitude observations). A discrete propagation is usually sufficient. Discrete propagation can be derived using a power series approach (Crassidis and Junkins, 2004).

$$\hat{\boldsymbol{q}}_I^c(t) = \boldsymbol{\Omega}(\hat{\boldsymbol{\omega}}^c)\hat{\boldsymbol{q}}_I^c(t_0) \quad (22)$$

where

$$\hat{\boldsymbol{\omega}}^c = \tilde{\boldsymbol{\omega}}^c - \hat{\boldsymbol{b}}_\omega^c \quad (23a)$$

$$\hat{\boldsymbol{\omega}}^c = \tilde{\boldsymbol{\omega}}^c - \hat{\boldsymbol{b}}_\omega^c$$

$$\boldsymbol{\Omega}(\hat{\boldsymbol{\omega}}^c) = \begin{bmatrix} cos\left(\frac{1}{2}\|\hat{\boldsymbol{\omega}}^c\|\Delta t\right)\boldsymbol{I}_{3\times3} - [\hat{\boldsymbol{\psi}}^c \times] & \hat{\boldsymbol{\psi}}^c \\ -\hat{\boldsymbol{\psi}}^{c\mathrm{T}} & cos\left(\frac{1}{2}\|\hat{\boldsymbol{\omega}}^c\|\Delta t\right)\end{bmatrix},$$

$$\hat{\boldsymbol{\psi}}^c = \frac{sin\left(\frac{1}{2}\|\hat{\boldsymbol{\omega}}^c\|\Delta t\right)\hat{\boldsymbol{\omega}}^c}{\|\hat{\boldsymbol{\omega}}^c\|} \quad (23b)$$

The propagation dynamic model for the error parameters is given by

$$\begin{bmatrix}\hat{\boldsymbol{b}}_\omega^c(t)\\ \hat{\boldsymbol{\epsilon}}_s^s(t)\\ \hat{\boldsymbol{\epsilon}}_l^l(t)\end{bmatrix} = \phi_{\mathrm{Markov}}(t, t_0)\begin{bmatrix}\hat{\boldsymbol{b}}_\omega^c(t_0)\\ \hat{\boldsymbol{\epsilon}}_s^s(t_0)\\ \hat{\boldsymbol{\epsilon}}_l^l(t_0)\end{bmatrix} \quad (24)$$

where ϕ_{Markov} is defined as follows:

$$\phi_{Markov} = \begin{bmatrix} e^{-\frac{\Delta t}{\tau_b^\omega}}\boldsymbol{I}_{3\times3} & \boldsymbol{0}_{3\times3} & \boldsymbol{0}_{3\times3} \\ \boldsymbol{0}_{3\times3} & e^{-\frac{\Delta t}{\tau^s}}\boldsymbol{I}_{3\times3} & \boldsymbol{0}_{3\times3} \\ \boldsymbol{0}_{3\times3} & \boldsymbol{0}_{3\times3} & e^{-\frac{\Delta t}{\tau^l}}\boldsymbol{I}_{3\times3}\end{bmatrix} \quad (25)$$

An extended Kalman filter is derived from the nonlinear models as illustrated in the equations below (Brown and Hawag, 1997).

$$\dot{\boldsymbol{x}} = \boldsymbol{f}(\boldsymbol{x}, \boldsymbol{u}, t) + \boldsymbol{w}(t), \qquad \boldsymbol{w}(t) \sim N(0, Q) \quad (26a)$$

$$\tilde{\boldsymbol{z}}_k = \boldsymbol{h}(\boldsymbol{x}_k, t) + \boldsymbol{v}_k, \quad \boldsymbol{v}_k \sim N(0, R_k) \quad (26b)$$

Here, the state vector x can represent relative position, velocity, and orientations of the chaser as well as other parameters that need to be estimated for the use by other flight algorithms. The time derivatives of the states $\dot{x}$ are a function of the states, inputs, time, and additive process noise w. This process noise is used to approximate unmodeled disturbances and other random disturbances to the dynamics. The measurements $\tilde{z}_k$ are modeled as a function of the states, time, and measurement noise v_k. The process noise and measurement noise are normally distributed with zero mean and covariance Q and R_k respectively.

The following steps summarize the Kalman filter equations, that are used to estimate the relative motion states and it is based on minimizing mean square of the error.

- Enter prior estimate of x_k^- and its error covariance P_k^- and compute the Kalman gain

$$K_k = P_k^- H_k^T (H_k P_k^- H_k^T + R_k)^{-1} \tag{27a}$$

- Update estimate by measurement $\tilde{z}_k$

$$\hat{\boldsymbol{z}}_k^- = \boldsymbol{h}(\hat{\boldsymbol{x}}_k^-) \tag{27b}$$

$$\hat{\boldsymbol{x}}_k = \hat{\boldsymbol{x}}_k^- + K_k(\tilde{\boldsymbol{z}}_k - \hat{\boldsymbol{z}}_k^-) \tag{27c}$$

- Compute error covariance for updated estimate

$$P_k = (\boldsymbol{I} - K_k H_k)P_k^- \tag{27d}$$

- Project ahead

$$\hat{\boldsymbol{x}}_{k+1}^- = \phi_k \hat{\boldsymbol{x}}_k \tag{27e}$$

$$P_{k+1}^- = \phi_k P_k \phi_k^T + Q_k \tag{27f}$$

The term ϕ_k is the state transition matrix, and H_k is the measurement partial matrix that represents the sensitivity of the measurements to changes in the states. The state vector of the Kalman filter is defined to be:

$$\boldsymbol{x} = [\boldsymbol{\rho} \quad \dot{\boldsymbol{\rho}} \quad \boldsymbol{\theta}_c \quad \boldsymbol{b}_\omega^c \quad \boldsymbol{\epsilon}_s^s \quad \boldsymbol{\epsilon}_l^l]^{\mathrm{T}} \tag{28}$$

and Kalman filter matrices are given by:

$$\phi_k = \begin{bmatrix} \phi_{\mathrm{LTV}} & \boldsymbol{0}_{6\times6} & \boldsymbol{0}_{6\times6} \\ \boldsymbol{0}_{6\times6} & \phi_{\mathrm{Attitude}} & \boldsymbol{0}_{6\times6} \\ \boldsymbol{0}_{6\times6} & \boldsymbol{0}_{6\times6} & \phi_{\mathrm{Markov}} \end{bmatrix} \tag{29a}$$

$$P_k = \begin{bmatrix} P_{\mathrm{LTV}} & \boldsymbol{0}_{6\times6} & \boldsymbol{0}_{6\times6} \\ \boldsymbol{0}_{6\times6} & P_{\mathrm{Attitude}} & \boldsymbol{0}_{6\times6} \\ \boldsymbol{0}_{6\times6} & \boldsymbol{0}_{6\times6} & P_{\mathrm{Markov}} \end{bmatrix} \tag{29b}$$

$$Q_k = \begin{bmatrix} Q_{\mathrm{LTV}} & \boldsymbol{0}_{6\times6} & \boldsymbol{0}_{6\times6} \\ \boldsymbol{0}_{6\times6} & Q_{\mathrm{Attitude}} & \boldsymbol{0}_{6\times6} \\ \boldsymbol{0}_{6\times6} & \boldsymbol{0}_{6\times6} & Q_{\mathrm{Markov}} \end{bmatrix} \tag{29c}$$

The state vector contains θ_c instead of q_I^c because the quaternion must obey a normalization constraint, which can be violated by the linear measurement updates associated with the filter. The most common approach to overcome this shortfall involves using a multiplicative error quaternion, where after neglecting higher order terms, the four component quaternion can effectively be replaced by a three component error vector θ_c (Crassidis and Junkins, 2004).Therefore, within first order, the quaternion update is given by:

$$\boldsymbol{q}_I^{c+} = \boldsymbol{\delta q}(\boldsymbol{\theta}_c) \otimes \boldsymbol{q}_I^{c-} \tag{30}$$

and the discrete attitude error state transition matrix can also be derived using a power series approach to be:

$$\phi_{\mathrm{Attitude}} = \begin{bmatrix} \phi_{11}^c & \phi_{12}^c \\ \phi_{21}^c & \phi_{22}^c \end{bmatrix} \tag{31}$$

where

$$P_{\mathrm{LTV}}^- = \begin{bmatrix} \sigma_x^2 & 0 & 0 & 0 & \varepsilon\sigma_x\sigma_{\dot{y}} & 0 \\ 0 & \sigma_y^2 & 0 & \varepsilon\sigma_{\dot{x}}\sigma_y & 0 & 0 \\ 0 & 0 & \sigma_z^2 & 0 & 0 & 0 \\ 0 & \varepsilon\sigma_{\dot{x}}\sigma_y & 0 & \sigma_{\dot{x}}^2 & 0 & 0 \\ \varepsilon\sigma_x\sigma_{\dot{y}} & 0 & 0 & 0 & \sigma_{\dot{y}}^2 & 0 \\ 0 & 0 & 0 & 0 & 0 & \sigma_{\dot{z}}^2 \end{bmatrix} \tag{32a}$$

$$\phi_{12}^c = [\hat{\boldsymbol{\omega}}^c \times] \frac{\{1 - \cos(\|\hat{\boldsymbol{\omega}}^c\|\Delta t)\}}{\|\hat{\boldsymbol{\omega}}^c\|^2} - \boldsymbol{I}_{3\times3}\Delta t - [\hat{\boldsymbol{\omega}}^c \times]^2 \frac{\{\|\hat{\boldsymbol{\omega}}^c\|\Delta t - \sin(\|\hat{\boldsymbol{\omega}}^c\|\Delta t)\}}{\|\hat{\boldsymbol{\omega}}^c\|^3} \tag{32b}$$

$$\phi_{21}^c = \boldsymbol{0}_{3\times3} \tag{32c}$$

$$\phi_{22}^c = \boldsymbol{I}_{3\times3} \tag{32d}$$

By following the line steps of Woffinden and Geller (2007), Woffinden (2004) and Lear (1985), the initial error covariance matrix P_o^-, which represents how accurate the initial states are known, is given below for the proposed linear relative model, attitude, and error parameters.

$$P_{\text{LTV}}^- = \begin{bmatrix} \sigma_x^2 & 0 & 0 & 0 & \varepsilon\sigma_x\sigma_{\dot{y}} & 0 \\ 0 & \sigma_y^2 & 0 & \varepsilon\sigma_{\dot{x}}\sigma_y & 0 & 0 \\ 0 & 0 & \sigma_z^2 & 0 & 0 & 0 \\ 0 & \varepsilon\sigma_{\dot{x}}\sigma_y & 0 & \sigma_{\dot{x}}^2 & 0 & 0 \\ \varepsilon\sigma_x\sigma_{\dot{y}} & 0 & 0 & 0 & \sigma_{\dot{y}}^2 & 0 \\ 0 & 0 & 0 & 0 & 0 & \sigma_{\dot{z}}^2 \end{bmatrix} \tag{33a}$$

$$P_{\text{Attitude}}^- = \begin{bmatrix} \sigma_{w_b^\theta}^2 & \mathbf{0}_{3\times3} \\ \mathbf{0}_{3\times3} & \sigma_{w_b^\omega}^2 \end{bmatrix} \tag{33b}$$

$$P_{\text{Markov}}^- = \begin{bmatrix} \sigma_{w^s}^2 \boldsymbol{I}_{3\times3} & \mathbf{0}_{3\times3} \\ \mathbf{0}_{3\times3} & \sigma_{w^l}^2 \boldsymbol{I}_{3\times3} \end{bmatrix} \tag{33c}$$

Parameters σ_x, σ_y and σ_z denote the standard deviation uncertainties of the relative position components, and $\sigma_{\dot{x}}$, $\sigma_{\dot{y}}$ and $\sigma_{\dot{z}}$ are for the relative velocity components. The coefficient ε refers to the uncertainty correlation coupling between relative position and velocity components in the LVLH coordinate frame, and it ranges between a positive and a negative one. The standard deviations $\sigma_{w_b^\theta}$, $\sigma_{w_b^\omega}$, σ_{w^s} and σ_{w^l} are referring to the uncertainties of initial attitude, gyro biases, star tracker misalignments, and LIDAR misalignments, respectively. The discrete process noise matrix components of the relative motion canbe approximated by:

$$Q_{\text{LTV}} = \begin{bmatrix} \sigma_{W_x}^2\left(\frac{\Delta t^3}{3}\right) & 0 & 0 & \sigma_{W_x}^2\left(\frac{\Delta t^2}{2}\right) & 0 & 0 \\ 0 & \sigma_{W_y}^2\left(\frac{\Delta t^3}{3}\right) & 0 & 0 & \sigma_{W_y}^2\left(\frac{\Delta t^2}{2}\right) & 0 \\ 0 & 0 & \sigma_{W_z}^2\left(\frac{\Delta t^3}{3}\right) & 0 & 0 & \sigma_{W_z}^2\left(\frac{\Delta t^2}{2}\right) \\ \sigma_{W_x}^2\left(\frac{\Delta t^2}{2}\right) & 0 & 0 & \sigma_{W_x}^2(\Delta t) & 0 & 0 \\ 0 & \sigma_{W_y}^2\left(\frac{\Delta t^2}{2}\right) & 0 & 0 & \sigma_{W_y}^2(\Delta t) & 0 \\ 0 & 0 & \sigma_{W_z}^2\left(\frac{\Delta t^2}{2}\right) & 0 & 0 & \sigma_{W_z}^2(\Delta t) \end{bmatrix} \tag{34a}$$

$$Q_{\text{Attitude}} = \begin{bmatrix} \left(\sigma_{v_\omega^c}^2 \Delta t + \frac{1}{3}\sigma_{v_b^\omega}^2 \Delta t^3\right)\boldsymbol{I}_{3\times3} & -\left(\frac{1}{2}\sigma_{v_b^\omega}^2 \Delta t^2\right)\boldsymbol{I}_{3\times3} \\ -\left(\frac{1}{2}\sigma_{v_b^\omega}^2 \Delta t^2\right)\boldsymbol{I}_{3\times3} & -\left(\sigma_{v_b^\omega}^2 \Delta t\right)\boldsymbol{I}_{3\times3} \end{bmatrix} \tag{34b}$$

$$Q_{\text{Markov}} = \begin{bmatrix} \sigma_{v^s}^2 \Delta t \boldsymbol{I}_{3\times3} & \mathbf{0}_{3\times3} \\ \mathbf{0}_{3\times3} & \sigma_{v^l}^2 \Delta t \boldsymbol{I}_{3\times3} \end{bmatrix} \tag{34c}$$

Here, σ_{W_x}, σ_{W_y} and σ_{W_x} are the standard deviations for the random unmodeled acceleration disturbances that act on the relative motion during the sample time period Δt and $\sigma_{V_\omega^c}$, $\sigma_{V_b^\omega}$, σ_{V^s} and σ_{V^l} are the random process uncertainty noises for gyros, gyro biases, star tracker misalignments, and LIDAR misalignments, respectively.

The measurements sensitivity matrices H_k and sensor measurements noise matrices R_k are defined for both star sensor and LIDAR as:

$$H_k = \begin{bmatrix} H_k^{lidar} \\ H_k^s \end{bmatrix} \tag{35}$$

$$R_k = \begin{bmatrix} R_k^{lidar} & \mathbf{0}_{3\times3} \\ \mathbf{0}_{3\times3} & R_k^s \end{bmatrix} \tag{36}$$

The measurement partials for the azimuth, elevation and range measurements are computed with the help of the LIDAR measurement range vector. Utilizing Eq. (13) and small angle approximations leads to the following equation for the relative range in terms of the navigation states:

$$\boldsymbol{\rho}^l = \rho\begin{pmatrix} c_\alpha c_\beta \\ s_\alpha c_\beta \\ s_\beta \end{pmatrix} = [\boldsymbol{I}_{3\times3} - \boldsymbol{\epsilon}_l^{l\times}][\boldsymbol{I}_{3\times3} - \boldsymbol{\epsilon}_s^{s\times}][\boldsymbol{I}_{3\times3} - \boldsymbol{\theta}_c{}^{\times}]\boldsymbol{T}^{\hat{s}I}(\boldsymbol{q}_I^{c-})\boldsymbol{T}^{lT}(-\boldsymbol{\rho}) \tag{37}$$

Using the chain rule, the partial of the range vector with respect to the navigation states can be expressed as (Woffinden and Geller, 2007):

$$\left.\frac{\partial\boldsymbol{\rho}^l(\boldsymbol{x})}{\partial\boldsymbol{x}}\right|_{\hat{x}} = \left[\frac{\partial\boldsymbol{\rho}^l(\boldsymbol{x})}{\partial\alpha}\frac{\partial\alpha(\boldsymbol{x})}{\partial\boldsymbol{x}} + \frac{\partial\boldsymbol{\rho}^l(\boldsymbol{x})}{\partial\beta}\frac{\partial\beta(\boldsymbol{x})}{\partial\boldsymbol{x}} + \frac{\partial\boldsymbol{\rho}^l(\boldsymbol{x})}{\partial\rho}\frac{\partial\rho(\boldsymbol{x})}{\partial\boldsymbol{x}}\right]_{\hat{x}} = \left[(\rho\boldsymbol{p}_\alpha^l)\boldsymbol{h}_\alpha + (\rho\boldsymbol{p}_\beta^l)\boldsymbol{h}_\beta + \boldsymbol{p}_\rho^l\boldsymbol{h}_\rho\right]_{\hat{x}} \tag{38}$$

$$\boldsymbol{p}_\alpha^l = \begin{bmatrix} -c_\beta s_\alpha \\ c_\beta c_\alpha \\ 0 \end{bmatrix}, \quad \boldsymbol{p}_\beta^l = \begin{bmatrix} -s_\beta c_\alpha \\ -s_\beta s_\alpha \\ c_\beta \end{bmatrix}, \quad \boldsymbol{p}_\rho^l = \begin{bmatrix} c_\beta c_\alpha \\ c_\beta s_\alpha \\ s_\beta \end{bmatrix} \tag{39}$$

The measurement geometry can now be computed by taking the advantages of the property that p_α^l, p_β^l and $p_p{}^l$ are orthogonal to each other and taking the dot product with respect to each of them.

$$\boldsymbol{h}_\alpha = \frac{(\boldsymbol{p}_\alpha^l)^{\mathrm{T}}}{\rho c_\beta^2}\left[\frac{\partial\boldsymbol{\rho}^l(\boldsymbol{x})}{\partial\boldsymbol{x}}\right], \quad \boldsymbol{h}_\beta = \frac{(\boldsymbol{p}_\beta^l)^{\mathrm{T}}}{\rho}\left[\frac{\partial\boldsymbol{\rho}^l(\boldsymbol{x})}{\partial\boldsymbol{x}}\right], \quad \boldsymbol{h}_\rho = (\boldsymbol{p}_\rho^l)^{\mathrm{T}}\left[\frac{\partial\boldsymbol{\rho}^l(\boldsymbol{x})}{\partial\boldsymbol{x}}\right] \tag{40}$$

The evaluation of the relative range vector with respect to the navigation states yields

$$\left[\frac{\partial \boldsymbol{\rho}^l(\boldsymbol{x})}{\partial \boldsymbol{x}}\right] = \left[-\boldsymbol{T}^{lT} \quad \mathbf{0}_{3\times 3} \quad -\boldsymbol{T}^{l\hat{l}}\boldsymbol{T}^{\hat{l}s}[\boldsymbol{\rho}^{\hat{s}}\times] \quad \mathbf{0}_{3\times 3} \quad -\boldsymbol{T}^{l\hat{l}}[\boldsymbol{\rho}^{s}\times] \quad -[\boldsymbol{\rho}^{\hat{l}}\times]\right] \tag{41}$$

Now, the LIDAR measurement sensitivity matrix and covariance matrix can be written as:

$$H_k^{lidar} = \left[\hat{\boldsymbol{h}}_\alpha \quad \hat{\boldsymbol{h}}_\beta \quad \hat{\boldsymbol{h}}_\rho\right]^{\mathrm{T}} \tag{42}$$

and

$$R_k^{lidar} = \begin{bmatrix} \sigma_\alpha^2 & 0 & 0 \\ 0 & \sigma_\beta^2 & 0 \\ 0 & 0 & \sigma_\rho^2 \end{bmatrix} \tag{43}$$

When processing star tracker data, a derived measurement is calculated (Woffinden and Geller, 2007). This quantity is effectively the residual to be processed by the filter.

$$\begin{bmatrix} \frac{1}{2}\tilde{\boldsymbol{z}}_S \\ 1 \end{bmatrix} = \tilde{\boldsymbol{q}}_I^S \otimes [\hat{\boldsymbol{q}}_c^I \otimes \hat{\boldsymbol{q}}_S^c \otimes \boldsymbol{\delta}\hat{\boldsymbol{q}}(\hat{\boldsymbol{\epsilon}}_S^S)] \tag{44}$$

The derived star tracker measurement can be written as a function of the navigation states as:

$$\tilde{\boldsymbol{z}}_s = \boldsymbol{h}_s(\boldsymbol{x}) + \boldsymbol{v}_s^s = \boldsymbol{\theta}_c^s + \boldsymbol{\epsilon}_s^s + \boldsymbol{v}_s^s \tag{45}$$

Therefore, the measurement sensitivity matrix for the star tracker can be derived to be

$$H_k^s = \left.\frac{\partial \boldsymbol{h}_s(\boldsymbol{x})}{\partial \boldsymbol{x}}\right|_{\hat{\boldsymbol{x}}} = [\mathbf{0}_{6\times 6} \quad \boldsymbol{I}_{3\times 3} \quad \mathbf{0}_{3\times 3} \quad \boldsymbol{I}_{3\times 3} \quad \mathbf{0}_{3\times 3}] \tag{46}$$

and the star tracker measurement covariance is

$$R_k^s = \sigma_s^2 \boldsymbol{I}_{3\times 3} \tag{47}$$

For close proximity operations, a propositional-derivative (PD) controller is employed for both the rotational and translational controls. The commanded torques for the chaser spacecraft to match its orientation with the target vehicle are computed as

$$\hat{\boldsymbol{\tau}}_{cc} = \boldsymbol{K}_q \boldsymbol{\delta}\hat{\boldsymbol{q}} + \boldsymbol{K}_\omega \boldsymbol{\delta}\hat{\boldsymbol{\omega}} \tag{48}$$

where

$$\boldsymbol{\delta}\hat{\boldsymbol{q}} = \boldsymbol{\delta}\hat{\boldsymbol{q}}_{e_v}\ \boldsymbol{\delta}\hat{\boldsymbol{q}}_{e_s} \tag{49a}$$

$$\boldsymbol{\delta}\hat{\boldsymbol{\omega}} = \hat{\boldsymbol{\omega}}_{des}^c - \hat{\boldsymbol{\omega}}^c \tag{49b}$$

and

$$\hat{\boldsymbol{q}}_{I_{des}}^c = \hat{\boldsymbol{q}}_I^t \tag{50a}$$

$$\hat{\boldsymbol{\omega}}_{des}^c = \boldsymbol{T}^{ct}\hat{\boldsymbol{\omega}}^t \tag{50b}$$

$$\hat{\boldsymbol{q}}_e = \hat{\boldsymbol{q}}_{I_{des}}^c \otimes \hat{\boldsymbol{q}}_I^c = \begin{pmatrix} \delta\hat{\boldsymbol{q}}_{e_v} \\ \delta\hat{\boldsymbol{q}}_{e_s} \end{pmatrix} \tag{50c}$$

$\hat{q}_{I_{desc}}^c$ and $\hat{\omega}_{desc}^c$ are the desired orientation and angular velocity, respectively, to be tracked by the chaser vehicle. The angular offset and angular rate offset between target and chaser are denoted by $\delta\hat{q}_e$ and $\delta\hat{\omega}$, respectively. The proportional and derivative control gains K_q and K_ω are determined based on the desired natural frequency ω_θ, damping ratio ζ_θ of the attitude control system, and the moment of inertia of the chaser spacecraft I_c (Wie, 1998).

$$\boldsymbol{K}_q = \omega_\theta^2 \boldsymbol{I}_c, \qquad \boldsymbol{K}_\omega = 2\zeta_\theta\omega_\theta \boldsymbol{I}_c \tag{51}$$

On the other hand, The translation control algorithm computes the required continuous thrust, f_c, based on the previous linear model, in order to track the desired trajectory specified by the following guidance algorithm:

$$\boldsymbol{U}_{\Delta V} = \mathbf{f_c} = \boldsymbol{K}_\rho \boldsymbol{\delta\rho} + \boldsymbol{K}_{\dot{\rho}} \boldsymbol{\delta\dot{\rho}} \tag{52a}$$

$$\boldsymbol{\delta\rho} = \boldsymbol{\rho}_{des} - \hat{\boldsymbol{\rho}} \tag{52b}$$

$$\boldsymbol{\delta\dot{\rho}} = \dot{\boldsymbol{\rho}}_{des} - \hat{\dot{\boldsymbol{\rho}}} \tag{52c}$$

The proportional and derivative control gains K_ρ and $K_{\dot{\rho}}$ are determined based on the desired natural frequency ω_ρ and damping ratio ζ_ρ of the translational control system.

$$\boldsymbol{K}_\rho = \omega_\rho^2 \boldsymbol{I}_{3\times 3} \boldsymbol{K}_{\dot{\rho}} = 2\zeta_\rho\omega_\rho \boldsymbol{I}_{3\times 3} \tag{53}$$

Variables ρ_{des} and $\dot{\rho}_{des}$ are, respectively, the desired relative position and relative velocity to be tracked by the chaser vehicle, and it is defined by the guidance algorithms. It is worth noting that the equivalent continuous velocity increment ΔV, based on the continuous thrust, can be approximated for small to be

$$\boldsymbol{\Delta V} \approx \boldsymbol{U}_{\Delta V}\Delta t \quad (54)$$

SIMULATION EXAMPLES

The key metrics of the analysis fall into three main categories. The first is navigation performance, which is how well the states are estimated by the filter. This metric is measured by the navigation error, the difference between the true states and the filter states. The second is trajectory control performance, which is a measure of how closely the chaser vehicle is able to follow the guidance algorithms. The third is fuel performance, or ΔV fuel usage, and it is computed based on the linear model developed in the previous section.

The preceding guidance and navigation algorithms are illustrated now through different examples. Initial conditions for simulation are listed in Tables 4 to 6.

A Simulink model is built using the MATLAB software to demonstrate the closed-loop guidance transfer of the chaser in order to approach and to depart from the target vehicle in any orbit, either circular or elliptic, given uncertain initial conditions, noisy measurements, and limited dynamics. This model consists of three main parts, guidance, navigation, and control. The proposed linear time varying model is used in designing the navigation filter and in maneuver targeting of the guidance system. The target is assumed to be in a passive nadir pointing mode andnot in maneuvering. The chaser uses star tracker data and gyro data to determine attitude and attitude rate. Momentum wheels and PD controller are used to point the chaser LIDAR at the target. The chaser uses LIDAR data to determine the relative position and velocity of the target. Maneuver targeting algorithms, based on PD controller, are used to compute commands in the chaser body frame as to track the desired trajectory.

The performance of the navigation system is shown in Figs. 5 to 7. In this case, the thrusters are off and both target and chaser vehicles are initially in the same neighborhood (Table 5). Figure 5 shows the relative position and relative velocity between the vehicles during simulation. Figure 6 depicts how accurately the navigation system can estimate the chaser's relative position and velocity. Form this figure, the filter is able to converge within few minutes and the relative position and velocity can be accurately estimatedwithin the accuracy of the sensors. The attitude navigation errors and the PD control tracking performance are shown in Fig. 7. As indicated by this figure, the chaser attitude navigation system is able to converge quickly and the chaser attitude PD controller can track the target attitude and angular velocity trajectories.

The basic glidelope rendezvous and close proximity operations scenario used to evaluate the performance of the entire closed-loop relative position and attitude control system with the navigation filter consists of two main segments: the inbound and theoutbound segments. Each segment of the glideslope is followed by 3 minutes of station keeping. First, the inbound segment: the chaser starts to approach the target form [58-580 0] m behind the target and ends at [0-100 0] m. After 3 minutes

Table 4. Navigation Filter Parameters.

Parameter	Value
Initial Relative Position and Velocity Uncertainties	$\sigma_x = \sigma_y = \sigma_z = 33.3$ m $\sigma_{\dot{x}} = \sigma_{\dot{y}} = \sigma_{\dot{z}} = 0.01$ m/s
Process Noise	$\sigma_{w_x} = \sigma_{w_y} = 5 \times 10^{-4}\text{m/s}^{3/2}$ $\sigma_{w_z} = 5 \times 10^{-6}\text{m/s}^{3/2}$
Measurements Noise	$\sigma_\alpha = \sigma_\beta = 0.06$ deg, $\sigma_\rho = 0.5$ m
Simulation Step	0.1 s
Measurements Update	1 Hz

Table 5. Vehicles Orbital Elements.

Parameter	Target	Chaser
a,km	6723.2576	6723.2576
e	0.1	0.1
i, deg	51.6467	51.6467
Ω,deg	188. 0147	188. 0147
ω,deg	174.3022	174.3022
f,deg	270.0882	270.0832

Table 6. Simulation Initial Conditions.

Initial Conditions		Parameter	Value
Inertia	Chaser	I_c	$\begin{bmatrix} 1.31 & 0 & 0 \\ 0 & 10.220 & 0 \\ 0 & 0 & 10.65 \end{bmatrix} \times 10^6$ kg-m^2
	Target	I_t	$\begin{bmatrix} 128 & 0 & 0 \\ 0 & 107 & 0 \\ 0 & 0 & 201 \end{bmatrix} \times 10^6$ kg-m^2
Initial Relative Attitude Errors		$\delta\phi, \delta\theta, \delta\psi$	[(7.5&-7.5&7.5)]deg
Control Parameters	Rotationalnatural frequency	ω_θ	1/30 s^{-1}
	Rotational damping ratio	ζ_θ	0.7
	Translational natural frequency	ω_p	1/50 s^{-1}
	Translational damping ratio	ζ_p	0.7
Unmodeled Disturbances	Rotational disturbances	τ_{td}	kg-km^2/s^2
	Translational disturbances	f_w	km/s^2
Sensors Errors	Gyro error (3)	Drift rate	3 deg/hr/axis
		Random walk	0.05 mrad/s$^{1/2}$
	Star Tracker error (3)	Misalignment	1 mrad/axis
		Noise	1 mrad/axis
		Measurements	1 Hz
	LIDAR error (3)	Misalignment	1 mrad/axis
		Noise	[1 mrad 1 mrad 0.5 m]
		Measurements	1 Hz

of station keeping at -100 m behind the target, the chaser starts to depart away from the target and leading to a new location -1000 m behind the target. The chase then stays at rest at that location for another 3 minutes. The results of this scenario are shown in Figs. 8 and 9. In all of these figures, different segments of the glideslope are shown, and the variations of in-plane relative motion of the chaser with respect to target vehicle are presented. Figure 8 shows the relative position and velocity plots of relative motion along with the required in order to achieve this trajectory maneuver, while Fig. 9 shows the error in relative position and velocity between the truth model and the navigation model. In all of the above glideslopes, the overall performance of the rendezvous and proximity operations are satisfactory.

The continuous thrust is calculated using the estimated relative position and velocity, either from the Kalman filter or from the knowledge of initial conditions, not from the true relative position and velocity of the chaser. As such, the chaser is not expected to reach its intended place exactly, but in the neighborhood thereof. Aided by the sensors, the initial estimation errors subside to an optimal level, determined by the ratio of the process noise matrix , and the measurement noise matrix ,earlier defined. Because of the active range and the angle measurements from the LIDAR system, and relatively small measurement errors, the true and the estimated relative position and velocity states are almost indistinguishable, as seen in previous figures during the steady state.

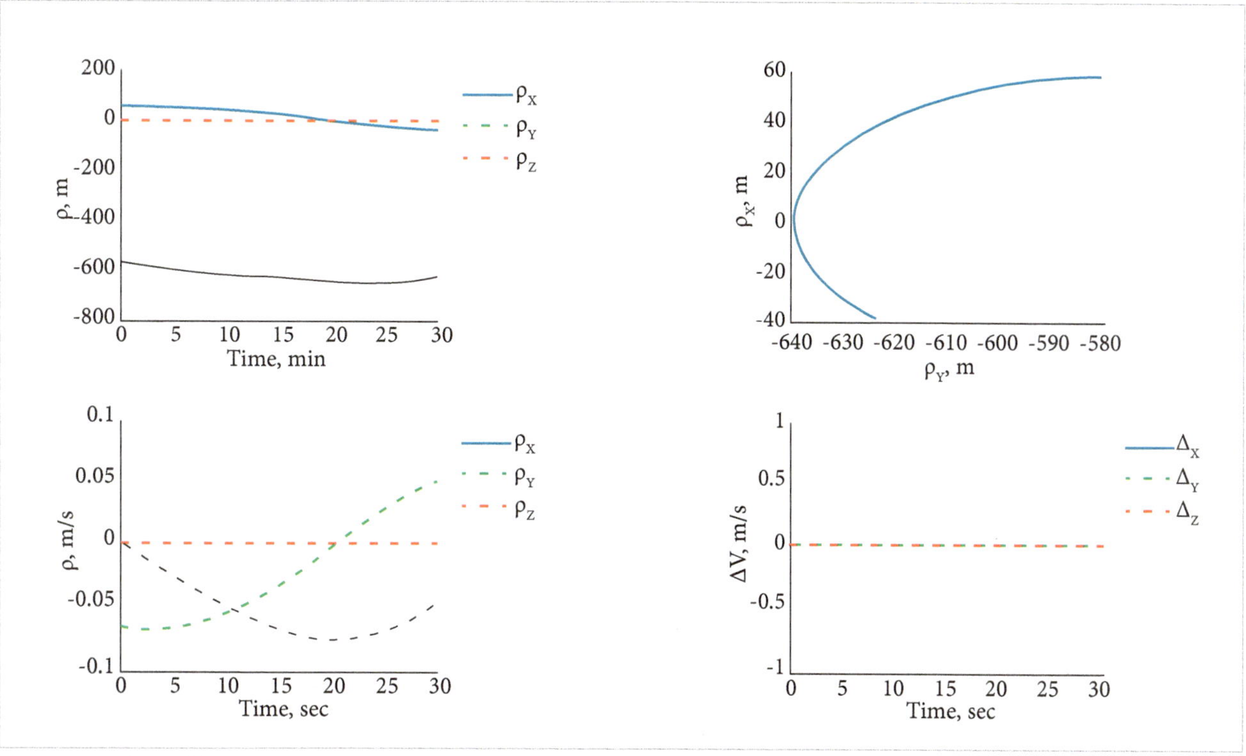

Figure 5. Relative Motion Without.

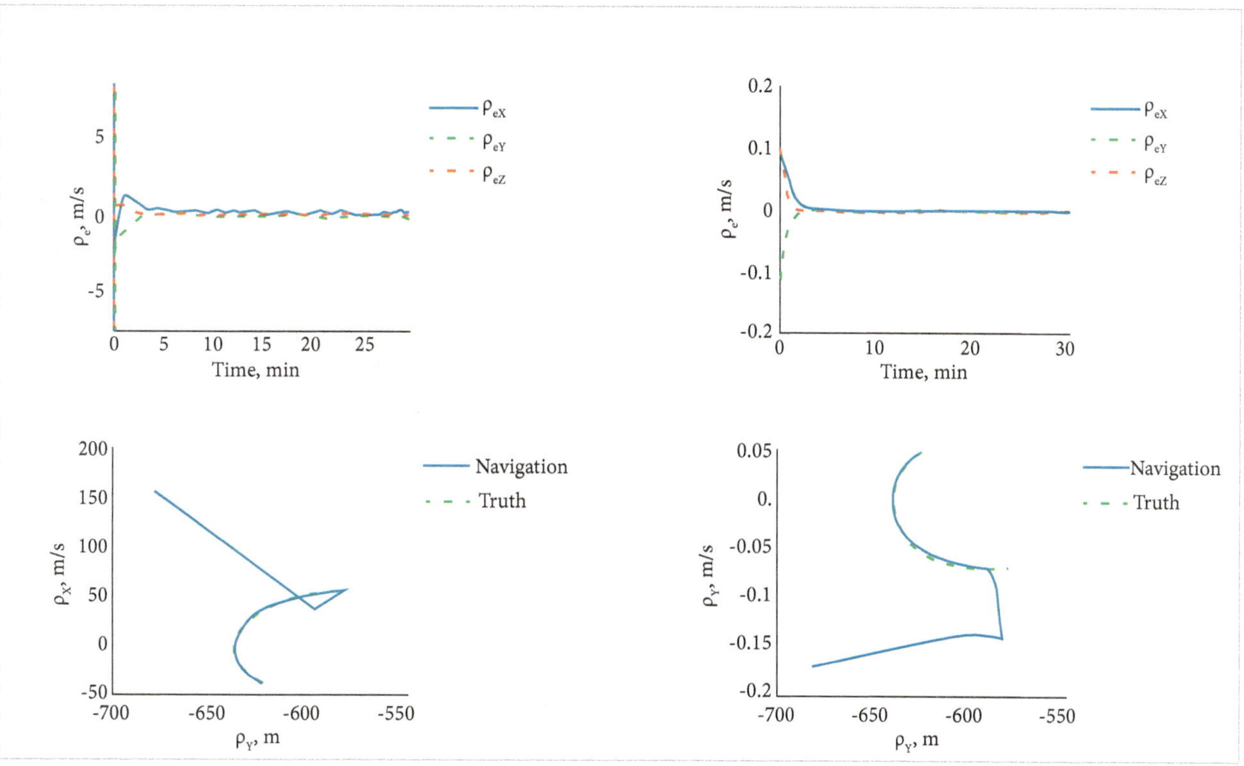

Figure 6. Relative Motion Navigation Performance Without.

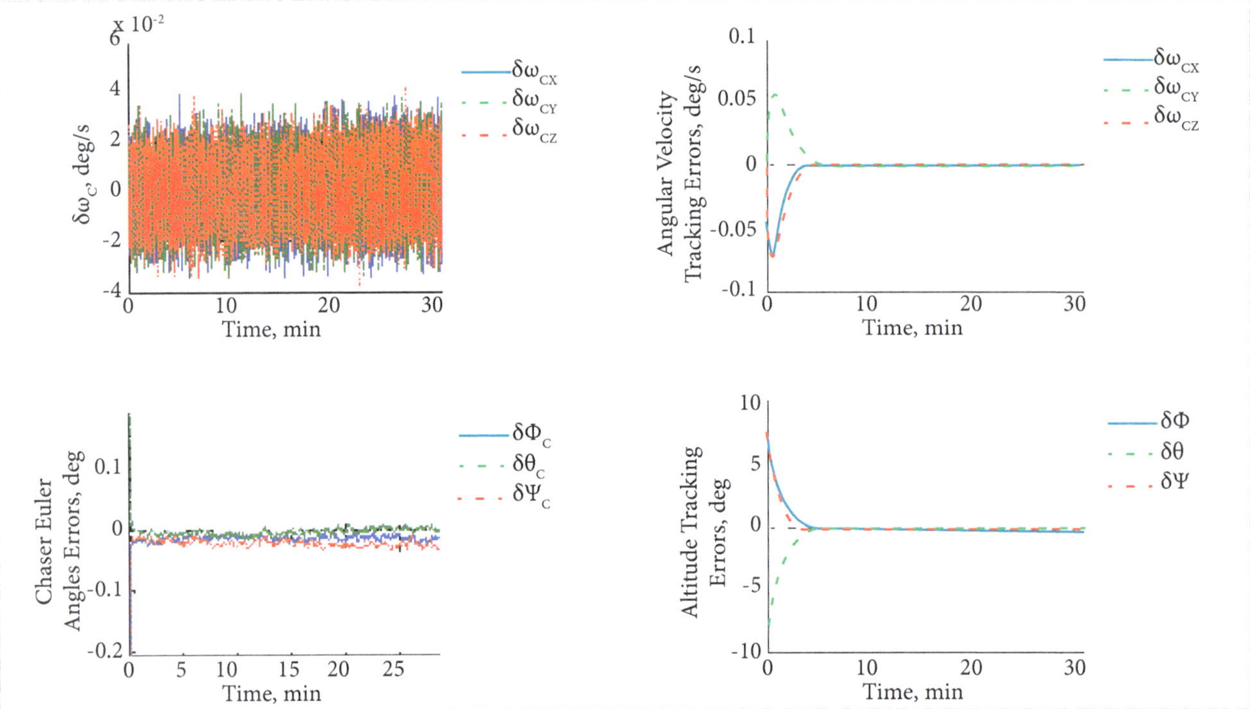

Figure 7. Chaser Attitude Navigation and Control Performance.

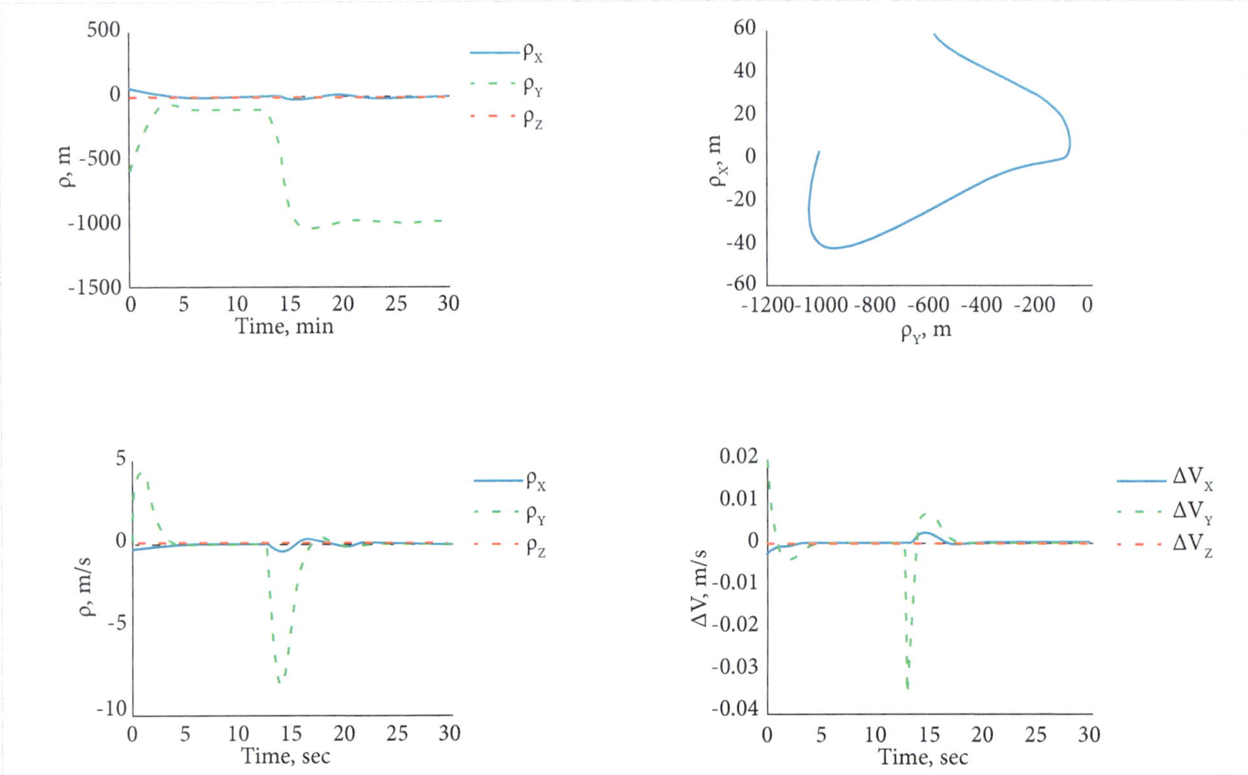

Figure 8. Relative Motion Inbound-Outbound Glideslope (Summary Scenario).

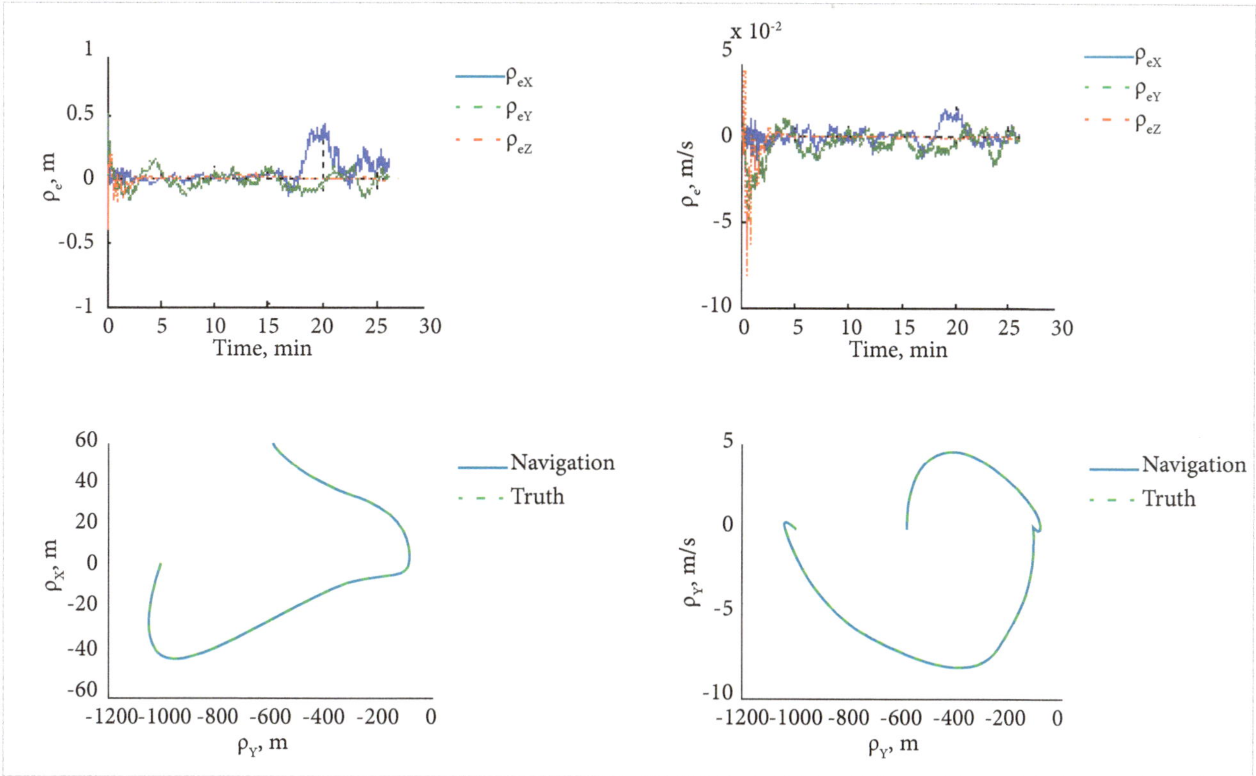

Figure 9. Relative Motion Navigation and Control Performance (Summary Scenario).

CONCLUSION

The results of this study indicate that the proposed linear model is clearly effective at estimating the relative position and velocity and controlling the relative trajectory. In addition, this model is not restricted to a circular orbit but it can be used as well for an eccentric orbit. Furthermore, by using this model, simple guidance algorithms for glideslope are developed to autonomously approach and depart form a target vehicle. The relative navigation in this study is utilizing range, azimuth, and elevation measurements of the target relative to the chaser froma simulated LIDAR system, along with the star tracker and gyro measurements of the chaser and an extended Kalman filter. The vehicle attitude dynamics, attitude tracking control, attitude determination, and uncertainties like measurement biases and sensor misalignments are considered in this study to fire the thrusters in the right direction and spin the momentum wheels at the proper rate in the chaser coordinate frame. The analyst must consider, in addition, off nominal situations, limitations and operational range of the sensors, and limitations of the actuators. These topics and others will be addressed in the future.

REFERENCES

Brown, R.G. and Hawag, P., 1997, "Introduction to Random Signals and Applied Kalman Filtering", 3rd Edition, John Wiley & Son Inc., United States.

Broucke, R.A., 2003, "Solution of the Elliptic Rendezvous Problem with the Time as Independent Variable", Journal of Guidance, Control, and Dynamics,Vol. 26, No. 4, pp. 615-621. doi: 10.2514/2.5089.

Carter,T.E., 1998, "State Transition Matrices for Terminal Rendezvous Studies: Brief Survey and New Example", Journal of Guidance, Control, and Dynamics, Vol. 21, No. 1, pp. 148-155.doi: 10.2514/2.4211.

Cho, H.C. and Park, S.Y., 2009, "Analytical Solution for Fuel-Optimal Reconfiguration in Relative Motion",Journal of Optimization Theory and Applications, Vol. 141, No.3, pp. 495-512.doi: 10.1007/s10957-008-9482-3.

Clohessy, W.H. and Wiltshire, R.S., 1960, "Terminal Guidance System for Satellite Rendezvous", Journal of the Aerospace Sciences, Vol. 27, No. 9, pp. 653–678.

Crassidis, J.L. and Junkins, J.L., 2004, "Optimal Estimation of Dynamic System", 1st Edition, CRC Press LLC, United States.

Fehse, W., 2003, "Automated Rendezvous and Docking of Spacecraft", 1st Edition, Cambridge University Press, United Kingdom.

Inalhan, G., Tillerson, M., and How, J. P., 2002, "Relative Dynamics and Control of Spacecraft Formation in Eccentric Orbits", Journal of Guidance, Control, and Dynamics,Vol. 25, No. 1, pp. 48-58.

Jenkins, S.C. and Geller, D.K., 2007, "State Estimation and Targeting For Autonomous Rendezvous and Proximity Operations", AAS 07-316, Proceedings of theAIAA/AASAstrodynamics Specialists Conference, Mackinac Island, MI.

Junkins, J.L., Kim, S., Crassidis, L., Cheng, Y., and Fosbury, A.M., 2005,"Kalman Filtering for Relative Spacecraft Attitude and Position Estimation", AIAA 2005-6087, Proceedings of the AIAA Guidance, Navigation, and Control Conference, San Francisco, California.

Lear, W.M., 1985, "Kalman Filtering Techniques", NASA Johnson Space Center: Mission Planning andAnalysis Division, Houston, TX, JSC-20688.

Melton, R.G., 2000, "Time-Explicit Representation of Relative Motion Between Elliptica lOrbits" Journal of Guidance, Control and Dynamics, Vol. 23, No. 4, pp. 604–610.doi: 10.2514/2.4605.

Sengupta, P. and Vadali, S.R., 2007, "Relative Motion and the Geometry of Formations in KeplerianElliptic Orbits with Arbitrary Eccentricity", Journal of Guidance, Control, and Dynamics,Vol. 30, No. 4, pp. 953-964.doi: 10.2514/1.25941.

Schaub, H. and Junkins, J.L., 2003, "Analytical Mechanics of Space Systems", American Institute of Aeronautics and Astronautics, Reston, Virginia, United States.

Tschauner, J. and Hempel, P., 1965, "Rendezvous zueinem in Elliptischer Bahn Umlaufenden Ziel", ActaAstronautica, Vol. 11, pp. 104-109.

Vallado, D.A., 2001, "Fundamentals of Astrodynamics and Applications", 2nd Edition, Microcosm Press, El Segundo, California.

Woffinden, D.C. and Geller, D.K., 2007, "Navigating the Road to Autonomous Orbital Rendezvous", Journal of Spacecraft and Rockets, Vol. 44, No. 4, 2007, pp. 898–909.doi: 10.2514/1.30734.

Woffinden, D.C. and Geller, D.K., 2007, "Relative Angles-Only Navigation and Pose Estimation for Autonomous Orbital Rendezvous", Journal of Guidance, Control and Dynamics, Vol. 30, No.5, pp. 1455-1469.doi: 10.2514/1.28216.

Woffinden, D.C., 2004, "On-Orbit Satellite Inspection Navigation and Analysis", M.S. Thesis, Aeronautics and Astronautics Department,Massachusetts Institute of Technology, Cambridge, MA, United States.

Wie, B., 1998,"Space Vehicle Dynamics and Control", American Institute of Aeronautics and Astronautics, Reston, Virginia, United States.

Yamanaka,K. and Ankersen,F., 2002, "New State Transition Matrix for Relative Motion on an Arbitrary Elliptical Orbit"Journal of Guidance, Control, and Dynamics, Vol. 25, No. 1, pp. 60–66.

4

Noise Source Distribution of Coaxial Subsonic Jet-Short-Cowl Nozzle

Odenir de Almeida[1], João Roberto Barbosa[2], Juan Battaner Moro[3], Rodney Harold Self[3]

ABSTRACT: The noise source distribution of a short-cowl coaxial jet operating at different velocity ratios is described in this work. This was motivated by an ongoing research about noise prediction of coaxial jets through Acoustic Analogy with purposes of industrial engine application. This research has been carried out between Universidade Federal de Uberlândia (UFU), Brazil and the Institute of Sound and Vibration Research (ISVR) at Southampton University, UK. The numerical approach employed is originally based on Lighthill Acoustic Analogy. This technique, although likely known, is associated with an improved energy transfer time-scale, used in the turbulence two-point correlation function, in order to enhance the source model. The source model is coupled with the aerodynamic calculation of flow through turbulence quantities evaluated by using a standard *k*-*ε* turbulence modeling. The Computational Fluid Dynamics data have also been used to provide complementary information about the coaxial jet noise production mechanisms. Experimental data were used in order to corroborate the results from the current model. Good agreement has been found, showing that high and low frequency contributors to the radiated noise for low velocity ratio are aggregated in a region about seven to ten secondary diameters downstream, while at higher velocity ratios sources are continuously spread from about one up to ten secondary diameters from the jet exit.

KEYWORDS: Aeroacoustics, Acoustic analogy, Turbulence, Noise, Nozzle.

INTRODUCTION

Measuring and locating noise sources in jet flows is doubtless a difficult task since it requires special equipments and refined techniques. However, the understanding of noise production and radiation mechanisms in these flows would perfectly hold up by such achievement. Indeed, knowing exactly all the patterns behind the sources of jet noise and their location would lead towards a new phase of developing noise suppression devices and optimization of different concepts for industrial applications, for example, aero-engines.

One of the most common methods for investigating the noise radiated by a jet flow is the determination of the source strength distribution along the jet axis. For single jet flows, this has been experimentally done by different techniques, using microphone phase array system and/or by using a highly directive microphone system like an elliptic mirror (Laufer *et al.*, 1976; Fisher *et al.*, 1977; Harper-Bourne, 1999).

The problem of noise source distribution intensifies when considering coaxial jets due to the nature or structure of the coaxial flow and mainly because the introduction of a number of additional variables, such as area ratio (*AR*), velocity ratio (*VR*) and temperature ratio (*TR*). Nevertheless, many experimental works were set in the last years in order to provide source location data of such flows (Battaner-Moro, 2003; Femi and Bridges, 2004). Concomitantly, numerical research has been evolved through some different approaches in order to overcome the lack of information in this field (Groschel *et al.*, 2006; Tinney *et al.*, 2006; Eschricht *et al.*, 2008).

One of the first works towards the use of acoustic analogy in order to evaluate noise sources is attributed to Ribner (1958).

1.Universidade Federal de Uberlândia – Uberlândia/MG – Brazil 2.Instituto Tecnológico de Aeronáutica – São José dos Campos/SP – Brazil 3.University of Southampton – Southampton/United Kingdom

Author for correspondence: Odenir de Almeida | Avenida João Naves de Ávila, 2.121 – Santa Mônica | CEP 38.408-100 Uberlândia/MG – Brazil | Email: odenir@mecanica.ufu.br

Despite the work was quite simple, the most important finding was that the overwhelming bulk of the jet noise is emitted in the region of eight up to ten nozzle diameters downstream the jet exit plane. Chen (1976), also presented a simple analytical model for predicting sound power spectra density of coaxial jets at ambient temperature. It is shown that for coaxial jets the sound source strength distribution along the jet axis is constant in the initial mixing region, decaying like x^{-7} in the fully turbulent region.

Based on the Lighthill (1952) acoustic analogy, Self (2004) introduces a jet noise source model in which a new feature is taken into account that is the inclusion of frequency dependence for the time and length scales used in the turbulence two-point correlation function. It is found that allowing for this dependence markedly improves the agreement of the prediction of the model with experimental data. This model has been utilized to describe the noise source distribution in a subsonic single jet with potential to other applications.

Thus, an extension of this model is presented in this work as part of an ongoing research about noise prediction of coaxial jet through acoustic analogy. Clearly, the source model can be generalized to use Reynolds Averaged Navier-Stokes (RANS) and other Computational Fluid Dynamics (CFD) data to predict jet noise, for single and coaxial jets and also for more novel nozzle geometries. Based on that, the main goal was to identify and quantify noise sources for coaxial jets for industrial application, through the development of a simpler model, which predicts the source noise distribution mainly based on simple turbulence scales quantities. Although the approach may be simplistic from the acoustic and flow interaction standpoint, since the predictions are limited to 90° to the jet axis, the source model shows promising results and contribute for the physical understanding of noise generation, being useful to address practical problems in industry.

The mathematical modeling is coupled and processed with the aerodynamic flow computation by turbulence quantities like turbulent kinetic energy (κ) and turbulent dissipation rate (ε). These quantities have been calculated by using CFD RANS approach with a standard κ-ε turbulence model – such details are given in sections; Acoustic model, The polar correlation technique, Test article and operating conditions and Computational fluid dynamics. The CFD simulations in this work were employed with two fold purposes: first, to provide turbulence quantities to the acoustic model and, second, to give support in understanding the jet noise production mechanism through the analysis of the pattern of the turbulence in the flow over different operating conditions. The numerical results were compared against experimental data collected from Coaxial and Jet Noise (CoJeN) project (European Union Research Programme – FP6), which undertook several measurements of various types of coaxial nozzles at different working conditions.

The experimental validation data is available from the work of Battaner-Moro, carried out in 2008 (not published yet), which undertook coaxial jet source location measurements using the Polar Correlation Technique (Battaner-Moro, 2003). These measurements were performed at QinetiQ's Jet Noise Facility in Farnborough, UK, and consist of source images and frequency distribution. This large facility (Fig. 1) is able to generate model-scale cold and hot jets in an anechoic environment.

ACOUSTIC MODEL

In this section, the acoustic model is described. The mathematical model for jet noise source distribution is based on the Lighthill acoustic analogy as shown in the following equations. In order to make predictions of the noise for coaxial jet flows, it is necessary to couple the acoustic model to CFD-turbulence results as an input for the source modeling. It is important to emphasize that although the Lighthill acoustic analogy includes convective amplification effects, such theory will not itself account for flow-acoustic interaction. This means that an accurate prediction of the far-field pressure distribution is limited to an angle of 90° to the jet axis, in which such effects are unimportant.

Figure 1. QinetiQ's Jet Noise Facility set up for the Coaxial and Jet Noise (CoJeN) test programs.

The model itself is quite simple and easy to be numerically implemented. The whole mathematical description will not be presented herein and additional details about the mathematical formulation can be seen in Self (2004). The overall intensity as an integral over the axial extent of the jet is considered according to Eq. 1:

$$I(R,\omega) = \int I_x(\omega)\,dx \tag{1}$$

where I_x is the overall sound intensity, R is the radial distance, ω is the angular frequency and x is the microphone distance.

Following Goldstein (1976), Eq. 1 defines the axial source distribution for a specific frequency (ω) at any radial distance (R). Since the aim is to identify the source distribution for a 90° observer, the following relation can be readily found from Lighthill's equation:

$$I_x(\omega) \propto \int_{\phi=0}^{2\pi}\int_{r=0}^{\infty} L_1 L_2 L_3 \tau u^4 \exp\left\{-\frac{\omega^2 L_1^2}{4c_0}\right\}\exp\left\{-\frac{\omega^2\tau^2}{4}\right\} r\,dr\,d\phi \tag{2}$$

where L_n are the characteristic correlation length scales in the three axial directions, characterizing the size of the eddies. In this case it is assumed $L_1 = L_2 = L_3$. The time-scale is represented by τ and υ is a velocity characteristic of the turbulence. The angular frequency, $2\pi f$, is represented by ω, c_0 is the mean sound speed, ϕ is the azimuthal location and r is the radial location.

As it has been pointed out by several authors, a crucial factor for describing the turbulence in a flow is τ. The definition of τ in the current model plays an important role for predicting the location and the contribution in the frequency range. It has been shown by experiments of Harper-Bourne (1998), that both length and time-scales are frequency dependent, which means that for a good spectrum prediction such dependence must be taken into account. The frequency dependence in both length and time-scales can control both the location of the peak-frequency and also the shape of the spectrum at higher frequencies.

In this work, an energy transfer time-scale (TET) has been used in order to enhance the predictions in the source model. This time-scale that is based on the transfer of turbulent energy between different wave numbers of the turbulent fluctuations is associated with the inertial sub range where the noise production mechanism is largely governed by the energy transfer phenomenon.

Thus, τ in Eq. 2 assumes the following relation (Azarpeyvand *et al.*, 2006):

$$\tau \approx \alpha^T \tau_d \left(\frac{\Lambda}{l}\right)^{2/3} \tag{3}$$

where α^T is an empirical constant that must be adjusted. This coefficient α^T is currently utilized to correct the shape of the spectrum and set the peak frequency location. A complete study about the influence of this factor α^T was presented in Almeida (2009). In this work, a range of values $\alpha^T = [0.38 - 0.45]$ was applied showing satisfactory results. These values are in agreement with the work of Almeida (2009). In Eq. 3, τ_d is the traditional time-scale based on the turbulence decay rate κ/ε. Λ denotes the size of the eddy which can be either found from experimental results or be estimated. In this work, a range of values of Λ=[1.8 – 2.0] has been used. Finally, l is the length-scale, as defined by:

$$l = \frac{k^{3/2}}{\varepsilon} \tag{4}$$

Equation 4 gives the values for $L_1 = L_2 = L_3$. It is important to observe at this time that a model for anisotropy could be taken into account to describe the relationship among the three axial directions. The use of the TET time-scale has provided a good agreement with the experimental results, showing a more physically realistic shape of source distribution.

The acoustic model presented herein is fed in with CFD RANS results. Both τ and l are evaluated by using κ and ε values provided by the aerodynamic simulation of the subsonic flow through a short-cowl nozzle as seen in aero-engines. Details of CFD simulations and analysis of results will be described in section Computational fluid dynamics.

THE POLAR CORRELATION TECHNIQUE

Phased microphone arrays have been the tool of choice to determine the spatial distribution of aircraft engine noise, both in static tests or in experimental benches (rigs) such as model scale jet noise facilities. A standard method used in aero-acoustic testing is the Polar Correlation Technique, developed by Fisher *et al.* (1977). Cross-spectrum measurements in the far field of a jet can be processed to form an equivalent line source image along the jet axis. The Polar Correlation Technique relies on the inverse Fourier relationship between the normalized cross-spectrum (C)

and an equivalent line-source strength distribution (S) along the jet axis y (Fig. 2). For an array of incoherent sources, the following relation holds:

$$C(\alpha,\omega) = \int_{-\infty}^{\infty} S(y,\omega)\exp(-jky\sin\alpha)dy \quad (5)$$

where α is the polar angle from the reference microphone at 90°. In principle,

$$S(y,\omega) = \frac{1}{2\pi}\int_{-\infty}^{\infty} C(\alpha,\omega)\exp(jky\sin\alpha)d(k\sin\alpha) \quad (6)$$

This Fourier relationship holds for uncorrelated far field sources. The following assumptions can then be taken:

1. The distances from each source to the jth microphone $r_j(y)$ and to the reference microphone $r_{ref}(y)$ are equal to the array radius, i.e. $r_{ref}(y) = r_j(y) = r$.
2. All the source/microphone angles are equal, i.e. for the directivity terms $d_i(\theta_j(y)) = d_i(\theta_{ref}(y))$ = constant.
3. The phase term, $r_{ref}(y) - r_j(y)$ may be approximated by $y \sin(\alpha_j)$.

It is important to emphasize that successful polar correlation imaging is based on careful consideration to the design of the microphone array. This design must take into account the characteristics of what are going to be measured and the frequency of interest. This is so because all practical microphone array techniques are subjected to geometric constraints which limit the available resolution and introduce undesired spatial aliasing. For the CoJeN nozzle, a ± 30° aperture microphone polar array with 63 microphones was used, as detailed in Table 1. Further details about the experimental setup can be found in Battaner-Moro (2003).

The microphones are positioned downstream the nozzle exit, covering a spherical arc, and numbered as described on Table 1 and Fig. 2.

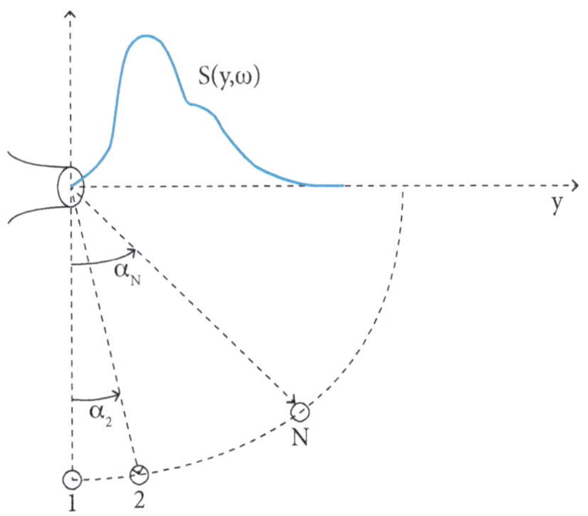

S: equivalent line-source strength distribution; *ω*: angular frequency; *N*: microphone number; *α*: polar angle.

Figure 2. Polar microphone array and equivalent line jet noise source distribution.

SOURCE IMAGES

The experimental approach by using Polar Correlation Technique provided source images assumed to be in the form of a line distribution and the noise intensity/meter of both the measured and the fitted data are plotted as a function of axial distance along the centerline of the jet. Figure 3 is an example, in which the measured image, obtained by taking a Fourier transform of the measured coherence data, is shown as a solid line. When source images of model coaxial exhaust jets were first studied, an important phenomenon was observed. Over a wide frequency range, the images were seen to display a double peak (Strange *et al.*, 1984). This double peak phenomenon was seen consistently over a range of coaxial nozzles of different size, showing the presence of a secondary source downstream the jet.

Further investigation, as presented by literature, showed that the appearance of this secondary downstream source was strictly related to the flow aerodynamics and the velocity ratio parameters. As presented by Strange *et al.* (1984), for a coaxial flow at velocity ratio equal to 1, only a single distributed source appears in the image. The emergence of the downstream source from its subdominant role, at a velocity ratio of 0.8, to the dominant source, at a velocity ratio of 0.62, gives a very clear indication that what is being observed in the source image is closely related to the flow conditions.

Table 1. Polar microphone array angular spacing (CoJeN).

Mic spacing	Mic positions	Mic units
0.25°	60° to 64° and 86° to 90°	32 mics
0.5°	64° to 68° and 82° to 86°	16 mics
1.0°	68° to 82°	15 mics

TEST ARTICLE AND OPERATING CONDITIONS

The test article consists of a short-cowl nozzle which is representative of a large modern high bypass ratio aero-engine, assuming a nominally $^{1}/_{10}{}^{th}$ scale. It has a secondary nozzle diameter of 274.4 mm, followed by a primary nozzle and a centerbody (Fig. 4).

The experimental acoustic data used in this section, for comparison purposes, are collected from the CoJeN project (European Union Research Programme - FP6) Contract no. AST3-CT-2003-502790. The current work has been processed at Institute of Sound and Vibration Research (ISVR) at Southampton University in 2009 and 2010.

Table 2 presents the flow conditions for the coaxial jet flows investigated in this work. Both primary and secondary streams where considered isothermal, i.e. the static temperature in the nozzle exit plane, T_j, is equal to the static temperature of the ambient air, T_0. The nozzle operating velocities have been changed to match three different velocity ratios *VR*=1.0, *VR*=0.75 and *VR*=0.63. The *VR* has been defined as the ratio between the secondary and primary stream (V_s/V_p).

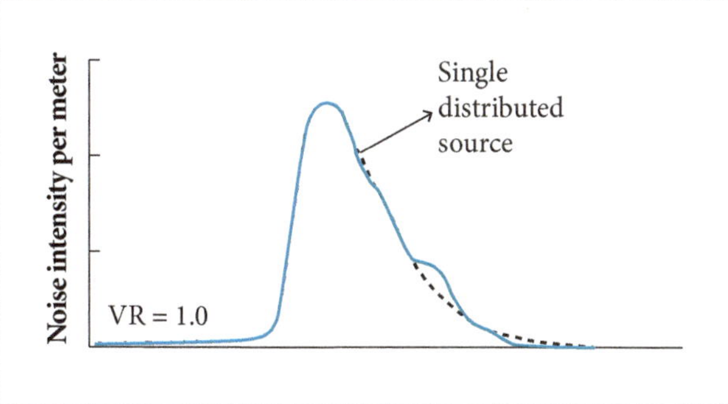

Figure 3. A typical coaxial jet source image (Strange *et al.* 1984).

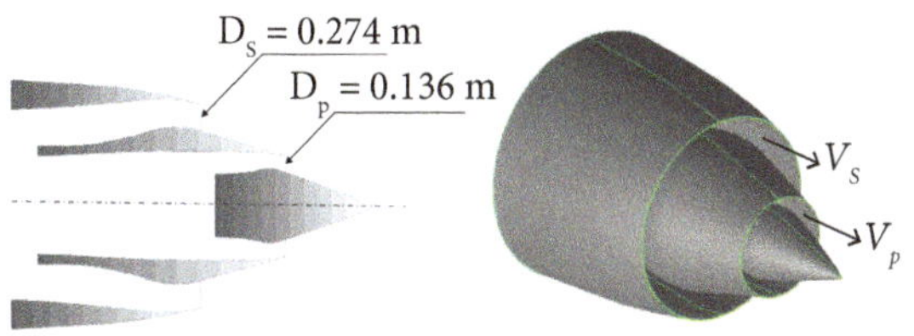

D_S: secondary diameter; D_p: primary diameter;
V_S: velocity at secondary stream; V_p: velocity at primary stream.

Figure 4. Short-cowl nozzle geometry.

Table 2. Operating conditions – isothermal flow.

Condition	V_p (m/s)	V_s (m/s)	*VR*
1	217.2	217.2	1.0
2	217.2	162.9	0.75
3	217.2	136.8	0.63

V_p: velocity at primary stream; V_s: velocity at secondary stream; *VR*: velocity ratio; P_0: ambient pressure =101,325 Pa); T_0: ambient temperature = 288,15K.

COMPUTATIONAL FLUID DYNAMICS

CFD through RANS methodology has been applied in this research in order to provide the mean flow quantities necessary to the acoustic model, as shown in section Acoustic model. The turbulence modeling is also a key point in the whole formulation, since quantities like κ and ε are used to compute the noise source strength, and is of great importance for understanding the underlying physics of the noise production mechanisms. It is clear that the CFD RANS alone does not reveal the frequency content of the sources. However, it is still possible to interpret CFD results by knowing that the high frequency sources are often aggregated in the close vicinity of the jet exit (between the potential core and the surrounding medium) and the low frequency sources further downstream (mixing region).

The dual-stream isothermal jets operating at the velocity ratios mentioned in Table 2 have been considered in this study and a RANS scheme using a finite volume scheme with a standard κ-ε turbulence model (Launder and Spalding, 1972) was used to calculate the mean and turbulence quantities. The dimensions and boundary conditions for the computational domain used in the RANS simulations are illustrated in Fig. 5. Figure 6 shows part of the short-cowl mesh used for the CFD calculations. For such a kind of approach, there is no need to process a 3D full model, since RANS is only able to provide an average flow-field for the parameters of interest quantities like κ and ε. A comparison between 3D and 2D simulations has shown a very good agreement, as presented in Almeida (2009). It must be remembered that the analysis is performed by a noise source distribution in a centerline at the exit plane of the nozzle. Such approach is useful for industry application since it requires less computational power and good results can be achieved. A well-defined axisymmetric 2D simulation is a compromise between time of processing and reasonable flow prediction. With modern hardware availability, such analysis could provide results around one and two days of work, being the mesh generation the most complex part of the process.

The final axisymmetric computational domain discretization consisted of a block structured mesh with 12 blocks and a total of 203,942 elements. The mesh points were concentrated to the shear layer region and clustered in the nozzle wall following a 7th- law turbulent boundary layer approach in order to provide a Y^+ of approximately 30 in the near wall regions. A linear growing law is used to increase the elements in axial and radial direction. A sharper mesh jump is avoided during block transitions. It is important to mention that, despite the fact that the Y^+ is not too large, some additional tests provided in Almeida (2009), showed the need to include near wall treatment for the short-cowl nozzles simulations.

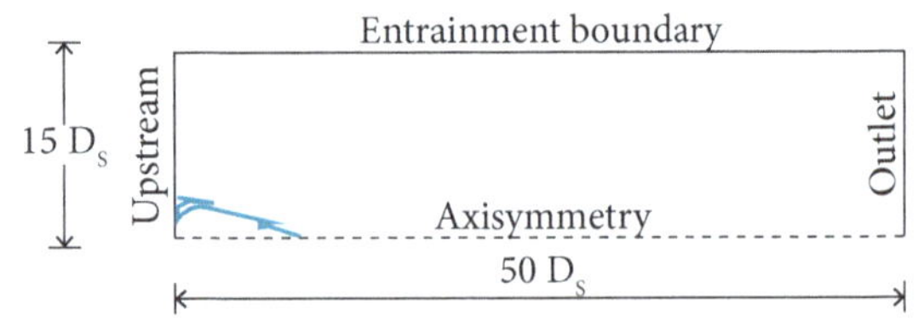

Inlet: total pressure and enthalpy; Entrainment: static pressure and temperature; Outlet: static pressure and temperature; Axis: axisymmetry.

Figure 5. Computational Fluid Dynamics domain and boundary conditions – short-cowl nozzle (application).

AERODYNAMICS AND SOURCE DISTRIBUTION RESULTS

The numerical results in this section are separated between the flow prediction through RANS simulations and the noise source distribution through the acoustic modeling. The section Aerodynamic Calculation describes the CFD results for different subsonic short-cowl coaxial jet flow operating at different *VR* of 1.0, 0.75 and 0.63, in accordance with Table 2. Following, the section Source distribution presents the predicted noise distribution at each frequency for the cases, respectively.

AERODYNAMIC CALCULATION

Aerodynamic calculations for the short-cowl nozzle jet operating at the *VR* mentioned in Table 1 have been considered in this study by employing a RANS scheme using a standard κ-ε turbulence model to evaluate the mean and turbulence quantities.

Figure 7 presents the contours of Mach and κ for the different *VR* investigated. It is possible to observe that as the velocity ratio is close to 1.0 the coaxial jet flow resembles a single jet flow with the noise source region (high kinetic energy) spread over two up to ten diameters. As the *VR* is decreased, it can be identified that the source region is moving downstream. At *VR*=0.63, the noise source is placed between $5 \leq x/D_S \leq 10$.

Figure 8 presents the axial distribution of κ, l and τ for three different radial positions (r/Dj= 0.0; r/Dj=0.5; r/Dj=1.0) comparing *VR*=1.0 and *VR*=0.63. As can be seen, the κ varies from an upstream distributed source at *VR*=1.0 to a downstream dominant source when *VR*=0.63. However, it can be seen that the source peaks at a different axial location, which for the short-cowl nozzle is approximately seven to eight secondary

Figure 6. Short-cowl mesh refinement over the domain – 203,942 quadrilateral elements.

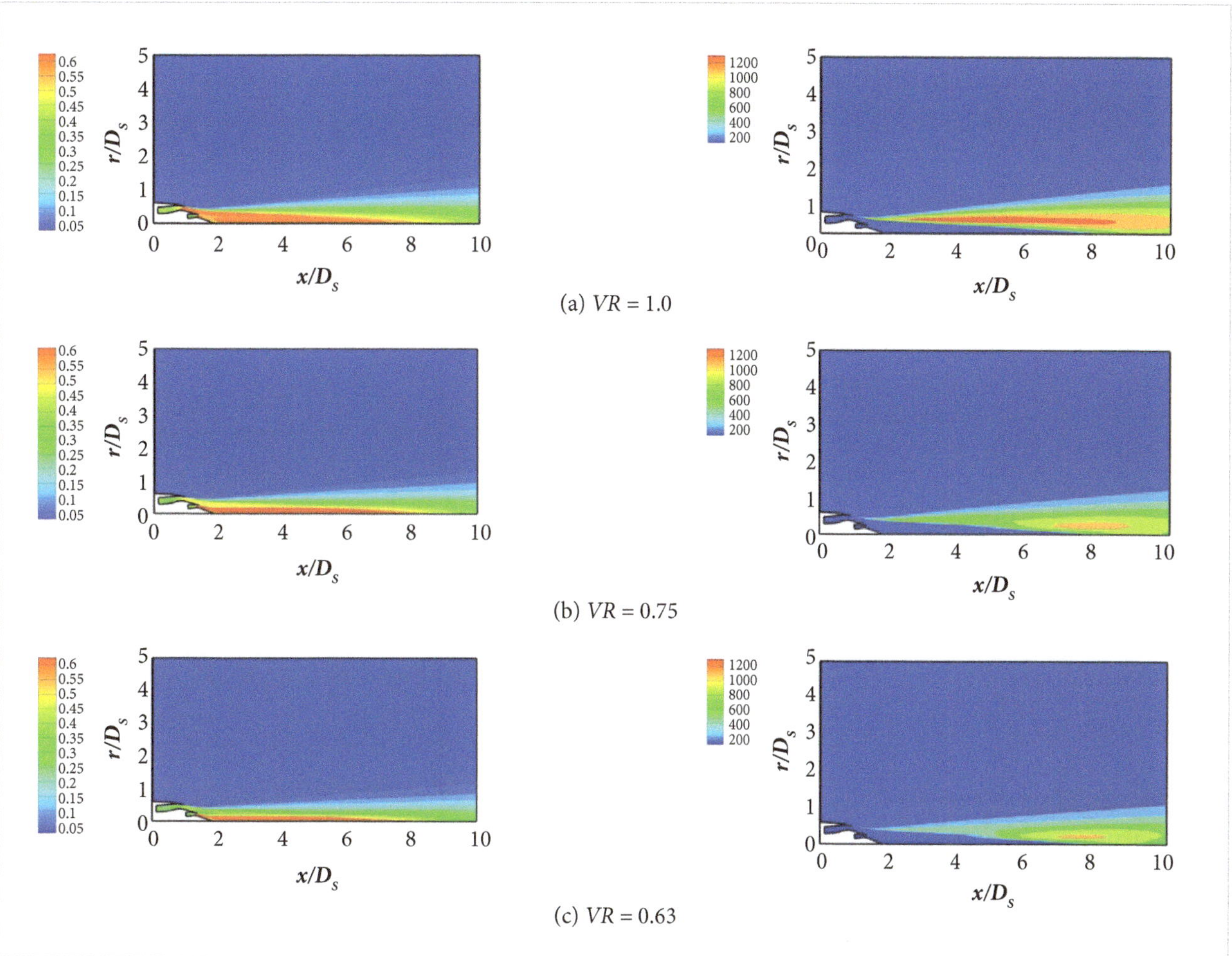

Figure 7. Mach number (M) (left) and turbulent kinetic energy (*k*) (right) – short-cowl nozzle (*AR* = 3).

diameters. Finally, based on τ and l, it can be inferred that the large structures (big eddies) also start to dominate after 15 D_S.

SOURCE DISTRIBUTION

Figure 9 shows the axial locations of peak of the I_X at different frequencies for an observer located at 90° and a radial distance of R=65.4 D_S. This result indicates a good agreement for the whole range of frequency (200 Hz up to 10 kHz) by using the source model associated with a TET time-scale.

In the left hand side, the contours (images) of the sources can be visualized. On the right hand side, it is illustrated a comparison of the source distribution over a transversal line placed at the origin – line plot. For *VR* equal to 1 and 0.75, the sources are not well distributed, peaking at different axial locations. The second downstream source, appearing for *VR*=0.75, seems to peak earlier and with a higher frequency. At *VR*=0.63, the downstream source appears to be more "compact" and peaks at approximately 6 D_S. For cases of *VR* equal to 1.0 and 0.63, the numerical and experimental data agreed satisfactorily. However, a close look on Fig. 9b reveals a discrepancy between experimental and numerical data. The reason for that discrepancy is associated to the presence of some "external" noise content coming either from internal sources in the rig or due to the presence of some solid surface close to the jet. Additional noise tests (experiments) may be necessary to confirm this observation.

Although the appearance of extra noise sources in the experimental data (it needs further investigation), the numerical location of the sources are reasonably predicted. Such approach is clearly important for study and investigation of regions of

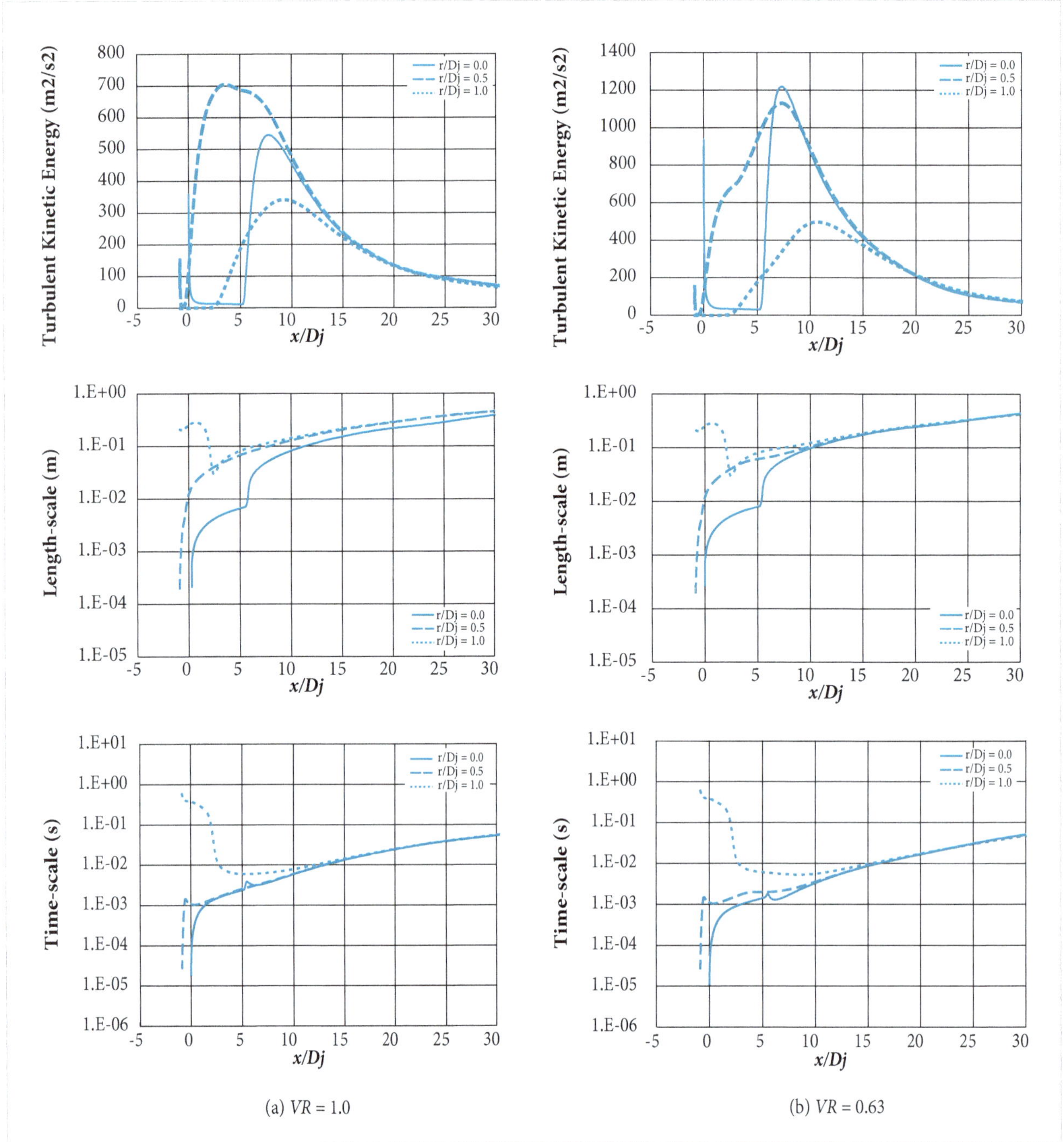

Figure 8. Flow quantities distribution at different radial locations – short-cowl nozzle (*AR* = 3; *VR* = 1.0 and *VR* = 0.63). In the plot, the origin (*x* = 0) is placed at the end of the plug.

noise source in coaxial jet flows. Especially for a *VR*=0.63, the numerical source location matched well the experimental results.

The final words in this section will be devoted to highlight the potential for using the acoustic model in the source location. However, as previously discussed, the exact characterization of the coaxial jet spreading and its levels of shear stresses may impose some restrictions in the accuracy of the predict spectra, for *VR* different than unity. Moreover, such failure in capturing, especially the high frequency content, may lead to an inexact source location matching.

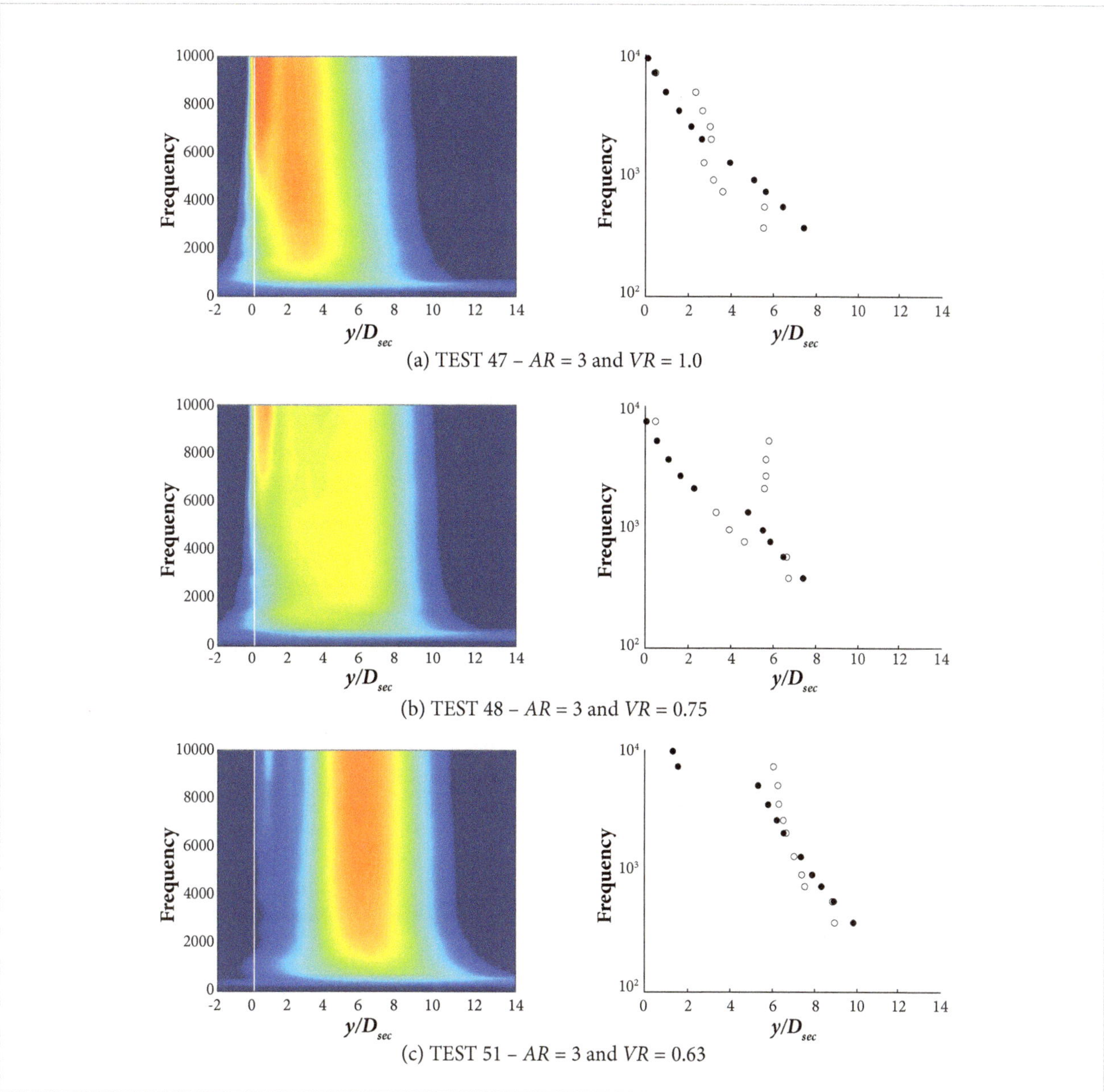

Figure 9. Coaxial source location – short-cowl nozzle; left: image; right: axial source strength/m: (○) experimental; (•) prediction.

Based on this affirmation and in all the results presented in this work, it is possible to affirm that the acoustic model is promising to deal with coaxial jet flows. However, the model seems to be much more dependent on the aerodynamic results (CFD) when the *VR* is decreased below 1.0. This is mainly due to the shifting in the noise source location inside the jet plume, which may not be correctly captured with a coarse mesh at downstream positions.

CONCLUSION

The noise source distribution of a short-cowl coaxial jet operating at different velocities ratios was investigated in this work. The numerical approach employed is originally based on Lighthill Acoustic Analogy. This technique is associated with an improved energy transfer time-scale (TET), used in the turbulence two-point correlation function, in order to enhance the source model. The main observations were registered as: a) high and low frequency contributors to the radiated noise for low velocity ratio are aggregated in a region about seven to ten secondary diameters downstream, while at higher velocity ratios sources are continuously spread from about one up to ten secondary diameters from the jet exit; b) as the velocity ratio is decreased, the noise sources move downstream the jet axis. This has been corroborated by experimental results. The numerical technique exposed herein is useful for industrial application and can be surely applied to investigate noise source regions in short-cowl nozzles, as seen in modern aircrafts.

ACKNOWLEDGMENTS

Prof. Odenir Almeida thanks the financial support provided by Coordenação de Aperfeiçoamento de Pessoal de Ensino Superior (CAPES), Brazil, for the development of this work. He also thanks for all the support received at the Institute of Sound and Vibration for the conclusion of his Ph.D thesis and the Fundação de Amparo à Pesquisa do Estado de Minas Gerais (FAPEMIG).

REFERENCES

Almeida, O., 2009, "Aeroacoustics of Dual-stream Jets with Application to Turbofan Engines", Ph.D. thesis, Instituto Tecnológico da Aeronáutica, São José dos Campos, Brazil.

Azarpeyvand, M., Self, R.H. and Golliard, J., 2006, "Improved Jet Noise Modeling Using a New Acoustic Time Scale, No. 2006-2598", 12th AIAA/CEAS Aeroacoustics Conference, Cambridge, MA, USA.

Battaner-Moro, J., 2003, "Report on Automated Source Breakdown for Coaxial and Single Jet Noise Measurements", ISVR Internal Report No. 03/10.

Chen, C.Y., 1976, "A Model for Predicting Aeroacoustic Characteristics of Coaxial Jets", AIAA Paper 76-4.

Eschricht, D., Yan, J., Michel, U. and Thiele, F., 2008, "Prediction of Jet Noise from a Coplanar Nozzle, No. 2008-2969", 14th AIAA/CEAS Aeroacoustics Conference, Vancouver, Canada.

Femi, A.A. and Bridges, J., 2004, "Jet Noise Source Localization Using Linear Phased Array", NASA TM-2004-213041.

Fisher, M.J., Harper-Bourne, M. and Glegg, S.A.L., 1977, "Jet Engine Noise Source Location: The Polar Correlation Technique", Journal of Sound and Vibration, Vol. 51, No. 1, pp. 23-54. doi: http://dx.doi.org/10.1016/S0022-460X(77)80111-9

Goldstein, M.E., 1976, "Aeroacoustics", McGraw-Hill, New York, USA.

Groschel, E., Schroder, W. and Meinke, M., 2006, "Noise Sources in Single and Coaxial Jets", ECCOMAS CFD Paper, T.U Delft, Netherlands.

Harper-Bourne, M., 1998, "Radial Distribution of Jet Noise Sources Using Farfield Microphones", 4th AIAA Aeroacoustics Conference, Toulouse, France (AIAA Paper 98-2357).

Harper-Bourne, M., 1999, "Jet near field noise prediction, No. 2002-2554", 5th AIAA/CEAS Aeroacoustics Conference, Seattle, WA, USA.

Laufer, J., Schlinker, R. and Kaplan, R.E., 1976, "Experiments on Supersonic Jet Noise", AIAA Journal, Vol. 14, No. 4, pp. 489-497.

Launder, B.E. and Spalding, D.B., 1972, "Lectures in Mathematical Models of Turbulence", Academic Press, London, England.

Lighthill, M.J., 1952, "On Sound Generated Aerodynamically. I. General Theory", Proceedings of the Royal Society, Vol. 211, No. 1107, pp. 564-587. doi: 10.1098/rspa.1952.0060

Ribner, H.S., 1958, "Strength Distribution of Noise Sources along a Jet", Journal of the Acoustical Society of America, Vol. 30, No. 9, pp. 876.

Self, R.H., 2004, "Jet noise prediction using the Lighthill acoustic analogy", Journal of Sound and Vibration, Vol. 275, No. 3-5, pp. 757-768. doi: 10.1016/j.jsv.2003.06.020

Strange, P.J.R., Podmore, G., Fisher, M.J. and Tester, B.J., 1984, "Coaxial jet noise source distributions", American Institute of Aeronautics and Astronautics, NASA, 9th Aeroacoustics Conference, New York, USA. (AIAA Paper N. 84-2361).

Tinney, C.E., Jordan, P., Guitton, A. and Delville, J., 2006, "A Study in the Near Pressure Field of Co-axial Subsonic Jets, No. 2006-2589", 12th AIAA/CEAS Aeroacoustics Conference, Cambridge, MA, USA.

5

Deterministic, Hybrid and Statistical Vibro-Acoustic Models - a Methodology to Determine the VLS Payload Fairing Acoustic Behavior

Rogério Pirk[1,2], Carlos d'Andrade Souto[1,2]

ABSTRACT: Finite element method (FEM) and boundary element method (BEM) are methods usually applied on low-frequency coupled fluid-structure analysis. On the other hand, statistical energy analysis (SEA) is a technique applied on small wavelength bandwidth predictions. However, in the mid-frequency, where deterministic calculations are costly and confidence intervals of SEA are not satisfied, hybrid FEM/SEA models can be used. This work presents a coupled vibro-acoustic analysis methodology of the Brazilian Satellite Launcher Vehicle payload fairing. FEM/FEM model accounts for the low-frequency internal responses. The high-frequency responses are computed by a SEA coupled elasto-acoustic model, and a hybrid FEM/SEA model calculates the payload fairing mid-frequency behavior.

KEYWORDS: Launcher, Payload fairing, Vibro-acoustics, Coupled analysis, Hybrid analysis.

INTRODUCTION

The acoustic noises generated during launching missions are typically described by three events as lift off, transonic and maximum dynamic pressure flights. According to Arenas and Margasahayam (2006), the launchers are submitted to excitations during the two-minute lift off and transonic climb phase through the dense atmosphere layer as well as the disturbances in the boundary layer during high-speed flight (Fig. 1).

The payload fairing (PLF) and equipment bays are the compartments where sensible electronics, inertial systems and payloads are embedded. In this way, external and internal acoustic and vibro-acoustic environments must be known, in order to establish the correct dimensioning and qualification tests for such elements, parts and subsystems. Many studies have been devoted to manage the vibro-acoustics inside the Brazilian Satellite Launcher Vehicle (VLS) PLF (Pirk *et al.*, 2002a, 2002b, 2005; Pirk and Góes, 2005, 2006; Pirk and Souto, 2008, 2010), where coupled low- and high-frequency techniques were applied. Deterministic coupled techniques as finite element method (FEM)/FEM (Pirk *et al.*, 2002a; Pirk and Góes, 2005) and FEM/boundary element method (BEM) (Pirk and Góes, 2005) were used to predict the low-frequency structural skin and acoustic responses, when the PLF structure is submitted to an acoustic diffuse field of 145 dB overall sound pressure level (OSPL). However, at higher frequencies, such deterministic methods are seldom practical (Shorter *et al.*, 2005). Firstly, because the number of degrees of freedom required describing

1.Instituto de Aeronáutica e Espaço – São José dos Campos/SP – Brazil 2.Instituto Tecnológico de Aeronáutica – São José dos Campos/SP – Brazil.
Author for correspondence: Rogério Pirk | Instituto de Aeronáutica e Espaço | Praça Marechal Eduardo Gomes, 50 – Vila das Acácias | CEP: 12.228-904 São José dos Campos/SP – Brazil | Email: rogeriorp@iae.cta.br

the response becomes intractable at higher frequencies, even with the evolution of the machines nowadays. The second problem is that the higher order modes of a system tend to be sensitive to small perturbations in the properties of the system.

In light of such practical limitations, a statistical description of systems becomes an essential tool to draw meaningful conclusions about high-frequency responses. Statistical energy analysis (SEA) (Lyon and DeJong, 1995) is a statistical technique that provides a good description of the response of large subsystems, compared with a wavelength (the response of such subsystems tends to be sensitive to perturbation). In this framework, coupled analyzes have also been done, in order to obtain the high-frequency PLF vibro-acoustic responses (Pirk *et al.*, 2002b; Pirk and Góes, 2005, 2006). In these works, the acoustic responses in the mid-frequency bandwidth or twilight-zone were characterized by extending the low- and high-frequency analysis bandwidths. However, despite the good results obtained by the applied prediction techniques (Pirk and Góes, 2006), an alternative hybrid FEM/SEA model can be proposed by including deterministic details in SEA models, aiming at improving the description of the medium-frequency dynamic responses. Then, the PLF analysis can be refined, once the statistical description employed in SEA is not appropriate for describing the response of "stiff" subsystems that are small compared with a wavelength.

This paper presents the procedures and coupled vibro-acoustic techniques, adopted to obtain the inner acoustic response of the VLS PLF. In view of having accurate models, well defined bandwidths are estabilished for the application of each coupled vibro-acoustic modeling technique as FEM/FEM deterministic, hybrid FEM/SEA and SEA. Then, the internal PLF acoustic environment could be predicted in the frequency range from 0 to 8,000 Hz.

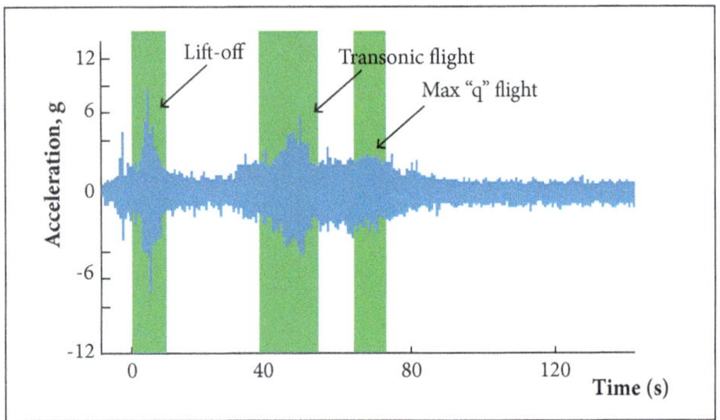

Figure 1. Typical vibration time history during a Space Shuttle Launch.

MODEL DESCRIPTION

BRAZILIAN SATELLITE LAUNCHER VEHICLE PAYLOAD FAIRING

Figure 2 shows the Brazilian VLS PLF structure, which is hammerhead-type geometry, with a maximum nominal diameter of 1.2 m and a height of 3.5 m. This compartment consists of aluminium shells, reinforced by beams. The exterior fairing surface is lined with cork and no acoustic lining is provided inside this cavity. The PLF weighs around 150 kg, including the aluminium structure and functional components as mechanisms, electronics and pyrotechnics that provide its halves separation as well as the exterior cork blanket heating protection.

Figure 2. Brazilian Satellite Launcher Vehicle payload fairing.

LOW-FREQUENCY MODELING METHODOLOGY

The FEM is the most appropriate numerical technique for the (low-frequency) dynamic analysis of these types of vibro-acoustic systems. Such problems are most commonly described in an Eulerian formulation, in which the fluid is described by a single scalar function, usually the acoustic pressure, while the structural components are described by a displacement vector. The resulting FE model in the unknown structural displacements and acoustic pressures at the nodes of, respectively, the structural and the acoustic finite element meshes are (Craggs, 1973).

$$\left(\begin{bmatrix} K_S & K_C \\ 0 & K_A \end{bmatrix} + j\omega \begin{bmatrix} C_S & 0 \\ 0 & C_A \end{bmatrix} - \omega^2 \begin{bmatrix} M_S & 0 \\ -\rho K_C^T & M_A \end{bmatrix}\right) \begin{Bmatrix} w_i \\ p_i \end{Bmatrix} = \begin{Bmatrix} F_{Si} \\ F_{Ai} \end{Bmatrix} \quad (1)$$

where K_S, K_C and K_A are the structural stiffness, the cross-coupling stiffness and the acoustic stiffness matrix, respectively; C_S and C_A are the structural and acoustic damping matrices, respectively; M_S and M_A are the structural and acoustic mass matrices, respectively; w_i is the unknown nodal displacement vector; p_i is the unknown nodal pressure vector; F_{Si} is the structural loading vector; F_{Ai} is the external acoustic force vector; ρ is the mass density of the acoustic fluid domain.

In comparison with a purely structural or purely acoustic FE model, the coupled stiffness and mass matrices are no longer symmetrical, since the loading of the fluid on the structure results in a cross-coupling term K_C in the coupled stiffness matrix, while the loading of the structure on the fluid results in a cross-coupling term $M_C = \rho K^T{}_C$ in the coupled mass matrix.

Structural wavelengths are usually much smaller than acoustic wavelengths, so that the structural mesh of the fairing should be finer than the acoustic mesh of the inner cavity. On the other hand, due to the continuity of the normal structural and fluid displacements along the fluid-structure interface, both structural and acoustic meshes should have comparable mesh densities, at least in the region of the fluid-structure coupling interface. In view of these two considerations and aiming at keeping the computational efforts within reasonable limits, the following modelling methodology has been adopted. A fine FE mesh of the PLF was used for the construction of the uncoupled structural modal data basis and the resulting modes were projected onto a coarse FE mesh. For the acoustic cavity mesh, the same mesh density was used along the fluid-structure coupling interface as the structural coarse mesh of the PLF, while the mesh density has been slightly decreased towards the inner central axis of the cavity. The uncoupled modes, resulting from this acoustic FE mesh, together with the projected structural modal base of the fairing, have then been used in a coupled FE model — Eq. 1 —, which used the modal superposition principle to calculate the acoustic and skin responses.

STRUCTURAL FINITE ELEMENT MODEL

The fairing was divided into four surfaces, discretized into 4-noded quadrilateral shell elements, while 2-noded beam elements were used to model the circumferential and the axial stiffeners. To account for the mass loading effect of the cork blanket on the exterior fairing surface, a distribution of concentrated mass elements was attached to the PLF surface nodes. Table 1 lists the properties of both the fine and the coarse structural meshes. Figure 3a shows the fine structural mesh.

Shell surface 1 has a thickness of 3 mm and is made of aluminum (elasticity modulus – E = 72 GPa, Poisson's ratio – $\nu = 0.29$, mass density – ρ = 2,750 kg/m^3), while the other three surfaces are 0.8 mm thick and made of an aluminum alloy (E = 72 GPa, $\nu = 0.29$, ρ = 7,000 kg/m^3). The fine mesh contains at least 6, 8 and 10 elements per wavelength in the frequency ranges up to, respectively, 350, 220 and 150 Hz, while the coarse mesh presents frequency limits of 200, 120 and 80 Hz. As a result, one could define the upper limit frequency analysis at 150 Hz. The PLF was assumed to be clamped at the bottom, since this boundary condition well represents the coupling with the fourth stage of the VLS launcher.

ACOUSTIC FINITE ELEMENT MODEL

The FE mesh of the fairing acoustic cavity consists of 119,577 nodes and 110,238 elements (106,050 8-noded hexahedral elements and 4,188 6-noded pentahedral elements). The internal fluid of the cavity is assumed as air with mass density ρ = 1.225 kg/m^3 and a speed of sound c = 340 m/s. The bottom and top faces of the cavity are assumed to be acoustically closed. Figure 3b illustrates the PLF cavity mesh.

Table 1. Mesh properties.

Mesh		# shell el.	# beam el.	# mass el.	# nodes
Fine	S1	4,000	240		
	S2	6,000	1,080		
	S3	2,000	360		
	S4	10,000	1,800		
	S5	12,000	1,672		
	Total	34,000	5,152	30,200	34,200
Coarse	S1	2,250			
	S2	3,000			
	S3	750			
	S4	6,000			
	S5	7,500			
	Total	19,500			19,650

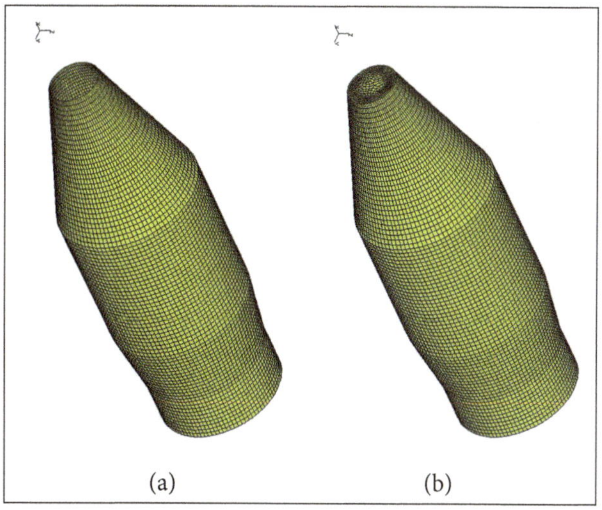

Figure 3. (a) Payload fairing structural mesh, (b) payload fairing acoustic mesh.

MODEL EXCITATION

According to Coyette *et al.* (1997) and Troclet (2000), the nature of the lift off acoustic pressure loading is close to a diffuse field excitation, having a (nearly) uniform pressure distribution.

Aiming at simulating the diffuse acoustic field generated at lift off, a uniform exterior pressure loading is applied as a normal point force on all nodes of the fairing shell elements. The force value is defined such that the total load is equivalent to a uniform pressure loading of 145 dB OSPL.

HIGH-FREQUENCY MODELING METHODOLOGY

Deterministic techniques have practical limitations for high-frequency analysis (Shorter *et al.*, 2005; Lyon and DeJong, 1995), since accurate models need to have increased the number of nodes and elements for a good spatial representation. This refinement imposes limitations, since the allocated memory and processing time increase, mainly when coupled vibro-acoustic analysis is done. In addition, the resonance frequencies and mode shapes show great sensitivity to small variations of geometry, construction and material properties, when higher frequency analysis is considered. In light of these uncertainties, the use of a statistical model of the dynamic parameters seems natural and appropriate. As an alternative for high-frequency analysis of the inner cavity of the VLS payload fairing, SEA approach is proposed.

The basic SEA equations express the energy balance of the different subsystems in the model. In general, subsystems can receive power, dissipate power and exchange power with other subsystems to which they are coupled. SEA fundamental hypotheses as dissipation losses in relation to the energy variable and modal energy proportionality from connected subsystems yield the SEA matrix. The distribution of the dynamical response in the system due to some excitation is obtained from the distribution of the energy among the mode groups, based on a set of power balance equations for the mode groups. As an example, Fig. 4 shows a three-subsystem SEA model (Lyon and DeJong, 1995).

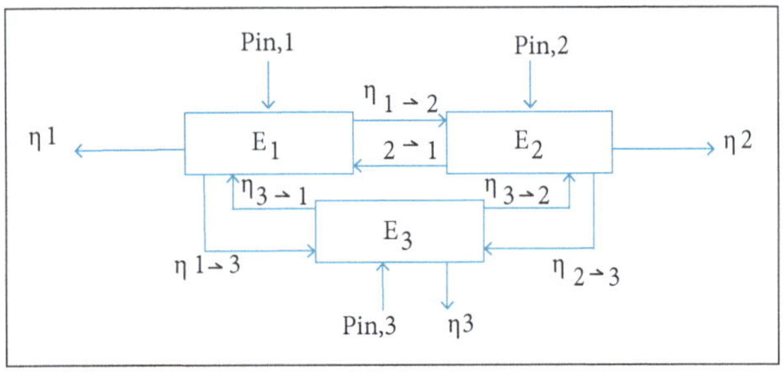

Figure 4. Statistical energy analysis model with three subsystems.

The corresponding SEA equation of the three-subsystem coupled model takes into account the input powers (P_{in}) for subsystems i, a vector with lumped total energies (E_i) and the SEA matrix [γ], which depends on the frequency and the SEA parameters, such as internal loss factors (η_{ii}), coupling loss factors (η_{ij}) and modal densities (n_i), as described in Eq. 2.

$$\begin{Bmatrix} Pin,1 \\ Pin,2 \\ Pin,3 \end{Bmatrix} = \omega \begin{bmatrix} \eta 1 + \sum_{i \neq 1}^{3} \eta 1i & -\eta 21 & -\eta 31 \\ -\eta 12 & \eta 2 + \sum_{i \neq 2}^{3} \eta 2i & -\eta 32 \\ -\eta 13 & -\eta 23 & \eta 3 + \sum_{i \neq 3}^{3} \eta 3i \end{bmatrix} \begin{Bmatrix} E1 \\ E2 \\ E3 \end{Bmatrix} \quad (2)$$

STRUCTURAL SEA MODEL

The PLF was divided into four surfaces, as shown in Fig. 5a. In order to account for the rib-stiffened plates of the surfaces 2, 3 and 4, the SEA structural fairing model considers connected plates and beams (longitudinal and circumferential). This structural FE mesh, previously generated in the Structural finite element model section, was imported and the SEA structural model was created. The plate subsystems were generated as singly curved shells and uniform plates, with damping loss factors of 1% (for flexure, extension and shear propagating waves). The external cork treatment on the surfaces 2, 3 and 4 was simulated as material addition. The layered area and the density of the cork were considered to assign this mass addition. A total of 72 beams (44 longitudinal and 28 circumferential) and 8 shells (2 singly curved shells of the adaptor, 2 singly curved shells — lower cone, 2 singly curved shells — main cylinder and 2 singly curved shells — upper cone) compose the structural SEA model (Fig. 5a).

ACOUSTIC STATISTICAL ENERGY ANALYSIS MODEL

The acoustic cavity was created considering the air as the fluid as well as the dimensional parameters of the PLF. As such, a mass density ρ = 1.225 kg/m³ and a speed of sound c = 340 m/s were assigned to create this subsystem. The top and bottom faces of the cavity were assumed to be acoustically closed. Figure 5b presents the PLF acoustic cavity.

CONNECTIONS

The coupling boundary between all the structural and acoustic subsystems is modelled to consider the transmission of power across the junctions. As such, all the subsystems that share common nodes are connected by point, line and area

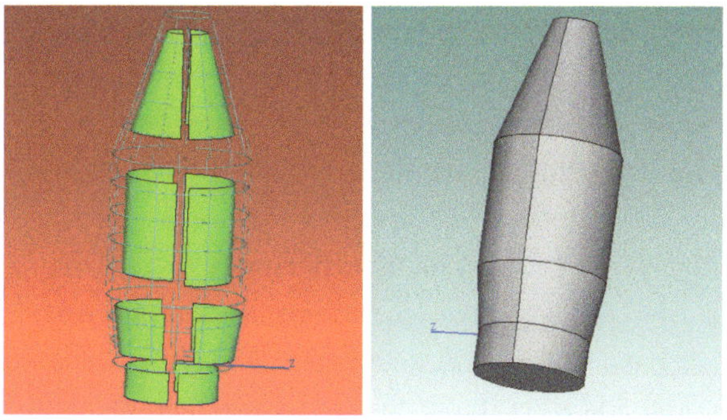

Figure 5. (a) Structural subsystems (shells, circumferential and longitudinal beams), (b) acoustic cavity of the fairing.

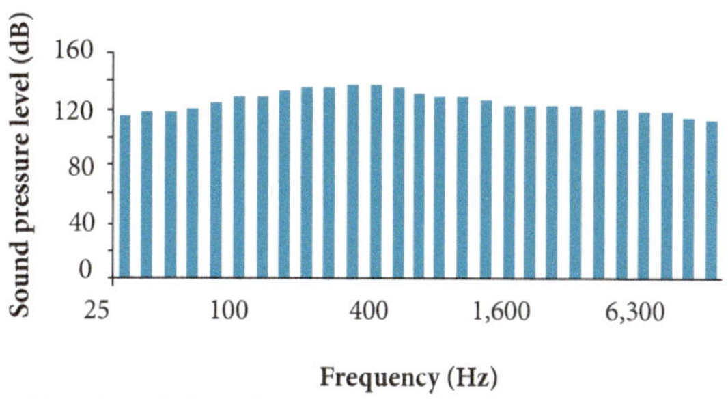

Figure 6. Diffuse pressure field at lift-off.

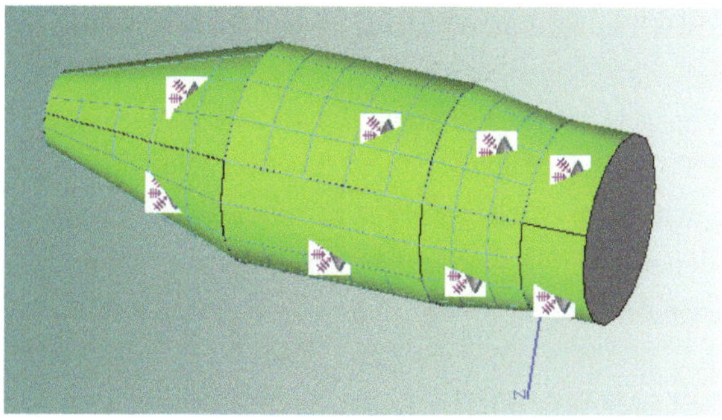

Figure 7. Complete model of the fairing.

junctions and all the appropriate wavefields are connected, as well as the corresponding coupling loss factors (CLF) between subsystems are created.

In the case of beam connections, point junctions are created to describe the transmission of the energy at a discrete point. The connections of beams and plates as well as plates and plates are provided by line connections, which describe the transmission of vibration energy along a line. Finally, the CLF of plates and acoustic cavities are accounted by area connections that represent the transmission of acoustic energy at this common bounding area.

EXCITATION

The estimated OSPL at 145 dB lift off (GEVS-SE, Rev A, June 1996) is considered. In contrast with the aerodynamic noise during flight ascent, the nature of the lift off acoustic pressure loading is close to a diffuse field excitation, having a (nearly) uniform pressure distribution (Coyette *et al.*, 1997). As stated by Defosse and Hamdi (2000), only elements with large surface areas, as plates and panels, are considered to be susceptible to acoustic excitation. Then, the referred diffuse pressure field was applied to the plates of the SEA fairing model, which simulates the power input into a structural plate. The spectral distribution of this air-borne excitation is shown in Fig. 6. Figure 7 shows the complete SEA fairing model, with 80 structural elements, 1 acoustic 3-D volume, and 8 diffuse pressure field excitations.

MID-FREQUENCY MODELING METHODOLOGY

A hybrid FE/SEA method ideally combines the low-frequency performance of the FEM with the high-frequency performance of SEA. This approach extends SEA to lower frequencies and the FEM to higher frequencies, allowing the complementary analysis in the mid-frequency or "twilight" zone, where the FEM model refinement imposes limitations and the acceptable confidence interval of the SEA model is not yet achieved.

In SEA technique, a system is modelled as a collection of subsystems, adopting the concept of universality (Langley *et al.*, 2005), which means that all subsystems are assumed to have a sufficiently high degree of complexity and randomness. However, in many practical applications, as aircraft and space structures, this condition is not fully met, once the skin panels may have many modes and a high degree of statistical overlap but the reinforcing frames do not present these characteristics. In such cases, it would be advantageous to combine the deterministic features of FEM with the SEA statistical technique. However, the coupling of these two methods, which differ in their nature (deterministic and statistical) as well as in their basic concepts (FEM is based on the system's dynamic equilibrium, while SEA is based on the conservation of energy flow), is a difficult task.

Shorter *et al.* (2004) have developed a way of providing this coupling, which is based on wave concepts. As stated by Shorter and Langley (2005), in the mid-frequency range, some components of a complex structure (for example, thin panels) display short wavelength vibrations and are sensitive to the effects of random uncertainties, while others (for example, beams) show little variation in their dynamic properties and are essentially deterministic. Therefore, in this hybrid FEM/SEA method, the deterministic components are modeled by using FEM, while the random components are modeled as SEA subsystems.

A key-feature of the coupling of this hybrid FE/SEA method is the concept of direct field or power absorbing dynamic stiffness matrix associated with each SEA subsystem. This matrix is defined as that resulting from the presence of the direct field waves, corresponding to power absorbing behavior computed for each subsystem, in the sense that the direct field waves propagate energy away from the boundaries (Langley *et al.*, 2005). It is important to note that the direct field dynamic stiffness matrix can also be viewed as the ensemble average of the full dynamic stiffness matrix when averaged over random boundary reflections.

FINITE ELEMENT/STATISTICAL ENERGY ANALYSIS EQUATIONS

One may consider that the bending motions of panels and cavities have a short wavelength of deformation and will be described using SEA subsystems and the bending degrees of freedom of these subsystems will be omitted from the FE model of the system, at all points other than the boundaries. The relevant "direct field" dynamic stiffness matrix is then added to the FE model at the boundaries and this augmented FE model is used in the subsequent analysis. If the responses (or degrees of freedom) of the deterministic part are called q, then the governing equations of motion (for harmonic vibration of frequency ω), considering the summation over all the SEA subsystems in the model, can be written as:

$$D_{tot}q = f + \sum_k f_{rev}^{(k)}, \tag{3}$$

$$D_{tot} = D_d + \sum_k D_{dir}^{(k)}, \tag{4}$$

where D_{tot} is the dynamic stiffness matrix of the FE model (excluding the SEA subsystem degrees of freedom), when augmented by the direct field dynamic stiffness matrix of each SEA subsystem; *f* is the set of external forces applied to this part of the system; $f_{rev}^{(k)}$ is the force arising from the reverberant field in subsystem k (not accounted in $D_{dir}^{(k)}$); D_d is the dynamic stiffness matrix given by the finite element model of the deterministic part of the system; $D_{dir}^{(k)}$ is the direct field dynamic stiffness matrix associated with subsystem k.

Notice that Eq. 3 and Eq. 4 are exact and that the forces arising from the SEA subsystems into a direct field part, which is accounted by $D_{dir}^{(k)}$, and a reverberant part, which is carried to the right hand side of Eq. 3, were split.

Equation 5 is a key-point to the development of the hybrid method, which shows that the cross-spectral matrix of the force exerted by the reverberant field is proportional to the resistive part of the direct field dynamic stiffness matrix, which is a form of diffuse field reciprocity statement (Shorter and Langley, 2005):

$$S_{ff}^{(k),rev} \equiv E\left[f_{rev}^{(k)} f_{rev}^{(k)*T}\right] = \frac{4E_k}{\omega \pi n_k} \mathrm{Im}\left\{D_{dir}^{(k)}\right\} \tag{5}$$

where E_k and n_k are the (ensemble average) vibrational energy and the modal density of the kth subsystem, respectively.

From Eq. 3, the responses of the deterministic part can be expanded as:

$$q = q_d + \sum_k q^{(k)} \tag{6}$$

$$q_d = D_{tot}^{-1} f \tag{7}$$

$$q^{(k)} = D_{tot}^{-1} f_{rev}^{k} \tag{8}$$

where q_d is the response considering dynamic stiffness matrix of the finite element model and the external forces.

The time averaged power input to the direct field of subsystem j can be written as Eq. 9, considering that the dynamic stiffness matrix is symmetric:

$$P_{in,j} = \left(\frac{\omega}{2}\right)\mathrm{Im}\left\{q^{*T} D_{dir}^{(j)} q\right\} = \left(\frac{\omega}{2}\right)\sum_{rs} \mathrm{Im}\left\{q^{*T} D_{dir,rs}^{(j)}\right\} S_{qq,rs} \tag{9}$$

where *rs* is the index referent to the symmetric matrix and S is the cross-spectral matrix.

If the various contributions $q^{(k)}$ of Eq. 8 are uncorrelated and of zero mean, then Eq. 5-9 yield:

$$P_{in,j} = P_{in,j}^{ext} + \sum_k \omega \eta_{jk} n_j \left(\frac{E_k}{n_k} \right) \tag{10}$$

where

$$P_{in,j}^{ext} = \left(\frac{\omega}{2} \right) \sum_{rs} \mathrm{Im}\{D_{dir,rs}^{(j)}\} \left(D_{tot}^{-1} S_{ff} D_{tot}^{-1*T} \right)_{rs} \tag{11}$$

$$\omega \eta_{jk} n_j = \left(\frac{2}{\pi} \right) \sum_{rs} \mathrm{Im}\{D_{dir,rs}^{(j)}\} \left(D_{tot}^{-1} \mathrm{Im}\{D_{dir}^{(k)}\} D_{tot}^{-1*T} \right)_{rs}$$

Since the dynamic stiffness matrices are symmetric, it is shown from Eq. 12 that reciprocity holds, in the sense that $\eta_{jk}\eta_j = \eta_{jk}\eta_k$. Moreover, it is shown that the terms η_{jk} are equivalent to the coupling loss factors that appear in SEA (Eq. 2).

The power output from the reverberant field in subsystem j can be written as the sum of the power dissipated through damping, the power transferred to the other subsystems, and the power dissipated in the deterministic system due to the response $q^{(j)}$, as Eq. 13 describes:

$$P_{out,j} = \omega \eta_j E_j + \sum_k \omega \eta_{kj} n_k \left(\frac{E_j}{n_j} \right) + \omega \eta_{dj} E_j \tag{13}$$

and

$$\omega \eta_{dj} = \left(\frac{\omega}{2E_j} \right) \mathrm{Im}\{q^{(j)*T} D_d q^{(j)}\} = \left(\frac{2}{\pi n_j} \right) \sum_{rs} \mathrm{Im}\{D_{d,rs}\} \left(D_{tot}^{-1} \mathrm{Im}\{D_{dir}^{(j)}\} D_{tot}^{-1*T} \right) \tag{14}$$

The energy balance equation for subsystem j is described by Eq. 10 and 14:

$$P_{in,j}^{ext} = \omega(\eta_j + \eta_{d,j}) E_j + \sum_k \omega \eta_{jk} n_j \left(\frac{E_j}{n_j} - \frac{E_k}{n_k} \right) + \omega \eta_{dj} E_j \tag{15}$$

Furthermore, the cross-spectral matrix of the response q can be derived from Eq. 5-8, which yields:

$$S_{qq} = D_{tot}^{-1} \left[S_{ff} + \sum_k \left(\frac{4E_k}{\omega \pi n_k} \right) \mathrm{Im}\{D_{dir}^{(k)}\} \right] D_{tot}^{-1*T} \tag{16}$$

Equations 15 and 16 form the two main equations of the hybrid method. It is clear that these equations couple FE and SEA methodologies: Eq. 15 has precisely the form of SEA, but the coupling loss factors η_{jk} and loss factors $\eta_{d,j}$ are calculated by using the FE model (augmented by the direct field dynamic stiffness matrices) via Eq. 12 and 14; furthermore, Eq. 16 has the form of a standard deterministic FE analysis, but additional forces arise from the reverberant energies in the subsystems. If no SEA subsystems are included, then the method becomes purely FE; on the other hand, if only the junctions between the SEA subsystems are modeled by FE, then the method becomes purely SEA, with a novel method of computing the coupling loss factors (Shorter and Langley, 2005).

HYBRID MODEL OF THE BRAZILIAN SATELLITE LAUNCHER VEHICLE PAYLOAD FAIRING

In the hybrid model proposed in this work, the subsystems of the PLF that present little variation in their dynamic properties, as the circular and longitudinal reinforcing beams, are considered as the deterministic parts of the FEM/SEA model. As a result, such beams are modelled using FEM. In order to obtain accurate analysis in the mid-frequency range (from 100 to 400 Hz), a discretization of 8 elements by wavelength was adopted. On the other side, the bending motions of the eight structural panels and the acoustic cavity of this vibro-acoustic system that present short wavelength of deformation and are sensitive to the effects of random uncertainties are modelled using SEA. The same excitation simulating the lift off acoustic diffuse field, shown in Fig. 6, was applied to this hybrid FEM/SEA model and clamped boundary condition was applied at the bottom part of the structural deterministic model. In order to obtain the FE modal data basis to be considered on the hybrid calculations, an intermediate analysis was done. Such a structural modal analysis yielded the modal data basis that is composed of 454 eigenvalues up to 536 Hz. Figure 8 shows the FE structural model with circular and longitudinal beams, used in this hybrid model. The SEA structural panels and cavity models are the same as shown in Figs. 5a and 5b.

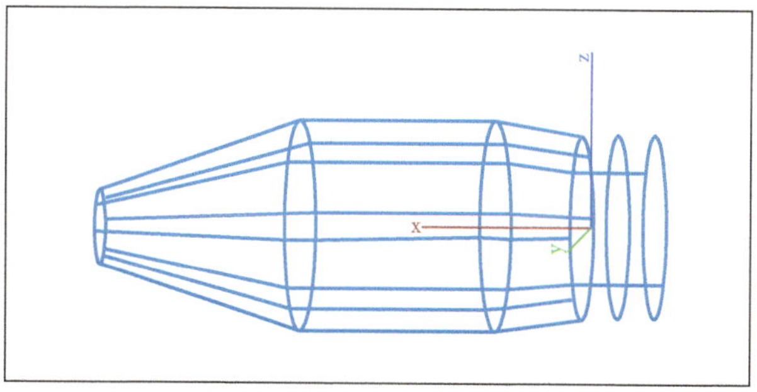

Figure 8. FE reinforcing beams model.

ANALYSIS AND RESULTS

In view of having the complete range of analysis from 0 to 8,000 Hz, the modelling procedures presented in sections Low-frequency modeling methodology, High-frequency modeling methodology and Mid-frequency modeling methodology were adopted. The limiting frequency range of the FEM/FEM model was found at 150 Hz (section Low-frequency modeling methodology). On the other side, the acceptable confidence interval of the SEA model is satisfied starting from 300 Hz, where the reliable vibro-acoustic predictions have 5 modes in the bandwidth and the modal overlap factor starts to be accurate (resonant mode controlled bandwidth). In practice, it is assumed that a modal overlap factor smaller than 1 does not present features of a SEA subsystem. On the other side, if this value is much bigger than unity, the modal peaks are not clearly apparent in the frequency response functions, which characterize the high modal density of the SEA subsystem. Finally, in the gap between 150 and 300 Hz, the "twilight" zone, a hybrid FEM/SEA model calculates the PLF acoustic responses. With these three coupled models, one can do the complete analysis for the low-, mid- and high-frequency bandwidths.

In the low-frequency range, a total of 174 structural modes in a frequency range up to 220 Hz have been identified using the fine PLF structural mesh (Fig. 3a). On the other hand, a total of 80 acoustic modes in a frequency range up to 566 Hz have been identified using the acoustic mesh (Fig. 3b). The uncoupled modal data bases were projected onto the FEM/FEM coupled vibro-acoustic model (Eq. 1) and the frequency response analysis was calculated, using the modal superposition principle. Figures 9a and 9b illustrate the 1st structural mode and the 1st acoustic mode, respectively, considered in the FEM/FEM coupled analysis.

In order to predict the mid-frequency acoustic response of the PLF, the hybrid FEM/SEA model has assumed the circular and longitudinal reinforcing beams of the fairing as the deterministic part of the vibro-acoustic model, while the structural panels and the acoustic cavity were modelled as the statistical part (section Hybrid model of the Brazilian Satellite Launcher Vehicle payload fairing). A total of 454 eigenvalues were considered in this hybrid calculation. As mentioned before, the bandwidth of interest in this analysis is from 100 to 400 Hz.

The energy levels and interactions of different subsystems were solved for the SEA vibro-acoustic model of the fairing (Eq. 2). Careful analyzes were done to determine the validity of the response calculations, and important parameters as modal density, mode count and modal overlap factor were considered to determine the SEA valid analysis bandwidth. Figure 10 displays the one-third octave bandwidth acoustic response of the VLS PLF acoustic cavity.

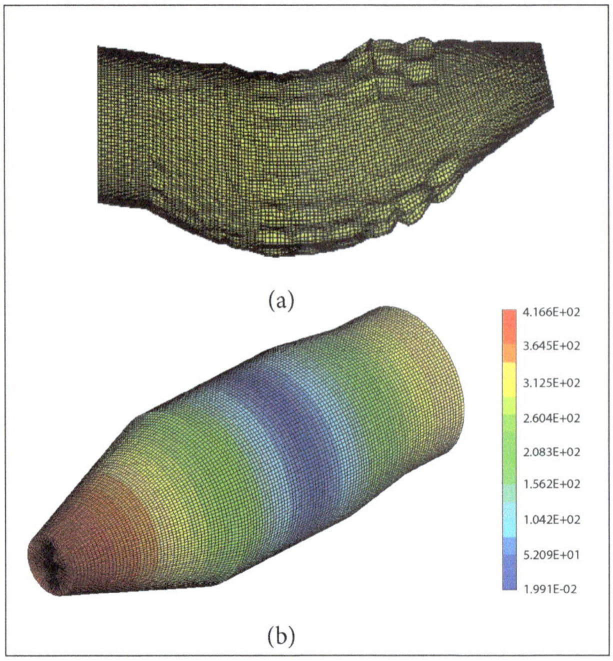

Figure 9. (a) Payload fairing 1st structural bending mode at 150.7 Hz, (b) payload fairing 1st acoustic longitudinal mode at 63.5 Hz.

In the low-frequency (up to 150 Hz), the PLF internal acoustic response presents discrete peaks around 63 and 112 Hz, related to the 1st and 2nd longitudinal acoustic modes of the referred cavity, while in the twilight zone, the responses do not present resonant peaks. This can be related to the "relative" high modal density in this frequency bandwidth due to the responses of the cylindrical and longitudinal beams, which influence the cavity responses. However, in this frequency band, one can see sound pressure levels next to

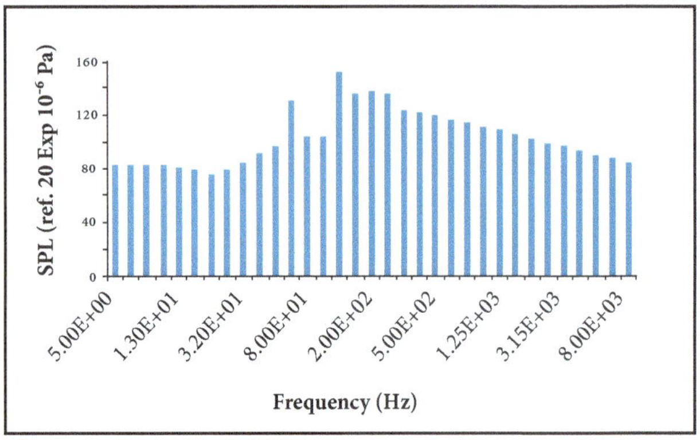

Figure 10. Payload fairing internal acoustic response.

140 dB. On the other side, the high frequency acoustic responses start from 120 dB at 350 Hz, with the reduction of the acoustic levels up to the higher frequencies (around 80 dB at 8,000 Hz).

COMMENTS AND CONCLUSIONS

The acoustic response of the VLS PLF was calculated by using coupled vibro-acoustic virtual prototypes, from 5 to 8,000 Hz. The low-, mid- and high-frequency bandwidths of PLF acoustic cavity were determined and three different vibro-acoustic models were adopted, being FEM/FEM for the low-frequency, hybrid FEM/SEA for mid-frequency and SEA for the high-frequency analysis. Parameters as model accuracy, calculation time and allocated memory were considered to define the low-frequency analysis' upper limit, at 150 Hz. The mid-frequency intervals, in the twilight zone, were assumed to be the gap between the FEM/FEM model's upper limit and the lower limit, defined for the SEA model, which starts to yield reliable results at 300 Hz.

ROAD MAP FOR THE FUTURE

Detailed analysis may be done, considering the radiation efficiency of the structural subsystems and structural FE mesh refinement results may also be accessed.

Model validations may be done with real size acoustic test results.

ACKNOWLEDGEMENTS

This work is supported by Instituto de Aeronáutica e Espaço (IAE).

REFERENCES

Arenas, J.P. and Margasahayam, R.N., 2006, "Noise and Vibration of Spacecraft Structures", Ingeniare. Revista chilena de ingeniería, Vol. 14, No. 3, pp. 251-264.

Coyette, J.P., Lecomte, C. and Meerbergen, K., 1997, "Treatment of Random Excitations Using SYSNOISE Rev. 5.3.1 - User's Manual and Validation Manual", LMS Numerical Technologies NV.

Craggs, A., 1973, "An Acoustic Finite Element Approach for Studying Boundary Flexibility and Sound Transmission Between Irregular Enclosures", Journal of Sound and Vibration, Vol. 30, No. 3, pp. 343-357. doi:10.1016/S0022-460X(73)80243-3

Defosse, H. and Handi, M.A., 2000, "Vibro-Acoustic Study of Ariane V Launcher During Liftoff", Proceedings of the Congress and Exposition on Noise Control Engineering, (INTERNOISE), Nice, France.

Langley, R.S., Shorter P.J. and Cotoni, V., 2005, "Predicting the Response Statistics of Uncertain Structures Using Extended Versions of SEA", Proceedings of the Congress and Exposition on Noise Control Engineering, Rio de Janeiro, Brazil.

Lyon, R.H. and DeJong, R.G., 1995, "Theory and Application of Statistical Energy Analysis", Second Ed., Butterworth-Heinemann, UK.

Pirk, R., Desmet, W., Pluymers, B., Sas, P. and Góes, L.C.S., 2002a, "Vibro-Acoustic Finite Element Analysis of the Brazilian Vehicle Satellite Launcher Fairing", Proceedings of the 9th International Congress on Sound and Vibration (ICSV9), Orlando, USA.

Pirk R., Desmet W., Pluymers B., Sas P. and Góes L.C.S., 2002b, "Vibro-acoustic Analysis of the Brazilian Vehicle Satellite Launcher (VLS) fairing", Proceedings of the International Conference on Noise and Vibration Engineering, ISMA, Leuven, Belgium.

Pirk, R. and Góes, L.C.S., 2005, "FEM/FEM Versus FEM/BEM Vibro-Acoustic Coupling Techniques Applied to the Brazilian Vehicle Satellite Launcher (VLS) Fairing Problem: Advantages and Drawbacks", Proceedings of the INTERNOISE, Rio de Janeiro, Brazil.

Pirk, R. and Góes, L.C.S., 2006, "Acoustic Theoretical x Experimental Comparison of the Brazilian Satellite Launcher Vehicle (VLS) Fairing", Proceedings of the International Conference on Advanced Acoustics and Vibration Engineering, ISMA, Leuven, Belgium.

Pirk, R., Góes, L.C.S., Sas, P. and Desmet, W., 2005, "Vibro-acoustic Analysis of the Brazilian Satellite Launcher Fairing, Using SEA", Proceedings of the XI DINAME, Ouro Preto, Brazil.

Pirk, R. and Souto, C.D.A., 2008, "Implementation of Acoustic Blankets to the VLS Fairing: a Sensitivity Analysis Using SEA", Proceedings of the International Conference on Noise and Vibration Engineering, ISMA, Leuven, Belgium.

Pirk, R. and Souto, C.D.A., 2010, "Implementation of Acoustic Blankets in the VLS Fairing, Using Multi- and Single-Layered Models: a Sensitivity Analysis Using SEA". Proceedings of the 17th International Congress on Sound and Vibration (ICSV), Cairo, Egypt.

Shorter, P.J., Cotoni, V. and Langley, 2004, "Numerical and Experimental Validation of Hybrid FE-SEA Method", Proceedings of the National Conference on Noise Control Engineering, Maryland, USA.

Shorter, P.J., Cotoni, V. and Langley, R.S., 2005, "A Hybrid FE-SEA Method for the Analysis of Complex Vibro-Acoustic Systems", Proceedings of the International Congress on Noise and Vibration Emerging Methods (NOVEM), St. Raphael, France.

Shorter, P.J. and Langley, R.S., 2005, "On the Reciprocity Relationship between Direct Field Radiation and Diffuse Reverberant Loading", Journal of the Acoustical Society of America, Vol. 117, No. 1, pp. 85-95.

Troclet, B., 2000, "Analysis of Vibroacoustic Response of Launchers in the Low and High Frequency Range", Proceedings of the International Congress on Noise and Vibration Emerging Methods, Lyon, France.

6

On Using Quasi-Newton Algorithms of the Broyden Class for Model-to-Test Correlation

Jan Klement[1]

ABSTRACT: The correlation of a model with test results is a common task in engineering. Often genetic algorithms or adaptive particle swarm algorithms are used for this task. In this paper another approach is presented using two quasi-Newton algorithms of the class defined by Broyden. A study was conducted with thermal models showing the performance of this approach. Comparing the results to other studies reveals that the approach reduces the number of iterations by several orders of magnitude; typical calculation times for model correlation times are reduced from the order of weeks and months to the order of hours and days.

KEYWORDS: Thermal vacuum test, Thermal analysis, Correlation, Broyden, ESATAN, Thermica.

INTRODUCTION

In every calculation, there are always discrepancies between the predictions of a mathematical model and their physical measurements. Refinement of the models can reduce the discrepancies, however, the exact values of many parameters are often unknown. Without perfect knowledge of the values of these parameters, a model refinement will never lead exactly to the measured results.

Determining the exact values of all parameters individually is too costly or sometimes even impossible. Testing is often only possible at the system level, and extracting parameter values from the test results is a difficult task which is done by correlating the mathematical model with the measurement data. This means that the model parameters are changed via feedback from the measurement results, so that the discrepancies between the measurements and the models are minimized.

Many methods have been developed and analyzed to perform model-to-measurement correlation (Jouffroy, 2007; De Palo *et al.*, 2011; Momayez *et al.*, 2009; Harvatine and De Mauro, 1994; Roscher, 2006; van Zijl, 2013; WenLong *et al.*, 2011; Mareschi *et al.*, 2005). Most methods are based on stochastic optimization algorithms and often require several hundred iterations to converge. In this paper, an approach is presented using Broyden's class of methods (Broyden, 1965), which use considerably less iterations for the majority of cases.

1. Tesat-Spacecom GmbH & Co – Backnang – Germany
Author for correspondence: Jan Klement | KG, Gerberstraße 49 | D-71522 Backnang – Germany | Email: Jan.Klement@tesat.de

THE THERMAL MODEL CORRELATION PROBLEM

The method will be presented with a typical thermal model correlation as used in the space industry, but the procedure can also be applied to many other models as used in other industries and areas.

Thermal models calculate temperatures at certain points depending on a set of parameters. Mathematically expressed, this means:

$$t_{mdl} = F(p), \quad (1)$$

where:
F is the mathematical model with function F:Rk→Rm (k is the number of parameters, m is the number of results);
t_{mdl} is a vector containing the temperatures at the selected points; and
p is a vector containing the parameters of the model, for example thermal conductivity, thermal capacity, thermal emissivity, and solar absorptivity.

During the tests, the temperatures are measured at the same points as described in the model t_{mdl}. The aim of the correlation is to find the set of parameters p_{corr} for which the root sum square of the differences between the measured temperatures t_{mes} and the calculated temperatures $F(p_{corr})$ is minimal.

$$||F(P_{corr}) - t_{mes}|| = \min\{||F(p) - t_{mes}|| | p \in P\} \quad (2)$$

where P is the solution space for the parameter vector.

THE METHOD

In contrast to most correlation methods developed (Jouffroy, 2007; De Palo *et al.*, 2011; Momayez *et al.*, 2009; Harvatine and DeMauro, 1994; Roscher, 2006; van Zijl, 2013; WenLong *et al.*, 2011; Mareschi *et al.*, 2005), this method will not attempt to minimize the length of the vector $||F(p) - t_{mes}||$, but searches a root of the equation system:

$$F(p_{corr}) - (t_{mes}) = 0. \quad (3)$$

The advantages of this method are that a whole vector of information, in this case temperature differences, can be evaluated instead of a single scalar and that a set of simple functions are used instead of a complex function. One of the best methods to solve this type of problem is the multidimensional Newton method, which applies the following formula iteratively:

$$p_{n+1} = p_n - J(p_n)^{-1} F(p_n) \quad (4)$$

where:
p_n is the vector of parameters;
$J(p_n)$ is the Jacobian matrix at p_n

A limitation of this method is that the Jacobian matrix can often only be calculated numerically, with the consequence that, for each individual parameter in p_n, a separate calculation is needed. Fortunately, Broyden developed a class of methods to approximate the Jacobian matrix based on the results from the previous step. The methods are described in Broyden (1965) and therefore will not be explained here. Broyden's first method (the "good Broyden method") is chosen for the algorithm used in this paper as it performed better for Broyden's practical problems.

In addition, a self-developed method of the Broyden class is tested. This method updates each element b_n^{ij} of the approximated Jacobian Matrix B_n with the following formula:

$$b_{n+1}^{ij} = (1 + k_n^i b_n^{ij} s_n^j)\, b_n^{ij} \quad (5)$$

where:

$$s_n := p_n - p_{n-1} \quad (6)$$

$$y_{ex,n} := B_n s_n \quad (7)$$

$$y_{is,n} := F(p_n) - F(p_{n-1}) \quad (8)$$

$$k^i = \frac{y_{is,n}^i - y_{ex,n}^i}{\sum_{j=1}^{m} (b_n^{ij} s_n^j)^2} \quad (9)$$

CONSTRAINTS FOR THE METHOD

Every optimization method has requirements which need to be fulfilled in order to be used effectively. Within this section, some of the more important requirements for practical model-to-test correlation, and the main practical consequences, are described. The requirements are important when choosing to use this method and also for the selection of the parameters and results to be used.

MONOTONY AND DIFFERENTIABILITY

An important restriction in the use of this method is that each result of the model is a differentiable monotone function of the parameters within the solution space.

Most thermal models are monotone and differentiable functions of the parameters, but there are a few exceptions such as controlled heating systems or when the parameter is, for example, the orbit position of a spacecraft. For small changes the model is neither monotone nor differentiable due to numerical noise.

If the function does not fulfill this requirement, within the solution space, the algorithm may be prevented from converging.

OBSERVABILITY OF THE PARAMETERS

Each of the parameters must have an effect on at least one result. This is not a specific issue to this particular method, but it is a general requirement for correlation. Although the requirement appears trivial, it is often the reason why the algorithm is unable to converge. For the presented method, a low degree of observability combined with numerical noise can lead to instability.

Often, the relevance of parameters can be deduced through an understanding of the model, but for complex systems it is sometimes difficult to estimate the relevance of an individual parameter. Fortunately, the Jacobian matrix addresses some of these limitations. If a column of the Jacobian matrix consists solely of zeros, the corresponding parameter cannot be observed. In practical applications there are often many values close to zero, mainly due to small effects or numerical noise. If the relevance of the parameter is smaller than the accuracy of the measurement data, the parameter can then be considered as not observable.

INFLUENCE OF THE PARAMETERS ON THE RESULTS

Each result has to be influenced by at least one parameter. This is also not a general requirement for correlation, but it prevents model correlation. For the presented method, numerical noise combined with a low level of influence on a result can lead to instability.

Again, the Jacobian matrix is a good tool for estimating the influence of a parameter on results. If a row of the Jacobian matrix has only zeros or very small values, the corresponding result cannot be influenced.

ACCURACY OF THE MEASUREMENT AND MODEL INACCURACIES

Due to measurement errors and model inaccuracies, it is unlikely that the result vector will converge towards zero. If the measurement error is larger than the effect of a parameter, then the relevant parameter may not converge to a reasonable result. Model inaccuracies may, in addition, make it impossible to correlate it within the parameter space.

CONSTRAINTS

In most instances, each parameter has certain limits defined by the physics of the model. For example, thermal conductivity will never be zero or below and never exceed a certain value. As the method defined by Broyden is for unconstrained problems, the following algorithm is used:

- Calculate p_{n+1} without considering the constraints;
- If a parameter of p_{n+1} is outside the defined boundaries, scale the change of the parameters so that all parameters are within the boundaries;
- If some parameters have already reached their constraints, calculate p_{n+1} for the condition that these parameters are fixed at their constraints.

LOAD CASES ANALYZED

Two load cases were analyzed: a constructed simple model and a complex model of real piece of hardware. These models where build in Systema/THERMICA, a thermal analysis software specialized to simulate spacecraft (mostly compatible to ESATAN and similar to SINDA). The standard syntax for these models was used.

THE CONSTRUCTED SIMPLE MODEL

The thermal model is a steady state model consisting of four diffusive nodes and an environment node at 0°C. Each node has a surface of 0.1 m² with an emissivity of 1, which radiates to the environment. 10 W are dissipated in each node. Node 1 is connected to the environment node with a thermal conductance of 1 W/K. Nodes 1 to 4 are all connected to each other by 6 thermal linear links (GL) as shown in Fig. 1.

In order for the temperatures at the nodes to correspond to the measured values, the parameter values of the thermal links need to be changed. Therefore, the equation system to be solved by the algorithm is:

$$T_{\text{calculated}}(GL) - T_{\text{measured}} = 0, \quad (10)$$

where:

$T_{\text{calculated}}$ is the function which represents the model;
GL is the vector containing GL_1 to GL_6; and
T_{measured} is the vector of the calculated temperatures for $GL_i = (0.1 + i/100)$W/K. This is assumed to be a measurement.

The initial estimates for all GL are 0.5 W/K.

Three load cases are derived from this problem:

- An underdetermined system: All links are variable. There are six parameters (links) for four values (temperatures);
- A determined system: Links 3 and 6 are constant ($GL_3 = 0.13W/K$, $GL_6 = 0.16W/K$). There are four parameters (links) for four values (temperatures);
- An overdetermined system: Links 3, 5, and 6 are constant ($GL_3 = 0.13W/K$, $GL_5 = 0.6W/K$, $GL_6 = 0.16W/K$). There are three parameters (links) for four values (temperatures), and GL_5 is chosen so that it is impossible to solve the system.

THE COMPLEX MODEL

The complex model is a model of real spacecraft hardware, which was tested in a thermo vacuum chamber. In total, 13 parameters where chosen; they consist of thermal boundary conditions, thermal conductances, and emissivities for the infrared spectrum.

Temperature readings were taken from 26 sensors at 6 test points over time, yielding 156 evaluated temperatures.

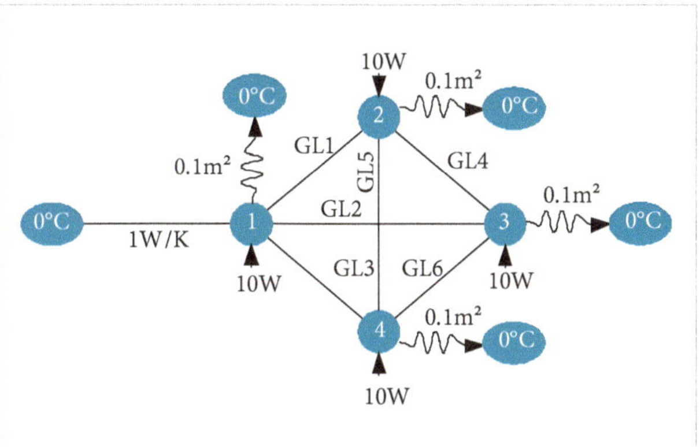

Figure 1. Thermal test model schematics.

RESULTS

RESULTS OF THE UNDERDETERMINED SIMPLE MODEL

Figure 2 shows the development of the root sum square of all temperature differences for both analyzed algorithms. The values of some significant points shown in this diagram are listed in Table 1. The model quickly converged toward 0, with both algorithms

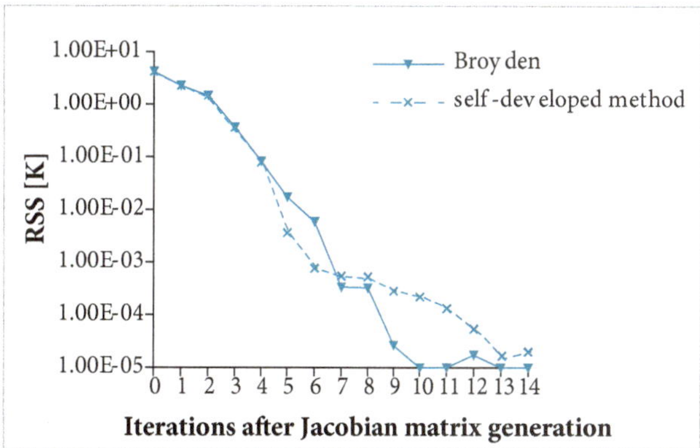

Figure 2. Root sum square of the temperature differences over iterations of the underdetermined simple model.

Table 1. Selected points of the algorithm convergence of the underdetermined simple model.

Algorithm	Iterations for Jacobian matrix generation	N° of iterations after Jacobian matrix generation	RSS/K
Broyden	6	0 (initial)	4.261
		6	5.921E-03
		9	2.646E-05
Self-developed method	6	0 (initial)	4.261
		6	7.652E-04
		12	5.385E-05

reducing the RSS by approximately one order of magnitude every two to three iterations, until the numerical accuracy limit of 1E-5 was reached. After the 4th iteration, the self-developed algorithm was faster, but after the 9th iteration, the Broyden algorithm converged faster. In conclusion, the performance of both algorithms can be considered comparable for this case.

RESULTS OF THE DETERMINED SIMPLE MODEL

In the same way as shown for the underdetermined simple model, Fig. 3 and Table 2 show the root sum square development for the determined simple model. The model converged slower than for the underdetermined problem. The Broyden algorithm needed an average of 11 steps per order of magnitude while the self-developed algorithm needed an average of only 5 of them.

RESULTS OF THE OVERDETERMINED SIMPLE MODEL

The overdetermined model converged, as expected, not to 0 but to a finite value (0.0375K), as is shown in Fig. 4 and Table 3. For this load case, the self-developed method converged considerably faster, with approximately 4 iterations per order of magnitude, while the Broyden algorithm used approximately 8 iterations per order of magnitude.

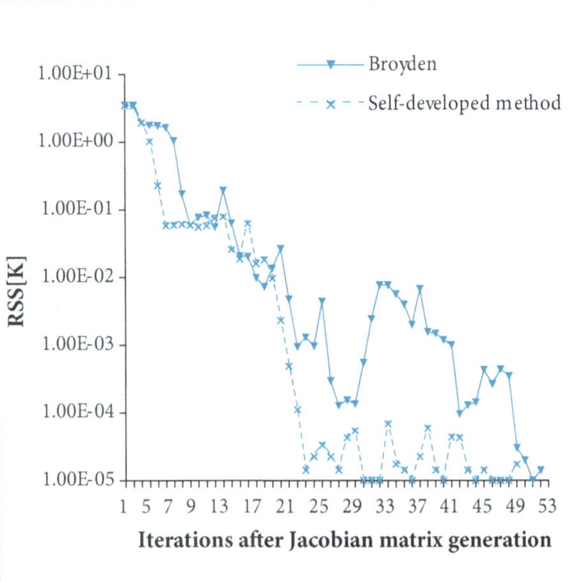

Figure 3. Root sum square of the temperature differences over iterations of the determined simple model.

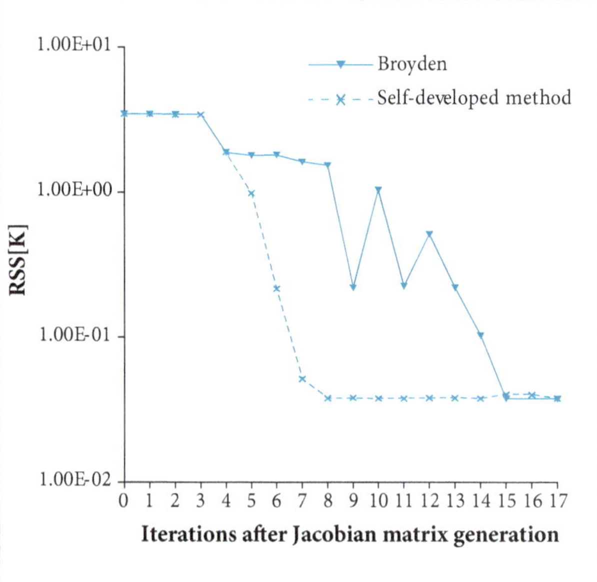

Figure 4. Root sum square of the temperature differences over iterations of the overdetermined simple model.

Table 2. Selected points of the algorithm convergence of the determined simple model.

Algorithm	Iterations for Jacobian matrix generation	N° iterations after Jacobian matrix generation	RSS/K
Broyden	4	0 (initial)	3.442
		6	1.024
		41	9.487E-05
Self-developed method	4	0 (initial)	3.442
		6	5.813E-02
		22	1.414E-05

Table 3. Selected points of the algorithm convergence of the overdetermined simple model.

Algorithm	Iterations for Jacobian matrix generation	N° iterations after Jacobian matrix generation	RSS/K
Broyden	3	0 (initial)	3.442
		4	1.869
		15	3.760E-02
Self-developed method	3	0 (initial)	3.442
		4	1.869
		8	3.757E-02

RESULTS OF THE COMPLEX MODEL

Figure 5 and Table 4 show the convergence of the algorithm for the complex model in a different way to the way that was shown for the simple model. The complex model is an overdetermined system and it is not expected to converge to 0. Therefore, the difference between the RSS of each iteration and the best RSS reached is used. This difference is standardized to the initial value. The best RSS was reached by restarting the algorithm near the optimum.

As can be seen in the following figure, the optimum is nearly reached after a few iterations. In this steep initial part, both algorithms required approximately 2 iterations, on average, to reduce the target value by one order of magnitude.

After reaching 0.25%, the Broyden algorithm stayed stable while the new algorithm converged slowly to a lower value.

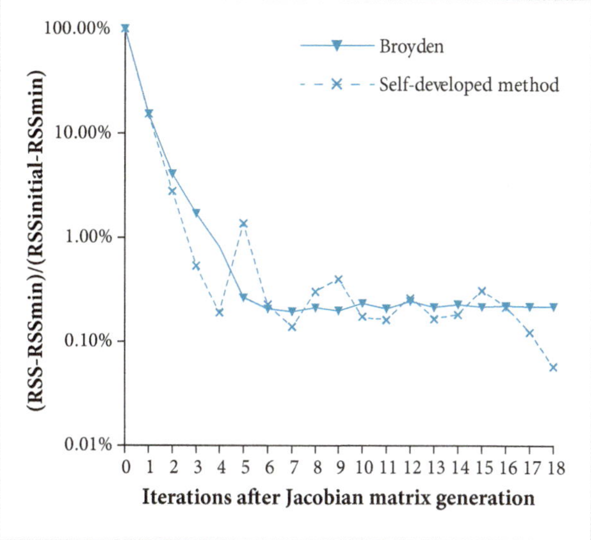

Figure 5. relative root sum square of the temperature differences over iterations of the complex model.

Table 4. Selected points of the algorithm convergence of the complex model.

Algorithm	Iterations for Jacobian matrix generation	N° iterations after Jacobian matrix generation	$(RSS-RSS_{min})/(RSS_{initial}-RSS_{min})$
Broyden	13	0 (initial)	100.00%
		4	0.81%
		18	0.22%
Self-developed method	13	0 (initial)	100.00%
		4	0.19%
		18	0.06%

DISCUSSION OF THE RESULTS

Both algorithms converged in all of the studied cases until either the numerical accuracy limit or a limit due to an over-constrained system was reached. The number of iterations can be approximated by the formula:

$$m = 1 + k + r * c \tag{11}$$

where:

m is the total number of iterations including the first;

k is the number of parameters (iterations of Jacobian matrix generation); and

r is a value specific to the problem; it depends mainly on the nonlinearity of the model and the interactions between the parameters. The total number of parameters is not very significant for this value. For the load cases studied, the value was between 2 and 10, on average, until a certain limit was reached. For similar load cases, it is not expected to increase significantly; c is the desired reduction in the order of magnitude of the difference of the root sum square and the minimal possible root sum square. It is defined by the following formula:

$$c = log_{10}(R_0 - R_{min}) - log_{10}(R_{final} - R_{min}) \tag{12}$$

where:

R_0 is the initial root sum square of the deviation vector $(F(p_0) - t_{mes})$;

R_{final} is the final root sum square of the deviation vector $(F(p_{final}) - t_{mes})$; and

R_{min} is the minimum possible root sum square of the deviation vector $(F(p_{final}) - t_{mes})$.

As can be seen in Fig. 6 the number of iterations needed for the presented algorithms (10 to 30) to the typical number of iterations needed for genetic (25600 (van Zijl, 2013), 843 to 33072 (Jouffroy, 2007)), adaptive particle swarm optimization (6000 (van Zijl, 2013)) or the stochastic design improvement (555, (Mareschi *et al.*, 2005)), this approach is much faster. The number of parameters of the models used in these evaluations is comparable to the complex model used in this paper, but the

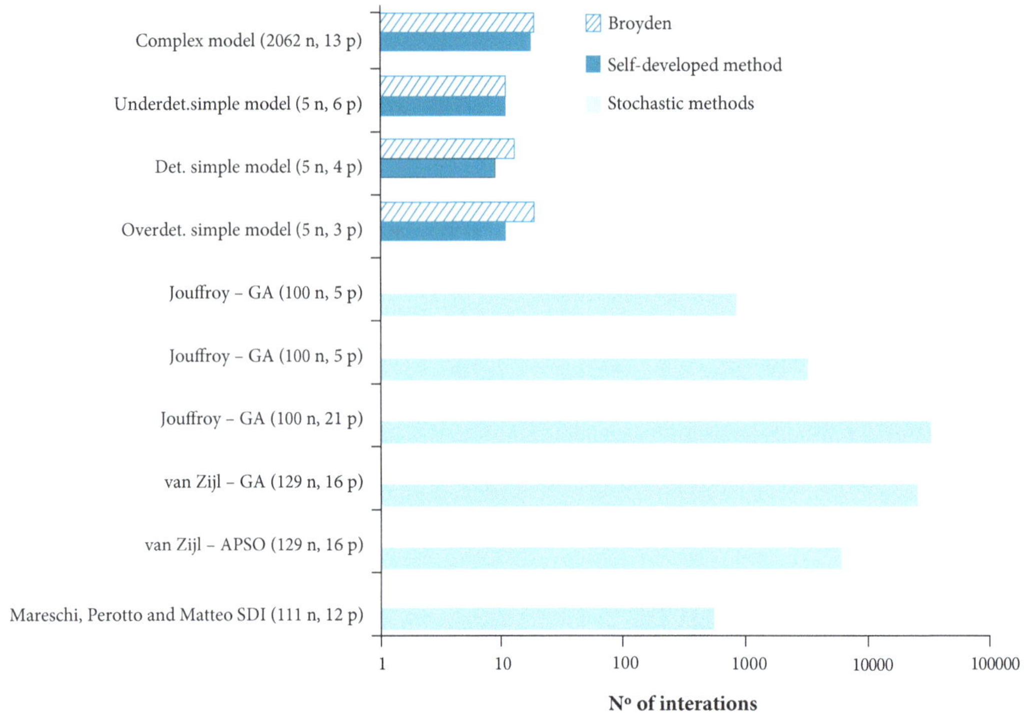

Figure 6. Comparison of the number of iterations needed for thermal model correlation (n=nodes of the model, p=parameters correlated).

number of nodes used is considerably lower. Unfortunately, the models and tools used for these methods are not accessible so that a direct comparison was not possible.

For each iteration, the CPU time required for generating a new set of parameters is only a few milliseconds and, therefore, it is negligible compared to the minutes or hours required for solving the model. It can be concluded that the total CPU time is nearly proportional to the number of iterations, independent from the algorithm.

Both algorithms studied showed similar performance. Therefore, both are considered suitable for model correlation. The new algorithm showed better performance for most situations.

CONCLUSION

It has been shown that using root finding algorithms of the Broyden class for thermal mathematical model-to-test correlation is feasible and efficient. Although a direct comparison of the analyzed methods and a genetic algorithm was not possible, the number of iterations needed for the tested algorithms is 20 to 1,000 times smaller than the number of iterations reported for genetic or adaptive particle swarm algorithms. In conclusion, this algorithm significantly reduces the cost of a thermal model correlation.

The "good Broyden method" and the self-developed method delivered satisfactory results and are, therefore, both considered suitable. The self-developed method delivered slightly better results for most of the cases analyzed.

ACKNOWLEDGMENTS

The laser communication terminal (LCT) project, which this publication is based on, is carried out on behalf of the Space Administration of the German Aerospace Center (DLR e.V.) with funds from the Federal Ministry of Economics and Technology, under the reference number 50-YH1330. The author is responsible for the contents of this publication and gratefully acknowledges their support.

REFERENCES

Broyden, C.G., 1965, "A Class of Methods for Solving Nonlinear Simultaneous Equations", AMS Mathematica of Computation, Vol. 19, No. 92, pp. 577-593. doi: 10.1090/S0025-5718-1965-0198670-6.

De Palo, S., Malost, T. and Filiddani, G., 2011, "Thermal Correlation of BepiColombo MOSIF 10 Solar Constants Simulation Test", 25th European Workshop on Thermal and ECLS Software.

Harvatine, F.J. and DeMauro, F., 1994, "Thermal Model Correlation Using Design Sensitivity and Optimization Techniques", 24th International Conference on Environmental Systems and 5th European Symposium on Space Environmental Control Systems.

Jouffroy, F., 2007, "Thermal model correlation using Genetic Algorithms", 21st European Workshop on Thermal and ECLS Software.

Mareschi, V., Perotto, V. and Gorlani, M., 2005, "Thermal Test Correlation with Stochastic Technique", 35th International Conference on Environmental Systems (ICES).

Momayez, L., Dupont, P., Popescu, B., Lottin, O. and Peerhossaini, H., 2009, "Genetic algorithm based correlations for heat transfer calculation on concave surfaces", Applied Thermal Engineering, Vol. 29, No. 17-18, pp. 3476-3481. doi: 10.1016/j.applthermaleng.2009.05.025.

Roscher, M., 2006, "Genetische Methoden zur Optimierung von Satelliten Thermalmodellen bei EADS ASTRIUM", Hochschule Wismar & EADS Astrium.

van Zijl, N., 2013, "Correlating thermal balance test results with a thermal mathematical model using evolutionary algorithms", Faculty of Aerospace Engineering, Delft University of Tecnology.

WenLong, C., Na, L., Zhi, L., Qi, Z., AiMing, W., ZhiMin, Z. and ZongBo, H., 2011, "Application study of a correction method for a spacecraft thermal model with a Monte-Carlo hybrid algorithm", Chinese Science Bulletin, Vol. 56, No. 13, pp. 1407-1412. doi: 10.1007/s11434-010-4053-z.

7

Trajectory Control During an Aeroassisted Maneuver Between Coplanar Circular Orbits

Willer Gomes dos Santos[1], Evandro Marconi Rocco[1], Valdemir Carrara[1]

ABSTRACT: This paper presents the simulation results of an aeroassisted maneuver around the Earth, between coplanar circular orbits, from a geostationary orbit to a low orbit. The simulator developed considers a reference trajectory and a trajectory perturbed by external disturbances combined with non-idealities of sensors and actuators. It is able to operate in closed loop, controlling the trajectory (drag-free control) at each instant of time using a Proportional-Integral-Derivative (PID) controller and propulsive jets. We adopted a spacecraft with a cubic body composed of two rectangular plates arranged perpendicular to the velocity vector of the vehicle. Propulsive jets are applied at the apogee of the transfer orbit in order to keep the perigee altitude and control the rate of heat transfer suffered by the vehicle during atmospheric passage. A PID controller is used to correct the deviation in the state vector and in the keplerian elements. The U.S. Standard Atmosphere is adopted as the atmospheric model. The results have shown that the aeroassisted transfer presents a smaller fuel consumption when compared to a Hohmann transfer or a bi-elliptic transfer.

KEYWORDS: Aeroassisted maneuvers, Orbital dynamic, Trajectory control.

INTRODUCTION

An orbital maneuver is the transfer of a satellite from one orbit to another by means of a change in velocity. To perform this change, the spacecraft has to engage the thrusters or use the natural forces of the environment. The Hohmann transfer and the bi-elliptic transfer are some alternatives to perform an orbital maneuver by propulsive means. In 1961, Howard London presented the approach of using aerodynamic forces to change the trajectory and velocity of a spacecraft, this new technique became known as aeroassisted maneuvers (Walberg, 1985). This type of orbital transfer can be accomplished in several layers of the atmosphere. The altitude reached by vehicle is linked to the mission's purpose and to the maximum thermal load supported by the vehicle structure. The main advantage of this type of maneuver is the fuel economy. According to Walberg (1985), many papers on aeroassisted orbital transfer have been made in recent decades and it has been shown that a significant reduction in fuel can be achieved using aeroassisted maneuvers instead of Hohmann transfer. Consequently, the reduction of fuel provides an increase in the payload capacity of the vehicle.

Orbit transfer between two circular and coplanar orbits is very common. The technique of using atmospheric drag to reduce the semi-major axis is known as aerobraking and it was first used on March 19th, 1991, by spacecraft Hiten. The launch was conducted by the Institute of Space and Astronautical Science of Japan (ISAS). The spacecraft passed through Earth's atmosphere at an altitude of 125.5 km over the Pacific Ocean at a speed of 11 km/s. The experience resulted in a decrease in apogee altitude of 8,665 km. In May 1993, an aerobraking maneuver was used on a mission to Venus by Magellan spacecraft, whose goal was to circularize the orbit of the spacecraft. In 1997, the probe U.S. Mars Global Surveyor (MGS)

1.Instituto Nacional de Pesquisas Espaciais – São José dos Campos/SP – Brazil

Author for correspondence: Willer Gomes dos Santos | Instituto Nacional de Pesquisas Espaciais | Avenida dos Astronautas 1758, São José dos Campos/SP | CEP: 12227-010 – Brazil | Email: willer.gomes@inpe.br used its solar panels as "wings" to control its passage through the tenuous upper atmosphere of Mars and lower its apoapsis.

There are several missions requirements that become feasible with the use of aeroassisted vehicles, as for example, to reconfigure orbital systems that are unable to perform an orbital maneuvering (such as replacing a malfunctioning satellite by a spare), to transfer space debris to a new orbit, to operate Space Transportation Systems (STS), to use the atmospheric drag as a brake force to provide orbit capturing of the vehicle, to assist the International Space Station with the transfer of cargo between geostationary orbit (GEO) and low earth orbit (LEO), among others.

Within the context of this paper, we can cite the scientific micro-satellite Franco-Brazilian (FBM) project, in partnership between the Brazilian and French space agencies (INPE and CNES), which would be released as piggy-back on an Ariane 5, and then would perform aerobraking maneuvers to transfer the satellite to the orbit service (Furlan, 1998). The Ariane 5 rocket has the capacity to carry up to eight microsatellites with a maximum individual weight of 120 kg through the Ariane Structure for Auxiliary Payload (ASAP). However, the rocket was designed to place satellites in geostationary transfer orbits. The propellant required to transfer the FBM to a low orbit (between 800 and 1,300 km) by means of chemical propellants, would exceed the allowed amount of mass. This question has led space agencies to study the concept of aerobraking as a workaround. However, CNES, in 2003, left the program which was subsequently discontinued (Brezun *et al.*, 2000). In another interesting study with the same context, Schulz (2001) developed an optimal control law that minimizes the fuel consumption during an aeroassisted maneuver, as well as analyzed orbital changes.

This paper will present the simulation of an aerobraking maneuver between GEO and low orbit. The study aims at examinining the effects that this kind of maneuver can cause in the orbital elements. It will also demonstrate the difference in fuel costs and the elapsed transfer time between an aeroassisted maneuver and a fully propulsive maneuver. The results show that the aeroassisted transfer has a propellant consumption lower than a Hohmann or a bi-elliptic transfer.

PROBLEM DEFINITION

In this paper, a spacecraft with a cubic body composed of two rectangular plates was adopted, called aerodynamic plates, placed in opposite sides of the vehicle's body. The inclination angle of the plates, regarding its molecular flow (attack angle), was fixed at 90 degrees, in order to maximize the projected area and the drag force. The spacecraft will be transferred from the GEO to a low orbit. The orbits are considered circular and coplanar. A multipass aeroabraking strategy is used to perform the transfer.

First, the spacecraft applies an impulse to take the vehicle out of the GEO and put it into an elliptical orbit with perigee within the limits of the atmosphere. After each passage at atmospheric region, a reduction of the apogee transfer orbit occurs. When the spacecraft reaches its final apogee altitude, then, a new impulse is applied to the vehicle to remove it from the transfer orbit and insert it into the final orbit. In order to control the rate of heat transfer suffered by the vehicle during the passage through atmosphere, propulsive jets are applied at the apogee, correcting the decay of perigee. This transfer strategy is shown in Fig. 1.

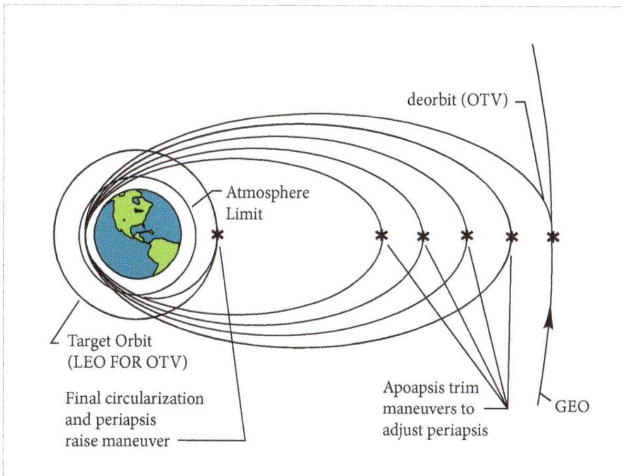

Figure 1. Multipass aerobraking (Walberg, 1985).

In this work, it was used the Aeroassisted Spacecraft Maneuver Simulator (SAMS) based on the Spacecraft Trajectory Simulator (STRS), which uses a drag-free control architecture to control the spacecraft trajectory (Rocco, 2006, 2008, 2009, 2010). Usually, a ground based open loop control is used for maneuver correction and orbit transfer. However, in some drag-free missions (Gravity Probe B, Hipparcos, GOCE, etc) the feedback control is mandatory. The STRS and SAMS consider a reference trajectory and a trajectory perturbed by external disturbances, including the aerodynamic effects, combined with non-idealities of sensors and actuators. The simulator works in closed loop controlling the trajectory at each instant of time, which is one of the input parameters, using a Proportional-Integral-Derivative (PID) controller and propulsive jets. Santos (2011) used the SAMS to study how the orbital elements can be changed by an aeroassisted maneuver and how much fuel is saved comparing with a propulsive maneuver. Figure 2 shows a basic diagram of the running logic of the aeroassisted maneuver simulator. In this work, noises and non-idealities in the actuator and in the sensor were not considered.

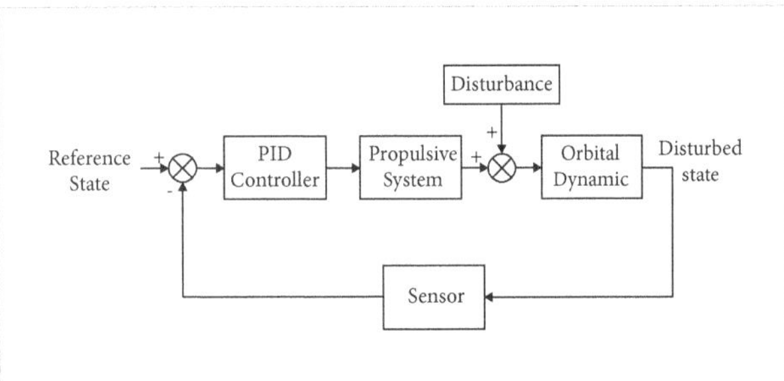

Figure 2. Basic diagram of the aeroassisted maneuver simulator.

BACKGROUND CONCEPTS

This section aims at presenting the main equations and concepts used in the development of this work. Firstly, the concepts about drag-free technology are presented, and then the equations of aeroassisted maneuver and trajectory control system are introduced.

THE DRAG-FREE SATELLITE

Drag-free satellites have a payload that follows a geodesic path through space. The satellite is affected by gravity and by non-gravitational forces. The control system for compensation of the non-gravitational forces is called drag-free control system. The drag-free device has an outer shell and an inner mass called proof mass. Inside the outer shell, the proof mass is floating freely and the distance between the outer shell and the proof mass is constantly measured. When a displacement of the proof mass regarding the outer shell is detected, this means that the outer shell has been influenced by non-gravitational forces and it has moved. Hence, thrusters will reposition the outer shell regarding proof mass so that it returns to the initial position.

According to Theil and Silas-Guilherme (2005), drag-free technology is essential for scientific missions which need a very low disturbance environment. Several missions have used this technology, such as Gravity Probe B (GP-B) to test the relativistic effects on a gyroscope; STEP and MICROSCOPE had the objective of testing the weak principle of equivalence; Laser Interferometer Space Antenna (LISA) for the detection of gravitational waves; and Gravity Field and Steady State Ocean Circulation Explorer (GOCE), launched in March 2009, to determine the gravity-field anomalies with high accuracy; among others.

AEROASSISTED MANEUVER

The main forces acting on a spacecraft in LEO are gravitational force (mg), thrusters force (T_s) and aerodynamic forces (F), caused by the interaction of the satellite with the atmosphere. The spacecraft position in space determines the magnitude of aerodynamic forces suffered by the spacecraft. The higher the planetary atmospheric density is, the stronger the aerodynamic forces are. The aerodynamic force can be divided into two: the drag force (F_D), whose direction is opposite to the velocity vector, and the lift force (F_L), perpendicular to the drag force. According to Vinh (1981), the magnitude of these forces is given by the following equations:

$$F_D = \frac{1}{2}\rho S V^2 C_D \tag{1}$$

$$F_L = \frac{1}{2}\rho S V^2 C_L \tag{2}$$

where r is the atmosphere density, C_D and C_L are, respectively, the drag and lift coefficients on the projected area S, and V is the velocity of the spacecraft in relation to the atmosphere. The lift can also be decomposed into altitude lift force (F_A) and lateral lift force (F_B). The attack angle (a) is measured between the longitudinal axis of the spacecraft and velocity in relation to the atmosphere. The magnitude of the aerodynamic force depends mainly on the attack angle, and its direction varies depending on the bank angle (s) between the lift plane and the plane, formed by the velocity vector in relation to the atmosphere and the vector position of the spacecraft, as shown in Fig. 3.

The direction and amplitude of these forces can be calculated by the following equations (Guedes, 1997):

$$\mathbf{F_D} = -F_D\hat{\mathbf{V}} \tag{3}$$

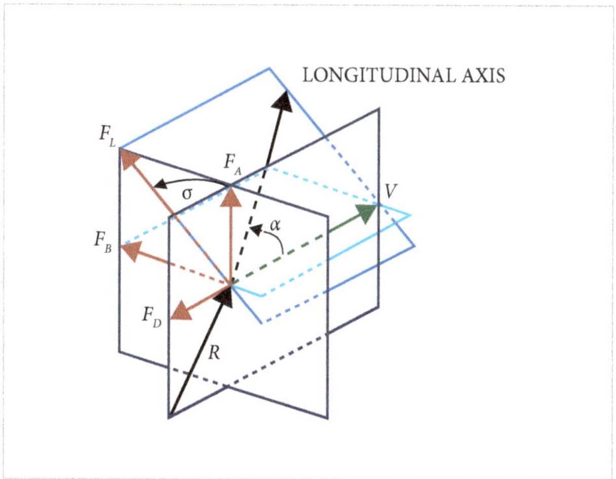

Figure 3. Components of the aerodynamic forces, attack angle and bank angle (Guedes, 1997).

$$\mathbf{F_L} = F_A\hat{\mathbf{N}} + F_B\hat{\mathbf{H}} \quad (4)$$

where $\hat{\mathbf{V}}$ is the velocity in relation to the atmosphere versor; $\hat{\mathbf{H}}$ is the angular momentum versor; and $\hat{\mathbf{N}}$ is the altitude versor. The altitude lift force, the lateral lift force, the angular momentum vector and the altitude vector are calculated according to the following equations:

$$F_A = \frac{1}{2}\rho S V^2 C_A \quad (5)$$

$$F_B = \frac{1}{2}\rho S V^2 C_B \quad (6)$$

$$\boldsymbol{H} = \boldsymbol{R}\times\boldsymbol{V} \quad (7)$$

$$\boldsymbol{N} = \boldsymbol{V}\times\boldsymbol{H} \quad (8)$$

The drag coefficient (C_D), altitude lift (C_A) and lateral lift (C_B), are calculated using the Impact Method (Regan and Anandakrishnan, 1993), according to the following equations:

$$C_D = 2sen^2\,\alpha \quad (9)$$

$$C_L = 2sen\alpha\, cos\alpha \quad (10)$$

$$C_A = C_L\, cos\sigma \quad (11)$$

$$C_B = C_L\, sen\sigma \quad (12)$$

The Impact Method, or Newtonian Impact Theory (Vinh *et al.*, 1970), is a simplified numerical technique to approximate the forces and torques acting on a body. It assumes an elastic reflection of particles in a specular surface. The normal component of impact velocity is reversed while the tangential component is unchanged. This model assumes that the particles have no random velocity component, usually associated to microscopic particles of gas (Regan and Anandakrishnan, 1993).

The U.S. Standard Atmosphere model provides the value of the atmospheric density, depending on the position of the vehicle, for the calculation of aerodynamic forces. The velocity of the spacecraft in relation to the atmosphere in the inertial system is calculated assuming that the atmosphere has the same rotation velocity of the Earth and its equation is given by Kuga *et al.* (2008):

$$\mathbf{V} = \dot{\mathbf{r}} - \omega\times\mathbf{r} = \begin{bmatrix} \dot{x}+\omega y \\ \dot{y}-\omega x \\ \dot{z} \end{bmatrix} \quad (13)$$

where $\dot{\mathbf{r}}$ is the velocity vector in relation to the inertial system and ω is the angular velocity vector of Earth's rotation.

Some of the major difficulties faced by the spacecraft during atmospheric maneuver are related to the heating rate and velocity deceleration. These quantities increase when the vehicle is submitted to high atmospheric densities and high velocities. In the upper atmosphere, it should be considered a form of heating known as free molecular heating. This phenomenon occurs due to the impact of free molecules against the vehicle. The rate of heat transfer as per area unit is given by the following equation (Gilmore, 1994):

$$\dot{Q} = \frac{1}{2}\alpha_c \rho V^3 \quad (14)$$

where α_c is the thermal accommodation coefficient (Gilmore (1994) recommends the use of $\alpha_c = 1$).

The orbital spacecraft state is described by the coordinates $\mathbf{X} = [\mathbf{r}\ \ \mathbf{V}]$, measured in an inertial frame centered on Earth, and the dynamic model of the spacecraft used in this paper is given by:

$$\ddot{\mathbf{r}} = -\frac{\mu}{r^3}\mathbf{r} - \frac{1}{2}C_D\rho SV\,\mathbf{V} + \Delta\mathbf{V}_p \quad (15)$$

where μ is the central body gravitational constant (product of the central body mass and the universal gravitational constant) and ΔV_p is the velocity variation caused by propulsive thrusters when activated. The lift force is null because the attack angle of the aerodynamic plates is perpendicular to the velocity vector.

TRAJECTORY CONTROL SYSTEM

A PID controller was used to correct the deviation of the spacecraft trajectory. Most of the industrial controllers are PID due to its flexibility, low cost and robustness. The PID control action is computed by:

$$c(t) = K_P e(t) + K_I \int e(t)dt + K_D \frac{de(t)}{dt} \quad (16)$$

where K_P, K_I and K_D are the proportional gain, integral gain and derivative gain respectively, and $e(t)$ is the position error. Most control systems today use digital computers. Hence, to implement the PID control law in a digital computer, it is necessary to discretize the PID equation $c(t)$. Several discretization methods can be consulted in Franklin *et al.* (1998). Using the discretization methodology proposed by Hemerly (2000), we can write the discrete PID control law equation, as

$$u(k) = u(k-1) + \left(K_P + \frac{K_D}{T} + K_I T\right)e(k) + \left(-K_P - 2\frac{K_D}{T}\right)e(k-1) - \frac{K_D}{T}e(k-2) \quad (17)$$

where T is the sample period.

RESULTS

This section aims at presenting the results of an aeroassisted maneuver simulation to transfer the spacecraft from a GEO to a low orbit of 1,000 km of altitude. The presented curves refer only to the aeroassisted transfer. Table 1 shows the initial conditions of the orbit.

The complete maneuver was performed in 58.93 days and, at the end of the period, there was a reduction of approximately 35,000 km in the apogee altitude, according to Fig. 4. The perigee altitude remained at an average of 115 km, with a variation of ± 0.5 km due to the application of jet propulsion at the apogee of the orbit, shown in Fig. 5.

There was no change in the orbit inclination because lift forces were not being applied to the vehicle. Figure 6 illustrates

Table 1. Initial condition of the transfer orbit.

Description	Value	Units
Apogee altitude	35786.14	km
Perigee altitude	115	km
Eccentricity	0.7332	-
Inclination	1	degrees
RAAN*	200	degrees
Perigee argument	10	degrees
Mean anomaly	180	degrees

*Right Ascension of Ascending Node.

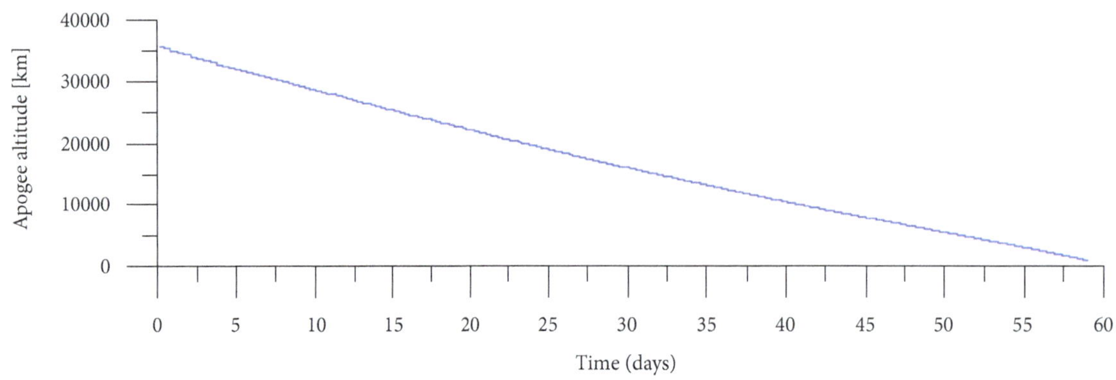

Figure 4. Apogee altitude as function of time.

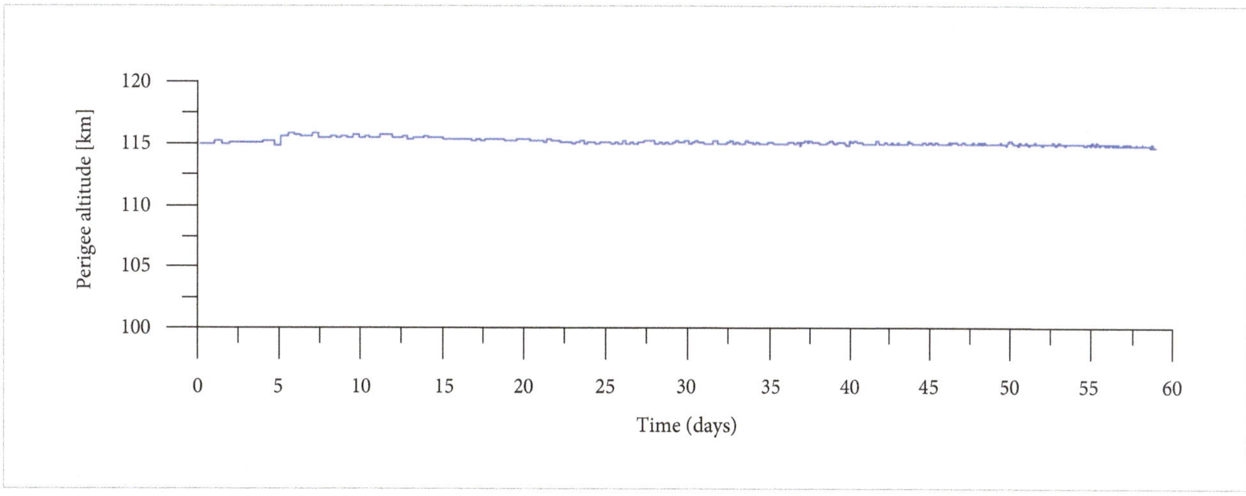

Figure 5. Perigee altitude as function of time.

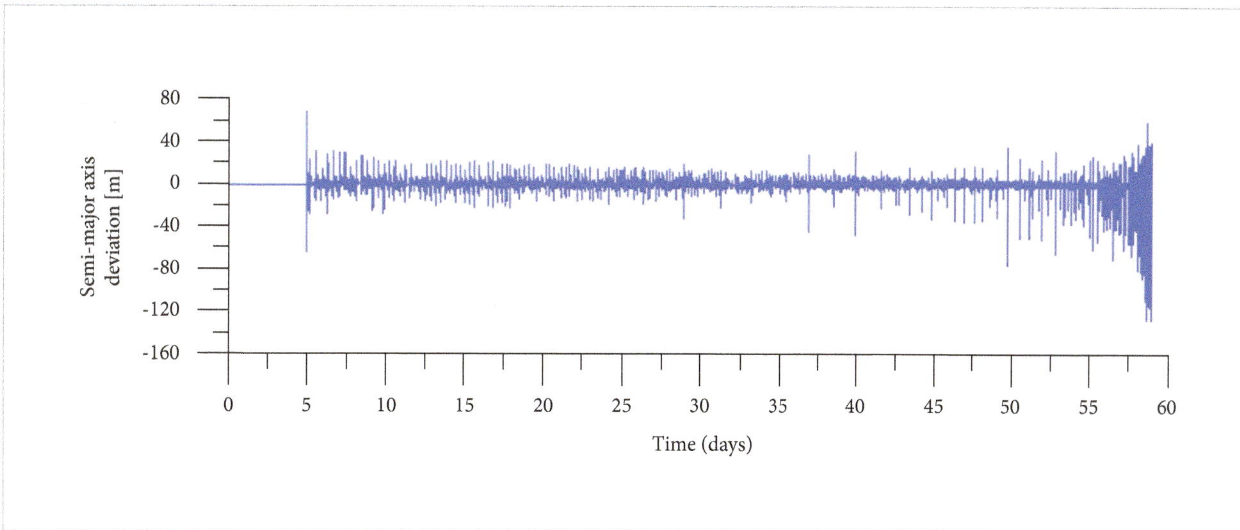

Figure 6. Semi-major axis deviation as function of time.

the deviation of the semi-major axis versus time. During the maneuver, the control system acts to reduce the error between the reference trajectory and the disturbed trajectory. The error appears when the first thrust is applied at the apogee.

Figure 7 shows the variation in the position components X, Y and Z, for the first five days of maneuvering. The variation is almost zero at the Z component due to the roughly equatorial orbit. The behavior X and Y components is related to the orbit eccentricity. As the vehicle approaches the perigee the orbital velocity increases, and vice versa.

The applied thrust at the orbit apogee versus time is shown in Fig. 8. Due to orbit circularization the trajectory path inside low atmosphere is increased, causing perigee decay. So, while low thrusters are applied at the beginning of the maneuvering process in order to adjust the trajectory derivations, high impulses are employed at the final orbit to correct the perigee height.

Figure 9 illustrates the drag force suffered by the vehicle along the atmospheric path. It can be observed a downward trend close to the final orbit. This behavior is related to the velocity reduction and with the circularization of the orbit.

The heat transfer rate suffered by the vehicle as function of time is shown in Fig. 10. The maximum value supported by the vehicle depends on the material structure. The values obtained are within the limit given by Kumar and Tewari (2005). However,

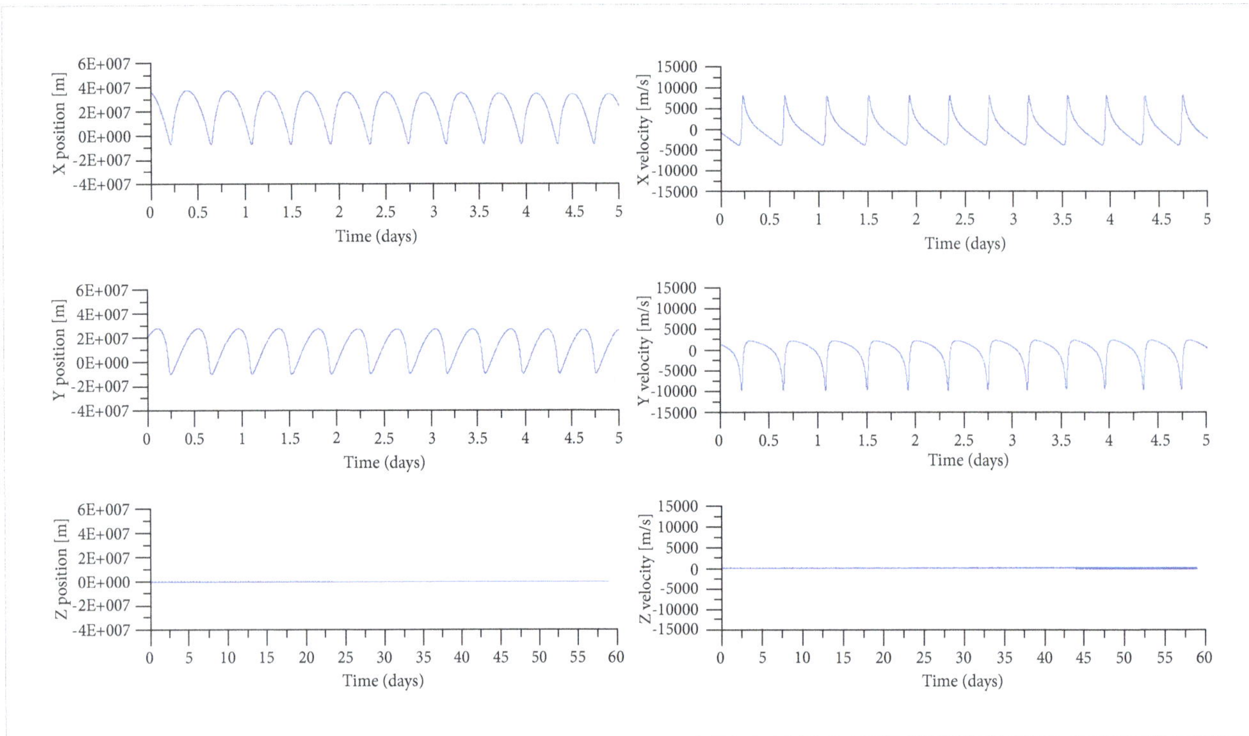

Figure 7. X, Y and Z components of the position and velocity vectors as function of time.

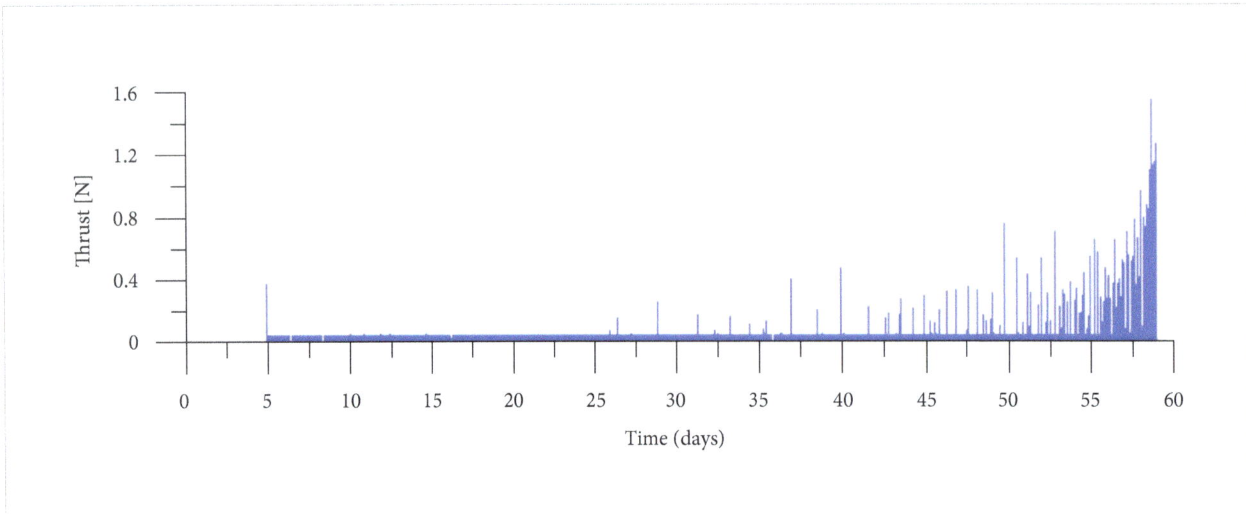

Figure 8. Propulsive thrust as function of time.

the heat rate experienced by the vehicle can be increased or decreased by controlling the perigee altitude; the lower the perigee height, the higher the heat suffered by the vehicle.

Regarding the consumption of propellant, two situations are considered. The first one illustrates a hypothetical situation: it is considered that the drag force suffered by the vehicle is provided by the propulsion system, whose results are presented in Fig. 11; and the second situation (Fig. 12), shows the case that considers only the thrust necessary to correct the perigee altitude.

Simulations show that up to 200 kg of fuel propellant should be needed in order to perform orbit adjusting using only the propulsion system, with application of thrust at perigee of the orbit

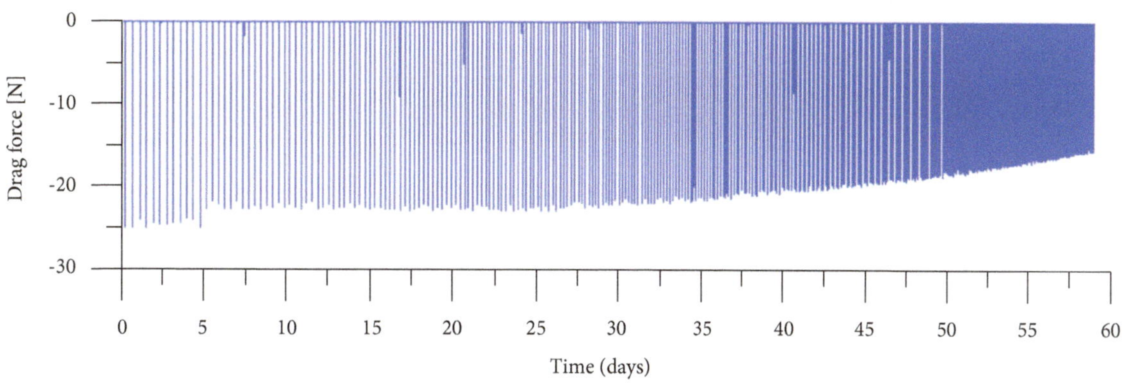

Figure 9. Drag force as function of time.

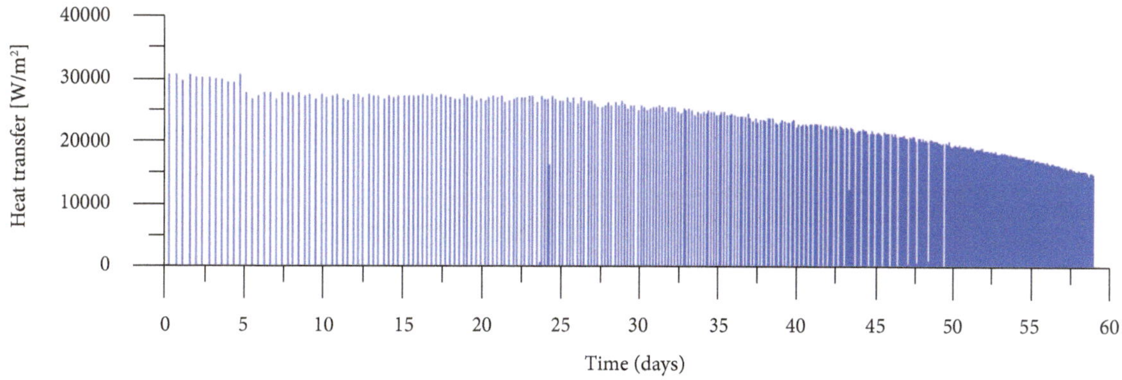

Figure 10. Rate of heat transfer as function of time.

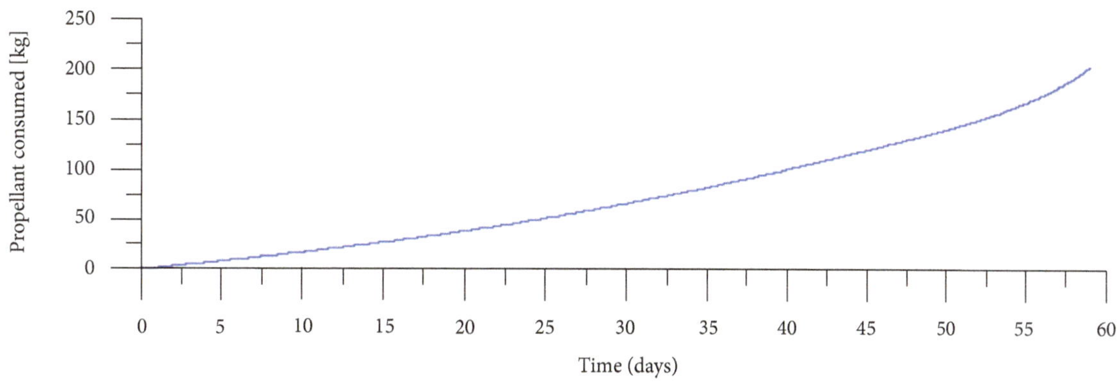

Figure 11. Hypothetical situation: propellant required for applying a thrust equal to drag force.

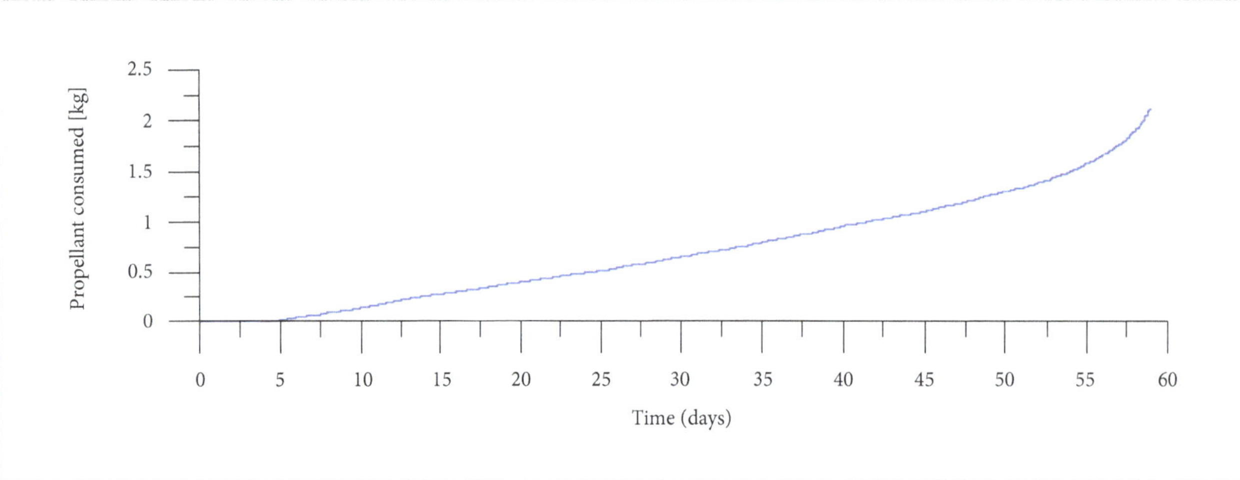

Figure 12. Aeroassisted situation: propellant necessary to correct the decay of perigee.

transfer, compared to 2.11 kg of fuel, necessary to the aeroassisted, plus PID control orbit maneuvering. Furthermore, it should be taken into account the propellant needed to enter to and exit from the transfer elliptical orbit. The propellant adopted was the liquid oxygen/liquid hydrogen, whose specific impulse is 460 s.

Table 2 shows a comparison of propellant consumption and transfer time among Hohmann transfer, Bi-elliptic transfer and aeroassisted maneuver. It was considered the consumption of propellant to enter and to exit from the elliptical orbit transfer. The aeroassisted maneuver spent 160.92 kg of propellant, where 2.11 kg were used to correct the decay of perigee, and 158.81 kg to get in and out of the transfer orbit. The fuel economy of aeroassisted maneuver related on the transfer of Hohmann was approximately 116 kg.

Table 2. Comparative table of propellant consumption and transfer time.

Maneuvers	Propellant Consumption (kg)	Transfer time (days)
Hohmann	276.62	0.22
Bi-elliptic	361.25	1.48
Aeroassisted	160.92	58.93

The aeroassisted maneuver took 58.93 days to reach the final orbit, while the transfer of Hohmann and bi-elliptical, required far less time, 0.22 and 1.48 days respectively. However, it was considered, in computations, the ideal case for calculating the transfer time of propulsive maneuvers, but when thrust is small, due to thruster size, and the number of propulsive maneuvers to reach the final orbit is increased, the time to perform the maneuver also increases.

CONCLUSIONS

It was presented the simulation of an aeroassisted maneuver to transfer a vehicle from a GEO to a low orbit altitude using propulsive jets to correct the decay in perigee altitude and to correct deviations between the reference trajectory and the disturbed trajectory. It can be concluded that the control system met expectations and that it has maintained the residual error in state vector within acceptable limits.

The control propulsive jets is able to maintain the perigee altitude within reasonable limits (115 km ± 0.5 km), avoiding the vehicle to suffer high thermal loads during the atmospheric path. This case can be compared to the mission of scientific FBM, in which there was a need to transfer the satellite from a GEO to a low orbit, by means of aeroassisted maneuvers. In general, the aeroassisted maneuvers were more advantageous in terms of fuel economy than the fully propulsive maneuvers. The closed loop control system was critical to the simulation success, without which it would not be possible to eliminate efficiently the residual errors in the trajectory.

REFERENCES

Brezun, E., Bondivenne, G. and Kell, P., 2000, "Aerobraking design and study applied to CNES microsatellite product line", 5th International Symposium of Small Satellites Systems and Services, La Baule, France. Proceedings... Paris: CNES, 2000. pp. 673-680.

Franklin, G.F., Powell, J.D. and Workman, M.L., 1998, "Digital Control of dynamic systems", 3rd Edition, Addison-Wesley Publishing Co., Massachusetts, USA

Furlan, B.M.P., 1998, "Several studies apply to the French-Brazilian mission", Instituto Nacional de Pesquisas Espaciais (INPE), São José dos Campos, Brazil (in Portuguese).

Gilmore, D.G., 1994, "Satellite thermal control handbook", 1st Edition, The Aerospace Corporation Press, California, USA.

Guedes, U.T.V., 1997, "Dispersion analysis of the reentry trajectory over the landing point, using geocentric inertial and lateral maneuvers", Ph.D. in Space Mechanics and Control Thesis, Instituto Nacional de Pesquisas Espaciais (INPE), São José dos Campos, Brazil (in Portuguese).

Hemerly, E.M., 2000, "Controle por computador de sistemas dinâmicos", 2nd Edition, Edgard Blücher Ltda, São Paulo, Brazil.

Kuga, H.K., Rao, K.R. and Carrara, V., 2008, "Introduction to Orbital Mechanics" 2nd Edition, Instituto Nacional de Pesquisas Espaciais (INPE), São José dos Campos, Brazil (in Portuguese).

Kumar, M. and Tewari, A., 2005, "Trajectory and attitude simulation for aerocapture and aerobraking", Jornal of Spacecraft and Rockets, Vol. 42, No. 4, pp. 684-693. doi: 10.2514/1.7117.

Regan, F.J. and Anandakrishnan, S.T., 1993, "Dynamics of atmosphere re-entry", 1st Edition, American Institute of Aeronautics and Astronautics, Washington, DC, USA.

Rocco, E.M., 2006, "Tools for analysis and simulation of spacecraft trajectories in keplerian orbit" Bremen: Center of Applied Space Technology and Microgravity ZARM, University of Bremen, Germany.

Rocco, E.M., 2008, "Perturbed orbital motion with a PID control system for the trajectory", XIV Colóquio Brasileiro de Dinâmica Orbital, Águas de Lindóia, Brazil.

Rocco, E.M., 2009, "Earth albedo model evaluation and analysis of the trajectory deviation for some drag-free missions", Proceedings of the 8th Brazilian Conference on Dynamics Control and Applications, Bauru, Brazil.

Rocco, E.M., 2010, "Evaluation of the terrestrial albedo applied to some scientific missions", Space Science Reviews, Vol. 151, No. 1-3, pp. 135-147. doi: 10.1007/s11214-009-9622-6, 2010.

Santos, W.G., 2011, "Simulação de manobras aeroassistidas de um veículo espacial controlado por placas aerodinâmicas reguláveis e sistema propulsivo", Dissertação de Mestrado em Mecânica Espacial e Controle, Instituto Nacional de Pesquisas Espaciais (INPE), São José dos Campos, Brazil, pp. 272.

Schulz, W., 2001, "Study of orbital transfers including aeroassisted maneuvers" Ph.D. in Space Mechanics and Control Thesis, Instituto Nacional de Pesquisas Espaciais (INPE), São José dos Campos, Brazil, pp. 178 (in Portuguese).

Theil, S. and Silas-Guilherme, M., 2005, "In-orbit calibration of drag-free satellites", Advances in Space Reseach, Vol. 36, pp. 504-514. doi: 10.1016/j.asr.2005.05.126.

Vinh, N.X., Busemann, A. and Culp, R.D., 1970, "Hypersonic and planetary entry flight mechanics" Elsevier, Amsterdam, Germany.

Vinh, N.X., 1981, "Optimal trajectories in atmospheric flight" Elsevier, Amsterdam, Germany.

Walberg, G.D., 1985, "A survey of aeroassisted orbit transfer", Journal of Spacecraft and Rockets, Vol. 22, No. 1, pp. 3-18.

8

A Numerical Study on Smart Material Selection for Flapped and Twisted Morphing Wing Configurations

Mauricio Vicente Donadon[1], Lorenzo Iannucci[2]

ABSTRACT: The developments of innovative adaptive structures on Unmanned Aerial Vehicles (UAVs), such as morphing wings, can potentially reduce system complexities by eliminating control surfaces and their auxiliary equipment. This technology has the potential of allowing a UAV to adapt to different mission requirements or to execute a particular mission more effectively by maintaining an optimum airfoil section over a range of speeds for different segments of a mission profile. Studies on a number of smart materials candidates are currently available in the open literature to achieve wing morphing. The material selection depends on several factors including fast dynamic response, low weight, capability to operate over a wide range of flight conditions and low power consumption. This paper presents a review on smart materials technologies for UAV morphing wings. A numerical study in terms of power requirements is also presented for two morphing wing concepts: flapped and twisted wing planforms. The energy calculations for both morphing configurations were based on a two-step procedure. The first step consists of computing the aerodynamic energy using an in-house Vortex-Lattice (VL) based program. Subsequently the pressure field obtained from the first step is then mapped into a finite element mesh and the structural strain energy is calculated. The numerical results indicated that flapped morphing wings have a better aerodynamic performance when compared to twisted wings and different morphing levels can be achieved using lighter smart materials with lower specific energy for this configuration.

KEYWORDS: Morphing wings, Smart materials, Wing design.

INTRODUCTION

The goal of multi-mission capability in military and civil air vehicle systems has created a need for technologies which allow drastic wing shape changes during flight. Since most current aircraft are fixed-geometry, they represent a design compromise between conflicting performance requirements in mission segments, such as high-speed cruise, low-speed loiter and low turn radius turn maneuver. If a hybrid aircraft is designed to combine several flights profiles, the wing design must maximize the overall efficiency of the anticipated mission. Through morphing, the aerodynamics of the aircraft can be adapted in order to optimize performance in each segment by changing shape features such as the camber of the airfoils and the twist distribution along the wing.

Adapting the shape of wings in flight allows an air vehicle to perform multiple, radically different tasks by dynamically varying its flight envelope. The wing can be adapted to different mission segments, such as cruise, loitering and high speed maneuvering by sweeping, twisting and changing its span, area and airfoil shape. Within this context, morphing wing technology is considered to be a key component in next-generation unmanned aeronautical vehicles (UAVs) for military and civil application. The design of UAVs demands a multidisciplinary integration of different engineering areas, including aerodynamics, structural elasticity, control and actuators/sensors dynamics as schematically shown in Fig. 1. The work presented in this paper is part of an ongoing international research program on UAVs between the Department of Aeronautics, at Imperial College

1.Instituto Tecnológico de Aeronáutica – São José dos Campos/SP – Brazil 2.Imperial College London – London – United Kingdom

Author for correspondence: Mauricio Vicente Donadon | Instituto Tecnológico de Aeronáutica | Praça Eduardo Gomes, 50 – Vila das Acácias | CEP: 12228-901 São José dos Campos/SP – Brazil | Email: donadon@ita.br

London-UK, and the Instituto Tecnológico de Aeronáutica-ITA, in Brazil. The paper is focused on a review on smart materials technologies for UAV morphing wings. A numerical study in terms of power requirements is also presented for two morphing wing concepts: flapped and twisted wing planforms. The energy calculation for both morphing configurations is based on a two-step procedure. The first step consists of computing the aerodynamic energy using an in-house Vortex-Lattice (VL) based program. Subsequently, the pressure field obtained from the first step is then mapped into a finite element mesh and the structural strain energy is calculated.

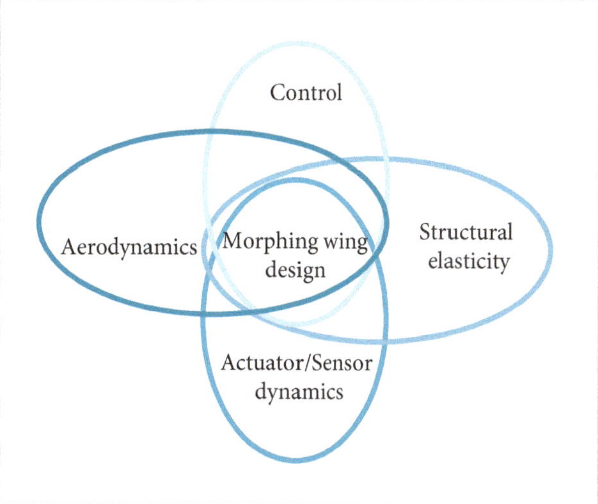

Figure 1. UAVs design methodology.

REVIEW ON MORPHING WING TECHNOLOGY

MORPHING WING CONCEPTS

The different concepts based on smart material technology, currently available in the open literature for UAVs flight control, can be classified into four distinct groups according to the adopted solution strategy (Fontanazza *et al.*, 2006):

- Wings with local morphing capabilities;
- Wings with global morphing capabilities;
- Composite wings with multi-stable structural behaviour;
- Wings with variable stiffness structural parts.

The solutions adopted for wings with local morphing capabilities rely on the deformation of compliant parts of the wing. Examples of wings with local morphing capabilities were presented by Lim *et al.* (2005), in which the authors proposed a compliant trailing edge configuration with lightweight piezo-composite actuator (LIPCA), bonded on the upper part of the skin. Kota *et al.* (2003) proved the effectiveness of novel compliant mechanisms to change the wing chamber of an airfoil to minimise drag without causing flow separation (Fig. 2)

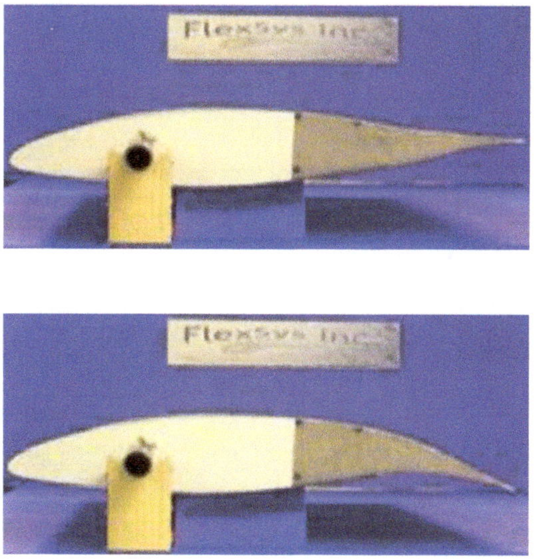

Figure 2. Trailing edge control (Flexus Inc.).

In this case, power is required in order to deform both the (compliant) structure and to generate the required aerodynamic forces. Because of the high chord-wise bending stiffness of a typical closed wing section, twisting the whole wing, or part of the wing, would be more effective (Barrett and Brozoski, 1996). The solution adopted for wings with global morphing capabilities implies in deforming the whole wing, such as twisting the wing along its entire span. This solution has been considered for flight control of fixed wing aircraft, rotorcraft, and missiles. Previous concepts made use of directionally attached piezoelectric actuators (DAP), embedded within the outer skin of high aspect ratio wings (Barrett and Brozoski, 1996). For low aspect ratio wings and missiles fins, designs with integral main spar and active torque plate were considered (Barrett, 1995). Later designs employed a bending element included into the wing, in order to achieve greater deflections, actively pitching the aerodynamic surface.

More recently, Cesnik and Brown (2003) studied a solution with anisotropic piezocomposites (AFCs) distributed along a high aspect ratio wing. It has been concluded that novel single crystals fibre composites may be capable of providing the required control capability. Composite wings with multi-stable structural behaviour exploit the possibility of changing the shape of an unsymmetric composite laminate from one stable position to another, supplying a small amount of energy (Schultz, 2005) to promote mode switching. The advantage is that no further power is required to keep the structure at an equilibrium configuration. The main drawback is that few (two or three) stable shapes are possible so that the resulting control system could not be used for manoeuvring over different range of speeds.

Wings with variable stiffness structural parts exploit the energy on the fluid (aerodynamic forces) rather than directly using the smart actuators to change the shape of the wing. Griffin and Hopkins (1997) suggested the Variable Stiffness Spar (VSS) concept in order to improve the manoeuvrability of flexible aircrafts (e.g. to counteract aileron reversal). The solution is based on the simultaneous actuation of a control surface and modification of the wing stiffness. In the VSS, a spar made of separated parts, linked with hinges, can be rotated in order to change its ability to react the shear forces (Chen *et al.*, 2000). Another design, the Torsion-Free (TF) wing concept, consists of two closely spaced very stiff spars which carry most of the shear. The stiffness of the other spars is reduced in order to produce a wing with low torsional stiffness. Two VSS, placed along the leading and trailing edge, are used to tune the wing torsional stiffness (Chen *et al.*, 2000). The TF concept was also investigated by Changho *et al.* (2002), employing variable stiffness Shape Memory Alloys (SMA) spars to increase roll effectiveness. Amprikidis *et al.* (2005) have recently developed an "adaptive internal structure" to twist a wing, by moving the position of the elastic axis. This can be obtained by rotating two spars or changing their chord-wise position. With this approach, a considerably lower amount of energy is required to twist the wing and keep it in the desired position. For the specific problem of UAV roll control, third and forth concepts seem to be the most promising ones. Smart materials can be employed either to twist the whole wing or to tune its stiffness. For the latter solution, in order to guarantee fast response and high efficiency, the use of "active materials" in an "intrinsically adaptive" mode is a requirement.

CANDIDATE MATERIALS FOR SMART MORPHING WINGS

Smart materials are able to respond to a stimulus in a useful and reproducible manner (Suleman, 2001). The materials themselves are not "smart", in the sense that they passively react to an input rather than making decisions or adapting themselves to the environment. A more accurate definition proposed by Kornbluh *et al.* (2004) classifies them into "Intrinsically adaptive materials" and "Active materials". Intrinsically adaptive materials are materials subjected to transformations in their molecular or microscopic structure due to a particular external stimulus (usually characterized by a small energy content regarding the deformation energy within the material), resulting in changes in mechanical properties. The SMA and shape memory polymers (SMP) are examples of intrinsically adaptive materials. Active materials act as transducers, converting some forms of energy (typically electrical, magnetic, and thermal) into mechanical energy. Electroactive polymers, piezoelectric ceramics, and magnetostrictive (Terfenol-D) are some examples of active materials. Active materials with high electromechanical coupling can also be used in an "intrinsically adaptive mode"; in this case, they require less power supply, but their performance are more limited.

The main advantages in using smart materials rather than conventional pneumatic or hydraulic actuators are the reduced complexity and improved reliability of the system. Similarly, the potential weight saving and the possibility of using active materials as both actuators and sensors within the structure are clear advantages. The 'best' materials and concepts to adopt depend on the specific morphing purpose. Since the aim is to change the shape of the wing for flight control, the morphing system should exhibit:

- Relatively fast dynamics;
- Capability to operate over a wide range of flight conditions;
- High reliability;
- Capability of repetitive actuations;
- Robustness against uncertainties and disturbances (e.g. gusts);
- Low power consumption;
- Insensitivity to environment variation.

Hence, the ideal material should respond quickly to the external stimuli, be capable of large and recoverable free strains, transform effectively the input energy into mechanical energy, and not to be affected by fatigue issues. Table 1 reports the main characteristics of the most common smart materials (maximum free strain, maximum stress, deformation energy density, efficiency, and relative speed of response). SMAs and SMPs can undergo large free strains and exhibit large blocking forces, but they have slow response and limited efficiency. Piezoelectric Ceramic (PZT) and single crystal piezoceramics, exhibit a much lower free strain, but they are electrically activated, capable of producing quite high blocking forces, and sensibly more efficient ones too. Electroactive polymers exhibit good properties, although they can produce low blocking stress.

ACTUATION ENERGY REQUIREMENTS ANALYSIS FOR MORPHING AIRFOILS

One of the most import issues and concerns in smart wing technology have been the actuation energy and power which have to be provided by the vehicle onboard power system. Naturally, a smart wing may almost always require the deformation of some, preferably secondary, wing structure with the actual power requirements heavily dependent on the wing structural realization and actuation scheme. For the present study, actuation energy of the deformable parts of the wing have been calculated on the basis of the work performed by the aerodynamic forces during the wing morphing in the aerodynamic flow field. The computation of the aerodynamic work has been carried out using an in-house computational program based on the VL method (Donadon and Iannucci, 2006a). The program enables the prediction of lift, pressure distribution, rolling and pitching moment calculations for flapped and twisted wing planforms.

AERODYNAMIC ENERGY COMPUTATION

The term aerodynamic energy defined here refers to the total energy induce by the pitching moment acting on the deformable parts of the wing. Thus, the expressions for the aerodynamic energy for flapped and twisted wings can be respectively written as follows

$$W_h = \int_0^{\theta_f} M_y^h(\theta) d\theta = \int_0^{\theta_f} \sum_{n=1}^{N} \Gamma_n \Delta y_n x_n^h d\theta \tag{1}$$

$$W_t = \int_0^{\varnothing_f} M_y^t(\varnothing) d\varnothing = \int_0^{\varnothing_f} \sum_{n=1}^{N} \Gamma_n \Delta y_n x_n^t d\varnothing \tag{2}$$

where $M_y^h(\theta)$ is the pitching moment around the flapping line, $M_y^t(\theta)$ is the pitching moment around the twisting line, θ is the flap tip deflection and ø is the wing tip twisting angle. Δy_n is the elemental spanwise length, M_n^h is the distance between the elemental leading vortex segment and the flapping line

Table 1. Most common smart materials (Fontanazza *et al.*, 2006).

Material	Max. Strain (%)	Max. Stress (MPa)	Elastic energy density (J/g)	Max. Efficiency (%)	Relative speed
Dielectric Polymer Acrylic	215	16.2	3.4	60-80	Medium
Silicone	63	3	0.75	90	Fast
Electrostrictor Polymer P(VDF-TrFE)	4	15	0.17	---	Fast
Piezoelectric Ceramic (PZT)	0.2	110	0.013	>90	Fast
Single Crystal (PZN-PT)	1.7	131	0.13	>90	Fast
Polymer (PVDF)	0.10	4.8	0.0013	n/a	Fast
SMA (TiNi)	>5	>200	>15	<10	Slow
SMP	100	4	2	<10	Slow
Terfenol-D	0.2	70	0.0027	60	Fast
Conducting polymer (Polyanaline)	10	450	23	<1	Slow

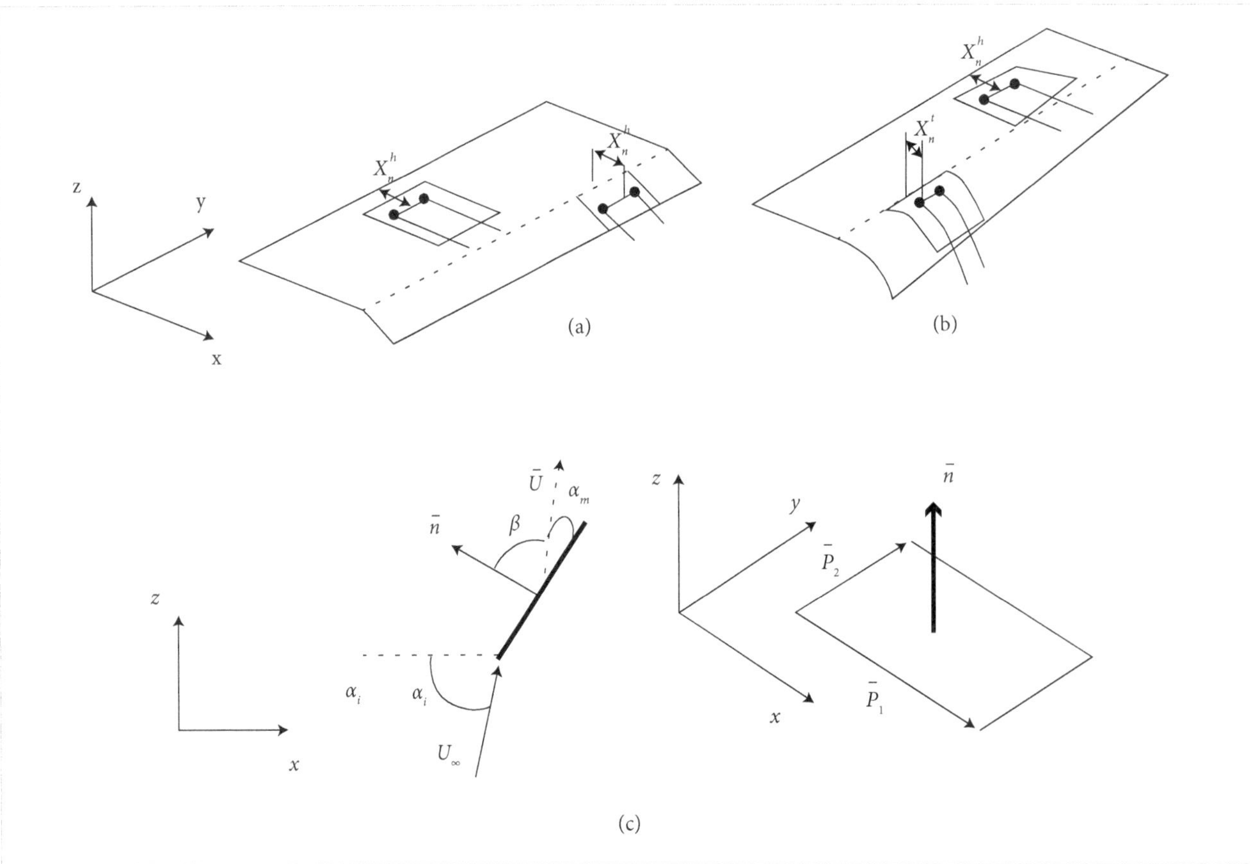

Figure 3. (a) reference line for pitching moment calculations for flapped wing, (b) reference line for pitching moments calculations for twisted wing and (c) elemental angle of attack.

and X_n^t is the distance between the elemental leading vortex segment and the twisting line, as shown in Fig. 3 (b). N is the total number of panels and Γ_n is the elemental vortex strength obtained by solving the following linear system of equations,

$$\{\Gamma_n\} = \left[\bar{C}^w_{m,n} - \bar{C}^w_{m,n} \tan(\Psi_n) \right]^{-1} 4\pi U_\infty \{\alpha_m\} \tag{3}$$

where $\bar{C}^w_{m,n}$ and $\bar{C}^w_{m,n}$ are the downwash and sidewash influence coefficients, respectively, computed according to the Biot-Savart Law (Bertin, 1989). Ψ_n is the elemental wing dihedral angle, U_∞ is the air flow velocity and α_m is the elemental angle of attack schematically illustrated in Fig. 3 (c).

INCREMENTAL PRESSURE COEFFICIENT CALCULATION

The incremental pressure coefficient for the *n-th* panel of the wing is given by (Lamar and Margason, 1971)

$$\Delta c_{p,n} = \frac{2\Gamma_n}{c_n U_\infty} \tag{4}$$

where c_n is the elemental chord. The resultant pressure acting on each panel of the wing can be written in terms of the incremental pressure coefficients as follows

$$\Delta P_n = \frac{\rho U_\infty^2}{2} \Delta c_{p,n} \tag{5}$$

WING STRUCTURE STRAIN ENERGY

The wing structure strain energy is given by

$$U = \frac{1}{2} \iiint \{\varepsilon\}^T \{\sigma\}\, dV \tag{6}$$

where V is the volume occupied by the structural elements of the wing, $\{\sigma\}$ and $\{\varepsilon\}$ are the stress and strain vectors, respectively. By using the finite element method, Eq. (6) can be rewritten in terms of the wing stiffness matrix and nodal displacement vector as follows

$$U=\frac{1}{2}\{\delta\}^T[K]\{\delta\} \tag{7}$$

where U is the total strain energy generated during the wing morphing process in the presence of the aerodynamic pressure field, together with the actuation forces provided by the smart materials to deform the wing.

NUMERICAL SIMULATIONS

This subsection presents a numerical study in terms of actuation energy requirements for both flapped and twisted morphing wing configurations. The chosen wing dimensions as well as flight conditions are typical of small UAVs and they are listed in Tables 2 and 3, respectively. A NACA 0012 airfoil section with dimensions shown in Fig. 4 was assumed for both wings. Both wings have flexible trailing edges made of elastomeric skins, starting at 70% of the chord (region indicated by red dashed line in Fig. 4) and extending up to the full-length chord dimension of the wings. The mechanical properties of the elastomeric skins are depicted in Table 4. In order to compute the pressure field as well as the aerodynamic work induced by the airflow, an in-house VL program (Donadon and Iannucci, 2006a) based on the formulation described in the previous sections was used. Full wing models were required for both wing configurations in order to obtain the resultant pressure field. Once the pressure field was determined, the aeroelastic problem was then solved by mapping the differential pressure field into a finite element model of the deformable parts of the wing and the required elastic energy density determined. The finite element models were developed using the rectangular four-node bi-linear Mindlin shell elements (S4R), with reduced integration available in ABAQUS/Standard finite element code. The structural analyses were carried out assuming that the trailing edge was clamped to the remaining part of the wing. The meshes used for the structural and aerodynamic simulations were the same in order to allow a direct pressure mapping between lattices and finite elements.

For the flapped wing, the hinging line was positioned at 70% of the chord and the final flap deflection was assumed to be 10°. A convergence study for the pressure field values indicated that a mesh density of 20 elements, spanwise by 10 elements chordwise, gives results within an accuracy of less

Table 2. Wing dimensions.

Spanwise length [m]	1.40
Root chord [m]	0.27
Tip chord [m]	0.27
Angle of Attack [Degrees]	3.0
Flap deflection (for the flapped wing) [Degrees]	10.0
Twisting angle at the wing tip (for the twisted wing) [Degrees]	10.0

Table 3. Flight conditions and air properties.

U_{∞} [m/s]	40.0
Altitude [m]	1000
ρ_{∞} [kg/m³]	1.117

Table 4. Mechanical properties for the elastomeric skin (Donadon and Iannucci, 2006b).

E [MPa]	6.90
V	0.30
ρ [kg/m³]	1080

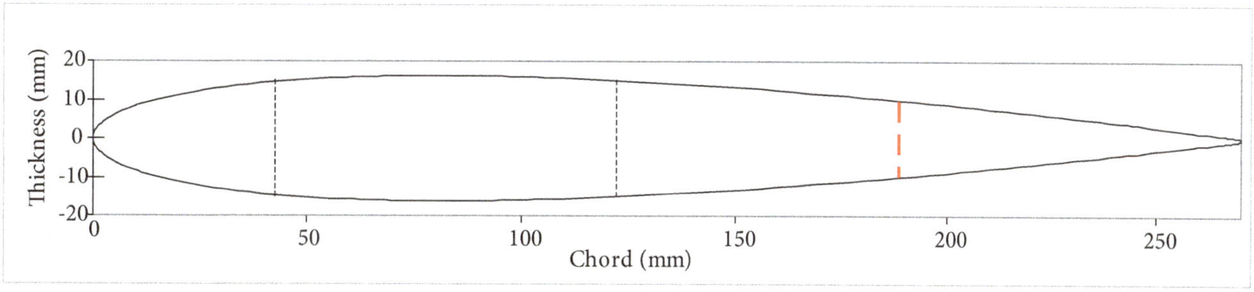

Figure 4. NACA 0012 airfoil section.

than 1% when compared to finer meshes and, for this reason, this mesh density was used throughout this work. A typical VL mesh for the flapped wing is shown in Fig. 5.

Figure 6 shows the numerical results in terms of pressure distribution for the flapped wing. It can be seen that there is a singularity in the pressure distribution around the hinging line, as expected. This singularity is due to the change in the local angle of attack, which increases the vorticity strength in that region, affecting both lift and pressure distributions.

The dimensions and flight conditions for the twisted wing were assumed to be the same as those defined for the flapped wing in order to provide a direct comparison between both wings planforms in terms of lift, pressure distribution, aerodynamic energy and required elastic energy density. The twisting line was placed at 70% of the chord extending throughout the wing span length. The VL mesh and the pressure distribution for the twisted wing are shown in Figs. 7 and 8, respectively.

It can be seen from Fig. 8 that there is an increase in both lift and pressure distributions towards the tip of the wing due to the change of the local angle of attack in that region. It also can be noticed that the lift generated by the twisted wing is lower than the lift generated by the flapped wing. The pitching moment about the twisting line is also lower than the one obtained for the flapped wing. The higher values of pitching moment for the flapped wing were expected, because in the twisted wing, just part of the trailing edge is deflected whilst in the flapped wing the whole trailing edge is deflected. Figure 9 shows a comparison in terms of lift generation between the flapped and twisted wings for flap deflection and the local wing twisting angles ranging from 0° up to 10°. A comparison in terms of aerodynamic energy generated by flapped and twisted wing planforms is presented in Fig. 10.

Figure 11 compares the aeroelastic strain field induced in the trailing edge portions of the flapped and twisted morphing wings.

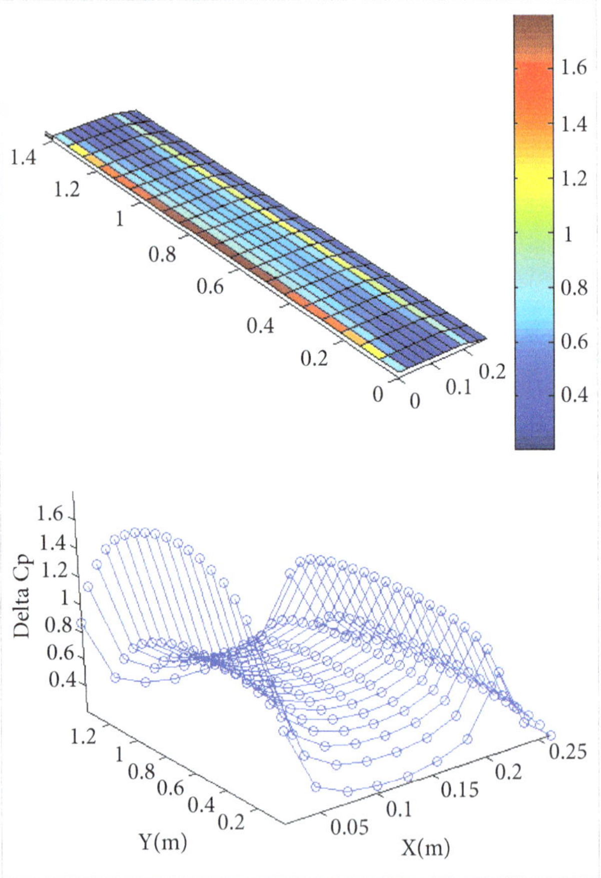

Figure 6. Pressure distribution for the flapped wing.

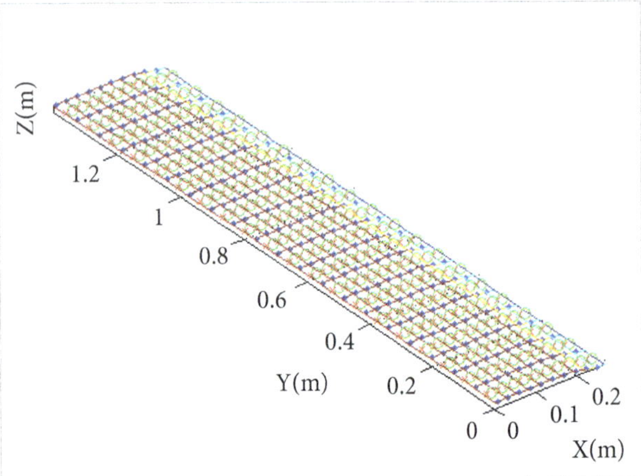

Figure 5. Vortex lattice mesh for the flapped wing.

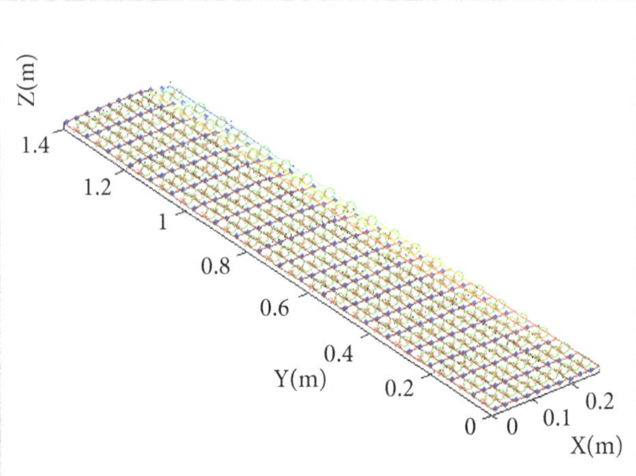

Figure 7. Vortex lattice mesh for the twisted wing.

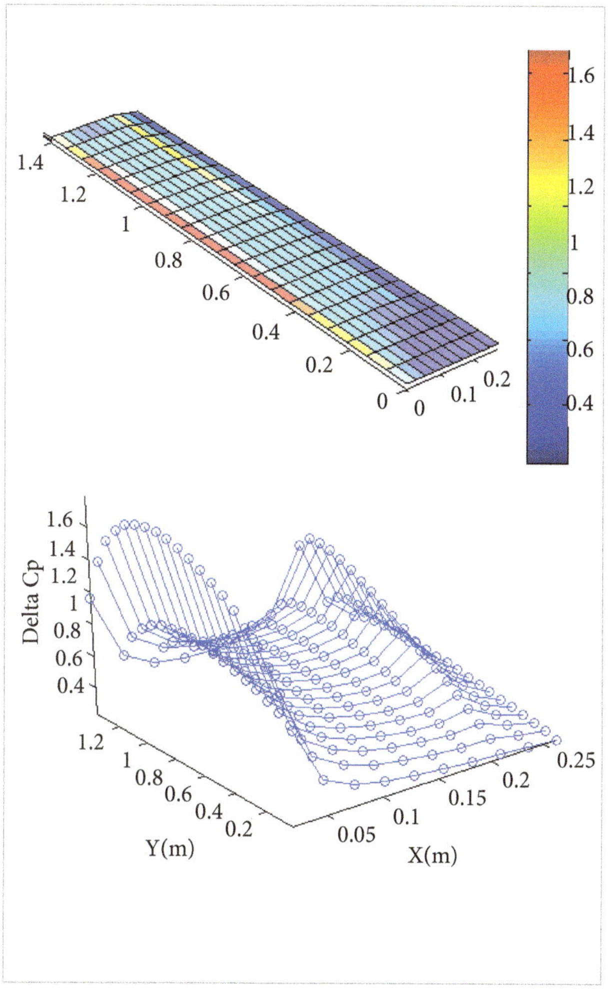

Figure 8. Pressure distribution for the twisted wing.

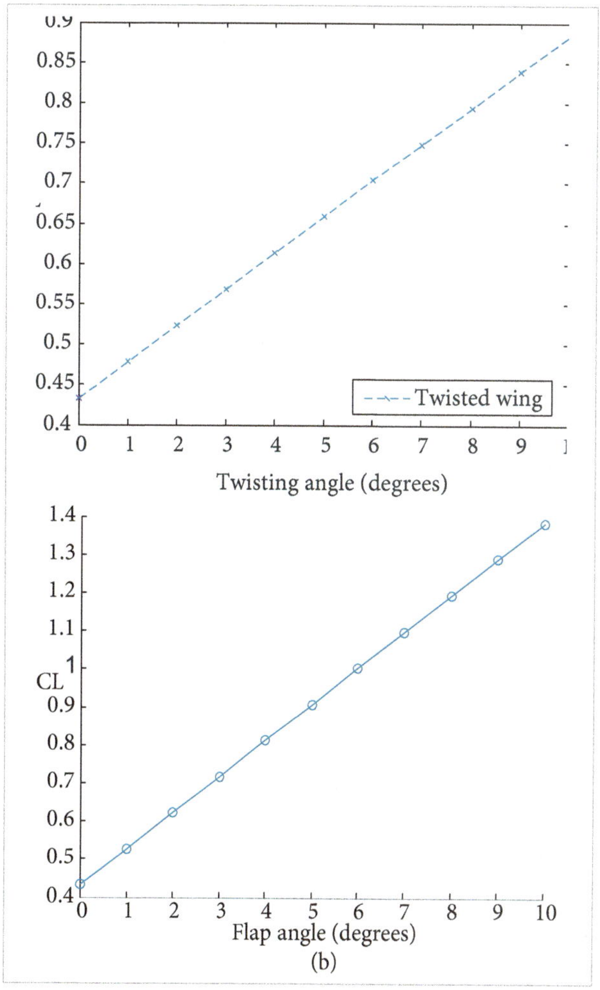

Figure 9. Lift generation comparison between twisted (a) and flapped (b) wings.

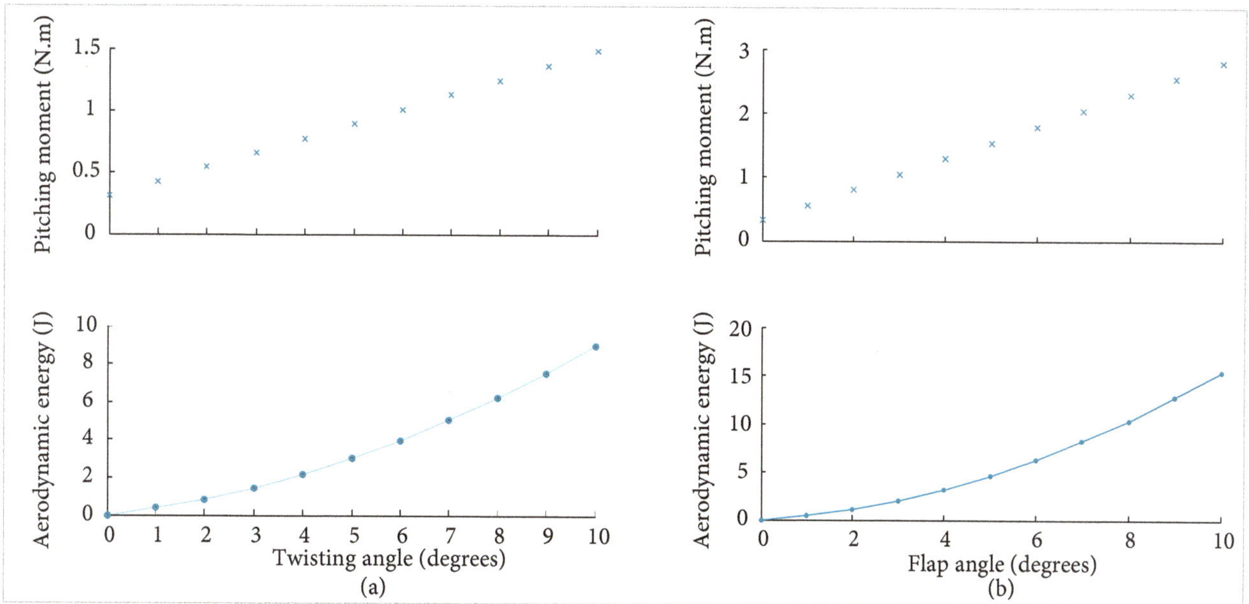

Figure 10. Comparison between aerodynamic energies for twisted (a) and flapped wings (b).

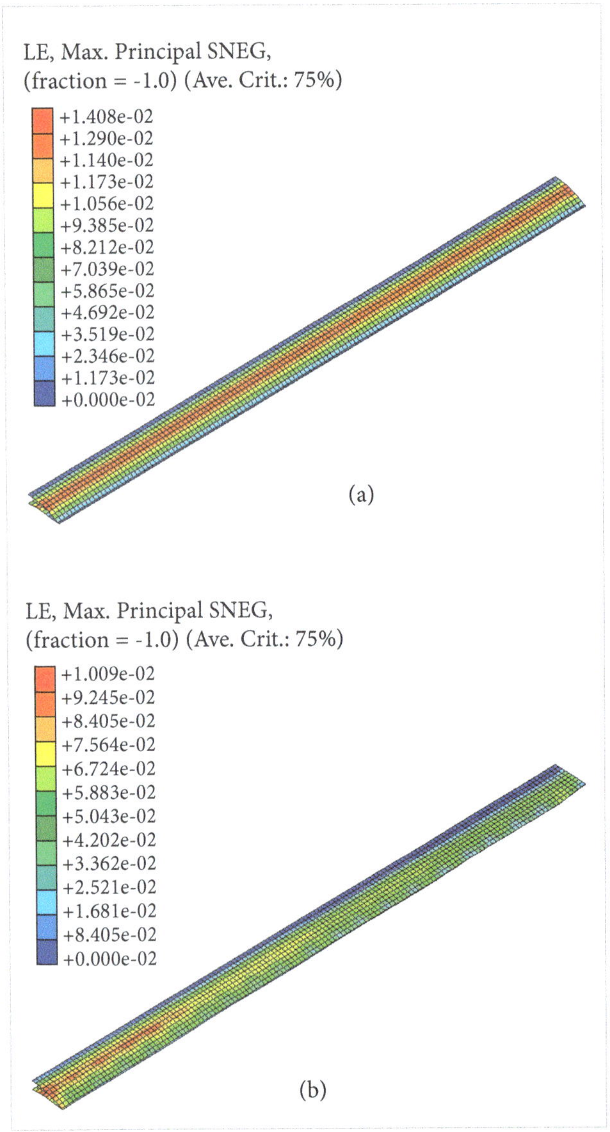

Figure 11. Aeroelastic strain field induced in the trailing edge portions flapped (a) and twisted morphing wings (b).

Table 5 compares the required elastic energy density for both morphing wing configurations. It can be seen that twisted configuration requires less actuation energy than flapped wing configuration. On the other hand, the aerodynamic performance in terms of lift generation is much better for the flapped morphing configuration. Comparing the required elastic energy density for both wing configurations with the values provided in Table 1, one can see that only materials 2, 3, 4, 5, 7, 8, and 10 are able to provide the amount of elastic energy density required to deform the wings and sustain them against the aerodynamic pressure field. However, only Single Crystal (PZN-PT) can provide a fast response with maximum efficiency for both morphing configurations.

Table 5. Energy quantities computed for flapped and twisted morphing wing configuration.

	Flapped	Twisted
Strain energy (J)	0.0760	0.0250
Aerodynamic energy (J)	15.00	9.00
Elastic energy density (J/g)	0.0060	0.0036
Max. Stress (MPa)	0.100	0.072
Max. Strain (%)	1.40	1.01

CONCLUSIONS

A review on smart materials technologies and concepts for morphing wing structures was presented and discussed in this paper. A formulation based on the Vortex Lattice Method (VLM) was proposed in order to compute the pressure distribution, lift generation and aerodynamic energy for both flapped and twisted morphing wing planforms. The proposed formulation has been implemented into MATLAB software. Numerical simulations were carried out for a typical small UAV, considering two morphing concepts: flapped and twisted wing configurations. The numerical results indicated that the flapped wing generated higher lift when compared to the twisted wing for the same deflection range. However, less aerodynamic power was required to sustain the twisted wing against the aerodynamic loads. These findings indicated that flapped wing configurations have a better aerodynamic performance when compared to the twisted wing, however, there is still a need of further investigation considering global twisting instead of twisting just part of the wing. A better aerodynamic performance means that the deformable parts of the wing can be made of lighter smart materials with lower specific energy, which allows the fabrication of lighter aircrafts with higher performance and less fuel consumption. The preliminary study presented in this paper suggests Single Crystal (PZN-PT) materials as potential candidates for smart morphing wing structures due to its fast response with maximum efficiency for both morphing configurations studied in this work.

ACKNOWLEDGEMENTS

The authors acknowledge the financial support received for this work from BAe Systems-UK through FLAVIIR SEEDCORN PROJECT.

REFERENCES

Amprikidis, M., Cooper, J.E., Rogerson, C. and Vio, G., 2005, "On The Use of Adaptive Internal Structures for Wing Shape Control", 46th AIAA/ASME/ASCE/AHS/ASC Structures, Structural Dynamics and Materials Conference, Austin, TX.

Barrett, R. and Brozoski, F., 1996, "Adaptive Flight Control Surfaces, Wings, Rotors and Active Aerodynamics", Proceedings of the SPIE – International Society for Optics and Photonics, San Diego, CA, v 2717.

Barrett, R., 1995, "All-Moving Active Aerodynamic Surface Research", Smart Materials and Structures, Vol. 4, No. 2, pp. 41-44. doi: 10.1088/0964-1726/4/2/001.

Bertin, J. J., 1989, "Aerodynamics for Engineers", 2nd Edition, Prentice-Hall Inc.

Cesnik, C. E. S. and Brown, E. L., 2003, "Active warping control of a joined-wing airplane configuration", 44th AIAA/ASME/ASCE/AHS/ASC Structures, Structural Dynamics and Materials Conference, Norfolk, VA.

Chen, P. C., Sarhaddi, D., Jha, R., Liu, D. D., Griffin, K. and Yurkovich, R., 2000, "Variable Stiffness Spar Approach for Aircraft Maneuver Enhancement Using ASTROS", Journal of Aircraft, Vol. 37, No. 5, pp. 865-871. doi: 10.2514/2.2682.

Changho, N., Chattopadhyay, A. and Youdan K., 2002, "Application of Shape Memory Alloy (SMA) Spars for Aircraft Maneuver Enhancement", Proceedings of the SPIE - International Society for Optics and Photonics, San Diego, CA, v 4701.

Donadon M. V. and Iannucci L., 2006a, "A Vortex Lattice Program to Compute Aerodynamic Loads in Flapped and Twisted Wing Planforms", Internal Report-Flaviir SeedCorn Project, Department of Aeronautics, Imperial College London, London, UK .

Donadon M. V. and Iannucci L., 2006b, "Morphing Wing Concepts and Smart Coupons Manufacturing", Internal Report- Flaviir SeedCorn Project, Department. of Aeronautics, Imperial College London, London, UK.

Fontanazza, A., Talling, R., Jackson, M., Dashwood, R., Dye, D. and Iannucci, L., 2006, "Morphing Wing Technologies Research", Proceedings of the 1st SEAS DTC Technical Conference, Edinburg, Scotland (CD ROM).

Griffin, K. E. and Hopkins, M. A., 1997, "Smart Stiffness for Improved Roll Control", Journal of Aircraft, Vol. 34, No. 3, pp. 445-447. doi: 10.2514/2.2191.

Kota, S., Hetrick, J., Osborn, R., Paul, D., Pendleton, E., Flick, P. and Tilmann, C., 2003, "Design and Application of Compliant Mechanisms for Morphing Aircraft Structures", Proceedings of the SPIE - International Society for Optics and Photonics, San Diego, CA, v 5054.

Kornbluh, R. D., Prahlad, H., Pelrine, R., Stanford, S., Rosenthal, M. A. and von Guggenberg, P. A., 2004, "Rubber to Rigid, Clamped to Undamped: Toward Composite Materials with Wide-Range Controllable Stiffness and Damping", Proceedings of the SPIE - International Society for Optics and Photonics, San Diego, CA, v 5388.

Lim, S. M., Lee, S., Park, H. C., Yoon, K. J. and Goo, N. S., 2005, "Design and Demonstration of a Biomimetic Wing Section Using a Lightweight Piezo-composite Actuator (LIPCA)", Smart Materials and Structures, Vol. 14, No. 4, pp. 496-503. doi: 10.1088/0964-1726/14/4/006.

Lamar, J. E. and Margason, J. R., 1971, "Vortex-lattice fortran program for estimating subsonic aerodynamic characteristics of complex planforms", NASA TN D-6142.

Schultz, M. R., 2005, "A New Concept for Active Bistable Twisting Structures", Proceedings of the SPIE - International Society for Optics and Photonics, San Diego, CA, v 5764.

Suleman, A., Ed., 2001, "Smart structures: application and related technologies", Springer Wien New York, NY.

9

Parametric Model Method and Deployment Simulation of Inflatable Antenna Structures

Xu Yan[1], Zheng Yao[1], Guan Fuling[2], Huang He[2], Xu Xian[2]

ABSTRACT: In this paper, we develop a parametric model method and simulate the deployment mechanism of inflatable antenna structures. Different folded methods are developed for the primary members of inflatable antenna structures, which include inflatable tubes, an inflatable torus, a reflector etc. The unstressed configuration and the folded configuration of these members are modeled parametrically using the developed folded methods. A simulation software is developed for the deployment mechanism of inflatable structures by the improved spring-mass system. The driving forces in the deployment process, i.e. the gas pressure and the moment of the fold hinges, are analyzed for each member. During the development process, self-contact or collision with the membrane occurs. A rule for identifying self-contact elements is applied, and a penalty function method is developed to solve this challenging problem. Finally, the equation of motion is solved using finite difference method. The developed simulation software is validated by simulating a cylindrical inflatable tube that is folded in half, and the simulation agrees well with the experiment. The deployment mechanism of the antenna model similar to that for the Inflatable Antenna Experiment (IAE) is modeled, analyzed and estimated. The deployed configurations and the dynamic parameters of each node are obtained. The numerical simulation results show that the simulation software for the deployment mechanism can correctly predict the deployment process of inflatable antenna structures.

KEYWORDS: Inflatable antenna, Improved spring-mass system, Folded methods, Deployment mechanism simulation.

INTRODUCTION

Large-size aerospace structures are difficult to construct because of the limitations of their weight and launch volume. Compared to conventional electromechanical deployable structures (Xu *et al.*, 2012), the inflatable structures offer many advantages including design flexibility, lightness in weight and compact launch volumes and are suitable for large-size aerospace structures. Since the 1990s, inflatable technologies have been used for antenna structures in NASA's Jet Propulsion Laboratory etc. Many ideas, concepts and models have been proposed, from the inflatable antenna (Cliff and Ron, 2001; David, 2003) to the inflatable aerodynamic decelerator (Kramer *et al.*, 2013) and the lunar habitat (Capua *et al.*, 2011). One of the key technologies of these structures is that the system can be folded into a compact launch volume and deployed in a smooth and orderly manner. Numerical simulations are needed to investigate the deployment mechanism in greater detail.

Deployment simulations and research studies of inflatable structures to date have primarily focused on inflatable tubes. Simulations have been based on two methods: multi-rigid-body dynamics theory and the non-linear finite element method. In the first simulation method, an inflatable membrane tube is modeled as several rigid struts or beam elements, and the deployment process of the tube can then readily be analyzed using multi-body dynamics theory (Fang *et al.*, 2006). A tube between a two-fold point as inflatable cantilever beams was modeled (Clem and

1. Zhejiang University – School of Aeronautics and Astronautics – Hangzhou – China. 2. Zhejiang University – College of Civil Engineering and Architecture – Hangzhou – China.
Author for correspondence: Xu Yan | Zhejiang University – School of Aeronautics and Astronautics | Hangzhou 310027 – China | Email: xyzs@zju.edu.cn

Smith, 2000), which was then analyzed using multi-rigid-body dynamic theory. The 3-D deployment mechanism of the inflatable structures must be analyzed using the non-linear finite element method (Wang and Johnson, 2002; Lampani and Gaudenzi, 2010) and some commercially available software was applied to this simulation work. PAM-CRASH software was used to simulate the deployment of an inflatable structure (Haug *et al.*, 1991; Katsumata *et al.*, 2014). LS-DYNA software was used to analyze the deployment of a tube and accounted for the reciprocity between the gas flow and the flexible tube shell (Salama *et al.*, 2001). The numerical and experimental investigations on the deployment of inflatable tubes with one fold point were conducted (Miyazaki and Uchiki, 2002; Bouzidi *et al.*, 2013). Also the deployment of an inflatable tube was validated by space experiment (Wei *et al.*, 2015). However, only a few research studies have been performed on the deployment mechanism of complicated inflatable structures, which should be investigated further.

For complicated inflatable structures — such as an inflatable antenna, which consists of inflatable tubes, an inflatable torus and a reflector —, it is difficult to model the folded configuration of the entire structure. In this paper, we report on recent efforts to develop a parametric model and deployment simulation software for inflatable antenna structures. The entire antenna structure is modeled parametrically and analyzed, and the deployment mechanism is simulated using an improved spring-mass system in which the self-contact of the membrane is solved using a penalty function method.

FOLDED METHODS AND PARAMETRIC MODEL

There are four configurations in the deployable simulation of inflatable structures: the unstressed configuration (prior to folding), the initial folded configuration, the configurations during deployment and the final fully deployed configuration. The purpose of the deployment simulation is to simulate the motion from the initial folded configuration to the fully deployed configuration using the known unstressed and the initial folded configurations. In this section, the unstressed and initial folded configurations of the primary members, such as the inflatable tubes, the inflatable torus and the reflector, are designed and modeled parametrically.

INFLATABLE TUBES

The unstressed configuration of the inflatable tubes is a cylinder shell. The length of the cylinder shell is L_1 and the section radius is R_1. The unstressed configuration of the section is a circle, and the folded configuration of the section is shown in Fig. 1, which includes two sections of circular curves. The radius of two circular curves is R_2 and the corresponding central angle is θ. The strain of the membrane material during the folding process is neglected, which results in an invariant circumference for the tube section. So, the central angle θ_1 in the folded configuration can be expressed as follows:

$$\theta_1 = (\pi R_1)/R_2 \tag{1}$$

The coordination of each node in the circular curves can then be calculated. There are typically two folded patterns for inflatable tubes along the length direction: the curve folded pattern and the zigzag folded pattern. The second folded pattern is used for the inflatable antenna structures (Katsumata *et al.*, 2014).

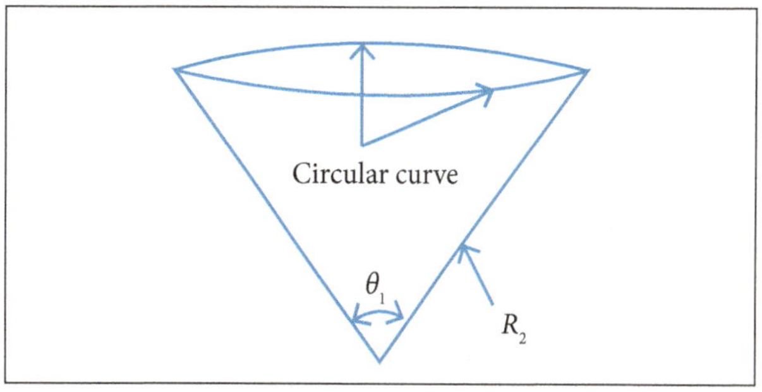

Figure 1. Folded configuration of the tube's section.

INFLATABLE TORUS

The unstressed configuration of the inflatable torus is a circular shell surface. This surface has a non-zero Gaussian curvature, which causes an initial stress to appear in the membrane after the folding process. Similarly to the inflatable tubes, the unstressed configuration of a section of a torus is a circle, which can be folded in the same way as for the inflatable tubes.

The axes curve of the inflatable torus is also a circle and the radius of the axes curve is denoted by R. The developed folded method is shown in Fig. 2a. A circular curve can be approximated by an equilateral polygon. The number of sides of the equilateral polygon is denoted by n. The length of each side of the polygon is given as follows:

$$L = 2\pi R/n \tag{2}$$

The projected shape in the 0XY plane is also a polygon after the inflatable torus is folded. θ is the folded angle of each polygon side along vertical direction. The length of each side is given as follows:

$$L_{xy} = (2\pi R/n)\cos\theta \tag{3}$$

The circumradius of the polygon in the folded configuration is:

$$r = L_{xy}(R/L) \tag{4}$$

The projected height of the folded configuration onto the 0XZ plane is:

$$L_{xy} = (2\pi R/n)\sin\theta \tag{5}$$

Thus, the coordination of each fold point is:

$$x_i = r\ cos\varphi_i,\ y_i = r\ sin\varphi_i,\ z_i = \pm\ L_{xz}/2 \tag{6}$$

where:

φ: the corresponding central angle, and the sign of the Z-coordinate alternates.

The folded methods for the axes curve are used for the torus section to calculate the coordination of each node in the folded configuration. The folded configuration of the inflatable torus is shown in Fig. 2b.

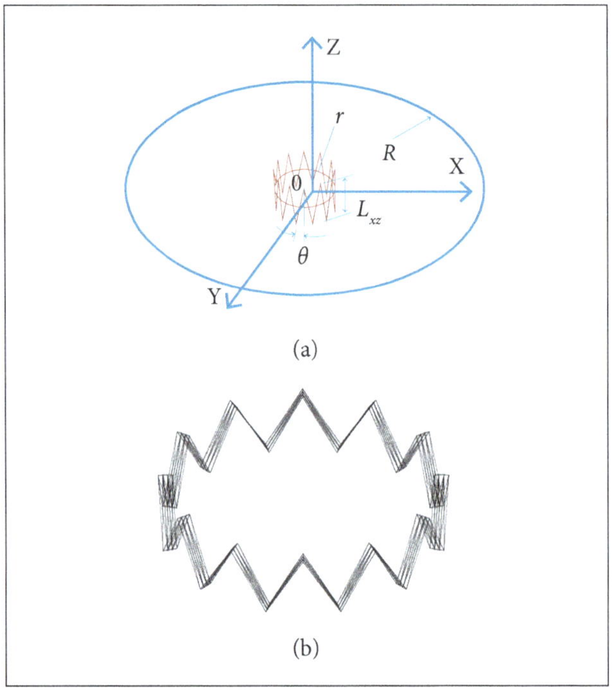

Figure 2. Parametric model for folded torus. (a) Folded method for axes curve; (b) Folded configuration.

INFLATABLE REFLECTOR

The inflatable reflector consists of a reflective surface and a canopy. The reflective surface assumes a parabolic shape under the inflation pressure. A parabolic surface is difficult to fold. The parabolic surface is a surface of revolution; thus, the surface is first folded along the longitudinal direction and then along the circumferential direction. The fold pattern along the latitudinal direction is similar to that along the axes direction of the inflatable torus. The fold pattern along the latitudinal direction is similar to that of the Z-folded method for inflatable tubes.

The fold pattern of the canopy is similar to that of the reflector surface. First, the canopy is placed near the reflector surface. The Z-coordination of each point on the canopy is 1 mm smaller than the corresponding point on the reflector surface. The canopy is then folded synchronously with the reflector surface. The parametric models for unstressed configuration and folded configuration of inflatable reflector are shown in Fig. 3.

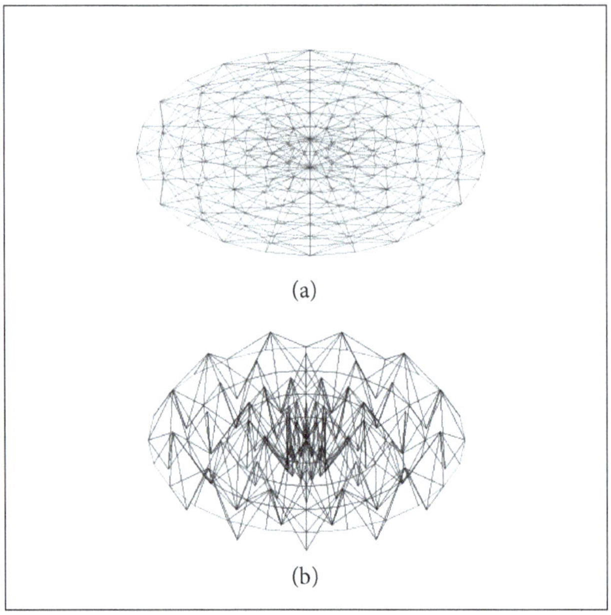

Figure 3. Parametric models of inflatable reflector. (a) Unstressed configuration; (b) Folded configuration.

INFLATABLE ANTENNA

A complete model of the inflatable antenna structure, such as that of the Inflatable Antenna Experiment (IAE), includes inflatable tubes, an inflatable torus, a reflector and tension cables. The support structure for the inflatable antenna structure consists of three tubes and a torus. There are several tension cables between the reflector and the support structure. Each primary member is folded using the developed method and is assembled to form the initial configuration of the entire structure. The unstressed configuration and the folded configuration of the entire antenna structure are shown in Fig. 4.

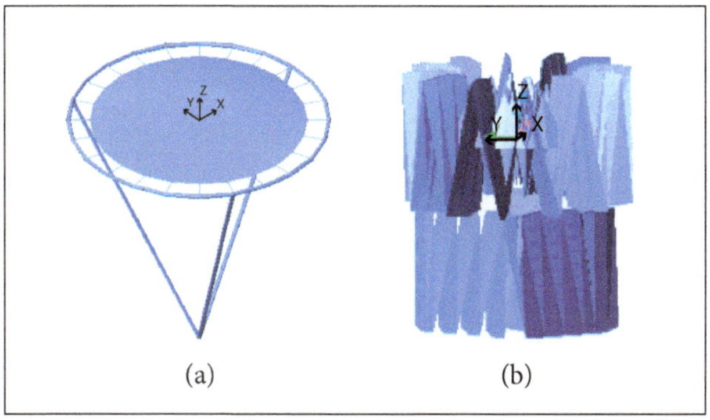

Figure 4. Parametric models of the entire antenna's structure. (a) Unstressed configuration; (b) Folded configuration.

DEPLOYMENT DYNAMICS SIMULATION METHOD

DESCRIPTION OF SPRING-MASS SYSTEM

A spring-mass system is developed to model the membrane material in the deployment simulation. The three vertexes of the triangular membrane element consist of masses, and the three borders consist of springs. The lengths of the springs in their initial state correspond to the original lengths of the springs. During the deployment process, the spring lengths do not equal their original lengths, *i.e.* internal forces exist in the springs. The three masses move under the combined action of the spring forces and the external forces to reach the final equilibrium configuration. The spring-mass system for an element is shown in Fig. 5.

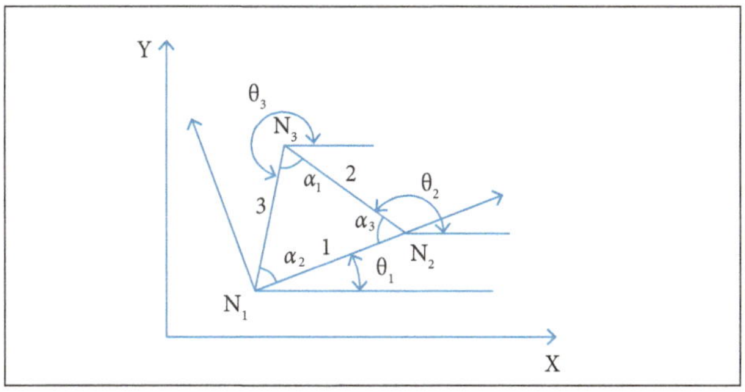

Figure 5. Spring-mass system.

The mass of point *i* is determined by the area of the membrane elements shown in Fig. 6 and is given as follows:

$$m_i = \frac{\alpha_i \cdot \rho}{2\pi} \sum_{k=1}^{s} A_k \tag{7}$$

where:

ρ: area density of the material; A_k: area of element *k* around the mass point *i*; *s*: number of elements around the mass *i*; α_i: internal angle for each vertex of the element.

The tension vector of the springs in the element can be obtained from Xu *et al.* (2007):

$$\mathbf{T} = \begin{Bmatrix} T_1 \\ T_2 \\ T_3 \end{Bmatrix} = \mathbf{B}^T\mathbf{D}\mathbf{B} \begin{Bmatrix} \delta_1 \\ \delta_2 \\ \delta_3 \end{Bmatrix} V \tag{8}$$

where:

δ: extension vector of the three springs; *V*: volume of the triangular element; D: elastic matrix of the membrane material; B: the relationship matrix between the strains of the element and the length deformations of three springs.

$$\mathbf{B} = \frac{1}{|\mathbf{A}|} \begin{bmatrix} (b_2c_3 - b_3c_2)l_1^{-1} & (b_3c_1 - b_1c_3)l_2^{-1} & (b_1c_2 - b_2c_1)l_3^{-1} \\ (a_3c_2 - a_2c_3)l_1^{-1} & (a_1c_3 - a_3c_1)l_2^{-1} & (a_2c_1 - a_1c_2)l_3^{-1} \\ (a_2b_3 - a_3b_2)l_1^{-1} & (a_3b_1 - a_1b_3)l_2^{-1} & (a_1b_2 - a_2b_1)l_3^{-1} \end{bmatrix} \tag{8a}$$

$$a_i = \cos^2\theta_i,\ b_i = \sin^2\theta_i,\ c_i = \sin\theta_i\cos\theta_i \tag{8b}$$

$$\mathbf{A} = \begin{bmatrix} a_1 & b_1 & c_1 \\ a_2 & b_2 & c_2 \\ a_3 & b_3 & c_3 \end{bmatrix} \tag{8c}$$

$$\mathbf{D} = \begin{bmatrix} \frac{E}{1-\nu^2} & \frac{\nu E}{1-\nu^2} & 0 \\ \frac{\nu E}{1-\nu^2} & \frac{E}{1-\nu^2} & 0 \\ 0 & 0 & \frac{E}{2(1+\nu)} \end{bmatrix} \tag{8d}$$

where:

E: elastic modulus; *ν*: Poisson's ratio; l_i: initial lengths of the springs; θ_i: angles that the sides of the element make with the X-axis.

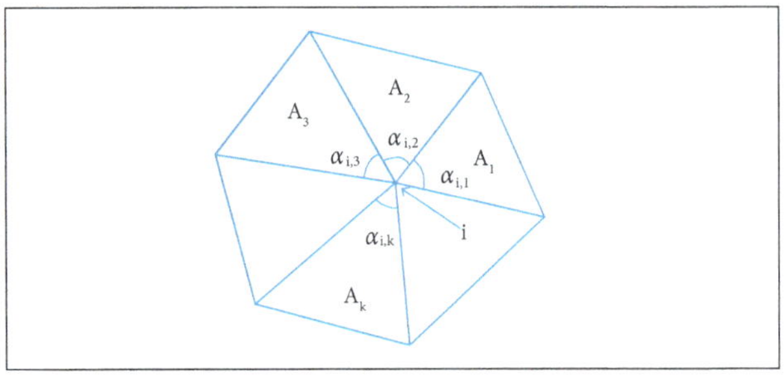

Figure 6. Triangular elements around mass *i*.

DEPLOYMENT DRIVING FORCES

In this section, the deployment driving forces that lead to the motion of the inflatable primary members are formulated. The driving forces in the inflatable antenna structures have two sources: the gas pressure and the moment of the fold hinge.

The internal chamber of the inflatable tubes and the inflatable torus are divided into a finite number of compartments (Katsumata *et al.*, 2014; Xiao *et al.*, 2010). The inflatable reflector can be considered to be a compartment. The variations in the gas state of each compartment are functions of the deployment time. the temperature and pressure of the inflation gas are also assumed to be uniform within the control volume of an individual compartment. For each compartment, a set of differential governing equations based on gas dynamic theory can be obtained. For three adjacent compartments, *i*, *j* and *k*, the differential governing equations for compartment *j* are given in Eq. 9. For brevity, the formulas are not derived here: further details can be found in Fang *et al.* (2006).

$$\begin{bmatrix} 0 & 0 & 0 & 0 & m_jV_j & V_jT_j & 0 & (k-1)m_jT_j \\ 0 & 0 & 0 & 0 & 0 & 1 & 0 & 0 \\ 0 & 0 & 0 & 0 & -m_jR & -RT_j & V_j & P_j \\ 0 & 0 & V_j & 0 & 0 & 0 & -V_j^* & (P_i - P_a) \end{bmatrix} \begin{Bmatrix} \dot{T}_i \\ \dot{m}_i \\ \dot{P}_i \\ \dot{V}_i \\ \dot{T}_j \\ \dot{m}_j \\ \dot{P}_j \\ \dot{V}_j \end{Bmatrix} = \begin{Bmatrix} 0 \\ 0 \\ 0 \\ 0 \\ kT_iV_j\dot{m}_{in} - kT_jV_j\dot{m}_{out} \\ \dot{m}_{in} - \dot{m}_{out} \\ 0 \\ 0 \end{Bmatrix} \quad (9)$$

where:

T_i, m_i, P_i, and V_i: temperature, mass, pressure and volume of compartment *i*; T_j, m_j, P_j, and V_j: temperature, mass, pressure and volume of compartment *j*; P_a: ambient gas pressure; $R = 8.3145$ J/molK; $k = c_p/c_v$: specific heat ratio of the gas; $\dot{m}_{in}$: mass rate of the gas flowing into compartment *j* from compartment *i*; $\dot{m}_{out}$: mass rate of gas flowing out of compartment *j* into compartment *k*.

The state of each compartment is then calculated by integrating this group of differential equations with respect to the deployment time. The Runge-Kutta method is used to integrate these equations numerically. The inflation pressure of each compartment is then obtained by performing a gas flow analysis and equals the nodal forces on the mass of the spring-mass system. The nodal forces $\mathbf{F}_a^i$ of a triangular element a in the shell of compartment *i* can be expressed in the following form:

$$\mathbf{F}_a^i = (1/3)A\{p_x\, p_y\, p_z\, p_x\, p_y\, p_z\, p_x\, p_y\, p_z\} \quad (10)$$

where:

$p_x = n_x \mathrm{x} P_i$, $p_y = n_y \mathrm{x} P_i$, $p_z = n_z \mathrm{x} P_i$; P_i: pressure of compartment *i*; n_x, n_y, n_z : normal vector components of the triangular element plane; *A*: area of the element.

A non-linear joint model and the Bernoulli-Euler method (Clem and Smith, 2000) were used to obtain the relationship between the restoring moment at a hinge point between two adjacent compartments and the folded angle θ:

$$M = \pi p r^3[1\text{-exp}(-3.12Et/p \tan \theta/2)] \quad (11)$$

where:

p: inflation pressure of the tube or torus near this fold joint; *r*: section radius of the member; *E* and *t* : Young's modulus and the thickness of the material, respectively.

The restoring moment can easily be transformed into the equivalent nodal forces of the mass point in the spring-mass system.

SELF-CONTACT PROBLEM

Membrane collision during the deployment process of the inflatable structures is commonly referred to as the self-contact problem. The self-contact problem is one of the major challenges in a deployment simulation because each thin film moves independently in a 3-D space. One solution for this problem is to apply the polygon intersection method, which identifies membrane collision by determining whether the boundary of one membrane film intersects the plane or boundary of another membrane film. If a collision is about to occur at a mass point, the motion of this mass is changed by altering the mass velocity such that the imminent collision is prevented.

In this study, a different method is used to solve the self-contact problem. To prevent self-contact and collision, a penalty force is added to the moving masses. Provot's method (Wang *et al.*, 2002) is adapted to apply a penalty force when the distance between one mass $A(x_A, y_A, z_A)$ and a membrane element *IJK* is less than a fixed value h_j^* (see Fig. 7). The penalty function is a vector function. When the mass *A* moves to a position that is too close to the element plane *IJK*, the penalty function is applied at the position of the mass to prevent collision.

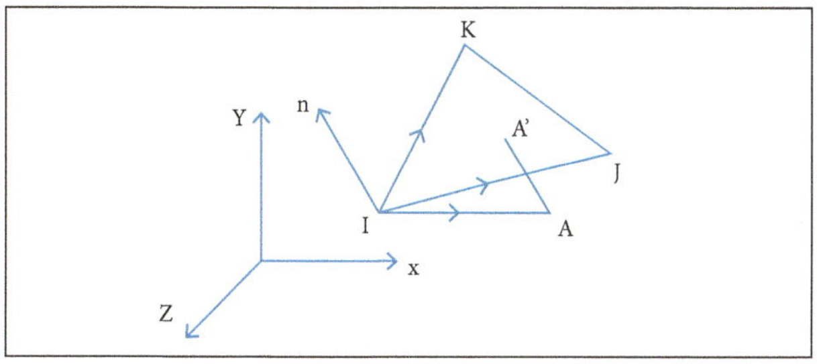

Figure 7. Self-contact element between the mass A and the element IJK.

Given the vectors $\mathbf{X}^I, \mathbf{X}^J, \mathbf{X}^K$ that describe the current position of the nodes in the global coordinate system, the boundary vectors of the triangular element can be defined as follows:

$$X^{JI} = X^J - X^I,\ X^{KI} = X^K - X^I,\ X^{KJ} = X^K - X^J \quad (12)$$

The normal vector of the element plane *IJK* can be defined as follows:

$$\mathbf{v}_3 = \mathbf{X}^{JI} \times \mathbf{X}^{KI},\ \mathbf{n} = \mathbf{v}_3/||\mathbf{v}_3|| = (x_n, y_n, z_n) \quad (13)$$

where:

n: unit normal vector.

The equation of the element plane *IJK* is given as follows:

$$\mathbf{nx} = (\mathbf{X} - \mathbf{X}^I) = 0 \quad (14)$$

where:

X: position vector of a random node in the plane.

First, we determine the side of the plane *IJK* on which the mass *A* lies using the following rules:

- if $\mathbf{n}.(\mathbf{X}^A - \mathbf{X}^I) > 0$, the mass *A* lies on the obverse side of the plane;

- if $\mathbf{n}.(\mathbf{X}^A - \mathbf{X}^I) < 0$, the mass *A* lies on the reverse side of the plane.

In Fig. 8, the mass *A* lies on the reverse side of the plane.

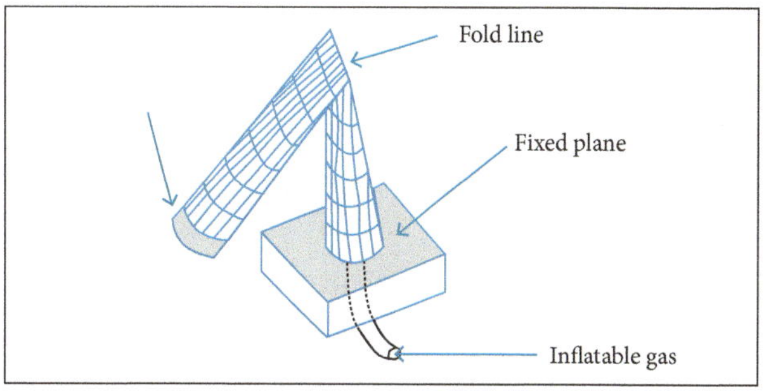

Figure 8. Components of the experimental model.

The projected node *A'* of mass *A* onto the plane is obtained by solving the following equations:

$$\begin{cases} \mathbf{n} \times (\mathbf{X} - \mathbf{X}^I) = 0 \\ \dfrac{x - x_A}{x_n} = \dfrac{y - y_A}{y_n} = \dfrac{z - z_A}{z_n} \end{cases} \quad (15)$$

We determine whether the projected node *A'* lies in the triangular field by the following method.

A radial line is drawn from node *A'* that is parallel to the *ox*-axis of the local coordinate system. The radial line intersects the boundary of the triangular element. If there is an odd number of intersection points, then the node *A'* lies within the triangular field; otherwise the node *A'* does not lie within the triangular field.

If the projected node *A'* lies within the triangular element field, there is a contact relationship between the mass *A* and the element *IJK*. When the distance from the mass *A* to the element *IJK* is less than the Euclidean distance, a penalty force is added to the mass *A*.

The penalty force for the mass *A* is defined next.

When the mass *A* is on the obverse side of the plane *IJK*, the penalty force is given as follows:

$$\mathbf{F}_p = \sum_{j=1}^{m} C_p \left| h_j - h_j^* \right| \mathbf{n} \quad (16)$$

When the mass A is on the reverse side of the plane IJK, the penalty force is given as follows:

$$\mathbf{F_p} = \sum_{j-1}^{m} C_p \left| h_j - h_j^* \right| \mathbf{n}$$

$$C_p = \begin{cases} k_{contact} & (h_j \le h_j^*) \\ 0 & (h_j > h_j^*) \end{cases} \tag{17}$$

where:

$\mathbf{F_p}$: penalty force vector for the mass A; C_p: penalty coefficient; $k_{contact}$: interface contact stiffness which depends on the membrane material property; h_j: current distance from the mass A to the element plane IJK; h_j^* : fixed value for the self-contact problem; n: unit normal vector.

If the projected node A' does not lie within the triangular element field, there is no contact between the mass A and the element IJK.

EQUATION OF MOTION

The equation of motion for the spring-mass system subjected to internal and external forces is given as follows:

$$\mathbf{M}\ddot{\mathbf{X}} + \mathbf{C}\dot{\mathbf{X}} = \mathbf{F_I} + \mathbf{F_E} \tag{18}$$

where:

$\mathbf{M}$ and $\mathbf{C}$: mass and damping matrices, respectively; $\mathbf{F_I}$: inertial force arising from the springs; $\mathbf{F_E}$: generalized external force associated with the degrees of freedom of the model; $\mathbf{x}$: position vector of all the mass points.

The following equation can be written for each mass:

$$m\frac{\partial^2 X}{\partial t^2} + c\frac{\partial X}{\partial t} = F_{ex} + F_{in} \tag{19}$$

where:

m: mass of each mass point; c: damping of each mass point; F_{ex}: inertial force subjected on each mass point; F_{in}:generalized external force subjected on each mass point.

The finite difference method can be used to simplify the two terms on the left-hand side of Eq. 19:

$$m\frac{\partial^2 X}{\partial t^2} = m\frac{1}{\Delta t^2}(X_{t+\Delta t} - 2X_t + X_{t-\Delta t}),$$
$$c\frac{\partial X}{\partial t} = c\frac{1}{2\Delta t}(X_{t+\Delta t} - X_{t-\Delta t}) \tag{20}$$

Substituting Eq. 20 into Eq. 19, one yields:

$$m\frac{1}{\Delta t^2}(X_{t+\Delta t} - 2X_t + X_{t-\Delta t}) +$$
$$c\frac{1}{2\Delta t}(X_{t+\Delta t} - X_{t-\Delta t}) =$$
$$F_{ex} + F_{in}\ (m\frac{1}{\Delta t^2} + c\frac{1}{2\Delta t})X_{t+\Delta t} = F_{ex} + F_{in} + \tag{21}$$
$$m\frac{2}{\Delta t^2}X_t - (m\frac{1}{\Delta t^2} - c\frac{1}{2\Delta t})X_{t-\Delta t}$$

Let us define the following terms:

$$M = m\frac{1}{\Delta t^2},\ C = c\frac{1}{2\Delta t}, F = F_{ex} + F_{in} \tag{22}$$

Then, we can write the following expression for the entire spring-mass system:

$$(\mathbf{M}+\mathbf{C})\mathbf{X}_{t+\Delta t} = \mathbf{F}_{ex} + \mathbf{F}_{in} + 2\mathbf{M}\mathbf{X}_t - (\mathbf{M}-\mathbf{C})\mathbf{X}_{t-\Delta t} \tag{23}$$

The initial configuration of the system corresponds to a static state. Thus, the displacement, the velocity and the acceleration at time 0 are zero.

$$\mathbf{X}_t\big|_{t=0} = 0, \mathbf{X}_{t-\Delta t}\big|_{t=0} = 0,$$
$$\left.\frac{\partial \mathbf{X}_t}{\partial t}\right|_{t=0} = 0, \left.\frac{\partial^2 \mathbf{X}_t}{\partial t^2}\right|_{t=0} = 0 \tag{24}$$

In the deployment process, the initial configuration is the folded configuration S_0 of the inflatable structures. The structures are deployed into the configuration S_i by a generalized external force. Eq. 8 is used to compute the distortion and the internal force of each spring. A penalty force is added to each of the masses in which self-contact is about to occur. Eqs. 20 and 23 are then used to calculate the values of the velocity, the acceleration and the displacement at this time; these values are added to the configuration S_i to obtain the new configuration S_{i+1}.

DEPLOYMENT SIMULATION SCHEME

A simple computational scheme is in the deployment mechanism simulation. This scheme is summarized next.

- The geometric parameters of the unstressed configuration and the material parameters of membranes and cables are inputted.
- Folded methods are designed, and the initial folded configuration of the entire structure is modeled parametrically.
- The spring-mass system is formulated to describe the membrane material.
- The deployment driving forces are calculated and applied to the spring-mass system.
- The self-contact between each mass and each membrane element is determined, and a penalty force is added to each of the masses at which self-contact is about to occur.
- The boundary conditions are applied, and the equation of motion is solved using the finite difference method.
- The configurations during the deployment process are obtained and saved as an AVI video file.
- The deployment mechanism of the entire process and the structural performance of the fully deployed configuration are evaluated.

The aforementioned simulation scheme is used to formulate the deployment simulation code for the inflatable antenna structures that is compiled in Fortran, and the software interface is compiled in MFC and OpenGL. The software consists of a pre-process module, a solver module and a post-process module. In the pre-process module, the material and geometric parameters are inputted, and the parametric model code is run. The numerical model of the folded configuration is then displayed graphically. After the numerical model of folded configuration is verified, the solver code is called, and the progress of the analysis process is shown. The analysis results can be opened in the post-process module. The formats of the analysis result files are .dxf, .avi and .txt. The deployed configurations and the fully deployed configuration are displayed by the visualization module and are subsequently analyzed to evaluate the deployment mechanism.

NUMERICAL EXAMPLES

Z-FOLD PATTERN

The developed simulation software is validated by simulating a cylindrical inflatable tube that is folded in half: the results are compared to the analysis and experimental results from Miyazaki and Uchiki (2002). Each part of the deployment experimental model for this tube is shown in Fig. 8. The bottom end of the tube is fixed, and a rigid column is attached to the tip of the tube. The initial folded configuration of the model in the simulation is shown in Fig. 9.

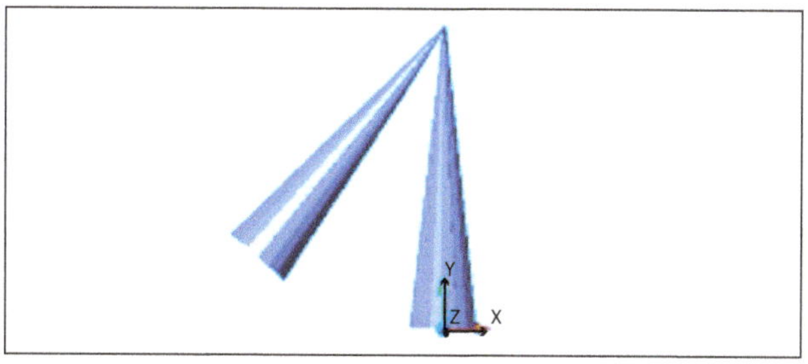

Figure 9. Folded configuration of the model.

The experiment is conducted under micro-gravity conditions, and the atmospheric pressure is maintained at 1.03 kPa. The tube has a section radius of 12 mm and a length of 200 mm. The membrane has a thickness of 104 μm, an elastic modulus of 1.08×10^8 N/m^2, a Poisson's ratio of 0.3 and a density of 910 kg/m^3. The inflatable gas is nitrogen, the experimental temperature is 300.68 K, and the gas constant is 296.798 J/kgK. The shell of the rigid column has a mass of 75.7 g, a section radius of 5 mm and a thickness of 30 mm.

The deployment mechanism of this tube model is simulated using the developed simulation software to obtain the configurations during the deployment process, which is illustrated in Fig. 10. The simulation results for the deployment process are in good agreement with the experimental results.

The internal chamber of the tube is divided into a bottom and a top compartment. The variations in the pressures in the two compartments with the deployment time are estimated using Eq. 9 and are shown in Fig. 11. Initially, the pressure in the bottom compartment increases, and the section between the two compartments is opened at a time of 0.9 s. The gas then enters the top compartment, and the pressure increases. Finally, the pressure in the two compartments reaches the design pressure of 31 kPa.

Figure 12 shows that the folded angle between the two compartments changes with the deployment time. The folded angle is initially 40 degrees and increases after 0.9 s. Finally, the tube straightens, and the folded angle reaches 180 degrees.

The x- and z-coordinates at the top end of the tube change during the deployment process and are shown in Fig. 13. The simulation and experimental results agree well with each other.

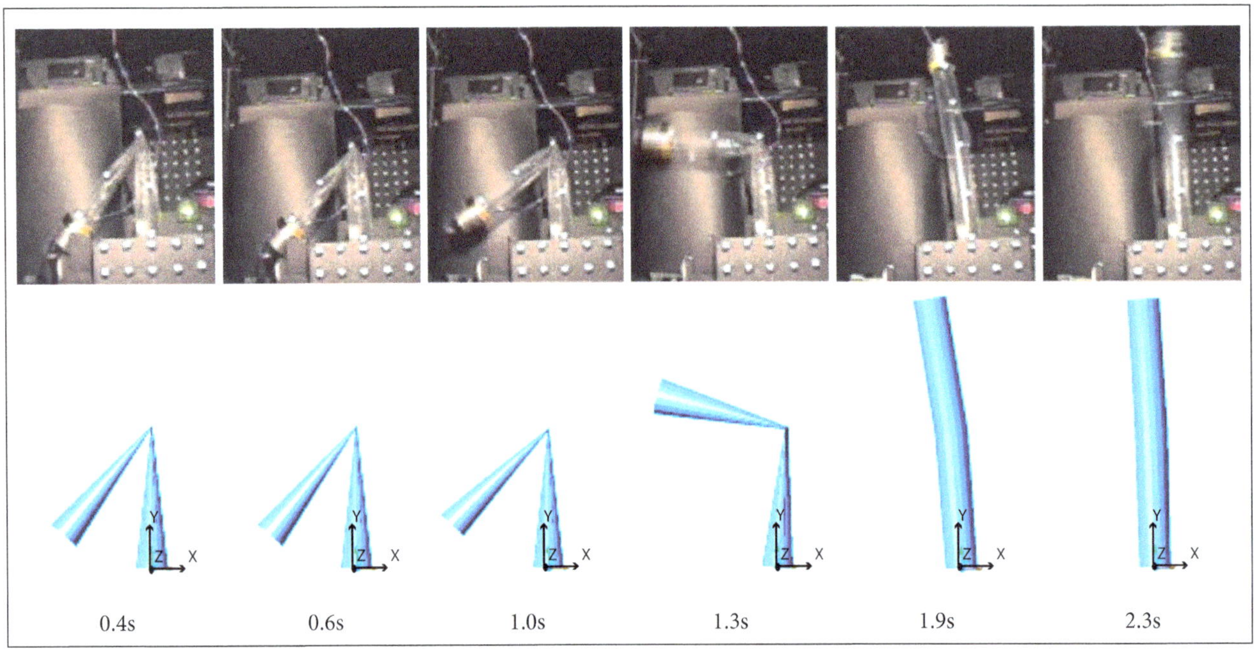

Figure 10. Comparison between experimental and simulation results.

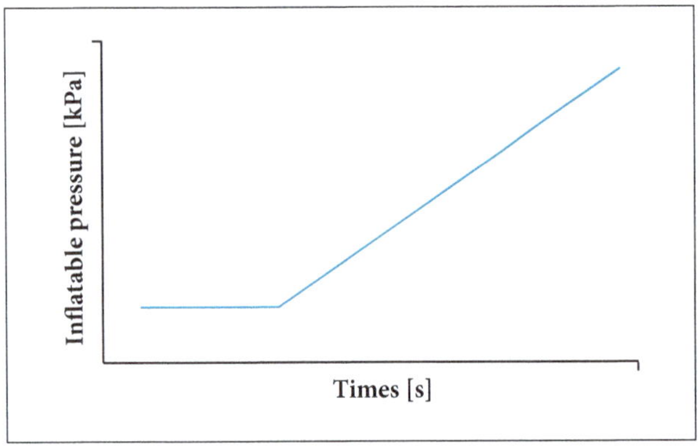

Figure 11. Changes in inflatable pressure.

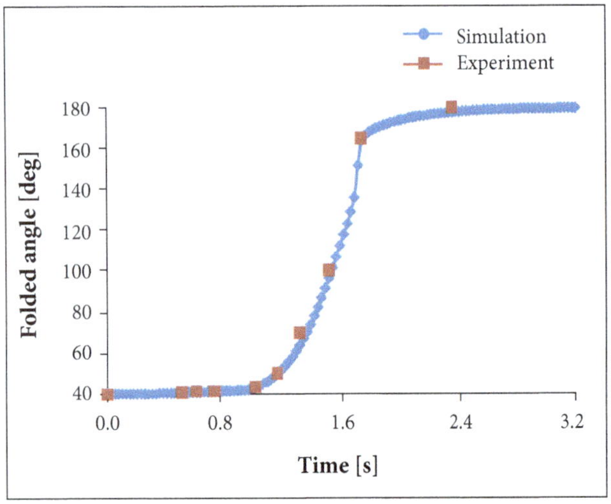

Figure 12. Folded angle between two compartments.

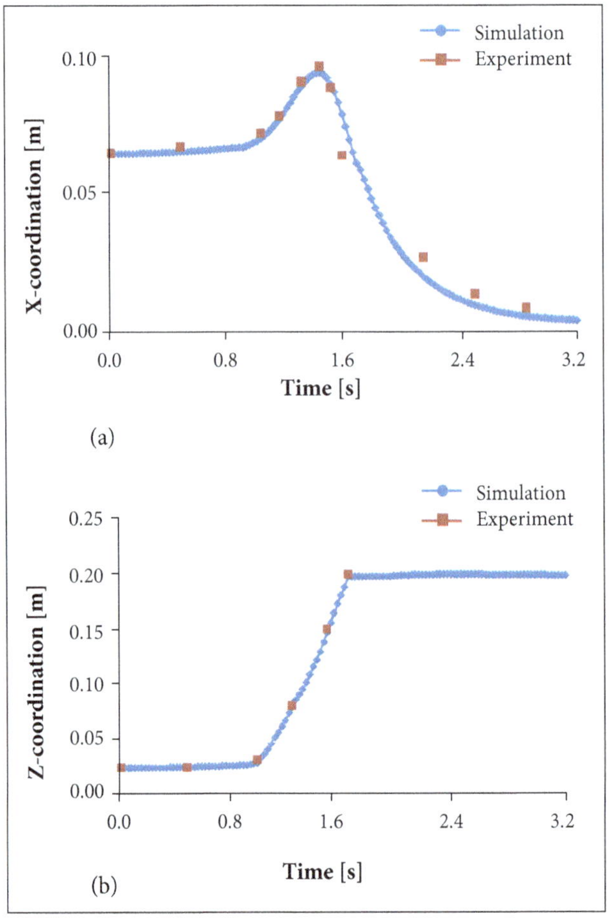

Figure 13. Coordinates at the top end of the tube. (a) X-coordinate; (b) Z-coordinate.

COMPLETE ANTENNA MODEL

A complex inflatable antenna structure similar to that in the IAE is analyzed and modeled. The model includes inflatable tubes, an inflatable torus, a reflector and tension cables. The support structure for the inflatable antenna structure consists of three tubes and a torus. There are 24 tension cables between the reflector and the support structure. The geometric parameters of the unstressed configuration are given here. The caliber D of the reflector is 6 m, and the focus length f is 7.5 m. The same equation is used for the reflector surface as for the canopy. The length of a single tube is 8.0498 m, and the section radius is 0.06 m. The outer diameter of the torus is 7.5 m, and the inner diameter is 7.0 m. The section radius of the torus is 0.125 m.

The membrane of the tubes, the torus and the reflector is made of Kapton. The membrane has an elastic modulus of 3.5 GPa, a Poisson's ratio of 0.35, a thickness of 0.127 mm, and a density of 1,450 kg/m^3. The tension cables are made of Kevlar and have a diameter of 1.2 mm. The elastic modulus of Kevlar is 131 GPa. The torus has 288 elements and 144 nodes. The reflector has 720 elements and 362 nodes. Each tube has 48 elements and 30 nodes. Thus, the entire model has 1,152 elements and 596 nodes.

The reflector has a folded ratio of 0.5 in the longitudinal direction. The circumferential direction of the reflector and the torus are divided into 24 parts, *i.e.* the fold number n = 24. The radius of the torus axes curve in the folded configuration is 1.8 m. The tube is divided into four parts lengthwise and is folded using a Z-fold pattern. The entire antenna is stowed into the volume of a cylinder with a radius of 2 m and a height of 2.6 m. The design pressures of the torus and the tubes are 23.6 kPa, and the design pressure of the reflector is 100 Pa. The tubes, torus and reflector have 4, 24 and 1 inflated compartments, respectively.

The bottom point of the three tubes is connected to the satellite and remains fixed during the deployment process.

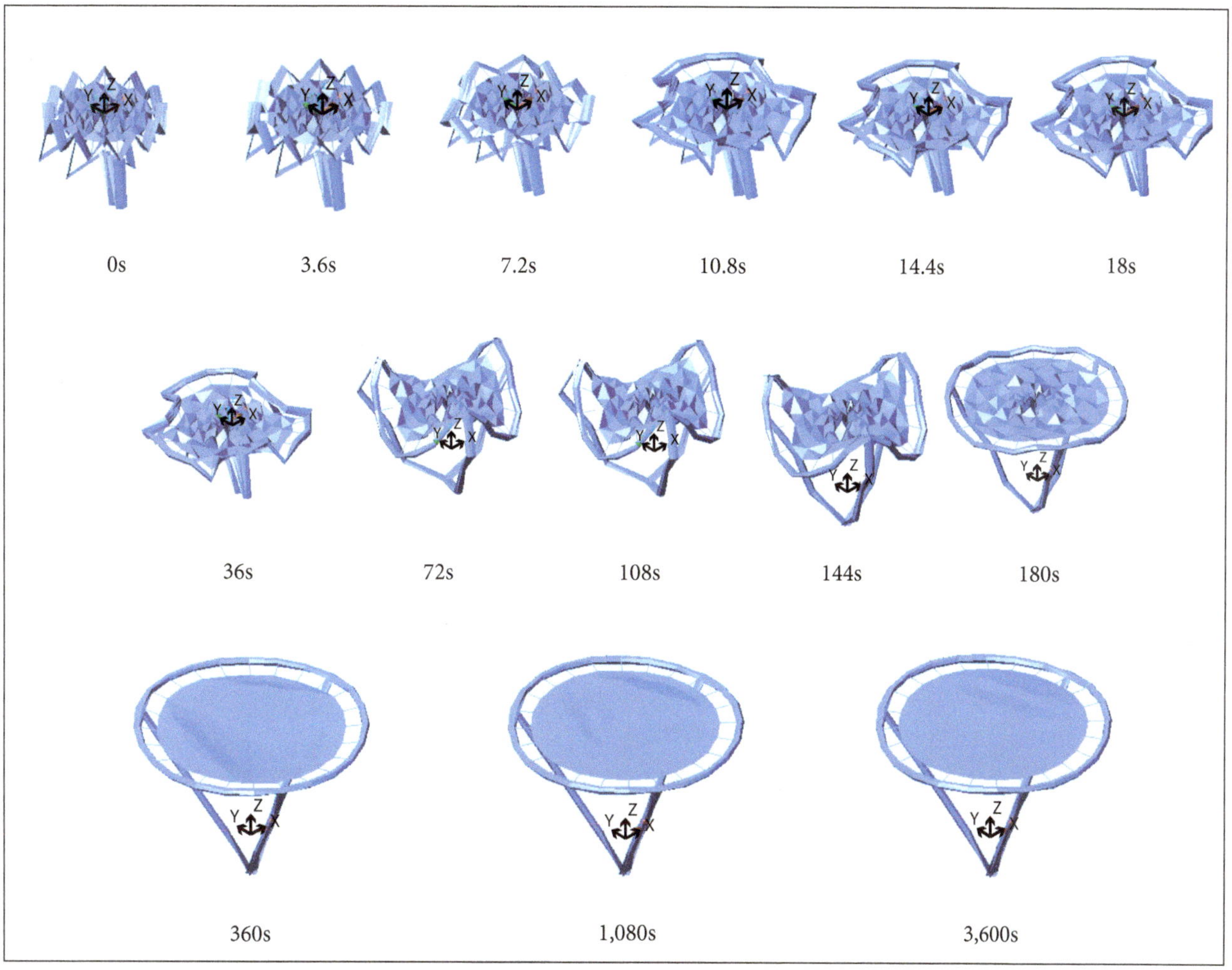

Figure 14. Individual deployment configurations.

Several configurations during the deployment process are shown in Fig. 14. First, the tube and torus are inflated. The tube straightens, which causes the torus and the reflector to move along the positive Z-coordinate. The cables cause the reflector to move. The axes of the torus tend to form a triangular shape, which is initially driven by the three tubes and then slowly deployed into a circle curve.

The results of the deployment dynamic simulation show that the three tubes and the axes of the torus are fully deployed at 360 s. The 3-D deployment velocities of the joint between the tube and the torus are shown in Fig. 15. At this time, wrinkles can be observed along the section direction of the torus. The reflector is pulled by the tension cables, but many wrinkled regions in the reflective surface remain. The reflector and the torus section continue to be deployed after 360 s. The torus section is fully deployed at 1,080 s. The reflector is slowly subjected to the design pressure and moves from its initial folded configuration to its final configuration. The internal force in the reflector surface is then transferred to the torus by the tension cables. The results show that the elastic deformation of the torus is very small.

The simulation results are used to evaluate the deployment dynamics of the antenna. The deployment process takes 3,600 s. The torus has a large diameter of 7.5052 m and a small diameter of 7.0855 m. The length of each tube is 8.0429 m. The fully deployed configuration has a high shape precision. Over the whole deployment process, the entire antenna structure moves smoothly from the folded configuration into the final design configuration. The results also validate the efficiency of the folded methods and the deployment mechanism simulation software of the inflatable structures.

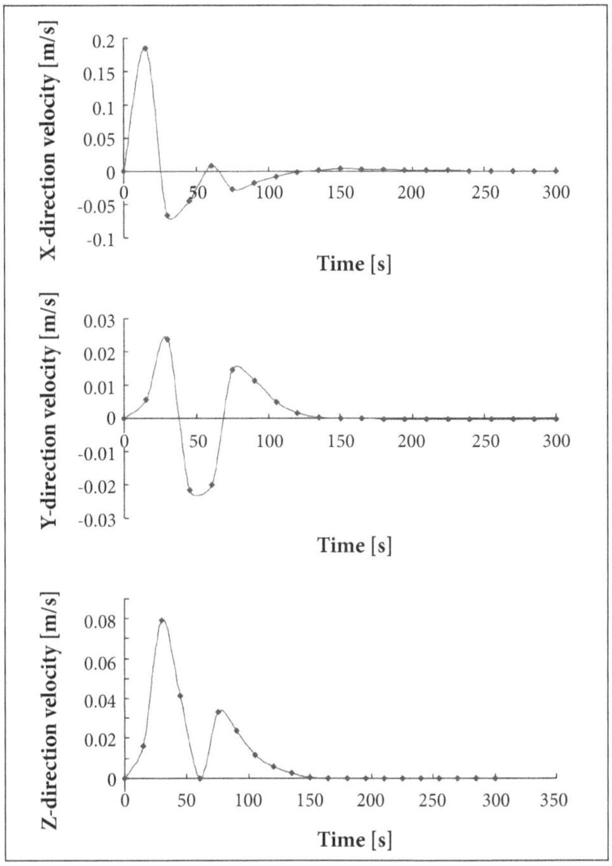

Figure 15. Deployment velocities of the joint between the tube and the torus.

CONCLUSION

Parametric model methods and simulation software for the deployment mechanism of inflatable antenna structures are investigated in this study. The unstressed and folded configurations of the entire structure are modeled parametrically. The time histories of the structural configuration during the deployment process and the dynamic parameters of each node are thus obtained by the simulation work. Numerical examples are used to validate the simulation software and the folded methods for inflatable structures. The results show that the entire antenna can be smoothly deployed from its initial folded configuration into the final fully deployed configuration. The shape of the fully deployed configuration is in good agreement with the design configuration.

In the present study, the number of elements used in the numerical model is not very large. The number of elements should be increased to verify the convergence of the numerical simulation. The primary members, such as the tubes and the torus, can be assembled to form complicated inflatable structures such as solar arrays and light shields. Assembly techniques for the parametric model design of the folded configuration should be studied in the future.

ACKNOWLEDGEMENTS

This work was supported by the National Natural Science Foundation of China (Grant No. 11402229, 51378458) and the Zhejiang Province Natural Science Foundation (Grant No. LQ14A020003, LY13E080002).

REFERENCES

Bouzidi, R., Buytet, S. and Van, A.L., 2013, "A Numerical and Experimental Study of the Quasi-Static Deployment of Membrane Tubes", International Journal of Solids and Structures, Vol. 50, No. 1, pp. 651-661. doi: 10.1016/j.ijsolstr.2012.10.027

Capua, M.D., Akin, D.L. and Davis, K., 2011, "Design, Development, and Testing of an Inflatable Habitat Element for NASA Lunar Analogue Studies", AIAA-2011-5044, pp. 1-20.

Clem, A.L. and Smith, S.W., 2000, "A Pressurized Deployment Model for Inflatable Space Structures", AIAA-2000-1808, pp. 1-11.

Cliff, E.W. and Ron, C.S., 2001, "A Hybrid Inflatable Dish Antenna System for Spacecraft", AIAA-2001-1258, pp. 1-9.

David, L., 2003, "Inflatable Deployed Membrane Waveguide Array Antenna for Space", AIAA-2003-1649, pp. 1-7.

Fang, H.F., Liu, M.C. and Hah, J., 2006, "Deployment Study of a Self-Rigidizable Inflatable Boom", Journal of Spacecraft and Rockets, Vol. 43, No. 1, pp. 25-30. doi: 10.2514/1.3283

Haug, E., Protard, J.B. and Gerren, A.M., 1991, "The Numerical Simulation of the Inflation Process of Space Rigidized Antenna Structure", Proceedings of the International Conference: Spacecraft Structures and Mechanical Testing, ESA, Zurich, Switzerland, pp. 862-868.

Katsumata, N., Natori, M.C. and Yamakawa, H., 2014, "Analysis of Dynamic Behavior of Inflatable Booms in Zigzag and Modified Zigzag Folding Patterns", Acta Astronautica, Vol. 93, No. 1, pp. 45-54. doi: 10.1016/j.actaastro.2013.06.008

Kramer, R.M.J., Cirak, F. and Pantano, C., 2013, "Fluid-Structure Interaction Simulations of a Tension-Cone Inflatable Aerodynamic Decelerator", AIAA Journal, Vol. 51, No. 7, pp. 1640-1656.

Lampani, L. and Gaudenzi, P., 2010, "Numerical Simulation of the Behavior of Inflatable Structures for Space", Acta Astronautica, Vol. 67, No. 3, pp. 362-368. doi: 10.1016/j.actaastro.2010.02.006

Miyazaki, Y. and Uchiki, M., 2002, "Deployment Dynamics of Inflatable Tube", AIAA-2002-1254, pp. 1-10.

Salama, M., Fang, H.F. and Lou, M.C., 2001, "Resistive Deployment of Inflatable Structures", AIAA-2001-1339, pp. 1-9.

Wang, C.C.L., Smith, S.S.F. and Yuen, M.M.F., 2002, "Surface Flattening Based on Energy Model", Computer-Aided Design, Vol. 34, No. 11, pp. 823-833.

Wang, J.T. and Johnson, A.R., 2002, "Deployment Simulation of Ultra-Lightweight Inflatable Structures", AIAA-2002-1261, pp. 1-14.

Wei, J.Z., Tan, H.F., Wang, W.Z. and Cao, X., 2015, "Deployable Dynamic Analysis and On-Orbit Experiment for Inflatable Gravity-Gradient Boom", Advances in Space Research, Vol. 55, No. 2, pp. 639-646. doi: 10.1016/j.asr.2014.10.024

Xiao, X., Guan, F.L. and Xu, Y., 2010, "Numerical Analysis and Experimental Study on Deployment Process of Coil-Folded Inflatable Tube" (in Chinese), Journal of Zhejiang University (Engineering Science), Vol. 44, No. 1, pp. 184-189.

Xu, Y., Guan, F.L., Chen, J.J. and Zheng, Y., 2012, "Structural Design and Static Analysis of a Double-Ring Deployable Truss for Mesh Antennas", Acta Astronautica, Vol. 81, No. 2, pp. 545-554. doi: 10.1016/j.actaastro.2012.09.004

Xu, Y., Guan, F.L. and Ma, Y.H., 2007, "Reflector Design and Precision Measurement of Inflatable Antenna" (in Chinese), Journal of Zhejiang University (Engineering Science), Vol. 41, No. 11, pp. 1921-1926.

The Sensitivity of Wind Forecasts with a Mesoscale Meteorological Model at the Centro de Lançamento de Alcântara

Elizabeth Diane de Jesus Reuter[1,2], Gilberto Fisch[3], Cleber Souza Correa[4]

ABSTRACT: The sensitivity of the planetary boundary layer (PBL) parameterizations is investigated using the hybrid Fifth Generation Penn State University/National Center for Atmospheric Research (NCAR) Mesoscale Model (MM5) and its results were compared with observations from two field campaigns held during the dry and wet seasons at the Centro de Lançamento de Alcântara (CLA). The comparisons were made using the integrated zonal and meridional components of observed and forecasted winds. Initially, three boundary layer parameterizations, in addition to the current parameterization, were selected for evaluation: Blackadar (BLK), Medium Range Forecast (MRF), Janjic (ETA) and Burk-Thompson (BT). The MRF and BLK schemes produced better results than the ETA and BT schemes. Nevertheless, MRF and BLK underestimate the zonal and meridional wind components by around 16% in the rainy season and overestimate them by on average 18% in the dry season.

KEYWORDS: Centro de Lançamento de Alcântara, MM5, Boundary layer parameterizations.

INTRODUCTION

Nowadays, in the field of Aerospace Meteorology, atmospheric modeling and numerical simulations are necessarily present in the operational planning and logistics of space vehicle launches, with the goals of guaranteeing higher security in the activities on the surface and in flight and thus contributing to the success of the launches. The Aerospace Meteorology is the application of atmospheric science towards the design, development, and operation of aerospace vehicles. At the Centro de Lançamento de Alcântara (CLA), it supports the observation missions and analysis of climate elements at the surface and altitude, as well as produces the weather forecast.

An aerospace vehicle's response to atmospheric disturbances, especially wind, must be carefully evaluated to ensure that the design will meet its operational requirements (Johnson, 2008).

A large number of different Numerical Weather Prediction models have been utilized in the operational sectors of Meteorology as a work tool. However, the limitations of each of them mean that, for each particular problem, it is necessary to develop or modify a code according to the specific desired requirements. In the case of rocket launches at the CLA, the search for a model capable of predicting the shear and wind profile with antecedence of hours in a large area of ocean-continent transition is very important given the operational necessity of understanding the winds for the determination of the vehicle trajectory and for studies of dispersion of the gases liberated by rocket combustion processes at the surface level at the launch pad.

1. Instituto Nacional de Pesquisas Espaciais – Centro de Previsão de Tempo e Estudos Climáticos – São José dos Campos/SP – Brazil. 2. Secretaria da Aviação Civil – Empresa Brasileira de Infraestrutura Aeroportuária – Guarulhos/SP – Brazil. 3. Departamento de Ciência e Tecnologia Aeroespacial – Instituto de Aeronáutica e Espaço – Divisão de Ciências Atmosféricas – São José dos Campos/SP – Brazil. 4. Departamento de Controle do Espaço Aéreo – Instituto de Controle de Tráfego Aéreo – Rio de Janeiro/RJ – Brazil.

Author for correspondence: Gilberto Fisch | Instituto de Aeronáutica e Espaço – Divisão de Ciências Atmosféricas | Praça Marechal Eduardo Gomes, 50 – Vila das Acácias | CEP: 12.228-901 – São José dos Campos/SP – Brazil | Email: gfisch@iae.cta.br

Although each physical parameterization package in a mesoscale model plays an important role in atmospheric simulations, the parameterizations of the planetary boundary layer (PBL) are formulations more simple than the complex physical processes that occur in the nearest layers of the Earth's surface. For better understanding, some definitions are important such as: mesoscale is the scale of meteorological phenomena that ranges in size from a few km to about 100 km. It includes local winds, thunderstorms, and tornadoes. PBL is the bottom layer of the troposphere that is in contact with the surface of the Earth. It is often turbulent and is capped by a statically stable layer or temperature inversion of air. Surface boundary layer (SBL) is the layer of air of order tens of meters thick adjacent to the ground where mechanical (shear) generation of turbulence exceeds buoyant generation or consumption.

The PBL parameterizations are mathematical formulations simpler than the complex physicals processes that occur in the layers closest to the Earth's surface. The principal process is turbulence, which is responsible for the vertical distribution of the properties related to heat surface fluxes, moisture and momentum within the PBL.

The realistic importance, in other words, the faithful representation of the existing processes in the PBL and SBL in meteorological models, is out of question, principally in mesoscale models. It should be taken into consideration because the majority of these parameterizations were developed based on measurements and studies conducted in Northern Hemisphere mid-latitude locations, principally in the United States and Europe. Thus, it is necessary to study the influence of these schemes on the simulation of meteorological variables in other regions, particularly in the equatorial region. There are relatively few validation studies of numerical weather models for the equatorial region: for instance, Oyama (2004), in a preliminary study, evaluated the sensitivity of the MM5 model to the deep convective parameterizations on reduced grids; Oyama and Giarolla (2006) simulated squall lines on the northern coast of Brazil also using the MM5 model. These simulations were performed in order to verify the ability of the model to represent the life cycle of squall lines (LI) initiated in the northern coast of Brazil. Squall line is a line of active thunderstorms, either continuous or with breaks, including contiguous precipitation areas resulting from the existence of the thunderstorms. The MM5 satisfactorily represented the life cycle of LI and, as a consequence, the weather; Pereira Neto and Oyama (2011) proposed adjustments in the Kain-Fritsch (KF2) deep convective parameterization scheme with the goal of improving precipitation prediction at the CLA. With the adjustments, a marked improvement in the representation of the total precipitation and of the monthly fraction of rainy days was found.

The spatial pattern of errors in the field, however, has not undergone many changes over the continent and, in general, rainfall is best represented over land than over ocean. Also using the MM5 regional model, Carvalho (2011) tested the sensitivity of simulated precipitation to different schemes of explicit convection, the activation of shallow convection (Grell, 1993) and adjustments in the scheme of deep convection KF2 scheme. Adjustments in KF2 comprehended changes in minimum depth of cloud required to activate the convection in the parameters "trigger" function and convective scales of advective and convective weather. The modeling study showed that the use of explicit convection "warm rain" and shallow convection schemes of Grell along with the "in line" in KF2 substantially reduced the overestimation of precipitation found in the simulations for the operating region of the Intertropical Convergence Zone — the axis, or a portion thereof, of the broad trade-wind current of the Tropics (ITCZ). All of these studies contributed to a better understanding of how to utilize atmospheric mesoscale modeling, in particular, the MM5 model, at the CLA region in order to support the launchings. However, they did not address the ability of the mesoscale model to forecast winds. The rainfall and the winds, both at surface and upper air, are the two key-parameters for the launchings of Brazilian rockets at CLA.

So, in order to use a numerical model to forecast the winds at the CLA, this study addressed the question of its evaluation using four different PBL parameterizations, to be described in section Description of the Modified MM5 Model, against data sets collected by radiosonde winds from the surface up to 5,000 m during wet and dry periods.

DATA

The experimental data collection was made at the CLA. The CLA is located on the coast of the state of Maranhão, with geographic coordinates of latitude 2°19' S, longitude 44°22' W and an altitude of 49 m, distant 30 km from the city of São Luís do Maranhão. Figure 1 shows a panoramic view of the CLA and the grid/domain utilized in the simulations with the MM5 model.

Source: Agência Espacial Brasileira.

Figure 1. Aerial view of the CLA and grids with horizontal resolutions of 18 km (larger domain) and 6 km (smaller domain).

The data were obtained from two field campaigns: Operation Murici 2008, during the dry season of 2008, and GPM 2010, wet season of 2010, and consist of systematic radiosounding launched at 00, 06, 12 and 18 UTC, during the periods from September 16–26, 2008, characterizing the dry season, to from March 19–25, 2010, characterizing the wet period. Meteorological parameters such as pressure, temperature and humidity were measured by the radiosonde in situ sensors while the direction and wind speed were determined by the GPS satellite navigation. The radiosonde soundings were performed using equipment model Vaisala OY (DigiCORA) using the sondes RS80-15G and RS92-SVP in experiments in 2008 and 2010, respectively.

DESCRIPTION OF THE MODIFIED MM5 MODEL

In this study, the hybrid Fifth Generation Penn State University/National Center for Atmospheric Research (NCAR) Mesoscale Model (MM5) (Grell *et al.*, 1994; Dudhia *et al.*, 2003) was utilized. The model is for numerical simulation of the atmosphere and was developed in the late 1970s by Penn State University, in conjunction with the National Center for Atmospheric Research (PSU/NCAR). In Brazil, the Laboratório de Prognósticos em Mesoescala (LPM) of the Departamento de Meteorologia of the Universidade Federal do Rio de Janeiro (UFRJ) provides predictions for the state of Rio de Janeiro with 20 km resolution. The Instituto de Controle do Tráfico Aéreo (ICEA), on the other hand, in partnership with the Centro Nacional de Meteorologia Aeronáutica (CNMA), both subordinated to the Departamento de Controle do Espaço Aéreo (DECEA) and the IAE of Departamento de Ciência e Tecnologia Aeroespacial (DCTA), uses the MM5 model with objectives of research and operational numerical modeling for aeronautical and space vehicle. The numerical simulation tests were conducted in the ICEA/DECEA research facility, which is located within the DCTA.

As the model was developed in order to simulate or predict atmospheric circulation, it is supported by pre- and post-processing programs. The MM5 model is a limited area model, which needs lateral boundary conditions to represent the real state of the atmosphere neighboring the simulation domain through the time of integration of the dynamic equations. These boundary conditions are obtained through a global scale atmospheric model, for example, the Global Forecast System (GFS), obtained from the address: http://www.ncdc.noaa.gov/data-access/model-data/model-datasets/global-forcast-system-gfs.

The modeling system consists of a set of components necessary to run the model: the Pre-processing System (WPS), the pre-processing package used in the Weather Research and Forecasting (WRF) Model, which horizontally interpolates the meteorological variables of geopotential height, wind, temperature and mixing ratio in each isobaric level, temperature and pressure on the surface and temperature and humidity of the soil layers; in this way, the version used MM5 can be considered hybrid format GRIB2. In this version of MM5, the REGRID package horizontally interpolates the analyzes and forecasts, the INTERPF vertically interpolates the pressure to sigma coordinates and also generates the initial and boundary conditions; in MM5, there are the parameter definitions, implementation of the model and periods of integration and assimilation of data and MM5toGrADS is done post-processing, which generates converted output to GrADS format. In this way, the version of MM5 used can be considered hybrid format GRIB2. Figure 2 depicted the flow chart of MM5.

This study considers the following parameterizations: Blackadar (BLK), described by Zhang and Anthes (1982), Burk-Thompson (BT; Burk and Thompson, 1989), ETA (Mellor and Yamada (1974); Janjic (1994) and Hong-Pan (Medium Range Forecast - MRF); Hong and Pan, 1996) PBL schemes available in the MM5 model, in agreement with Table 1.

A high-resolution, one-dimensional, moist PBL model is developed following Blackadar (1976) and verified using the 10 April 1979 SESAME data set (Zhang and Anthes, 1982). The model consists of two modules to predict the time-dependent behavior of the PBL under various surface characteristics. Under stable conditions, turbulent fluxes are related to a local Richardson number. In contrast, unstable conditions commonly said to free convection, the heat exchange, humidity and momentum occur through mixing between elements originating at the surface and the environmental air PBL. The stable regime, also known as nighttime, and the unstable regime are treated differently: in the nighttime, the atmosphere is stable or lightly unstable and turbulence is the result of mechanical processes, while during the daytime, or free convection regime, the atmosphere is unstable and the turbulence is the result of free convection of thermals of rising hot air, associated with the mechanical processes wind shear. In the nocturnal regime, a first order closure based on K theory approximation is used to determine the turbulent fluxes. K theory approximation, also called mixing-length theory, is a method of describing the movement of trace species on the turbulent or subgrid scale; the theory relates the fluxes of the trace species to the gradient of the mean quantities via eddy diffusivity denoted K. Considering the nocturnal regime as a local scheme, the mixture is assumed to occur only between adjacent layers in the model. The free convection regime employs a non-local approximation in which buoyancy plumes of hot air are assumed for the mixing of heat, humidity and momentum for each level in the mixing layer.

The B-T scheme available in MM5 is based on its initial implementation in NORAPS (Burk and Thompson, 1989). The PBL turbulence parameterization scheme with local second order closure is based on level 3 of Mellor and Yamada (1974). In the MM5 model, this scheme is used to predict the vertical mixing of horizontal wind, potential temperature, water vapor mixing ratio, cloud water and rain water through prognostic equations for the turbulent kinetic energy (TKE), temperature, humidity and a covariance of temperature and humidity. All other flows are obtained diagnostically. Their inclusion in the prognostic equations for larger statistical moments permits the simulation of better mixed PBLs, but with a large computational increase. It is a local closure PBL scheme of order 1 and ½ with a prognostic equation for turbulent kinetic energy.

The parameterization based on modeling developed by Janjic (1994), which will be called from here only as ETA, uses equations defined by Mellor and Yamada (1974), who consider the closure of second-order turbulence for the TKE and calculate the turbulent fluxes from it. According to Janjic (1994), this model assumes two distinct layers in PBL: a thin viscous layer above the surface, where the vertical transport

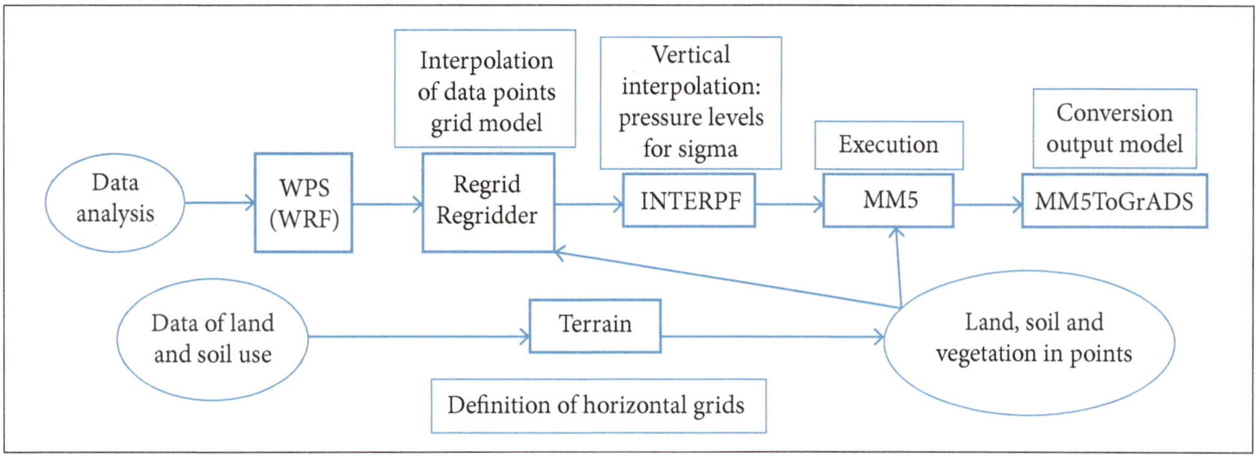

Figure 2. Flowchart of operation of the hybrid mesoscale model MM5.

Table 1. PBL parametrization schemes tested in MM5.

Name	Abbreviation	Original scheme	Closure order	Theory for vertical turbulent diffusion	Vertical diffusion	Surface layer similarity
Blackadar	BLK	Blackadar	First order	K-theory	non-local	Monin-Obhukov
Burk-Thompson	B-T	Mellor-Yamada	Second order	TKE	local	Louis
Mellor-Yamada	ETA	Mellor-Yamada	Order 1.5	TKE	local	Monin-Obhukov
Hong-Pan	MRF	Troen-Mahrt	First order	K-theory	non-local	Monin-Obhukov

Source: Adapted from Camillo (2011).

is determined by molecular diffusion, and a turbulent layer, where the vertical transport is dominated by turbulent flows. The thickness of the viscous layer is different relative to the temperature, specific humidity and wind speed and depends on the friction of its molecular diffusivity of each variable (Reynolds number). Additionally, viscous layers for temperature and humidity respectively depend on the Prandtl and the Schmidt numbers. These layers are much thinner than the surface layer. The turbulent flows of the different variables in the surface layer above the viscous layer are equal to those of the viscous layer and function of the gradient of these variables and coefficients of heat exchange and momentum defined by Mellor and Yamada. In MM5, the scheme is used to predict the vertical mixing of horizontal wind, potential temperature and mixing ratio.

An efficient scheme named MRF based on the Troen-Mahrt representation of the contragradient term and non-local K approximation (Hong and Pan, 1996) is a non-local scheme of first order based on results from large eddy simulations by Wyngaard and Brost (1984). In this scheme, there is a flow parametrization contrary to the gradient, which depends on the convective velocity and the flow to the surface. It has been used in general circulation and numerical weather prediction models due to its computational efficiency. In the MM5 model, the scheme is used to predict the vertical mixing of horizontal wind, potential temperature, water vapor mixing ratio, cloud water and cloud ice.

DESCRIPTION OF THE EXPERIMENTS

The initial and boundary conditions for the simulations were based on data from the Global Forecast System (GFS) global model. The model integrations were initialized using the 12 UTC analyses and extended for 72 hours. The first 12 hours of the simulation were not evaluated and the results were discarded as spin-up. The zonal (u) and meridional (v) wind components simulated were obtained for the grid point location closest to the CLA, whose coordinates are 2.23039° S and 44.4617° W. These variables are compared with the observations derived from radiosoundings. The wind analysis is done in the lower atmosphere; in this case, from the surface to 700 hPa, this is equivalent to a height up to 5,000 meters, because these are the altitudes where there are the largest rocket trajectory corrections.

Also, other additional input data are:

- Frequency of boundary conditions is that established by analysis and forecast files obtained from the National Centers for Environmental Prediction (NCEP) site. For MM5, files were obtained every 6 hours.
- Atmospheric information to the initial conditions: GFS analysis and forecasts. The program WPS, the WRF system, is used to read files in GFS format GRIB2 because the program responsible for this task only reads the MM5 format GRIB1.
- Sea surface temperature (SST) to the boundary conditions: the issue of SST is one aspect that still remains to be implemented in hybrid MM5. Therefore, the surface temperature was used in place of SST, given that this comes from the GFS file and is processed by the WPS.
- Moisture and soil temperature to the boundary conditions: these conditions were obtained from the GFS file.

The tested configuration of the MM5 can be summarized as: two domains nested; horizontal grid spacing of 18.6 km; number of the points of the domain size — 130 x 130, 121 x 121; 26 vertical levels and initialization at 12:00 UTC using GFS data. For the comparison of results, the parameterization MRF PBL was considered as reference or control and three other PBL parameterization options like BLK, ETA and BT were varied.

The physical parameterizations of the model are as follows: cloud radiation scheme for radiation; processes for multilayer surface; Grell for convection implied (Grell, 1993), which considers only a cloud-sized two-dimensional parameterization of vertical downward and upward movements and considers that there is only air mixture saturated with air at the top and the cloud base; simple ice (Dudhia, 1989) for cloud microphysics, which states that processes associated with the ice phase are more important in a layer between 0° and -20°C, where the vertical movements are forced to rise by releasing latent heat that results from deposition of water vapor in the snow.

METRIC FOR THE EVALUATION OF THE MM5 MODEL

In order to analyze the correlation of the simulation data in relation to the observations, the Willmott coefficient $\boldsymbol{d_r}$ (Willmott *et al.*, 2011) was utilized according to Eqs. 1 and 2. It can be calculated as:

If

$$\sum_{i=1}^{n}|P_i - O_i| \leq c \sum_{i=1}^{n}|O_i - \bar{O}| \tag{1a}$$

Then

$$d_r = 1 - \frac{\sum_{i=1}^{n}|P_i - O_i|}{c\ \sum_{i=1}^{n}|O_i - \bar{O}|} \quad (1b)$$

If

$$\sum_{i=1}^{n}|P_i - O_i| > c\ \sum_{i=1}^{n}|O_i - \bar{O}| \quad (2a)$$

Then

$$d_r = \frac{c\sum_{i=1}^{n}|O_i - \bar{O}|}{\sum_{i=1}^{n}|P_i - O_i|} \quad (2b)$$

where:

P_i: estimated value of a variable of order *i*; O_i: observed value of a variable of order *i*; $\bar{O}$: mean of the estimated values of a variable; *n*: number of events.

The value of *c* was assumed as 2, according to Willmott *et al.* (2011).

The interpretation of d_r is relatively simple: it indicates the sum of the magnitudes of the differences between the deviations of the model and observed values about the observed mean in relation to the sum of the magnitudes of the perfect model ($P_i = O_i$, for all *i*) and the observed deviations about the observed mean. Since the Willmott index can vary in the interval of -1.0 to 1.0, here the evaluation criteria presented in Table 2 were considered.

Table 2. Evaluation criteria adopted by Willmott.

Interval	Performance
1.0 to 0.75	Excellent
0.75 to 0.5	Good
0.5 to 0.25	Reasonable
0.25 to -0.25	Fair
-0.25 to 0.5	Unsatisfactory
-0.5 to -0.75	Poor
-0.75 to 1.0	Very poor

RESULTS

In this section, we will try to identify the main differences between the parameterizations of PBL with respect to the ability of the model to predict 60 hours of wind in terms of zonal (*u*) and meridional (*v*) components. The first 12 hours are not assessed and they are discarded as spin-up time effect. The two periods were analyzed as dry period of 2008 since 17 September at 12:00 UTC until 21 September at 12:00 UTC and as rainy period of 2010 since 19 March at 12:00 UTC until 25 March at 12:00 UTC.

The evaluation of the simulation is divided into two parts. First, a purely graphical comparison was made between the results obtained from the simulations, so considering each type of CLP parametrization with the observed value and also comparing with the NCEP reanalysis (Figs. 3 to 6). In the second part, the statistical metric Willmott index of agreement was used in the analysis, considering every 12 hours in advance prediction regarding the observation, namely, 60, 48, 36, 24 and 12 hours before the event.

GRAPHIC ANALYSIS

Year 2008

Considering for September 17, 2008 a booting from 12:00 UTC (Fig. 3), the ETA simulations (Fig. 3c), BLK (Fig. 3d), BT (Fig. 3e) and MRF (Fig. 3f) tend to overestimate, especially ETA and BT parameterizations, the zonal wind above 925 hPa. The startup from 12:00 UTC did not result in better forecasting for the meridional component (Fig. 4). It was noticed a change in the sign of the meridional component from 1,000 to 700 hPa, where, in the lower levels, the value

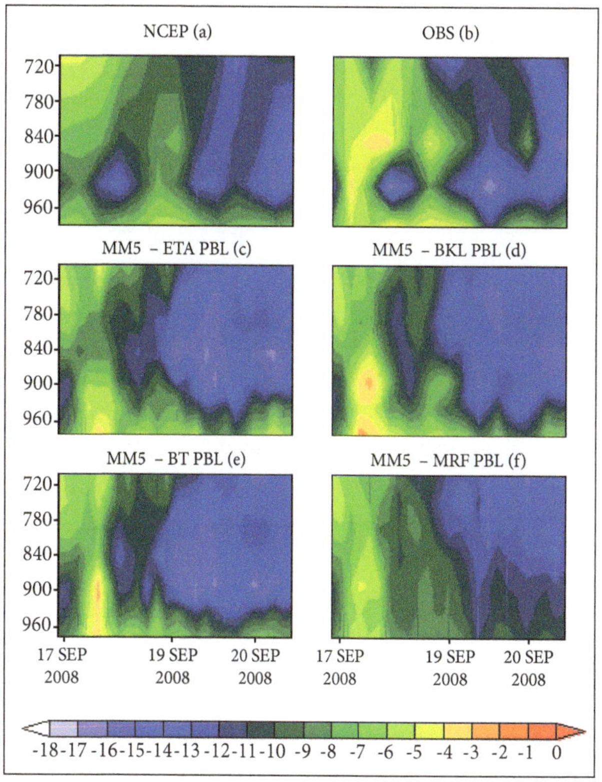

Figure 3. Example of comparisons of the zonal wind component (m/s) between 1,000 and 700 hPa. (a) Reanalyses - NCEP; (b) Radiosoundings at CLA; MM5 simulations utilizing the ETA (c), BLK (d), BT (e) and MRF (f) parametrizations.

is negative; from 850 hPa, it turned positive (Fig. 4b). Simulations (Figs. 4c, 4d, 4e and 4f) follow this change of sign of the meridional component, but with less fluctuation.

Concerning Figs. 3 and 4 for the zonal wind, simulation results near to 1,000 hPa suggested that ETA, BLK and BT parametrizations had better performance than the reference parameter, in this case, the MRF parametrization. From layer 925 to 700 hPa, there was not much difference between all parameterizations, which is best for predictions of 60 and 72 hours. In this case, there was no difference between the boots 00:00 (not shown) and 12:00 UTC. That is, the parameter did not appear to be important, either initialization. Possibly, this result is related to cloud microphysics.

It was found that, for the first levels of the atmosphere, between 1,000 and 925 hPa, the meridional component was best predicted by MM5 when it defined the BLK and MRF PBL parameterization as options. The intermediate layer, between 925 and 850 hPa, was not well-provided for any of the times, from the parameterization options studied, even by reference parameter, i.e. it may be that the type of PBL parameterization chosen do not interfere in the prediction of this wind component of the atmosphere at these levels. Between 850 and 700 hPa, wind component was well-provided forecasts of 24, 36 and 48 hours for all the parameterizations, being the BLK parameterization the foremost among them. For the meridional component of the wind, there was no difference between the boots from 00:00 (not shown) to 12:00 UTC; the boots from 12:00 UTC were closer to the observation, mainly from 1,000 to 925 hPa layer.

Year 2010

Figures 5 and 6 show the zonal and meridional components provided from the day March 19 at 12:00 UTC until the 22th at 12:00 UTC. ETA (Fig. 5c), BLK (Fig. 5d) and BT (Fig. 5e) simulations and also data from the NCEP reanalysis (Fig. 5a) underestimated the observed zonal. Only the prediction MRF (Fig. 5f) was treated in the observation of the entire length of the air layer and analyzed throughout the forecast.

From radiosonde launches at 12:00 UTC (Fig. 6), it was noted again the permanence of negative sign of the simulated meridional component for all forecasts (Figs. 6c, 6d, 6e and 6f)

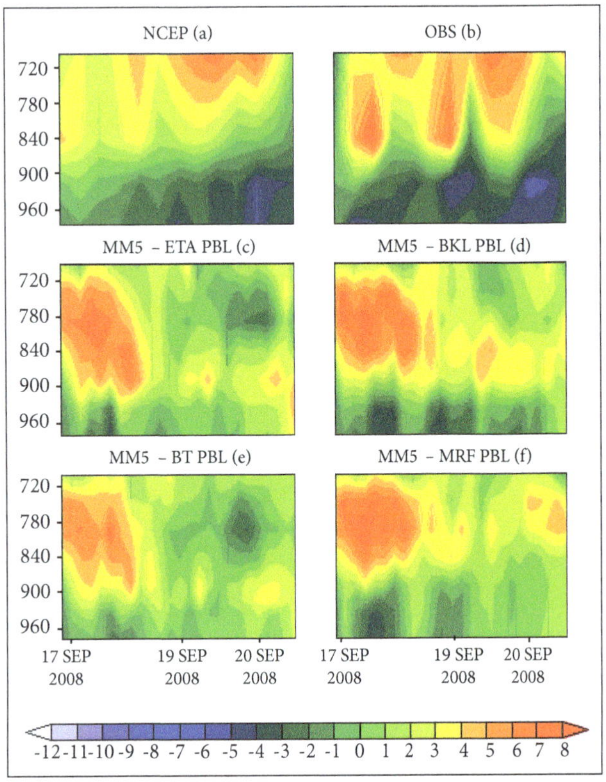

Figure 4. Example of comparisons of the meridional wind component (m/s) between 1,000 and 700 hPa. (a) Reanalyses - NCEP. (b) Radiosoundings at CLA; MM5 simulations utilizing the ETA (c), BLK (d), BT (e) and MRF (f) parametrizations.

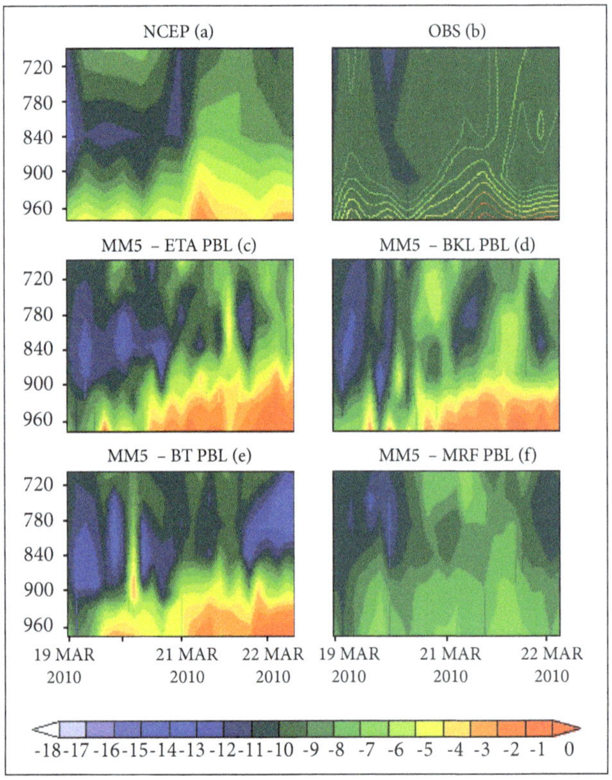

Figure 5. Example of comparisons of the zonal wind component (m/s) between 1,000 and 700 hPa. (a) Reanalyses - NCEP. (b) Radiosoundings at CLA; MM5 simulations utilizing the ETA (c), BLK (d), BT (e) and MRF (f) parametrizations.

above 925 hPa, unlike the observed (Fig. 6b). Nevertheless, both are close to the value observed 24 hours after the start of simulation from 1,000 to 850 hPa. The MRF (Fig. 6f) simulation was a little closer observation, with only 1 - 2 m/s apart (Fig. 6b).

The graphic result of 2010 simulations showed that when the simulations were initialized at 00:00 UTC (not shown) between levels 1,000 and 925 hPa and between 700 and 850 hPa, the MRF parametrization was not surpassed by any other, that is, the prediction of zonal values obtained better results using the MRF; however, at levels between 925 and 850 hPa, the ETA, BLK, BT the parametrizations were better than the MRF, although underestimated the observed values. Whereas, when the simulations were initialized at 12:00 UTC, no parametrization was more close to the observation than the MRF parametrization, considering the whole period of 72 hours for any layer of the atmosphere studied.

According to the results obtained for the meridional component, this was not well-simulated by the PBL parameterization when initialized at 00:00 UTC (not shown)

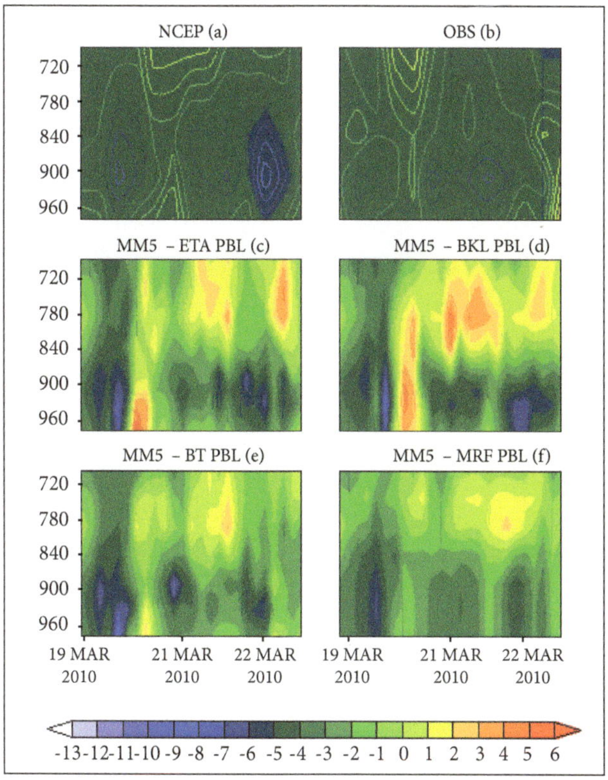

Figure 6. Example of comparisons of the meridional wind component (m/s) between 1,000 and 700 hPa. (a) Reanalyses - NCEP. (b) Radiosoundings at CLA; MM5 simulations utilizing the ETA (c), BLK (d), BT (e) and MRF (f) parametrizations.

and, when subjected to commence at 12:00 UTC, the ETA, BT and BLK parameterizations approached, but not exceed, the performance of MRF, which was better in 1,000 to 850 hPa layer.

VERTICAL PROFILES

Vertical Profile for the Dry Season (2008)

Figure 7 shows the forecasts with 24, 36, 48, 60 and 72 hours in advance of vertical profiles of wind speed and direction for the day September 20, 2008 from simulation initialized on September 17, 2008 at 00:00 UTC. For wind speed, the performance of the model estimates varied schedules. In some cases, the result was better forecasts for 24 and 72 hours of model integration also depending on when the simulation is initiated: 00:00 and 12:00 UTC (not shown).

The wind direction seemed to be better predicted by the model than the speed compared to observation. The parameterizations used predicted values of wind direction very close to the observations, but at certain times of the simulations with the MM5 MRF parameterization set up closer to the real value. As the wind speed, the time (integration time of the model) in which the prediction of wind direction is better was not defined. The rates differ from 1.0 to 5.0 m/s approximately, and the worst predictions of wind direction reached on average around 40 degrees of difference relative to the observed.

Vertical Profile for the Rainy Season (2010)

The vertical profiles of wind speed and direction for March 22 are shown in Fig. 8. In the same way as observed in the dry season (2008), the rainy season was not well-defined times when the predictions are better. The analysis results revealed some differences of up to 5.0 m/s of wind intensity between the simulated and the actual value, as in the simulation for March 22, 2010 at 00:00 UTC (Fig. 8), which was well-simulated by the model considering all parameterizations, almost throughout the integration period. However, for the final time (over 60 hours), this did not happen, mainly for ETA parameterization levels of 925 and 850 hPa. It was noted that the simulations from the MRF parameterization stood out among the others also in predicting the speed and the prediction of wind direction. An observed result is that forecasts of 24, 60 and 72 hours were better for the wind direction during the rainy season.

The results indicated the considerable potential of PBL parameterizations to improve the quality of wind forecasts in the short-term generated by MM5. Furthermore, these results show that the impact of parameterization is also not felt throughout the length of the atmosphere studied here between the levels of 700 and 1,000 hPa, that is, there are limitations in the wind forecast, in terms of u and v components; in parts of this layer, sometimes a particular parameterization performs better near the surface and, at other times, at higher levels. Basically, considering all tested options, it was noted that the model presented errors in wind velocity field mainly in 1,000 hPa. This represents that the wind field near the surface did not respond well to main features of CLA relief or suggests that we should adjust not only the parameterization of PBL, but also the physics of clouds parametrization. Considering the entire atmosphere layer studied, between 1,000 and 700 hPa, the differences were smaller in the dominant sectors of the wind, in this case, NE and SE, which somehow reduces this limitation since the most important sectors are reasonably simulated by the model.

Analysis of Concordance of Willmott Index

The impact in defining the parameterization of the MM5 PBL ideal for CLA, obtained by the wind forecast, has been reported for the dry and rainy seasons. The Willmott agreement index signaled some important results in search for the best parameterization of PBL. On average, the best results were Willmott index forecasts for 60 hours in advance, so during the dry period and for the rainy one.

Table 3 presents the comparison of the simulation results obtained for the different PBL parameterizations for the period from September 16, 2008 at 12:00 UTC to September 21, 2008 at 12:00 UTC for the zonal and meridional components. The best result for the Willmott index regarding the zonal component was for the MRF parametrization on both domains with values of -0.07 or fair performance for the 60-hour forecasts. The best result obtained for the meridional component of the Willmott value was -0.28 or unsatisfactory for BLK on domain 1 and -0.32 or unsatisfactory for MRF on domain 2.

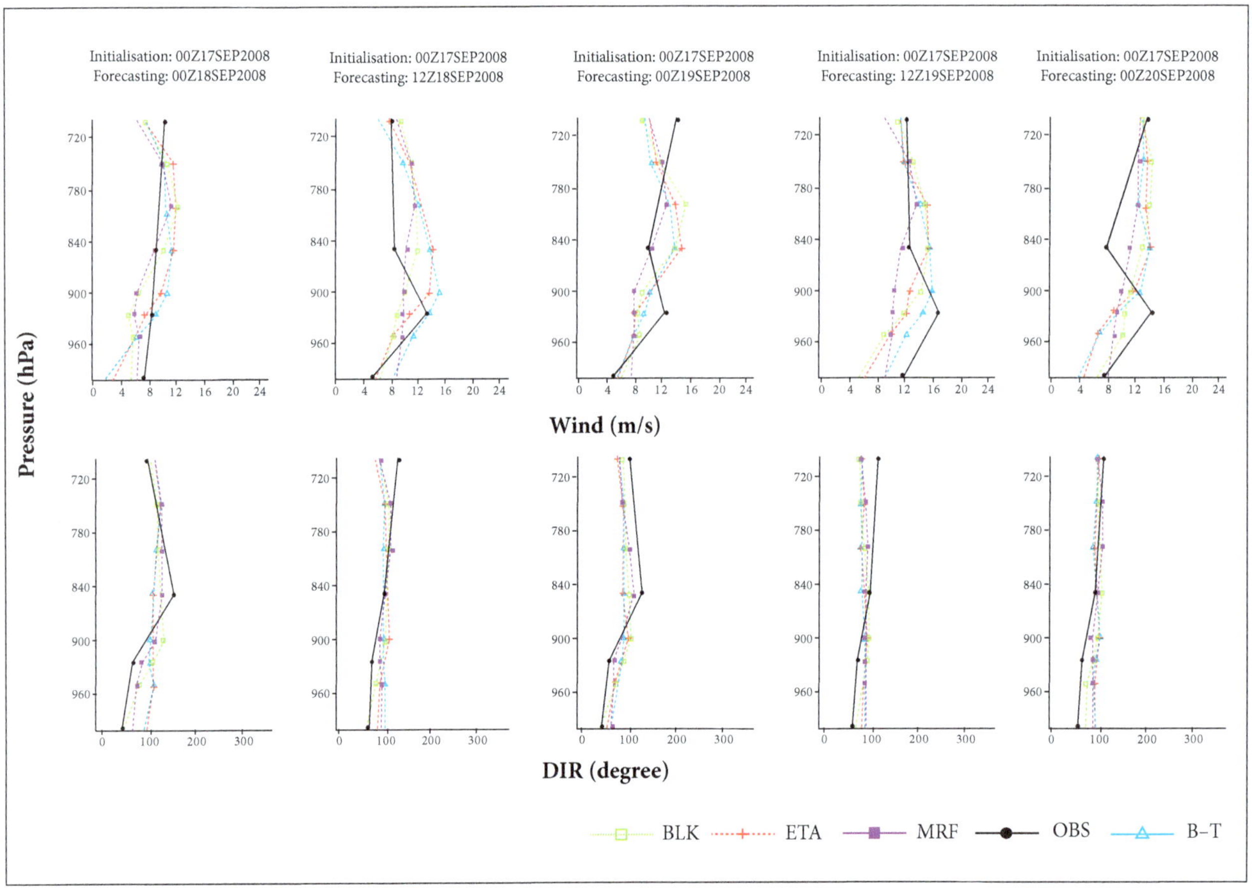

Figure 7. Example of the comparison of vertical wind profiles (m/s) predicted with the ETA, BLK, BT and MRF parametrizations and the radiosounding observations (OBS) in the layer from 1,000 to 700 hPa for forecast times of 24, 36, 48, 60 and 72 hours.

Table 4 shows results for the zonal and meridional components from the simulations initialized at 12:00 UTC for the period from March 20, 2010 at 12:00 UTC to March 25, 2010 at 12:00 UTC, in this case, wet season. The MRF parameterization did not obtain a good result; nevertheless, it was still better than the other parameterizations, both on domain 1, where it was observed the value of 0.17, as on domain 2, where it was observed the value of 0.14. That is, for both domains, it was considered fair according to the Willmott index.

Based on the results obtained, it was identified that, on average, the best results for the Willmott index were related to forecasts of 60 hours. In addition, it was perceived that the ETA and BT parametrizations are not adequate for the CLA or for the north coast of NE Brazil. Considering the dry (2008) and wet (2010) seasons, in general, the MRF parameterization, which is considered the MM5 default parametrization, still presented the best results. In addition to this reference parameterization, the BLK was the other that stood out.

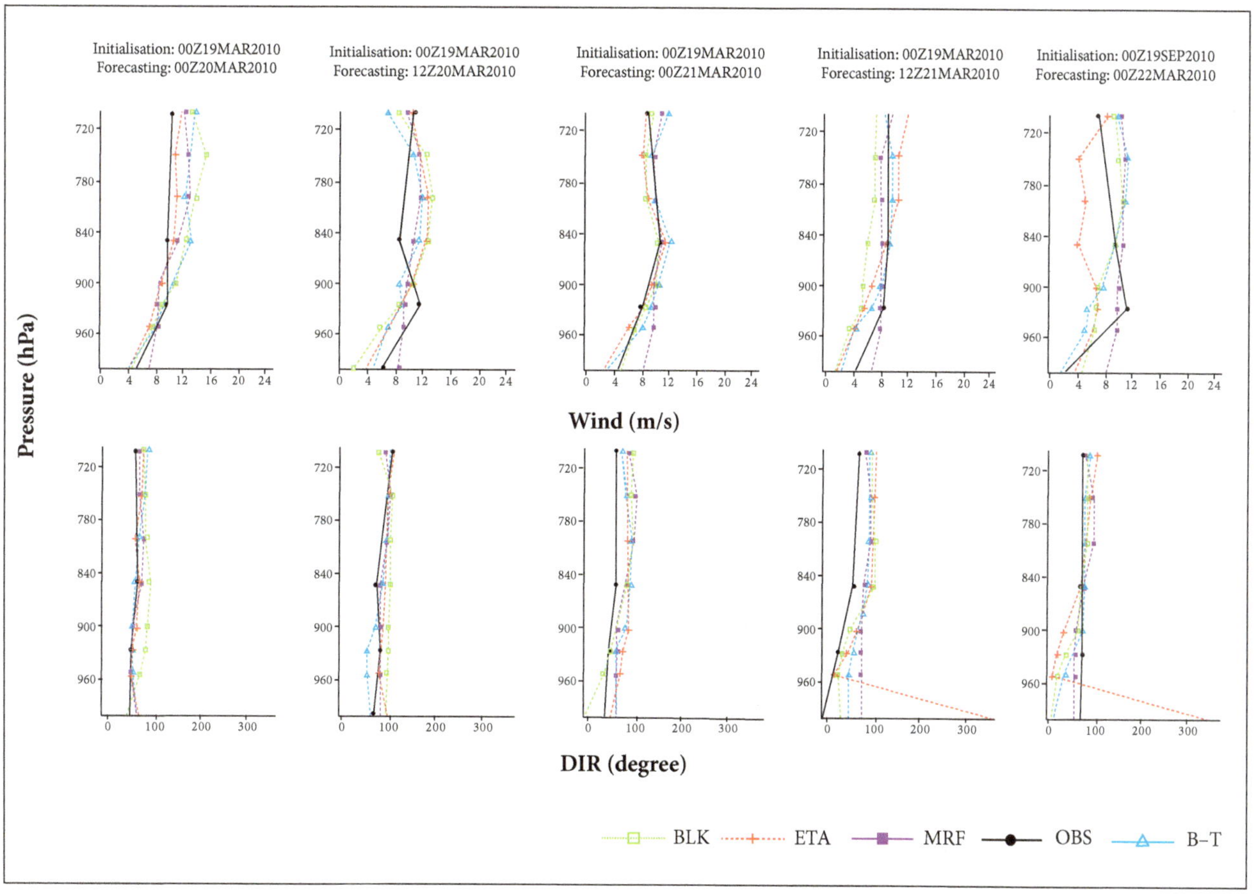

Figure 8. Example of the comparison of vertical wind profiles (m/s) predicted with the ETA, BLK, BT and MRF parametrizations and the radiosounding observations (OBS) in the layer from 1,000 to 700 hPa for forecast times of 24, 36, 48, 60 and 72 hours.

Table 3. Statistical indices obtained for the zonal and meridional components from 1,000 to 700 hPa for the period of the dry season.

Zonal component – 2008						Meridional component – 2008					
Domain 1						Domain 1					
Statistical index	Forecast	MRF	BLK	ETA	BT	Statistical index	Forecast	MRF	BLK	ETA	BT
Willmott	60	-0.07	-0.08	-0.11	-0.09	Willmott	24	-0.31	-0.28	-0.35	-0.33
Domain 2						Domain 2					
Willmott	60	-0.07	-0.07	-0.11	-0.11	Willmott	24	-0.32	-0.36	-0.38	-0.36

Table 4. Statistical indices obtained for the zonal and meridional components for the atmospheric layer from 1,000 to 700 hPa for the period of rainy season.

Zonal component – 2010						Meridional component – 2010					
		Domain 1						Domain 1			
Statistical index	Forecast	MRF	BLK	ETA	BT	Statistical index	Forecast	MRF	BLK	ETA	BT
Willmott	60	0.17	0.14	0.07	0.12	Willmott	60	0.28	0.22	0.19	0.24
		Domain 2						Domain 2			
Willmott	60	0.14	0.10	0.02	0.06	Willmott	60	0.29	0.22	0.16	0.22

CONCLUSION

It was noticed in this experiment that ETA and BT parameterizations are not suitable for the CLA. Considering the available data set, in general, both default MRF and BLK parameterization presented suitable schemes. The former is better for the wet season and the latter gave the best results for the dry season. An important factor that must be emphasized is that both parameterization schemes are non-local schemes. In this case, the non-local scheme is related to vertical diffusion, which is an important process that considers the effects of eddies moving to larger distances according to the physical processes that occur in turbulent atmosphere. The larger eddies are more important as they are more sensitive to the environment. This can be the cause of the MM5 model subtly better predict the wind in PBL considering these schemes, given that non-local schemes better simulate the deep and large eddies PBLs.

It was also found that the initializations at 12:00 UTC are better than those at 00:00 UTC. Elleman *et al.* (2003) assert that many of the parameterizations of the boundary layer MM5, including the MRF, tend to promote an excessive vertical mixing, which means that the errors they produce are higher during the night, that is, when there is the smallest vertical mixing. Possibly overnight, the main process responsible for errors generated is the cloudiness provided, which controls the radiation of long radiation lost to space.

REFERENCES

Blackadar, A.K., 1976, "Modeling the Nocturnal Boundary Layer". Proceedings of the 3rd Symposium on Atmospheric Turbulence, Diffusion and Air Quality, Boston, USA.

Burk, S.D. and Thompson, W.T., 1989, "A Vertically Nested Regional Numerical Weather Prediction Model with Second-Order Closure Physics", Monthly Weather Review, Vol. 117, pp. 2305-2324.

Camillo, G.L., 2011, "Impacto das Parametrizações de Camada Limite Planetária do MM5 na Previsão de Ventos em Baixos Níveis", Revista da Universidade da Força Aérea, Vol. 23, No. 28, pp. 67-79.

Carvalho, M.A.V., 2011, "Variabilidade da Largura e Intensidade da Zona de Convergência Intertropical Atlântica: Aspectos Observacionais e de Modelagem", Master's Dissertation, Instituto Nacional de Pesquisas Espaciais, São José dos Campos, Brazil.

Dudhia, J., 1989, "Numerical Study of Convection Observed during the Winter Monsoon Experiment Using a Mesoscale Two-Dimensional Model", Journal of Atmospheric Sciences, Vol. 46, No. 20, pp. 3077-3107. doi: 10.1175/1520-0469(1989)046<3077:NSOCOD>2.0.CO;2

Dudhia, J., Gill, D., Guo, Y., Manning, K., Wang, W., and Chiszar, J., 2003, "Mesoscale Modeling System Tutorial Class Notes and User's Guide: MM5 Modeling System Version 3", PSU/NCAR.

Elleman, R.A., Covert, D.S. and Mass, C.F., 2003, "Comparison of ACM and MRF Boundary Layer Parameterizations in MM5", Proceedings of the Models-3 Workshop, Raleigh, USA.

Grell, G.A., 1993, "Prognostic Evaluation of Assumptions Used by Cumulus Parameterizations", Monthly Weather Review, Vol. 121, pp. 764-787. doi: 10.1175/1520-0493(1993)121<0764:PEOAUB>2.0.CO;2

Grell, G.A., Dudhia, J., and Stauffer, D.R., 1994, "A Description of the Fifth-Generation Penn State/NCAR Mesoscale Model (MM5)", NCAR Technical Note, NCAR/TN-398+STR, Boulder, USA, 117 p.

Hong, S.-Y. and Pan, H.-L., 1996, "Nonlocal Boundary Layer Vertical Diffusion in a Medium-Range Forecast Model", Monthly Weather Review, Vol. 124, No. 10, pp. 2322-2339. doi: 10.1175/1520-0493(1996)124<2322:NBLVDI>2.0.CO;2

Janjic, Z., 1994, "The Step-Mountain Eta Coordinate Model: Further Developments of the Convection, Viscous Sublayer, and Turbulence Closure Schemes", Monthly Weather Review, Vol. 122, No. 5, pp. 927-945. doi: 10.1175/1520-0493(1994)122<0927:TSMECM>2.0.CO;2

Johnson, D.L., 2008, "Terrestrial Environment (Climatic) Criteria Guidelines for Use in Aerospace Vehicle Development, 2008 Revision", NASA/TM 2008-215633, Marshall Space Flight Center, Alabama, USA.

Mellor, G.L. and Yamada, T., 1974, "A Hierarchy of Turbulence Closure Models for Planetary Boundary Layers", Journal of Atmospheric Sciences, Vol. 31, No. 7, pp. 1791-1806.

Oyama, M.D., 2004, "Precipitação do Modelo de Mesoescala MM5: Avaliação Preliminar da Sensibilidade às Parametrizações de Convecção Profunda em Grade Reduzida", Proceedings of the XIII Brazilian Congress of Meteorology, Fortaleza, Brazil.

Oyama, M.D. and Giarolla, E., 2006, "Simulação de Linhas de Instabilidade na Costa Norte do Brasil de 6 a 9 de Julho de 2005 com o Uso do Modelo Regional MM5: Resultados Preliminares", Proceedings of the XIV Brazilian Congress of Meteorology, Florianópolis, Brazil.

Pereira Neto, A.V. and Oyama, M.D., 2011, "Mudanças do Esquema de Convecção Profunda Kain-Fritsch para a Região do Centro de Lançamento de Alcântara", Revista Brasileira de Meteorologia, Vol. 26, p. 579-590.

Willmott, C.J., Robeson, S.M. and Matsuura, K., 2011, "A Refined Index of Model Performance", International Journal of Climatology, Vol. 32, No. 13, pp. 2088-2094. doi: 10.1002/joc.2419

Wyngaard, J.C. and Brost, R.A., 1984, "Top-Down and Bottom-Up Diffusion of a Scalar in the Convective Boundary Layer", Journal of Atmospheric Sciences, Vol. 41, p. 102-112. doi: 10.1175/1520-0469(1984)041<0102:TDABUD>2.0.CO;2

Zhang, D. and Anthes, R. A., 1982, "A High-Resolution Model of the Planetary Boundary Layer Sensitivity Tests and Comparisons with SESAME-79 data", Journal of Applied Meteorology, Vol. 21, No. 11,, pp. 1594-1609. doi: 10.1175/1520-0450(1982)021<1594:AHRMOT>2.0.CO;2

11

An Approach to Outlier Detection and Smoothing Applied to a Trajectography Radar Data

Aguinaldo Bezerra Batista Júnior[1], Paulo Sérgio da Motta Pires[2]

ABSTRACT: The tracking of aerospace engines is reasonably achieved through a trajectography radar system that generally yields a disperse cloud of samples on tridimensional space, which roughly describes the engine trajectory. It is proposed an approach on cleaning radar data to yield a well-behaved and smooth output curve that could be used as basis for instant and further analysis by radar specialists. This approach consists on outlier detection and smoothing phases based on established techniques such as Hampel filter and local regression (LOESS). To prove the effectiveness of the approach, both filtered and unfiltered data are submitted to an extrapolation method, and the results are compared.

KEYWORDS: Trajectory, Radar, Filtering, Smoothing, Outlier detection.

INTRODUCTION

Trajectography radar systems play an important role on the tracking process of an aerospace engine. During the whole flight of a target, a radar system is able to retrieve linear distance, azimuth and elevation data of the flying engine to radar operators and trajectography subsystems.

The acquisition processes of a trajectory radar system deliver to data analysts a sparse cloud of samples, which is used to sketch, albeit crudely, the trajectory of the engine through the three-dimensional space. This occurs because radar systems suffer from a broad variety of perturbations that negatively influences the acquisition process and cannot be easily modeled, such as atmospheric issues, refraction and reflection, Earth's curvature, presence or absence of transponder, device calibration, antenna servomechanism interactions, SNR, target volume and position, etc. These intrinsic characteristics of trajectory radar systems may adversely affect any subsequent data analysis.

In some cases, the presence of outliers (aberrant or implausible values) in retrieved data, the high dispersion in collected samples, and the inherent noise of the process may hinder the assimilation of the actual trajectory of an engine. As a consequence, further assumptions from this raw data may lead to inaccuracies and misunderstandings related to the actual trajectory of the flying engine.

To address this problem, it is proposed an approach on filtering acquired data in order to produce as outcome a smooth and outlier-free curve, that is expected to be a better approximation of the real trajectory of the aerospace engine.

1. Centro de Lançamento da Barreira do Inferno – Natal/RN – Brazil 2. Universidade Federal do Rio Grande do Norte – Natal/RN – Brazil

Author for correspondence: Aguinaldo Bezerra Batista Júnior | Rodovia RN 063 - Km 11 PO Box 54 | CEP: 59.140-970 | Parnamirim/RN – Brazil | Email: aguinaldoabbj@clbi.cta.br

The proposal consists, basically, on the implementation of this approach to a moving-window filter which will be applied to incoming radar data or to offline data. Once interest data became outlier-free and smoothed, further analysis and processing will become more accurate, it will be possible to apply curve fitting techniques to get the trajectory's function parameters, driving out to more accurate calculations of speed and acceleration of the target, probable impact area, and to make more concise extrapolations when signal retrieval is somehow interrupted.

The core of the proposed filter is based on modern implementations of classical statistic methods with proven effectiveness, in particular, in the fields of outlier detection and data smoothing.

The remainder of the paper is structured as follows. The next Section outlines the background concepts of interest in this research. The subsequent Section presents our approach on radar data filtering. Then the accomplished results are discussed and some derived discussion is proposed on the final Section.

PAPER BACKGROUND

The task of excerpting meaningful information from imperfect data has always been a very common problem and topic of interest on data analysis. There are several families of techniques and approaches to accomplish it, such as time series and statistical analysis, digital signal processing, artificial neural networks, reinforcement learning, to name a few. The approach based on statistical analysis caught most attention in preliminary study phases due to the large amount of flexible and effective techniques suitable for general data cleaning problems. In this context, data smoothing and outlier detection techniques, especially when combined, seemed well suited for use our radar data filtering problem.

DATA SMOOTHING

A trajectography plot can be viewed as the evolution of tracking signal measurements as a function of time. Seen as a physical process, the trajectory of an spatial engine, like many other experiments, may be viewed as a discrete signal whose amplitudes change rather smoothly as a function of time, whereas many sorts of noise are noted as rapid, random, abrupt, and sometimes aberrant changes in amplitude from point to point within the signal. For some aerospace engines like rockets, the trajectory data is not supposed to fit models because the trajectory may change during flight. Nonetheless, it may be convenient not to force data into models but just to attempt to moderate frantic data by using a classic and straight-forward statistical approach known as data smoothing.

In few words, smoothing is a process where the samples of a signal are adjusted so that individual points that are higher than the immediately adjacent points are moderated, and points that are lower than the adjacent points are raised, leading to a naturally smoother signal output. Considering that the actual signal is smooth in nature, it will not be much distorted, though some noise could be substantially lessened.

Such a rough approach is sometimes evaluated as poorly suited because it may be close to data cooking (falsifying or selectively conditioning data in an attempt to prove an hypothesis), as it may inadvertently suppress some important information on data stream (Wilson, 2006). However, as we are interested mainly on short terms trends of an engine's trajectory and uncovering data, minor variations in target's position are not a major concern.

Literature shows that smoothing may be distinguished from the closely related and partly overlapping concept of curve fitting. It happens because the former outcomes an uppermost idea of relatively slow changes of values, provided by a smooth function which approximately fits the data with little concern to the close matching of values, while the latter concentrates on achieving as exact a match as possible (best fit) by using an explicit function form for the result. The main goal of most smoothing methods is not to specify a parametric model for the mean function but to afford a more flexible approach that allows the data points themselves to suggest the appropriate functional form of the smoothed curve. Smoothing methods provide a bridge between making no assumptions on a formal structure (a purely non-parametric approach) and making very strong assumptions (a parametric approach) by making only a relatively weak assumption: the targeted data might represent a smooth curve (Simonoff, 1998).

There are several ways to achieve data smoothing, especially through statistical techniques. Among these approaches based on regression analysis, more precisely local regression, and the employment of splines, more incisively smoothing splines are great stand outs, since they are classic, well established, simple to understand and interpret and proven effective when applied to general smoothing problems.

Local Regression

Local regression is an approach to fitting curves and surfaces to data in which a smooth function may be well approximated by a low degree polynomial in the neighbourhood of a point (Loader, 2012).

Early work on using the underlying principles of local regression to smooth data are dated from the late 19th century and continued in a reticent fashion until mid 20th century, mainly because they were too computationally intensive for that time. Hopefully, from the 1970s on, hitchhiking on great advancements in computer hardware/software and scientific computing, the local regression subject enjoyed a reborn. Since then, several relevant works on extending, modernizing and generalizing local regression have been developed, including the use of the method in other branches of scientific literature (Cleveland and Loader, 1996).

A modern, proficient and widely used local regression algorithm is the LOESS (short for LOcal RegrESSion) procedure. LOESS is a nonparametric local regression method pioneered by Cleveland (1979), and further developed by Cleveland and Devlin (1988), in which a smooth function may be properly fitted by a low degree polynomial in a chosen neighborhood (subset) of any point of a dataset in a moving window fashion. This method employs weighted least squares (WLS) to fit linear or quadratic polynomial functions of the predictors at the centers of neighborhoods in order to build up a curve which describes the deterministic part of the variation in the data, point by point. LOESS is a weighted polynomial regression procedure where more weight is given to points near the target point and less weight is given to points further away. The radius of each neighborhood contains a specified fraction of data points, known as smoothing parameter or bandwidth, which is the main parameter of the method and controls the smoothness of the estimated surface in each local surroundings. (Cohen, 1999; NIST/SEMATECH, 2012).

A chief advantage of this method is that the data analyst is not compelled to specify a global function to fit a model to the data, but only to fit pre-defined low-order polynomials to small segments of the data. LOESS is considered a versatile and coherent choice when it is demanded to model complex processes for which there are no theoretical models.

In most implementations of LOESS, there are few knobs to deal with. In general, the smoothing parameter value and the degree of the local polynomial are the user-specified inputs, though in some cases the weight function may also be a flexible parameter. The traditional weight function used by LOESS is the popular tri-cube weight function (NIST/SEMATECH, 2012).

The smoothing parameter or bandwidth, q, is a number between $(d+1)/n$ and 1, with d denoting the degree of the local polynomial and n denoting the number of data points. The value of q is the proportion of data used in each fit and the subset of data used in each WLS fit is comprised of the nq (rounded to the next largest integer) points, whose explanatory variables values are closest to the point at which the response is being estimated (NIST/SEMATECH, 2012). The smoothing parameter controls the flexibility of the LOESS regression, then large values of q yields smoother functions which would soften fluctuations in the data, while a smaller q value will make the regression function more conforming to the data and may eventually capture undesirable data oscillations. The selection of these parameters is normally an empirical task that depends on the dataset, but typical values lie in the range of 0.25 to 0.5, for most applications. Although fixed selection of the bandwidth may provide good fits in many cases (Cleveland and Loader, 1996), several criteria and procedures for intelligent, automated and adaptive bandwidth selection have been developed (Cleveland and Devlin, 1988; Cleveland and Loader, 1996).

The local polynomial degree d is either locally linear or locally quadratic. A zero degree polynomial turns LOESS into a weighted moving average while the use of higher-degree polynomials is unrecompensed, because they tend to overfit local data and are numerically unstable. Besides, one of the main goals of LOESS is to approximate any function in a small neighbourhood by fitting low-order polynomials (NIST/SEMATECH, 2012). In fact, locally quadratic fit is recommended in the early smoothing literature because it may provide a sufficiently good approximation when there are rapid shifts in the slope, peaks and valleys (Cleveland and Loader, 1996). It is a well known fact that the LOESS smoothing procedure, like other statistical analyses based on least squares fitting, is considerably sensitive to the presence of even a small proportion of outliers in the dataset. Depending on their location and deviation level, such inconsistent observations may fairly distort regression coefficients and corrode regression analysis, making the data fitting not representative of the bulk of the data. A good approach to verify how suspected outlying data may influence the results is to perform the targeted processing (in our case the application of LOESS) on data, both with and without these outliers, in order to examine their specific impact on the results and finally conclude if these outlying points should be worked out or not.

In Fig. 1, the tridimensional plot (a) shows 2,177 reasonably well behaved trajectory data points (with no extreme outliers) obtained from radar tracking of the International Space Station (ISS) while plot (b) shows the result of a 5% artificial outlier contamination procedure applied to the original data. Plot (c) shows the resulting smoothed curve from LOESS application over the original points from plot (a) while plot (d) shows the distorting effects caused by these outliers to the resulting LOESS smoothed curve, using the same smoothing parameters as in plot (c). In this work, all original test datasets were purposely artificially contaminated for better illustration of outlier effects and effectiveness of the techniques used in this work.

To overcome the influence of the outliers, a robust version of the LOESS has been referenced in the literature. The robustification consists, basically, in iterative reweighting processes that involve building robust weights with a specified robustness weight function by using current residuals and updating those on each iteration, until the residuals remain unchanged. However, these additional steps add much more computation complexity to the LOESS and dramatically decrease its performance (Cleveland, 1979; Cleveland *et al.*, 1990; Garcia, 2010). In our preliminary tests, robust LOESS was proved unattractive because it took too long to converge and neglected to several extreme outliers.

Smoothing Splines (Penalized Least Squares)

Another popular and established non-parametric regression is smoothing splines, which is based on the optimization of a penalized least squares criterion whose solution is a piecewise polynomial or a spline function (Loader, 2012). This approach employs fitting a spline with knots at every data point, so it could potentially fit perfectly into data, but the function parameters are estimated by minimizing the usual sum of squares plus a roughness penalty defined by the penalized sum of squares criterion (Garcia, 2010). An amount of penalty is imposed according to the magnitude of the tuning parameter (also known as degree of freedom) of the method, so that the lower is the parameter the closer is the data fit, which could lead to a noisy curve, as it follows every detail in data. The higher is the parameter, the smoother is the solution curve, which could end up in a very poor fit to data.

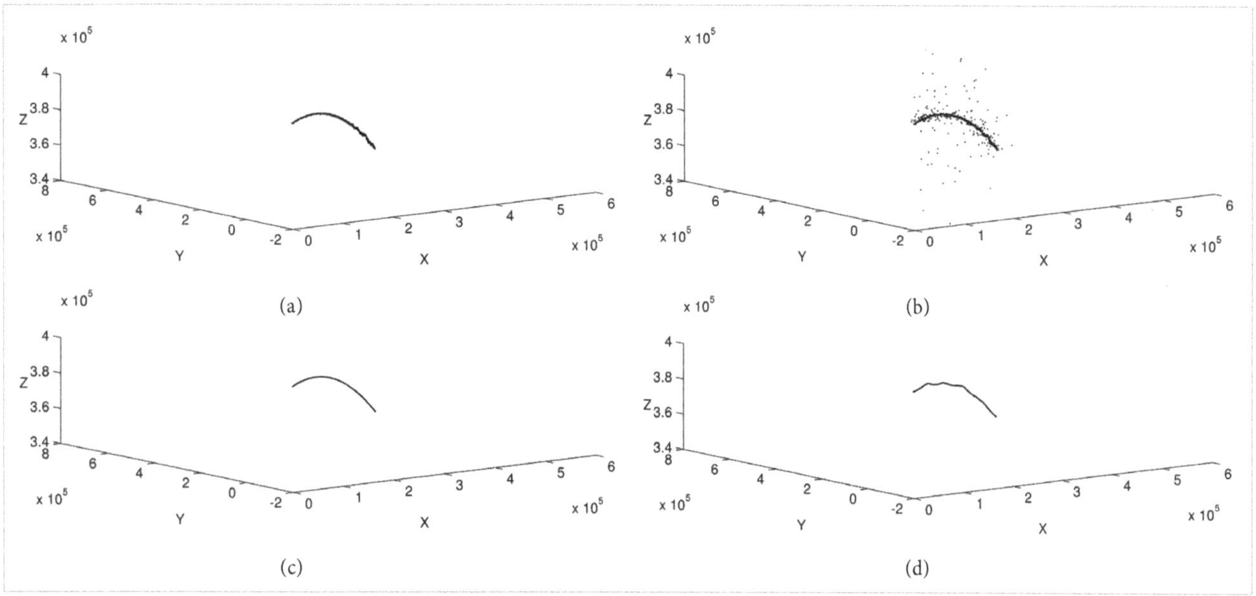

Figure 1. Influence of outliers on LOESS smoothing procedure. (a) Original data; (b) Original data with 5% artificial outliers contamination; (c) LOESS applied to original data; (d) LOESS applied to outlier contamined data.

It is a well known fact that the results obtained with smoothing splines are very similar to LOESS results. Moreover, the main drawback of smoothing splines is the proved sensitivity to outliers (Garcia, 2010). These facts were confirmed in preliminary tests with our trajectory datasets. Generally, even using a higher tuning parameter, the method tends to overfit misbehaving data (dispersed and highly-contaminated), always leading to a less smoothed output than the LOESS' output for the same scenario. Although it is a good, fast and well-referenced method, it proved to be less effective than LOESS when considering our trajectory datasets.

OUTLIER DETECTION

Although there is no single, unanimously accepted or rigid mathematical definition of what an outlier is, there is a consensus on referring to outliers as a statistical term for an observation that is numerically much deviant from the behaviour observed in the majority of the data. In statistics, there is a wide and classic discussion regarding the characterisation and categorisation of these unusual observations in outliers, high-leverage points and influential points. However, since this kind of analysis is not the focus of this work, we will not go into the merits of these differentiations.

Outliers are perhaps the simplest and best-known type of data anomaly (Pearson, 2005), highly common in most applied and scientific scenarios involving data collection and analysis. In terms of our trajectography dataset, outliers can be seen as points that disobey the general pattern of smooth variation seen in the data sequence, which represents the flight of an aerospace engine which is bound to the laws of physics.

These data anomalies demand close attention, because they are observations that do not follow the statistical distribution of the bulk of the data, and consequently, may lead to erroneous results regarding statistical analysis (Liu *et al.*, 2004). The presence of even a few of these anomalies in a large dataset can have a disproportional influence on analytical results (Pearson, 2005) and may cause general distortions, estimation biasing and inflated error rates that could lead to false alarms, improper decision making, faulty conclusions, model misspecification, etc.

In general, outliers may be treated merely as an extreme manifestation of the random variability inherent of the data, hence, they have to be retained and processed in the same manner as the other observations (ASTM International, 1980), or as a representation of some disorder or unexpected conditions in the system, such as gross measurement, sampling, computing or recording errors, transient malfunctioning, noise, missing data, human errors, etc. In the latter case, the identified implausible values may eventually be rejected, statistically adjusted and/or held for further analysis. Discussion regarding what to do with identified outliers is a common and controversial topic in outlier detection literature, but there is a consensus that outliers should not be simply discarded, since they may carry important (or key) information and insights about the process. After all, "one person's noise could be another person's signal".

Outlier detection has been suggested to detect implausible behaviour points for numerous applications such as business transactions, clinical trials, voting, network intrusion, weather prediction, geographic information systems, chemical data processing, industrial process monitoring, and so forth (Pearson, 2005; Nurunnabi and Nasser, 2008). Although dealing with outliers is an old and well-known problematic, there is no ultimate outlier identification procedure which is able to cover all kinds of outlier scenarios. Even so, since there are different types of outliers emanating from various sources and influencing data analysis in different ways, as well as there is no rigid formalisation of what constitutes an outlier, the identification of these doubtful observations is an arduous and ultimately a matter of interpretation, or at least previous knowledge of the data.

Several outlier detection criteria, procedures and guidelines have been actively developed for centuries (since the 19th century), using different diagnostic statistics. Most of them attempt to segregate the data into an outlier-free subset and a supplementary subset containing all potential outliers. These procedures can be grouped across a wide taxonomy of detection approaches, which include univariate, multivariate, parametric, non-parametric, distribution-based, distance-based, density-based, deviation-based and cluster-based methods (Pearson, 2005; Lee, 2008; Nurunnabi and Nasser, 2008). For the sake of time and study limitations, a historical review or a comparative study of outlier detection procedures diverges from the goal of this work. Thus, we briefly describe only some of the most known and referenced methods. Before that, it is crucial to highlight two concepts which are closely related to outlier detection: masking and swamping. Masking is the inability of a procedure to detect actual outliers while the swamping is the detection of a legit observation as an outlier. Literature shows that these opposing effects are faced by all outlier identification methods in a complementary fashion: a procedure which performs well with respect to masking is more susceptible to swamping and vice-versa (Pearson, 2005; Pearson, 2011).

Cook's distance is a general method for assessing the local influence of single data point against least squares regression analysis. The goal of the method is to detect influential points in regression coefficients, but it cannot be considered a conclusive test to detect outliers because it is very prone to both masking and swamping effects (Rancel and Sierra, 2000; Adnan *et al.*, 2003; Nurunnabi and Nasser, 2008).

Other widely known method is the Dixon's Q for outlier detection, mainly because of its simplicity. It is based on a comparison between the suspect value and its direct or close neighbour with the overall or modified range. A point is flagged as outlier if its calculated Q value exceeds the critical Q-value presented in a static table at the chosen significance level. Although it was originally recommended by ISO for inter-laboratorial tests, this test should be used just once in a dataset to detect a unique outlier, because it is highly prone to masking effects (Massart *et al.*, 1997).

The Grubbs' test for outliers is a commonly used procedure which replaced Dixon's Q test on ISO recommendations (Horwitz, 1995). It is suggested in order to detect outliers in a univariate dataset assumed to come from an approximately normally distributed population. The test is based on the difference of the mean of the sample and the farthermost data, considering the standard deviation. Although it has been modified and adapted (Horwitz, 1995; NIST/SEMATECH, 2012), this procedure still suffer from series of limitations, like the normality assumption, the use of non-robust characterisation tools, such as mean and standard deviation as core statistics, and the inability of reasonably detecting multiple outliers (Zhang *et al.*, 2004; Solak, 2009; NIST/SEMATECH, 2012).

Another popular approach to outlier identification is the well-known 3σ edit rule, based on the idea that, if data sequence is assumed to be approximately normally distributed, the probability of observing a point farther than three standard deviations from the mean is only about 0.3% (Pearson, 2002). Despite its historical importance and intuitive appeal, this outlier detection procedure tends to be ineffective in practice. The basic weakness is that the presence of outliers in the dataset can cause substantial errors in both estimated mean and standard deviation in which the procedure is based. It makes outliers harder to point out and, consequently, too few outliers are detected (Pearson, 2005).

From the described so far, it is noticeable that there is the need for a versatile outlier detection procedure, which could work independently from the distribution type assumption, capable of sanely detecting multiple outliers without a process model or assumptions and based on robust (outlier-resistant) statistical tools. The Hampel filter, regarded as one of the most robust and efficient outlier identifier (Liu *et al.*, 2004), satisfies these requirements by running through data a moving window cleaner centred at the current data point, using robust core statistics. A robust adaptation of the 3σ edit rule, named Hampel identifier, is then applied to this window to characterise each point regarding a local neighbourhood of preceding and subsequent samples, producing the replacement of the data point declared to be an outlier with a more representative value, according to other data points in the immediate vicinity, otherwise the data point is unchanged (Pearson 2005; Pearson, 2011).

The Hampel identifier, base processing of the Hampel filter, consists of replacing the original data location and data standard deviation estimates in the 3-σ edit rule. The mean is replaced with the median and the standard deviation is substituted with the MAD (Median Absolute Deviation). Because median and MAD are both less sensitive to outliers than the mean and standard deviation, respectively, the Hampel identifier behaves much more effectively than the 3 σ edit rule in a majority of outlier scenarios (Pearson, 2005; Pearson, 2011). The main drawback of these replacements is that the overall outlier detection procedure becomes more aggressive and, consequently, legit data may be declared as outliers. Then, this identifier is naturally more sensitive to the swamping effect than to masking effects. Davies and Gather (1993) reasonably described in details the overall Hampel functioning, including the employed criteria used in order to evaluate if a sample is an outlier.

The Hampel filter has only two tuning parameters: the half-width K of the window and the threshold parameter t (Pearson, 2005). The former defines the bandwidth for the cleaning window while the latter determines the aggressiveness of the filter in considering the suspect points as outliers. It is important to note that the filter remains well defined for $t = 0$, for which it is reduced to the median filter, since the target sample will always be replaced with the median of the window. In the other extreme, if t is large enough, the target sample will stay unmodified unless the MAD scale estimate is zero. This condition only occurs when a majority of values in the moving window are exactly the same, for example, when processing a coarsely quantised dataset (Pearson, 2005), which is not the case in our experiments.

The behaviour of the Hampel filter, applied to our trajectory datasets, was very satisfactory. For example, when the filter was processed over a test trajectory dataset, this procedure was able to detect about 90% of those artificial outliers, which purposely contaminated the original dataset (10% of the total samples). It also pointed out some evident abehant samples within the original data. Figure 2 plots illustrates the Hampel filter response to the previously contaminated ISS trajectory data, using K=3 and t=3. Plot (c) shows that most outlying samples were detected and replaced with the median of their surroundings. This is a good example of small but beneficial distortions introduced by this filter to an outlier contaminated dataset.

RADAR DATA FILTERING

The main interest in this research is to clean up tracking data of aerospace engines collected from a trajectory radar system, producing as result a smooth and outlier free curve that would afford much more accurate results regarding online and offline analyses. Due to the proven effectiveness and popularity of LOESS and the Hampel filter in their respective fields, they were chosen to integrate our proposal as base statistics.

PROPOSAL

The proposal consists on submitting trajectory radar data to a data filtering system founded on outlier detection/substitution and smoothing phases. The outlier detection pre-processing phase is borne to the Hampel filter while LOESS takes the charge of the main process of smoothing. It is important to highlight that, for illustration purposes, all original tracking datasets were submitted to an outlier contamination process based mostly in the application of additive Gaussian white noise to some randomly chosen samples, in order to simulate a slight distortion of the signal. These signal distortions could be caused, for example, by adverse weather conditions, malfunctioning of some radar subsystem or channel interferences.

The first approach to address the issue of cleaning radar data is to run both procedures sequentially on offline radar data. This would yield an after flight quality plot of the just tracked aerospace target to radar experts, helping them on reporting clean plots, extracting meaningful trends from radar data, detecting disagreements regarding nominal trajectories, modelling trajectories, etc. The second approach

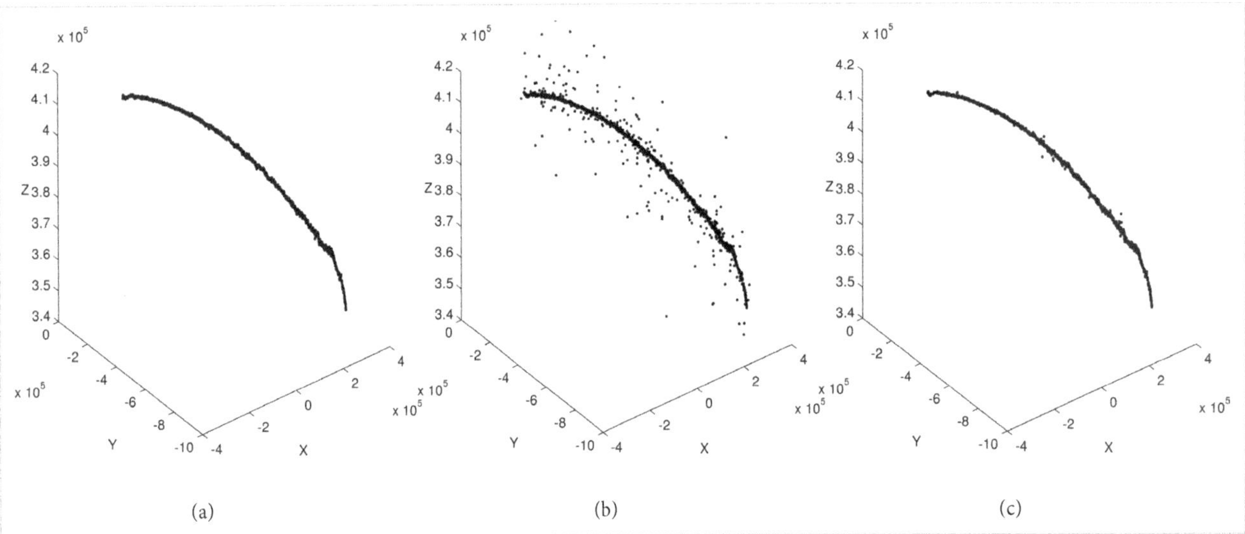

Figure 2. Hampel filter work over outlier-contaminated trajectory data. (a) Original data; (b) 10% Outlier contamination e (c) Hampel filtered data.

is supposed to be online and it involves establishing a moving window with parameterized bandwidth to be applied to incoming radar data, producing an on flight outlier-free and smoothed curve, which could be used for an accurate calculation of instant speed and acceleration of the target, determination of a less dispersed area of impact and establish concise extrapolations on the trajectory in case of radar malfunctioning or tracking loss.

IMPLEMENTATION

In these early research phases, it is important to prioritize a proper evaluation of the effectiveness of the methods under study over the achieved performance. After the validation of the effectiveness of the methodology, performance considerations should be taken into account. Given that, it is convenient in study and prototype phases to use numerical computing environments for the sake of simplicity, abstraction and flexibility, despite the well-known performance issues inherent to these environments. Several environments such as Matlab, R and Scilab were used in our experiments since most building blocks used in our approach are already satisfactorily implemented in these numerical environments.

The dataset is basically composed by tridimensional Cartesian coordinates converted from azimuth, elevation and distance coordinates of trajectory data generated by the radar during the tracking of a target. As they are obtained from different radar subsystems, they can be considered independent measures. Then, each trajectory coordinate is separately treated as a univariate time series, which is conjoined with the others only for plotting purposes.

The offline approach was implemented in a straightforward way. It consists on the sequential submission of the target dataset to both Hampel filter and LOESS procedures with predefined parameter values to get as result an outlier-free and smoothed curve which may better reflect the trajectory of the aerospace engine.

The online approach rendered somewhat more work, since a moving window to be applied to the incoming data needs to be established. To ensure the online scenario, a 20 Hz sampling rate to the already existent trajectory datasets is simulated. Thus, a sliding window is established, taken into account a fixed proportion of the full dataset points to set the window size. To step the window through data, it is required to fill a buffer with the latest incoming points. Thus, at each fulfilment of the buffer, the window moves a step and the overall processing is applied to it. The step size could be the whole window bandwidth or a fixed proportion of it. The overall processing has to be accomplished as fast as data arrives, once the online approach is supposed to support radar analysts' decisions and derivative processing during the flight of the engine.

RESULTS

EFFECTIVE COMBINATION

In most tests, the filtering process combining outlier detection and smoothing successfully delivered an outlier-free and smooth curve that, according to radar specialists, does not affront rocket models and nominal trajectories. The results are quite satisfactory as both procedures on which our approach is based performed very well in their actuation fields. The Hampel filter played an important pre-processing role as the routine preceding the LOESS smoother, especially in scenarios of high contamination. In most tested datasets, Hampel filter could show an identification rate exceeding 80% in relation to artificial outlying data. It was also able to ensure that the most significant outliers would not render much interference in local regression. Figure 3 shows a comparative plot set for offline and online filtering approaches. Figure 3 (a) presents a 10% artificial outlier contamination on a certain rocket trajectory data and Fig. 3 (b) reveals the offline results regarding the application of the filtering processes, using the parameters K=3, t=3 (Hampel filter) and g=0.5, d=2 (LOESS). A locally quadratic fit (d=2) was used in all LOESS tests.

Regarding the smoothing phase, it was observed significant differences between offline and online results. Offline results were quite impressive, since the output curve was indeed smooth, reflecting better the actual trajectory of the target flying engine. On the other hand, the results from the online implementation pointed out that there is a serious trade-off related to the size of the moving window and the effectiveness of the LOESS smoothing procedure.

In general, online outputs reveal a drop in performance and effectiveness of the method when the bandwidth of the moving window is narrow, when compared to the size of the whole dataset. In comparison with the offline method, the online approach always leads to a less smooth curve, obtained from a temporally and computationally more costly process.

It happens because LOESS requires a fairly large and densely sampled datasets in order to produce good models. That is, it needs good empirical information on the local structure of the process in order to perform the local fitting (NIST/SEMATECH, 2012). When the dataset is partitioned in windows that move as data arrives, much less data is delivered to each iteration of the LOESS procedure, making the smoothing process much more local when considering the whole dataset. Besides, as the LOESS procedure is invoked more often (at least in each window iteration), a performance drop is also justifiable. Substantially increasing the moving window bandwidth is not the solution to this problem, because this would make this online approach gradually closer to the offline approach. Since the methods are computationally intensive, limited computational capabilities becomes a crucial barrier in the online approach.

Figure 3 (c) also shows the results when applying the online method on the contaminated dataset, using the same parameters used on offline experiments (for Hampel filter and LOESS) and a window size w=100 (one hundred samples) for the moving window applied to incoming data. The online method outputs a slightly less smooth curve than the offline results and it is much more computationally intensive.

EXTRAPOLATING FILTERED DATA

Because of the good results after applying the filtering processes, it was expected that any further processing applied to this smooth and well-behaved data would render better results when compared to the processing of raw radar data. From existing filtered source data, extrapolation could play an important role by completing the missing portion of the signal in a tracking loss scenario or by predicting with better accuracy the impact point of a falling aerospace engine.

To prove this hypothesis, we decided to extrapolate both filtered and unfiltered data using a well known and effective extrapolation method and comparing the results. A proven effective and established method that could fit well to the problem was chosen. A linear predictive strategy using autoregressive modeling was preferred as its attractiveness stems, among others, from the fact that the numerical algorithms involved in the processing are rather simple and it depends on a limited number of parameters which are estimated from the already measured data (De Hoon *et al.*, 1996). However, this strategy produces coefficients which are not well suited for numerical computation and models which are not always stable. To overcome these constraints, the preferable algorithm to estimate the autoregressive parameters is the Burg's method, due to its reliability and accuracy on parameter estimates and because the estimated autoregressive model is guaranteed to be stable (De Hoon *et al.*, 1996).

To sum up very briefly, we used the strategy mentioned above to extrapolate a time series (each of the three

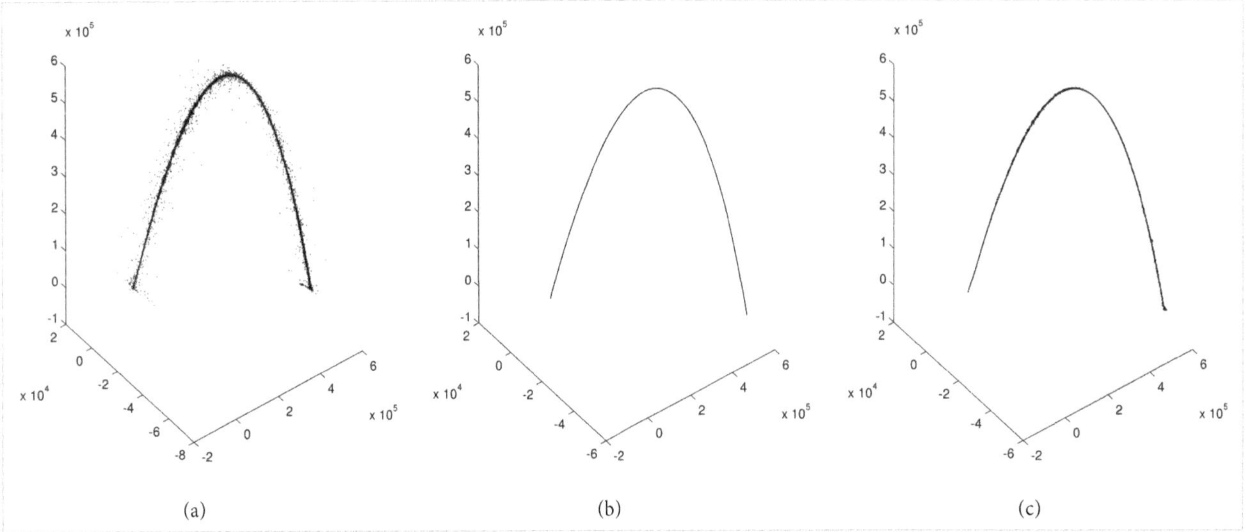

Figure 3. A comparative plot set for offline and online filtering approaches. (a) Original data with 10% outlier contamination; (b) Offline filtered curve; (c) Online filltered curve.

signal components) by fitting a linear model to the time series, in which each sample is assumed to be a linear combination of previously observed samples. Hence, the predicted samples are the next time samples of the input time segment.

The following figures illustrate the results of applying Burg's extrapolation over both unfiltered and filtered radar data, obtained from the tracking of a certain rocket. Again, for illustration purposes, the base data used in these extrapolation experiments is the original tracking data with 5% of samples artificially contaminated by outliers. Figure 4 shows a tridimensional plot comparing the entire base data to its extrapolation, starting from the 3501st of a 5820 sample total. A noticeable deviation between the unfiltered samples curve and the extrapolated samples curve can be noted. On the other hand, the tridimensional plot of Fig. 5 clearly shows that the deviation between the filtered data and its extrapolation is much lower.

DISCUSSION AND FUTURE WORK

This work represents just a primer approach to trajectory radar data cleaning, supported by a long-standing demand brought about by radar specialists. Preliminary results were considered quite satisfactory, mainly regarding the offline processing. The problem of outlying samples was reasonably surpassed and data analysts can rely on a smoother and more representative curve for subsequent analysis. Online processing demands more work regarding the optimal strategy on processing incoming radar data as the efficient use of a stepping moving window limits the amount of data delivered to smoothing processes and, consequently, degrades the quality of the smoothed curve. Besides, for effectiveness and performance reasons, it is essential to implement online processing methods in low-level programming languages, since the online approach needs several iterations of already costly processes, and the processed results should be instantly available.

As for what to be done with the filtered data, the possibilities are vast. It was shown that, from a smooth and well behaved trajectory curve, extrapolation strategies could be readily applied in case of radar malfunctioning and target loss. Also, it is possible to adjust parametric models to the curve which can propitiate accurate speed and acceleration calculations on specific points of the trajectory. These approaches could also be used in a scenario where it is needed to determine precisely the area of impact of a tracked flying engine.

The parameters of the processes used in this study were chosen based on literature indications and empiricism. A future work could comprise the use of an automated

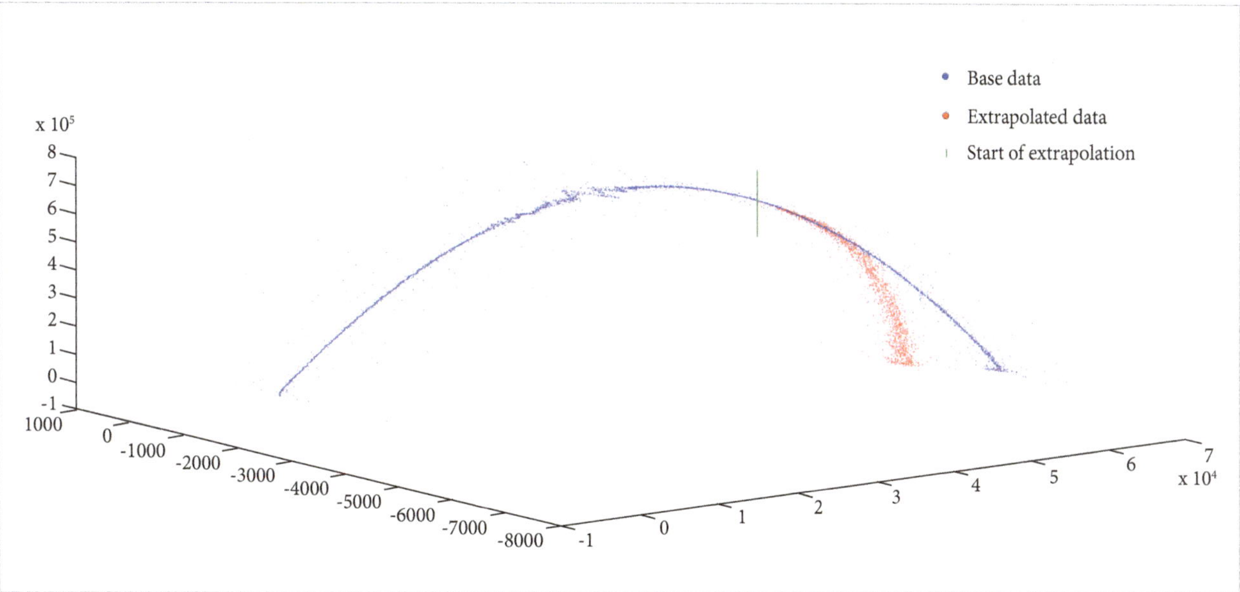

Figure 4. Extrapolation of unfiltered radar data.

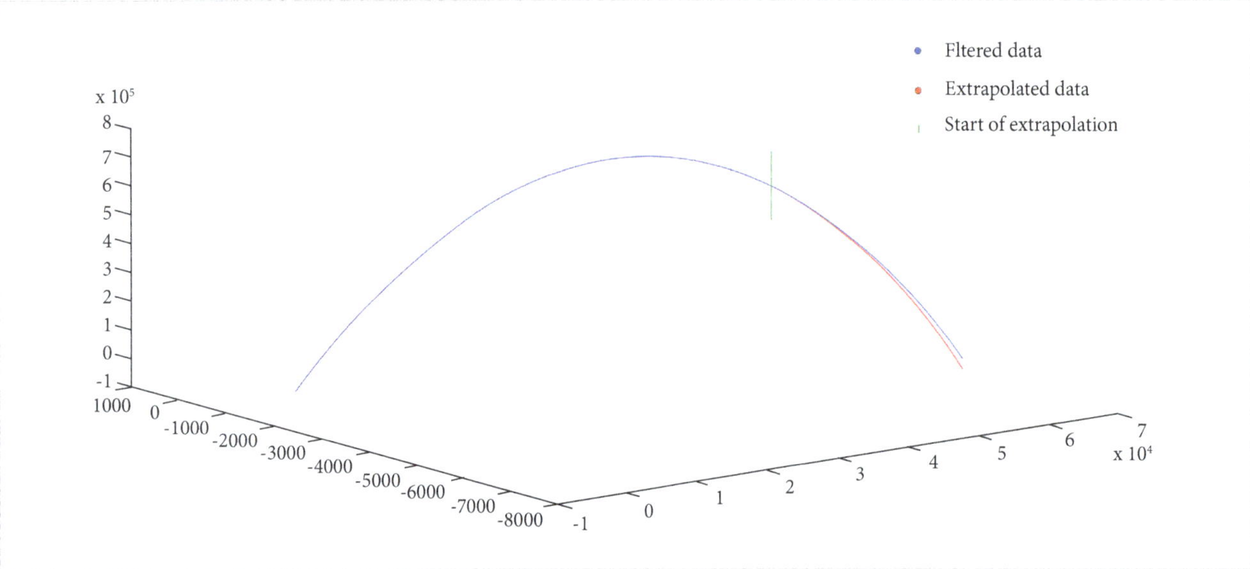

Figure 5. Extrapolation of filtered radar data.

parameter tuning algorithm to select the best parameters given the trajectory shape and quality of radar data. Also, the independent trajectory coordinates problematic is likely to be a good candidate for parallel or distributed processing approach in order to speed up the overall process, especially for the online filtering process.

A lot of validation work has to be done in order to verify if the output model really befits the actual trajectory of a tacked aerospace engine. This validation work may consider past tracking datasets, analytic physical analyses data and nominal trajectory behavior data for a specific engine.

ACKNOWLEDGMENT

We would like to thank *Eng.* Edson Duarte, from *Centro de Lançamento da Barreira do Inferno* (CLBI), for providing several radar tracking datasets and invaluable explanations about trajectory radar functioning.

REFERENCES

Adnan, R., Mohamad, M.N. and Setan, H., 2003, "Multiple Outliers Detection Procedures in Linear Regression", Matematika, Vol. 19, No 1, pp. 29-45.

ASTM International, 1980, "Standard Practice for Dealing with Outlying Observations", Active Standard E-178-08.

Cleveland, R.B., Cleveland, W.S., McRae, J.E. and Terpenning I., 1990, "STL: A seasonal-trend decomposition procedure based on LOESS", Journal of Official Statistics, Vol. 6, No 1, pp.3–73.

Cleveland, W.S. and Devlin, S.J., 1988, "Locally Weighted Regression: An Approach to Regression Analysis by Local Fitting", Journal of the American Statistical Association, Vol. 83, No 403, pp. 596-610.

Cleveland, W.S. and Loader, C.L., 1996, "Smoothing by Local Regression: Principles and Methods", Statistical Theory and Computational Aspects of Smoothing, pp. 10-49, Haerdle and M. G. Schimek, Springer, NY.

Cleveland, W.S., 1979, "Robust Locally Weighted Regression and Smoothing Scatterplots", Journal of the American Statistical Association, Vol. 74, No 368, pp. 829–836. doi:10.1080/01621459.1979.10481038.

Cohen, R. A., 1999, "An Introduction to PROC LOESS for Local Regression", Proceedings of the 24th SAS Users Group International Conference, Paper 273.

Davies, L. and Gather, U., 1993, "The identification of multiple outliers", Journal of the American Statistical Association, Vol. 88, No 423, pp. 782–792. doi:10.1080/01621459.1993.10476339.

De Hoon, M.J.L., van der Hagen, T.H.J.J., Schoonewelle, H. and van Dam, H., 1996, "Why Yule-Walker should not be used for autoregressive modeling", Annals of Nuclear Energy, Vol. 23, No 15, pp. 1219–1228. doi:10.1016/0306-4549(95)00126-3.

Garcia, D, 2010, "Robust smoothing of gridded data in one and higher dimensions with missing values", Computational Statistics & Data Analysis, Vol. 54, No 4, pp. 1167-1178. doi:10.1016/j.csda.2009.09.020.

Horwitz, W., 1995, "Protocol for the Design, Conduct and Interpretation of Method-performance Studies", Pure & Applied Chemistry, Vol. 67, No 2, pp. 331-343. doi:10.1351/pac199567020331.

Lee, J., Han, J. and Li, X., 2008, "Trajectory Outlier Detection: A Partition-and-Detect Framework", Proceedings of the IEEE 24th International Conference on Data Engineering (ICDE '08), pp. 140-149.

Liu, H., Shah, S. and Jiang, W., 2004, "On-line outlier detection and data cleaning", Computers & Chemical Engineering, Vol. 28, Issue 9, pp. 1635–1647. doi:10.1016/j.compchemeng.2004.01.009.

Loader, C., 2012, "Smoothing: Local Regression Techniques", Handbook of Computational Statistics, Ch. 20, pp. 571-596. doi:10.1007/978-3-642-21551-3_20.

Massart, D. L., Vandeginste, B.G.M., Buydens, L.M.C., De Jong, S., Lewi, P.J. and Smeyers-Verbek, J., 1997, "Handbook of Chemometrics and Qualimetrics: Part A", Data Handling in Science and Technology, Elsevier Science, Vol. 20A.

NIST/SEMATECH, 2012, "e-Handbook of Statistical Methods", Retrieved in May 27, 2013, from http://www.itl.nist.gov/div898/handbook/.

Nurunnabi, A.A.M. and Nasser, M., 2008, "Multiple Outliers Detection: Application To Research & Development Spending and Productivity Growth", BRAC University Journal, Vol. V, No 2, pp. 31-39.

Pearson, R. K., 2002, "Outliers in process modeling and identification", IEEE Transactions on Control Systems Technology, Vol. 10, No 1, pp. 55-63. doi:10.1109/87.974338.

Pearson, R.K., 2005, "Mining Imperfect Data: Dealing with Contamination and Incomplete Records", SIAM.

Pearson, R.K., 2011, "Exploring Data in Engineering, the Sciences, and Medicine", Oxford University Press, 2011.

Rancel, M.M.S. and Sierra, M.A.G., 2000, "Procedures for the Identification of Multiple Influential Observations Using Local Influence", The Indian Journal of Statistics (Sankhyā), Series A, Vol. 62, No 1, pp. 135-143.

Simonoff, J.S., 1998, "Smoothing Methods in Statistics", 2nd edition, Springer.

Solak, M.K., 2009, "Detection of Multiple Outliers in Univariate Data Sets", Schering-Plough Research Institute, Summit, NJ, paper SP06-2009, Retrieved in May 27, 2013, from http://www.lexjansen.com/pharmasug/2009/sp/SP06.pdf.

Wilson, D.I., 2006, "The Black Art of Smoothing", Electrical and Automation Technology, June/July Issue.

Zhang, M.H., Luypaert, J., Pierna, J.A.F., Xu, Q.S. and Massart, D.L., 2004, "Determination of Total Antioxidant Capacity in Green Tea by Near-infrared Spectroscopy and Multivariate Calibration", Talanta, Vol. 62, Issue 1, pp. 25-35. doi: 10.1016/S0039-9140(03)00397-7.

12

Characterization and Modeling of the Viscoelastic Behavior of a Self-Adhesive Rubber Using Dynamic Mechanical Analysis Tests

Lucie Rouleau[1], Rogério Pirk[1,2], Bert Pluymers[1], Wim Desmet[1]

ABSTRACT: The goal of this study is twofold. The first one is to assess the applicability of approaches based on dynamic-mechanical analysis to investigate the viscoelastic properties of a self-adhesive synthetic rubber. The second goal is to identify the parameters of a viscoelastic model which accurately represents the frequency-dependent mechanical properties. For that purpose, the time-temperature superposition principle is successfully applied to build the master curves of the material up to 1 MHz. The thickness of the samples and the thermal expansion effects are found to have a negligible influence on the mechanical properties measured by dynamic-mechanical analysis. The parameters of a generalized Maxwell model and a fractional derivative model are identified from the obtained master curves and lead to an accurate representation of the frequency-dependent mechanical properties of the rubber.

KEYWORDS: Dynamic mechanical analysis, Viscoelastic material, Generalized Maxwell model, Fractional derivative model.

INTRODUCTION

Structural vibration reduction is a fundamental concern in several industries to avoid failure due to mechanical fatigue, to comply with increasingly stringent Noise, Vibration and Harshness (NVH) standards, or to ensure optimal operating conditions for on-board electronics. There are several methods that may be used in order to control vibrations (Nakra, 1998), which can be divided in two categories: active and passive damping treatments. The former rely on the effect of active mechanisms, using for example piezoelectric actuators, while the latter takes advantage of the inherent energy dissipation effects of systems, using for example viscoelastic materials (VEM).

Passive damping treatments based on VEM have been widely used in a variety of industrial applications and especially in aircraft and automotive systems (Rao, 2003). Since their molecular structure allows the internal dissipation of part of the vibration energy (Ferry, 1980), VEM are usually applied on the surface of vibrating structures, as a constrained or unconstrained layer of material. Some add-on passive damping treatments are available commercially in the form of self-adhesive constrained layers, which offer the advantage of introducing damping with a minimum mass addition, while also being easily applied to structures with complex geometries.

The characterization and modeling of VEM mechanical properties are essential in the design and analysis of damped structures, to provide solutions with adequate damping properties. However, the VEM material properties depend on many factors,

1.Katholieke Universiteit Leuven – Department of Mechanical Engineering – Heverlee – Belgium. 2.Departamento de Ciência e Tecnologia Aeroespacial – Instituto de Aeronáutica e Espaço – São José dos Campos/SP – Brazil.

Author for correspondence: Lucie Rouleau | Cnam Paris – Structural Mechanics and Coupled Systems Laboratory | 2 rue Conté, 75003 | Paris – France
Email: lucie.rouleau@cnam.fr

such as temperature, excitation frequency, amplitude, pre-stress or relative humidity (Sjöberg, 2002; Kergourlay, 2004). Temperature and frequency being the most relevant factors, it is indispensable to determine the material properties over the range of operational temperatures and frequencies.

Various existing techniques are available in the literature to characterize VEM as a function of temperature and frequency and are reviewed, among others, in Lakes (2004). Among them, dynamical mechanical analysis (DMA) is one of the most common and standard methods used to measure the dynamic properties of VEM (ASTM D4065-12; Menard, 2008). In this method, the response of VEM samples to a sinusoidal excitation at different frequencies and temperatures is measured. The dynamic properties over a wide frequency range are then extrapolated by means of the time-temperature superposition principle (TTSP). This method has been successfully applied to the characterization of many polymers (Ferry, 1980; Emri, 2005; Dealy and Plazek, 2009). However, for some materials such as polymeric blends (Van Gurp and Palmen, 1998) or polymers with long chain branching (Hatzikiriakos, 2000), the TTSP may not be applicable anymore. Moreover, in the case of adhesive polymers, the nonlinearity of the adhesive behavior may have an influence on the mechanical properties measured by DMA, as evidenced in García-Barruetabeña *et al.* (2011). These observations motivate this study, of which the first goal is to assess the applicability of the DMA-based approach to investigate the viscoelastic properties of a self-adhesive synthetic rubber. In particular, since self-adhesive materials (also called pressure-sensitive adhesives) are a type of polymer blends (Creton, 2003), both the applicability of the TTSP and the nonlinearity of the adhesive behavior should be studied.

Once the frequency-dependent properties of the VEM are determined, one needs to model viscoelasticity in order to obtain accurate results using finite element models and thus assess the performance of the passive damping treatment. Several models exist in the literature to describe the viscoelastic behavior of VEM (Bert, 1973; Vasques *et al.*, 2010). The capacity of a viscoelastic model to represent the frequency-dependent mechanical properties depends on the material. For instance, a fractional derivative model may lead to a good fit of the viscoelastic properties of some polymers (Pritz, 2003), but a poor one for others (Corn *et al.*, 2009). The second goal is then to identify the parameters of a viscoelastic model which accurately represents the frequency-dependent mechanical properties. In this study, two viscoelastic models are investigated: the generalized Maxwell model and the fractional derivative model.

THEORY OF LINEAR VISCOELASTICITY

Considering an isotropic viscoelastic material under small isothermal and small deformation conditions, the theory of linear viscoelasticity assumes the existence of a constitutive relation between stress and strain histories, which for a one-dimensional system is given by a convolution integral (Christensen, 1982):

$$\sigma(t) = \int_{-\infty}^{t} E(t-\tau)\varepsilon(t)d\tau \tag{1}$$

where $\sigma(t)$ and $\varepsilon(t)$: respectively, the stress and strain at a given time t; $E(t)$: stress relaxation modulus.

A convenient way of representing the viscoelastic behavior in the frequency domain is through the complex modulus approximation. Let us assume a stationary harmonic strain $\varepsilon(t)$:

$$\varepsilon(t) = \varepsilon_0 e^{i\omega t} \tag{2}$$

and a stationary harmonic stress $\sigma(t)$:

$$\sigma(t) = \sigma_0 e^{i(\omega t + \varphi)} \tag{3}$$

where i : imaginary unit, defined by its property $i^2 = -1$; σ_0 and ε_0: respectively, the stress and strain amplitudes; ω: excitation frequency; φ: phase delay between the stress and strain.

The constitutive equation linking stress and strain in the frequency domain is given by:

$$\sigma(\omega) = E^*(\omega)\varepsilon(\omega) \tag{4}$$

where the complex modulus $E^*(\omega)$ can be written as:

$$E^*(\omega) = E'(\omega) + iE''(\omega) = E'(\omega)[1 + i\eta(\omega)] \tag{5}$$

where $E'(\omega)$: storage modulus; $E''(\omega)$: loss factor.

The loss factor $\eta(\omega) = E''(\omega)/E'(\omega)$ is a measure of the energy dissipated in the material, expressed in terms of recoverable energy. The storage and loss moduli, as well as the loss factor, are related to the phase and amplitude differences between stress and strain:

$$E'(\omega) \quad \frac{\sigma_0(\omega)}{\varepsilon_0(\omega)} \cos[\varphi(\omega)] \tag{6a}$$

$$E''(\omega) = \frac{\sigma_0(\omega)}{\varepsilon_0(\omega)} \sin[\varphi(\omega)] \tag{6b}$$

$$\eta(\omega) = \tan\,[\varphi(\omega)] \tag{6c}$$

Therefore, by measuring the time response of a sample of material to an imposed strain, DMA testing allows the measurement of the complex modulus defined in Eq. 5. However, the measurements which can be performed on most commercial DMA equipment are limited to the low-frequency domain (typically up to a few hundred Hz). In order to characterize the material on a broader frequency range, the TTSP is applied (Ferry, 1980). This principle relates the material response at a given time t (or frequency ω) and at a given temperature T to that at other conditions (denoted by subscript r):

$$\omega_r = a_T(T, T_r)\omega \tag{7a}$$

$$E'(\omega_r, T_r) = b_T(T, T_r)E'(\omega, T) \tag{7b}$$

$$E''(\omega_r, T_r) = b_T(T, T_r)E''(\omega, T) \tag{7c}$$

where $a_T(T, T_r)$ and $b_T(T, T_r)$: coefficients which indicate the amount of horizontal and vertical shifting (respectively) to be applied to isotherms of storage and loss moduli measured at a temperature T in order to estimate the material properties at a reference temperature T_r (see Fig. 1).

The horizontal shift factors $a_T(T, T_r)$ describe the temperature dependence of the relaxation time and usually follow the empirical Williams-Landel-Ferry (WLF) equation (Williams *et al.*, 1955):

$$\log[a_T(T, T_r)] = \frac{C_1(T - T_r)}{C_2 + T - T_r} \tag{8}$$

where C_1 and C_2: empirical constants whose order of magnitude is about 10 and 100 K, respectively.

The vertical shift factors $b_T(T, T_r)$ are related to thermal expansion effects, which for most polymers can be neglected. In that case, only horizontal shift factors are applied.

It is clear from Fig. 1 that by performing measurements at different temperatures, the TTSP allows to extend the frequency range on which the material properties are known. However, the TTSP holds only for materials which are called thermo-rheologically simple (Dealy and Plazek, 2009). The Black diagram (Van Gurp and Palmen 1998) (plot of the phase angle *versus* the amplitude of the complex modulus) and the Cole-Cole diagram (Han and Kim, 1993) (plot of the storage modulus *versus* the loss modulus) are two indicators commonly used to assess the applicability of the TTSP.

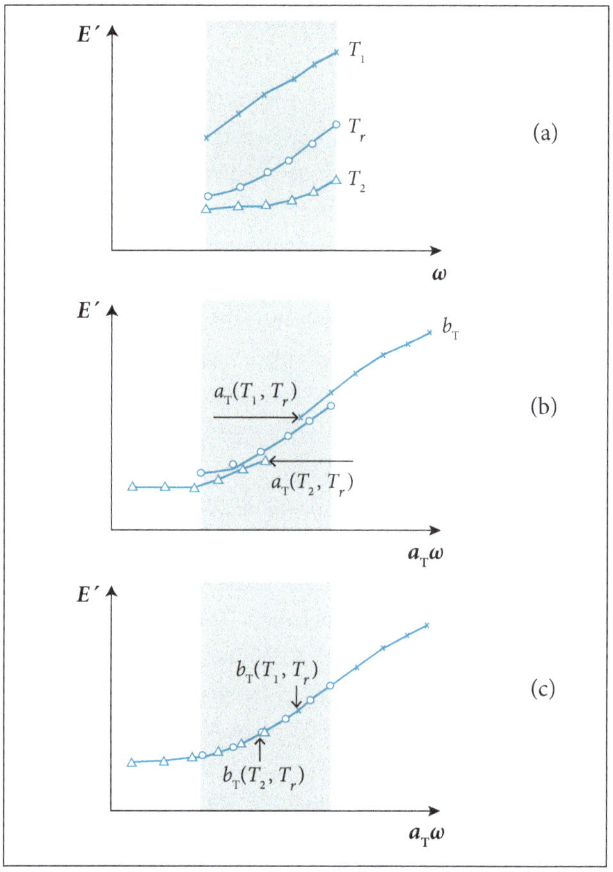

Figure 1. TTSP. (a) Isotherms of storage modulus on the frequency range measurable by DMA, at temperatures T_1, T_2 and T_r, with $T_1 < T_r < T_2$; (b) Isotherms of storage modulus after application of the horizontal shift factors, taking T_r as the reference temperature; (c) Isotherms of storage modulus after application of both horizontal and vertical shift factors, taking T_r as the reference temperature.

METHODS

DYNAMICAL MECHANICAL ANALYSIS EXPERIMENTS

The material studied is a self-adhesive synthetic rubber (SR) from an automotive TIER supplier. Measurements of its frequency-dependent shear modulus are carried out on a DMA Q800 dynamical mechanical analyzer (produced by TA Instrument Corporation), available at the Department of Metallurgy and Material Engineering of Katholieke Universiteit of Leuven (see Fig. 2a). For that purpose, the testing configuration is set in shear mode, *i.e.* two equal-size samples are sheared between a fixed and a moveable plate, as depicted in Fig. 2b. The lateral dimensions of the samples are 10 mm x 10 mm. When clamping the samples on the DMA, a compressive pre-strain γ_0~ 10% is applied in order to ensure a good adhesion between the samples and the clamps.

In order to study the linearity of the adhesive behavior, samples of different thicknesses (1.5 and 2.2 mm) are tested. The linearity of the viscoelastic behavior was studied for both samples by inducing a dynamic displacement varying from 1 to 10 μm, at a reference temperature of 20°C and a reference frequency of excitation of 100 Hz.

The dynamic behavior of the self-adhesive rubber is then studied by performing a multi-frequency test with temperature ramp. The glass transition temperature was estimated at -60°C by Differential Scan Calorimetry (DSC) on Q2000 DSC equipment (produced by TA Instrument Corporation) from the Department of Chemical Engineering of Katholieke Universiteit of Leuven. Considering the rheological characteristics of the tested material and the operational bands of DMA Q800, 15 frequency sampling points ranging from 1 to 200 Hz are selected, and the temperature ramp is done at a heating rate of 2°C/min from −70°C to 40°C. A dynamic displacement of 10 μm is applied to the samples.

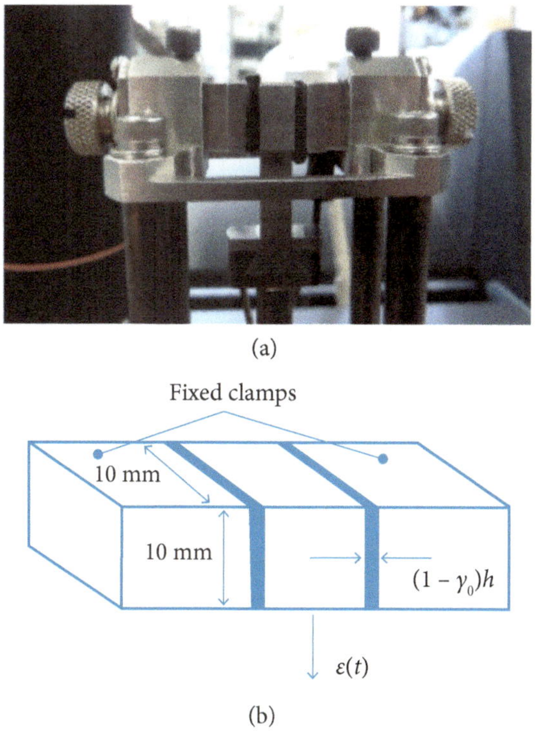

Figure 2. (a) Synthetic rubber samples installed on the DMA Q800 in shear test configuration. (b) Scheme of the test configuration: two equal-size material samples, of thickness *h*, are sheared between a fixed and an oscillating clamp. A compressive pre-strain γ_0 is applied during clamping to ensure a good adhesion between the clamps and the samples.

APPLICATION OF THE TIME-TEMPERATURE SUPERPOSITION PRINCIPLE

The TTSP consists in applying shift factors to isotherms in order to build the master curves of the material which represent the dynamic complex modulus as a function of frequency. Once the validity of the TTSP is checked, the master curves are built by using the method described in (Rouleau *et al.* 2013). This method aims at determining both horizontal and vertical shift factors to be applied to isotherms by means of a least square method which requires the fulfillment of the Kramers-Kronig relations. In this way, the master curves are built in accordance with the causality principle, while also accounting for possible thermal expansion effects.

This methodology is applied to at least two samples of each dimension (1.5 and 2.2 mm), to comply with standard norms (ISO, 1996).

RESULTS AND DISCUSSION

LINEARITY OF THE BEHAVIOR

While varying the dynamic displacement applied to samples, no variation of the measured modulus is observed, confirming the assumption of linear viscoelasticity. Following this result, a dynamic displacement of 10 μm is considered for the remainder of the study.

The storage modulus and loss factor measured for two samples of different thicknesses are plotted against temperature at different frequencies in Fig. 3. It can be noted that the storage modulus of the 2.2 mm thick sample is slightly higher than that of the other sample, at very low temperatures (below −30°C). A higher peak of loss factor is also observed for the 2.2 mm thick sample (about 5% higher than for the 1.5 mm thick sample). However, considering the inherent dispersion expected for this type of measurements, the influence of the sample thickness can be neglected (see the next section).

In Fig. 3, two regions of viscoelastic behavior can be identified. The first one is the glassy region, corresponding to low temperature (below –30°C), where the storage modulus is roughly constant and the loss factor is close to zero. In this region, the material has an elastic behavior. The second region is the glass transition region, for temperatures above –30°C, where the dynamic modulus undergoes strong variations with frequency and temperatures. This is the region of interest for damping applications since it is characterized by a peak of loss factor. Therefore, only the measurement points within the temperature range [–30°C, 30°C] are considered for the application of the TTSP.

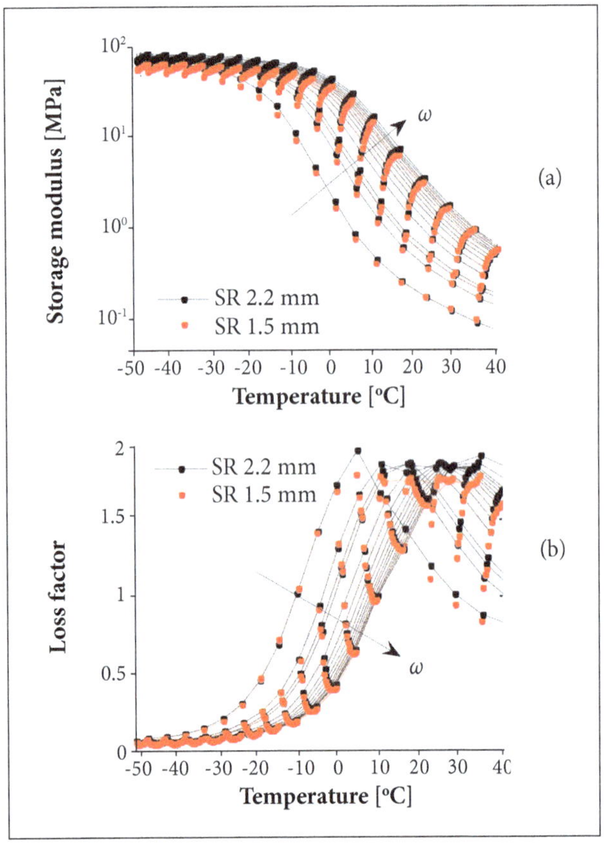

Figure 3. Dynamic modulus variation with temperature at frequencies from 1 to 200 Hz for 1.5 and 2.2 mm thick samples. (a) Storage modulus. (b) Loss factor.

VALIDITY OF THE TIME-TEMPERATURE SUPERPOSITION PRINCIPLE

The validity of the TTSP is checked by means of the Cole-Cole diagram, shown in Fig. 4a, and the Black diagram, shown in Fig. 4b. The G' *versus* G" and η *versus* |G*| curves in double logarithmic scales at different temperatures coincide in one continuous curve for most points, thus meeting the assumption of a thermo-rheologically simple material. Points that deviate from the curves correspond to measurements above 60 Hz. Since they may be related to resonance phenomena in the DMA (Placet and Foltête, 2010), they are removed.

Figure 5 shows the isotherms corresponding to the 2.2 mm thick sample which are used for the application of the TTSP. They are obtained by carrying out a polynomial curve fitting of the measured isochrones (shown in Fig. 3).

The horizontal and vertical shift factors applied to isotherms of Fig. 5 are plotted against temperature in Fig. 6. It can be noted that the vertical shift coefficients are very close to unity, which implies that thermal expansion effects can be neglected. The horizontal shift factors a_T are fitted by the WLF equation (Eq. 8): the parameters C_1 = 9.23 and C_2 = 141.2 K are found to lead to a very good fit. Those values are in accordance with previous studies, which give confidence in the built master curves.

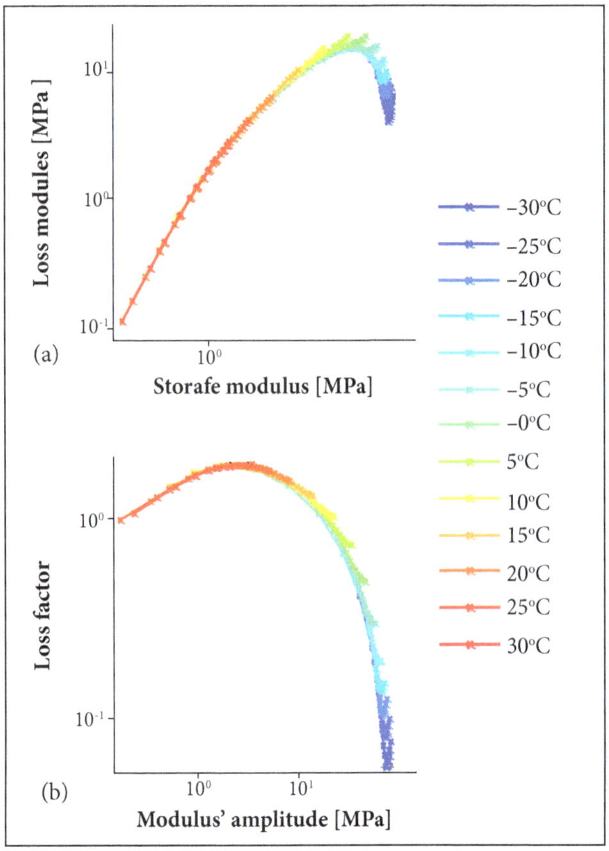

Figure 4. (a) Cole-Cole diagram: loss modulus *G" versus* storage modulus *G'*. (b) Black diagram: loss factor *η versus* the modulus' amplitude |*G**| for a 2.2 mm thick material sample.

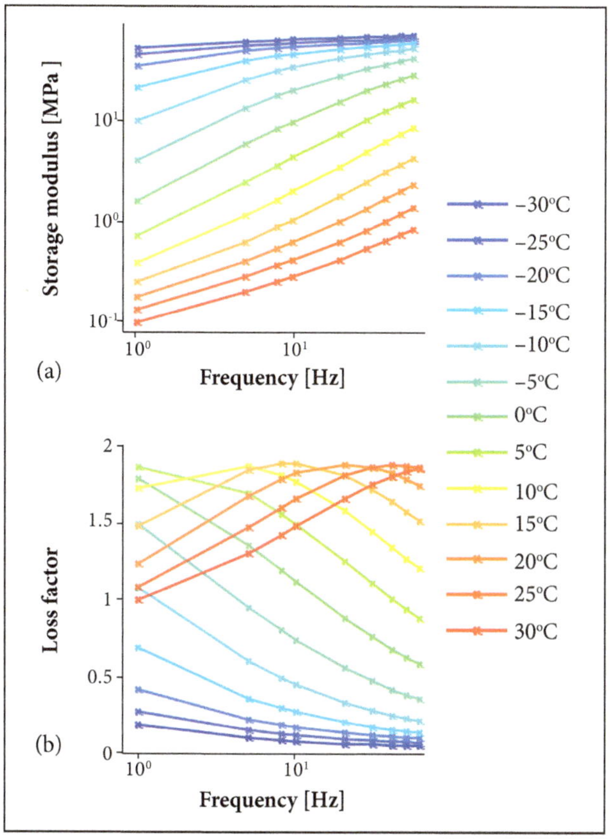

Figure 5. Isotherms of (a) storage modulus and (b) loss factor for a 2.2 mm thick material sample.

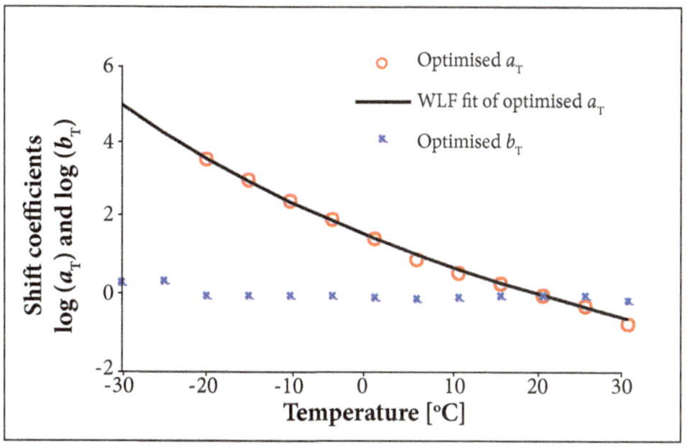

Figure 6. Horizontal and vertical shift factors $\log(a_T)$ and $\log(b_T)$ applied to the isotherms of Fig. 5. The horizontal shift coefficients are fitted by the WLF equation (Eq. 8).

The master curves obtained from the application of the TTSP at a reference temperature of 20°C for all the tested samples are given in Fig. 7. Little dispersion is observed in the mechanical properties of the material. The peak of loss factor appears around 20 – 30 Hz, so this material should be used for damping low-frequency vibrations.

VISCOELASTIC MODELS

Generalized Maxwell Model

In the Maxwell model, springs and dashpots are used to describe the elastic and viscous components of the stress/strain response of a viscoelastic material. The generalized Maxwell model is comprised of several Maxwell elements in parallel, and the corresponding complex modulus representation is:

$$G^*(\omega) = G_0 + \sum_{k=1}^{m} G_k \frac{i\omega\tau_k}{1 + i\omega\tau_k} \tag{9}$$

where G_0: relaxed modulus, *i.e.* $G_0 = \lim_{\omega \to 0} G^*(\omega)$; m: number of Maxwell components; G_k and τ_k: respectively, the relaxation strength and relaxation time of the kth Maxwell component.

The parameters of a generalized Maxwell model are identified from the master curves obtained for a SR sample of 2.2 mm thickness. The enclosing curve technique (Vinh, 1967; Renaud *et al.*, 2011), a graphical technique based on the Pole-Zero formulation of the generalized Maxwell model, is used for that purpose. Twelve Maxwell components are found to lead to a good fit of the frequency-dependent shear modulus of the SR, as shown in Fig. 8. Since the graphical identification method uses only the characteristics of the modulus' amplitude, the modeling of the phase is less accurate, but still reasonable. The identified parameters are given in Table 1.

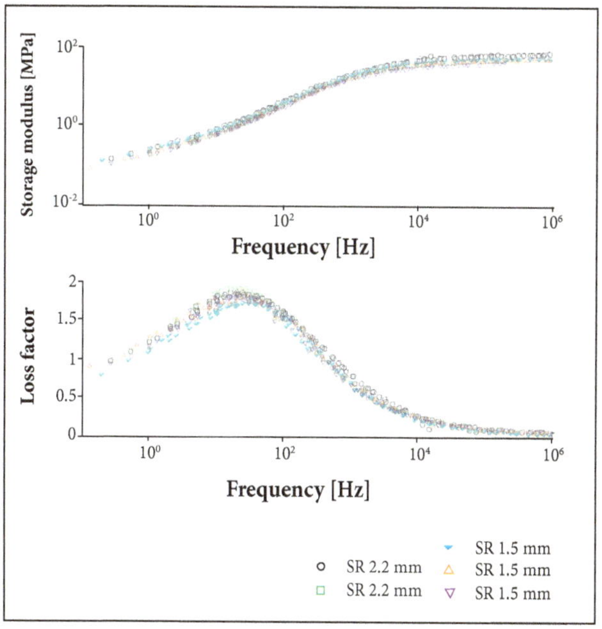

Figure 7. Master curves of the self-adhesive synthetic rubber at a reference temperature of 20°C obtained from DMA measurements of 5 samples of different thicknesses.

Fractional Derivative Model

The fractional derivative model used in this study is a Maxwell model where the dashpot is replaced by a spring-pot element. The corresponding complex modulus representation is:

$$G^*(\omega) = G_0 + (G_\infty - G_0)\frac{(i\omega\tau)^\alpha}{[1 + (i\omega\tau)^\alpha]} \quad (10)$$

where G_∞: unrelaxed modulus, *i.e.* $G_\infty = \lim_{\omega \to \infty} G^*(\omega)$; τ: relaxation time; α: order of derivation which is related to internal relaxation processes.

The four parameters constituting the fractional derivative model are identified from the master curves of a 2.2 mm thick material sample, by using a least square method. The identified parameters are listed in Table 2. Good correlation is obtained between the experimental master curves and the model, as shown in Fig. 9.

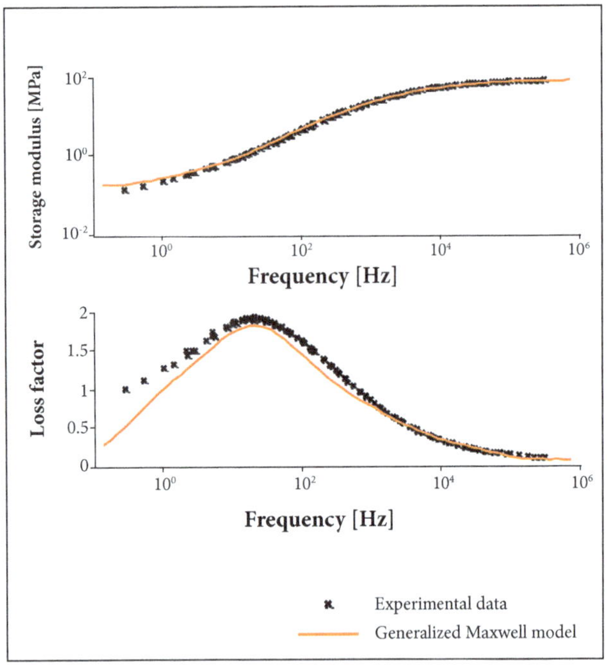

Figure 8. Master curves of the self-adhesive synthetic rubber (2.2 mm sample) at a reference temperature of 20°C fitted by a generalized Maxwell model.

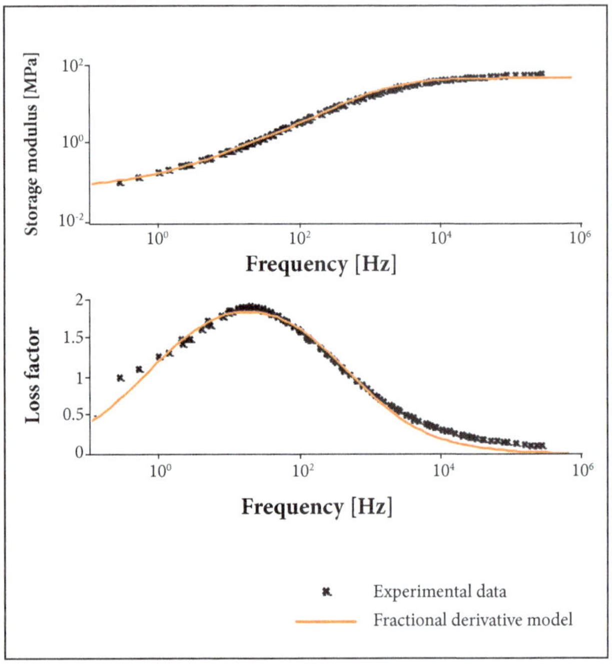

Figure 9. Master curves of the self-adhesive synthetic rubber (2.2 mm sample) at a reference temperature of 20°C fitted by a fractional derivative model.

Table 1. Parameters of the generalized Maxwell model identified by the enclosing curve technique.

Relaxed modulus G_0					
0.15 MPa					
Series (k)	Relaxation strength (G_k) (MPa)	Relaxation time (τ_k) (s)	Series (k)	Relaxation strength (G_k) (MPa)	Relaxation time (τ_k) (s)
1	0.043	$3.25e^{-1}$	7	5.795	$5.89e^{-4}$
2	0.080	$1.29e^{-1}$	8	6.282	$2.27e^{-4}$
3	0.157	$4.97e^{-2}$	9	13.89	$8.84e^{-5}$
4	0.416	$1.43e^{-2}$	10	15.67	$2.53e^{-5}$
5	1.522	$3.50e^{-3}$	11	15.40	$4.55e^{-6}$
6	2.282	$1.30e^{-3}$	12	8.18	$2.04e^{-7}$

Table 2. Parameters of the fractional derivative model identified by the least square method.

Relaxed modulus (G_0) (Pa)	Unrelaxed modulus (G_∞) (Pa)	Relaxation time (ρ) (s)	Order of derivation (α)
$8.59e^4$	$5.34e^7$	$1.0e^{-4}$	0.72

CONCLUSION

The viscoelastic behavior of a self-adhesive synthetic rubber, used in automotive applications, is investigated in this study by means of DMA measurements. It is found that the mechanical properties identified by this method are not influenced by the thickness of the sample tested. The viscoelastic material is proven as thermo-rheologically simple, and the TTSP is successfully applied to build master curves up to 1 MHz. The analysis of the horizontal and vertical shift factors determined when applying the TTSP shows that the thermal expansion effects can be neglected. Pre-strain effect has not been studied here and may require further investigation since in Martinez-Agirre *et al.* (2014) it was found to have an influence on the mechanical properties measured by DMA.

Following the obtaining of the material's master curves, two viscoelastic models are identified: a generalized Maxwell model, classically used to model viscoelastic materials in industrial software, and a fractional derivative model, which relies on only four parameters. Both models allow an accurate representation of the frequency-dependent properties of the material. The parameters identified can then be used as inputs in finite element calculations to optimize the NVH performance of damped systems.

ACKNOWLEDGEMENTS

The IWT Flanders within the ASTRA project and the European Project INTERACTIVE Marie Curie Initial IAPP are gratefully acknowledged for their support. The Brazilian Program *Ciência sem Fronteiras* is also acknowledged by the second author.

REFERENCES

Rouleau, L., Deü, J.-F., Legay, A. and Le Lay, F., 2013, "Application of the Kramers-Kronig Relations to Time-Temperature Superposition for Viscoelastic Materials", Mechanics of Materials, Vol. 65, pp. 66-75. doi: 10.1016/j.mechmat.2013.06.001

Bert, C.W., 1973, "Material Damping: an Introductory Review of Mathematical Measures and Experimental Techniques", Journal of Sound and Vibration, Vol. 29, No. 2, pp. 129-153. doi: 10.1016/S0022-460X(73)80131-2

Christensen, R.M., 1982, "Theory of Viscoelasticity", Dover Publications, Mineola, USA.

Corn, S., Ienny, P., Dupuy, J.S. and Daridon, L., 2009, "Identification des Propriétés Viscoélastiques d'un PMMA par Analyse Vibratoire: Comparaison entre Différentes Méthodes Expérimentales", Proceedings of the 19ème Congrès Français de Mécanique, Marseille, France.

Creton, C., 2003, "Pressure-Sensitive Adhesives: an Introductory Course", Material Research Society Bulletin, Vol. 28, No. 6, pp. 434-439. doi: 10.1557/mrs2003.124

Dealy, J. and Plazek, D., 2009, "Time-Temperature Superposition – a User's Guide", Rheology Bulletin, Vol. 78, pp. 16-31.

Emri, I., 2005, "Rheology of Solid Polymers", Rheology Reviews, Vol. 3, pp. 49-100.

Ferry, J.D., 1980, "Viscoelastic Properties of Polymers", John Wiley & Sons, New Jersey, USA.

García-Barruetabeña, J., Cortés, F., Abete, J.M., Fernández, P., Lamela, M.J. and Fernández-Canteli, A., 2011, "Experimental Characterization and Modelization of the Relaxation and Complex Moduli of a Flexible Adhesive", Materials and Design, Vol. 32, No. 5, pp. 2783-2796. doi: 10.1016/j.matdes.2011.01.005

Han, C.D. and Kim, J.K., 1993, "On the Use of Time-Temperature Superposition in Multicomponent/Multiphase Polymer Systems", Polymer, Vol. 34, No. 12, pp. 2533-2539. doi:10.1016/0032-3861(93)90585-X

Hatzikiriakos, S., 2000, "Long Chain Branching and Polydiversity Effects on the Rheological Properties of Polyethylenes", Polymer Engineering and Science, Vol. 40, No. 11, pp. 2279-2287. doi: 10.1002/pen.11360

Kergourlay, G., 2004, "Mesure et Prediction Vibro-Acoustique de Structures Viscoélastiques. Application à une Enceinte Acoustique", Ph.D. Thesis, École Centrale de Paris, Paris, France.

Lakes, R., 2004, "Viscoelastic Measurement Techniques", Review of Scientific Instruments, Vol. 75, No. 4, pp. 797-810. doi: 10.1063/1.1651639

Martinez-Agirre, M., Illescas, S., and Elejabarrieta, M.J., 2014, "Characterisation and Modelling of Prestrained Viscoelastic Films", International Journal of Adhesion and Adhesives, Vol. 50, pp. 183-190. doi: 10.1016/j.ijadhadh.2014.01.029

Menard, K.P., 2008, "Dynamic Mechanical Analysis: a Practical Introduction", CRC Press, Boca Raton, USA.

Nakra, B., 1998, "Vibration Control in Machines and Structures Using Viscoelastic Damping", Journal of Sound and Vibration, Vol. 211, No. 3, pp. 449-466. doi: 10.1006/jsvi.1997.1317

Placet, V. and Foltête, E., 2010, "Is Dynamic Mechanical Analysis (DMA) a Non-Resonance Technique?", EPJ Web of Conference, Vol. 6, pp. 41004. doi: 10.1051/epjconf/20100641004

Pritz, T., 2003, "Five-Parameter Fractional Derivative Model for Polymeric Damping Materials", Journal of Sound and Vibration, Vol. 265, No. 5, pp. 935-952. doi: 10.1016/S0022-460X(02)01530-4

Rao, D.K., 2003, "Recent Applications of Viscoelastic Damping for Noise Control in Automobiles and Commercial Airplanes", Journal of Sound and Vibration, Vol. 262, No. 3, pp. 457-474. doi: 10.1016/S0022-460X(03)00106-8

Renaud, F., Dion, J.-L., Chevallier, G., Tawfiq, I. and Lemaire, R., 2011, "A New Identification Method of Viscoelastic Behavior: Application to the Generalized Maxwell Model", Mechanical Systems and Signal Processing, Vol. 25, No. 3, pp. 991-1010. doi: 10.1016/j.ymssp.2010.09.002

Sjöberg, M., 2002, "On Dynamic Properties of Rubber Isolators", Ph.D. Thesis, Royal Institute of Technology (KTH), Stockholm, Sweden.

Van Gurp, M. and Palmen, J., 1998, "Time-Temperature Superposition for Polymeric Blends", Rheology Bulletin, Vol. 67, No. 1, pp. 5-8.

Vasques, C., Moreira, R. and Rodrigues, J., 2010, "Viscoelastic Damping Technologies – Part I: Modeling and Finite Element Implementation", Journal of Advances Research in Mechanical Engineering, Vol. 1, No. 2, pp. 76-95.

Vinh, T., 1967, "Sur le Passage du Régime Harmonique au Régime Transitoire Viscoélastique", Vol. 3, Mémorial de L'artillerie Française, Saint-Ouen, France.

Williams, M.L., Landel, R.F. and Ferry, J.D., 1955, "The Temperature Dependence of Relaxation Mechanisms in Amorphous Polymers and Other Glass-Forming Liquids", Journal of the American Chemical Society, Vol. 77, pp. 3701-3707. doi: 10.1021/ja01619a008

13

A Six Degrees-of-Freedom Flight Dynamics Simulation Tool of Launch Vehicles

Guilherme da Silveira[1], Valdemir Carrara[2]

ABSTRACT: The use of digital simulation has become an essential activity during the development and operation of launch vehicles, due to the complexity of such systems. Of particular interest is the flight dynamics simulation, which investigates the behavior of the vehicle in flight subjected to forces and moments. This work presents a simulation tool suited to perform six degrees-of-freedom flight dynamics investigations of launch vehicles. Developed at the Instituto Nacional de Pesquisas Espaciais (INPE) and the Instituto de Aeronáutica e Espaço (IAE) in Brazil, the tool was implemented following the requirement for flexibility, so that it can be used to simulate different types of launch vehicles. The assessment of the vehicle performance and the vehicle payload capacity are some examples of analysis that can be performed with the tool. A modular programming strategy was employed to assure the tool flexibility. Therefore, the models presented in the tool were implemented as separate modules. The combination of these models can originate flight models of different launch vehicles. Two flight scenarios of Brazilian rockets were simulated and the results were verified against simulation tools already employed by aerospace community. The developed tool showed good agreement with respect to the simulators used to perform the comparison.

KEYWORDS: Simulation, Flight dynamics, Trajectory, Launch vehicles, Rockets.

INTRODUCTION

With the ever-increasing power of computers, digital simulation has become an essential tool in today's engineering development. The use of digital simulation allows the reduction of both risks and costs associated with a project, as well as the assessment of different configurations of the product in a relative easy way (Steele *et al.*, 2002). Once the simulation tool is validated, several activities can be performed, such as: definition of performance requirements; assessment of different configurations; test support; reduction of testing costs; investigation of inaccessible environments and analysis of subsystems interactions (Zipfel, 2007).

Launch vehicles are objects designed to launch into space instruments like probes and satellites. Due to their great complexity, the use of digital simulation during the design and operation of these systems is essential. A very important subject associated with launch vehicles and generally addressed with digital simulation is the so-called flight dynamics, which studies the motion of the vehicle in space. Flight dynamics investigations are present during all the life cycle of a launch vehicle, assisting actions from the vehicle design until the analysis of actual flight data (Sarma *et al.*, 1978).

The simulation tool used to perform flight dynamics analysis should be able to predict the way the vehicle moves in space as a function of the forces and moments acting on it. This objective is achieved with the use of an adequate dynamic model (a set of equations of motion capable of evaluating the vehicle's position and velocity) and suitable techniques to predict the efforts acting on the vehicle during the flight. The dynamic model is generally described by a set of differential equations and, due to their complexity, these equations are normally solved numerically.

1.Departamento de Ciência e Tecnologia Aeroespacial - Instituto de Aeronáutica e Espaço – Divisão de Sistemas Espaciais - São José dos Campos/SP – Brazil
2.Instituto Nacional de Pesquisas Espaciais – Coordenação Geral de Engenharia e Tecnologia Espacial – Divisão de Mecânica Espacial e Controle – São José dos Campos/SP – Brazil.

Author for correspondence: Guilherme da Silveira | Instituto de Aeronáutica e Espaço – Divisão de Sistemas Espaciais | Praça Marechal Eduardo Gomes, 50 – Vila das Acácias | CEP: 12.228-901 – São José dos Campos/SP – Brazil | Email: guilhermegs@iae.cta.br

The flight dynamics investigations may have different levels of sophistication, depending on the vehicle's characteristics that are taken into consideration when deriving the dynamic model and the amount of information available about the vehicle's design and properties, like mass, inertia, drag coefficients and so on. As an example, the vehicle can be considered a point mass, where only the translational dynamics is included; a rigid body, where the relative motion of its parts are neglected; a body with a rigid structure and variable mass; and a body with flexible structure and variable mass.

Another aspect that influences the choice of the model is related to the type of analysis it is suited for. Generally, the flight dynamics investigations can be divided into two related major groups (Greensite, 1967; Sarma *et al.*, 1978). The long-period dynamics is concerned with factors such as vehicle's capability to accomplish a specific mission, payload capacity and trajectory dispersions. In this case, effects of non-spherical rotating Earth and variable mass of vehicle should be included. The short-period dynamics, in turn, is concerned with phenomena with relatively small occurrence interval, like the separation process between stages or the vehicle's oscillation about the center of mass considering fuel sloshing and bending.

The development of a computational tool for flight dynamics simulation usually requires a substantial amount of time. In order to reduce time and cost associated with this activity, it has become a common practice to implement generic and flexible simulators, which could be used during the design and operation of different launch vehicles (Ippolito and Pritchett, 2000; Steele *et al.*, 2002).

There are several examples of launch vehicle simulators developed in space agencies, like NASA and ESA, and also in companies around the world. The Program to Optimize Simulated Trajectories (POST) and the Optimization of Trajectories by Implicit Simulation (OTIS) are examples of powerful and flexible tools used by NASA to optimize and simulate the flight trajectory of launchers, and some of its applications can be found in the works of Albertson *et al.* (2012) and Falck and Gefert (2007). The Marshall Aerospace Vehicle Representation in C II (MAVERIC II), also developed by NASA, is a modularized high-fidelity simulation software and appears in the work of Lu and Rao (2004). ESA, in turn, has developed a generic multibody flight simulator, capable of performing extensive launcher dynamics analysis (Baldesi and Toso, 2012). As an example of a commercial simulation tool, the company Astos Solutions has been developing the AeroSpace Trajectory Optimization Software (ASTOS®), a modularized software capable of performing trajectory simulation and optimization of launch vehicles (Cremaschi *et al.*, 2010). Finally, in the work of Betts *et al.* (2007), it can be found an example of a simulator constructed using the modularized characteristic of Simulink®, which facilitates modifications of the vehicle model as necessary.

This work presents a launch vehicle simulation tool developed at INPE and IAE, two active space research institutes in Brazil. The tool was developed in order to enhance Brazilian capability and autonomy in launch vehicles simulation. The requirement for developing a flexible tool capable of simulating different vehicles and different missions was adopted. The tool is referred to as Rocket Trajectory Simulator (RTS). The results obtained with RTS were verified against other simulation tools already used by aerospace community.

RTS FEATURES

RTS was implemented using MATLAB®. It simulates the launch vehicle flight in six degrees-of-freedom, from launch to reentry or orbit insertion. The vehicle is considered a body with rigid structure and variable mass. Several phenomena that affect the vehicle's long-period dynamics can be considered in the simulation, such as the variation of mass and inertia, the aerodynamic and propulsive efforts that act on the vehicle, the control systems dynamics, the Earth's geometry and rotational motion, the presence of wind etc.

The RTS simulated trajectory is divided in phases. The definition of the phases is arbitrary, and phase changing is generally related to some abrupt alteration in vehicle properties, such as a change in vehicle mass due to jettisoning of part of its structure, or a change in propulsion thrust due to the ignition of a motor, as well as some change in environment, for example, during the vehicle exiting from the atmosphere. The different events that define the end of the phases are called "sequence of events". In RTS, the sequence of events corresponds to a set of time instants (all the events must have a known instant of occurrence).

RTS was implemented to be a generic and flexible flight simulation tool, allowing the simulation of different types of launch vehicles. This flexibility is ensured by a modular programming strategy, where each model is implemented as a separate module. The choice of different models can originate flight models for different launch vehicles.

Basically, RTS comprises a library of models, with all the available models that can be used to compose the vehicle flight model to be simulated and the main module, responsible for performing the trajectory integration. Figure 1 shows these components. A graphical user interface was also implemented to allow the creation of the whole vehicle flight model in an easy way.

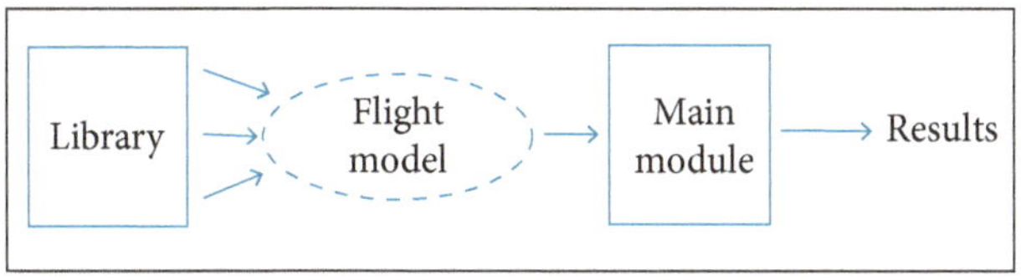

Figure 1. RTS library and main module.

LIBRARY OF MODELS

RTS library comprises adequate models to simulate the flight dynamics of launch vehicles. Several aspects of this dynamics, mainly those related to the long-period dynamics, can be investigated using these models. The library is divided into four categories: Dynamics, Vehicle Subsystems, Auxiliary Tools and Environment Models.

DYNAMICS

This category comprises mathematical models of the vehicle dynamics, which corresponds to sets of differential equations that describe the motion of the vehicle in space. Its objective is to determine the derivatives of the state variables which will be numerically integrated, resulting in the position and velocity of the vehicle along the trajectory. There are three mathematical models of the dynamics implemented in RTS: one for the simulation of the vehicle motion in space, and two for the launch phase.

The motion of the vehicle in space is described by 12 state variables. The translational motion is described by the vehicle velocity vector and position with respect to the surface of the Earth. The rotational motion is described by the vehicle angular velocity vector with respect to an inertial frame and by the vehicle orientation, or attitude, with respect to the local vertical reference frame (to be defined shortly).

The dynamic equation of the translational motion with respect to a non-inertial reference system fixed on Earth, stated in Cornelisse *et al.* (1979) and Mooij (1997), is:

$$\left.\frac{d\vec{V}_{rel}}{dt}\right|_B = \frac{\vec{F}_{total}}{M} - \vec{\Omega}\times\left(\vec{\Omega}\times\vec{R}_{ec}\right) - \left(\vec{\omega}+\vec{\Omega}\right)\times\vec{V}_{rel} \quad (1)$$

where:

$\vec{V}_{rel}$: vehicle velocity with respect to Earth; $\vec{R}_{ec}$: vehicle position with respect to the center of the Earth; $\vec{\Omega}$: Earth's angular velocity; $\vec{\omega}$: vehicle angular velocity with respect to an inertial frame; M: vehicle mass; $\vec{F}_{total}$: total force that acts on the vehicle; $|_B$: denotes the derivative as viewed by an observer in the rotating system attached to the vehicle.

The total force results from gravitational, aerodynamic and propulsive effects, and its value depends on the models considered in the simulation.

The vehicle position is specified in polar coordinates: its distance from the center of the Earth, its longitude and its geocentric latitude. The kinematic equations for the variation of these parameters, presented in Mooij (1997), are:

$$\frac{dR_{ec}}{dt} = -V^V_{relz} \quad (2)$$

$$\frac{d\mu}{dt} = \frac{1}{R_{ec}\cos\lambda_{ec}} V^V_{rely} \quad (3)$$

$$\frac{d\lambda_{ec}}{dt} = \frac{1}{R_{ec}} V^V_{relx} \quad (4)$$

where:

R_{ec}: magnitude of R_{ec}; μ: vehicle longitude on Earth; λ_{ec}: geocentric latitude; V^V_{relx}, V^V_{rely} and V^V_{relz}: components of the relative velocity $\vec{V}_{rel}$ written in the local vertical coordinate system F_V.

This system, shown in Fig. 2, has its origin at the vehicle's center of mass. The X^V axis is located at the local horizontal plane, pointing north, the Y^V axis is located at the local horizontal plane, pointing east, and Z^V axis points to the center of the Earth.

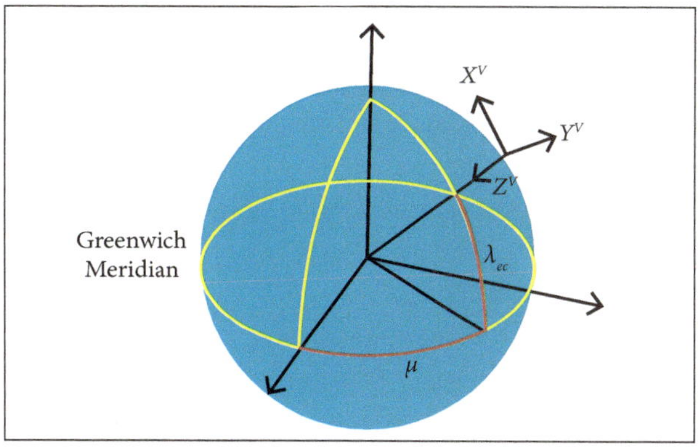

Figure 2. Local vertical coordinate system.

The dynamic equation for the rotational motion is stated in Cornelisse *et al.* (1979) and Mooij (1997) as:

$$I \cdot \left.\frac{d\vec{\omega}}{dt}\right|_B = \vec{M}_{total} - \vec{\omega} \times (I \cdot \vec{\omega}) \quad (5)$$

where:

I: vehicle inertia; $\vec{M}_{total}$: total moment that acts on the vehicle, resulting from aerodynamic and propulsive effects, depending on the models considered in the simulation.

Finally, the vehicle attitude is described by three Euler angles, and their kinematic equations, presented in Hughes (2004), are:

$$\frac{d\theta}{dt} = q_l \frac{\cos\phi}{\cos\psi} - r_l \frac{\sin\phi}{\cos\psi} \quad (6)$$

$$\frac{d\psi}{dt} = q_l \sin\phi + r_l \cos\phi \quad (7)$$

$$\frac{d\phi}{dt} = p_l - q_l \tan\psi \cos\phi + r_l \tan\psi \sin\phi \quad (8)$$

where:

θ, ψ and ϕ: angles that relate the body frame F_B to the local vertical frame F_V. These angles are defined by a sequence of rotation 2-3-1, where θ is the first rotation angle, ψ is the second and ϕ is the third angle; p_l, q_l and r_l: components of the angular velocity of the vehicle with respect to the local frame F_V which are obtained from the subtraction of the inertial angular velocity of frame F_V, which depends on the velocity of the vehicle with respect to the Earth, from the inertial angular velocity of the vehicle.

Besides the model for the motion in space, RTS has two additional dynamic models: one for rail launching simulation, in which the vehicle has one degree-of-freedom while in contact with the rail, and the other for launching simulation from launch pad, in which the relative velocity remains zero while the vehicle is in contact with the pad.

VEHICLE SUBSYSTEMS

The Vehicle Subsystems models describe the vehicle properties during the flight. This category is subdivided into mass properties, propulsive, aerodynamic and control system models.

The **mass properties models** simulate the way the mass properties of the vehicle vary during the trajectory, including the determination of the mass, the inertia tensor, the position of the center of mass, the mass flow and the vehicle inertia variation rate. RTS has a model for constant mass properties, which can be used when there are no active propulsion system and no jettisoning of structural mass, and one model for variable mass properties, for the phases of the trajectory when the vehicle loses mass due to an active propulsion system. Usually, the jettisoning of some structural part of the vehicle is considered an event that separates two consecutive phases of the trajectory.

The **propulsive models** have the objective of simulating the vehicle propulsion systems, by computing the forces and moments generated by these systems. RTS has models for rocket motors with fixed and gimbaled thrust, spin-up motors for inducing a roll velocity to the vehicle, and a generic model which determines the propulsive force and moment directly from user table interpolation as a function of time.

The **aerodynamic models** have the objective of calculating the aerodynamic force and moment that act on the vehicle. A mathematical model which computes these efforts using aerodynamic coefficients was implemented in RTS, a common approach used in several simulators. The model is linear (therefore valid only for low angles of attack) and the longitudinal axis of the vehicle is considered a symmetry axis.

Finally, the **control system models** are responsible for simulating the behavior of the vehicle control systems, by calculating signals or efforts produced by these systems. RTS has control models for pitch, yaw and roll attitude controls using a PID control architecture. Both continuous and discrete control models can be used.

AUXILIARY TOOLS

Auxiliary Tools have no specific objective and can be used to determine parameters not covered by any existing model, such as the position of the vehicle with respect to a radar, or the impact point of a jettisoned part of the vehicle's structure.

RTS has auxiliary models to predict the impact point of a jettisoned part of the vehicle's structure, to calculate the orbital elements of the vehicle or the payload after orbit injection, and to simulate the behavior of a roll velocity dumper system, known as yo-yo.

ENVIRONMENT MODELS

The objective of the Environment Models is to calculate the environment properties that affect the flight. This category is subdivided into Earth, atmospheric and wind models.

The **Earth models** describe the geometry of the Earth's surface and calculate the gravitational field strength at the vehicle location. RTS has a model for a spherical Earth with homogenous mass distribution and a model for a spheroidal Earth with axisymmetric mass distribution.

The **atmospheric models** calculate the atmosphere properties that affect the flight, including air temperature, pressure and density, the sound speed etc. The U.S. Standard Atmosphere, 1976, was implemented in RTS.

Finally, the **wind models** have the objective of simulating the behavior of the wind as the vehicle moves through the atmosphere. The model implemented in RTS considers a horizontal wind with varying strength and direction as a function of vehicle altitude.

RTS ARCHITECTURE

The creation of a flight model in RTS comprises basically the selection of each model to be used, the definition of a sequence of events, which establishes the several phases of the trajectory, and the definition of the models that will be active in each phase. This process is illustrated in Fig. 3. Once the flight model is ready, the integration of the trajectory equations can be performed by the main module, as shown in Fig. 4. The output results are the several flight parameters like position, velocity and acceleration of the vehicle, the forces and moments from propulsion and aerodynamics etc.

The architecture of RTS was defined to fulfill the requirement for flexibility. During the integration of the trajectory, the successive calls to the models comprised in the flight model obey a predefined sequential order, as shown in Fig. 5. This order follows a logical calculation sequence of the several flight parameters of the simulation, in a way that the information needed by a model has already been calculated by the models executed previously.

The first executed models belong to the environment category and compute the gravitational, atmospheric and wind properties at the vehicle position. Next, the subsystems

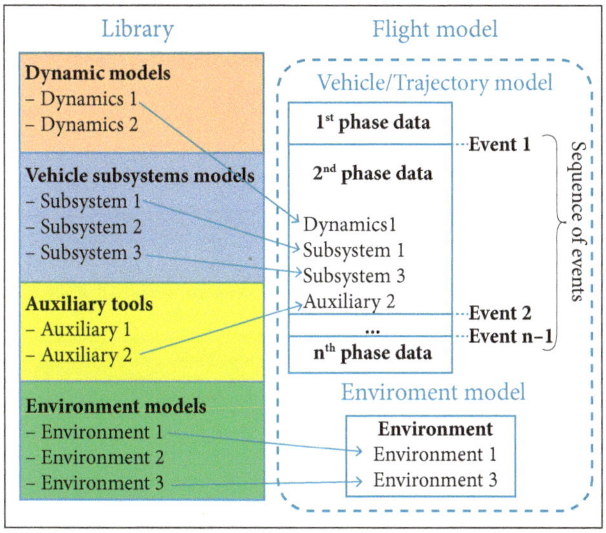

Figure 3. Flight model creation in RTS.

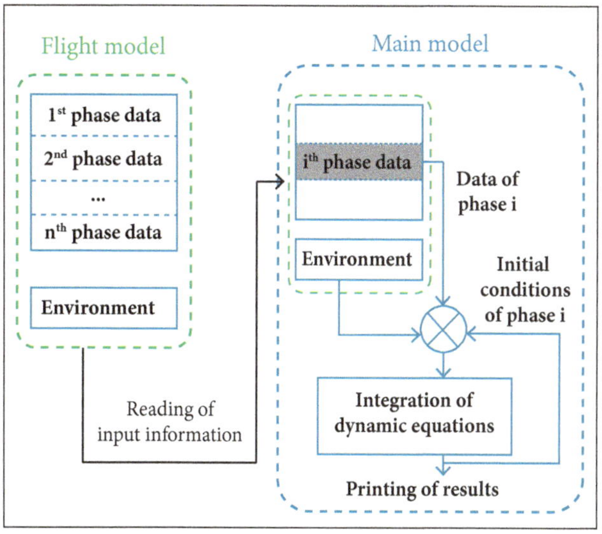

Figure 4. Trajectory integration in RTS.

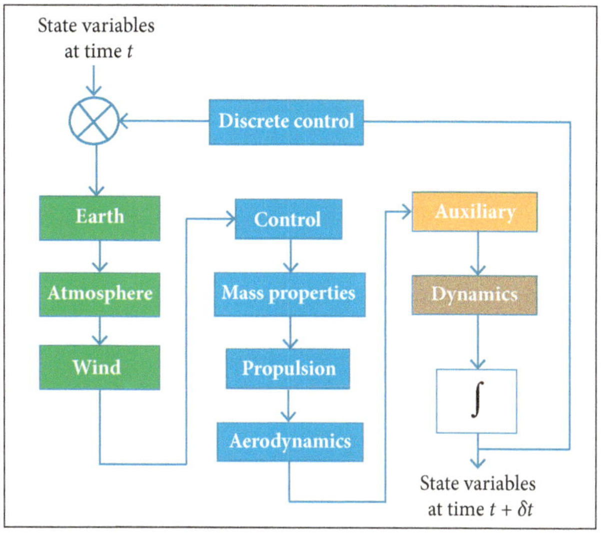

Figure 5. RTS architecture.

models are called, and several vehicle properties are calculated, including control actions and efforts, like nozzle deflection, mass properties, and also aerodynamic and propulsion parameters and efforts. Next, the auxiliary tools can be used to calculate the parameters not comprised by the previous models, like the vehicle position with respect to a radar, for instance. Finally, with all efforts that act on the vehicle, the dynamic model is executed, and the state variables that describe the vehicle motion are integrated. If a discrete control model is included, it is executed outside the integration loop. Except for the dynamic model, more than one model of a specific category can be executed. This is useful when, for example, two kinds of controllers or propulsive systems are active at the same time.

The integration process of the state variables is performed by a MATLAB® build-in function, the ode45. This is an initial value problem solver based on an explicit Runge-Kutta (4,5) formula. It is a one-step solver, meaning that the solution at a time instant depends only on the solution at the immediately preceding instant. The time step, represented by δt in Fig. 5, varies according to the solver capability to find a solution that satisfies the error tolerance criteria, and, for the present application, it generally lies between 10^{-3} and 10^{0}.

The architecture of RTS enables the flight simulation of a variety of launch vehicles, since the appropriate models are present in the library. This flexibility is enhanced with the possibility of the user to add models to the library in a relatively easy way. A key-characteristic of RTS that enables the addition of new models is that all parameters computed during a time step of the integration are stored in a common structure where the access to the contents of the variables is made by their name. By using this structure as input and output arguments to all models, including those added by the user, it permits the models to access all the needed variables and also to store the information calculated so far, making this information available to all other models executed later. This characteristic is illustrated in Fig. 6, where a nozzle deflection signal, β_{pitch}, is calculated as a function of the vehicle attitude, θ, and angular speed, q, and is stored in the same structure containing these parameters. In the figure, k_θ and k_q are constants of the controller.

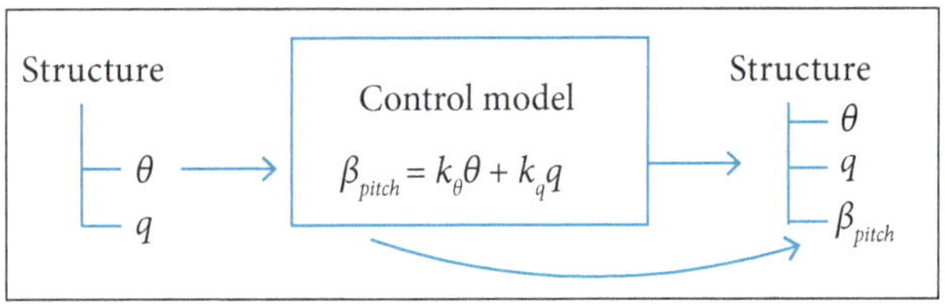

Figure 6. Access and storage of parameters in RTS.

RESULTS AND DISCUSSION

In order to verify the results of RTS, two different flight scenarios were simulated. The results were compared with two different simulation tools available at IAE. The simulations were performed in an Intel® Core 2 Quad 2.66 GHz, with 1.94 GB of RAM.

The first scenario was the simulation of VSB-30, an unguided and aerodynamically stabilized Brazilian sounding rocket composed by two solid propellant stages and a payload. The vehicle is shown in Fig. 7. Launched from rail, it has the capacity of boosting a payload of around 350 kg to an altitude of 300 km (Garcia *et al.*, 2011). The results of RTS were compared with those obtained with the tool Rocket Simulation (ROSI), a six degrees-of-freedom simulator used for flight simulation of Brazilian rockets and already validated by actual flight data. For this scenario, ROSI runtime was about 1 s and RTS runtime was about 59 s.

Figure 8 shows the altitude of the vehicle as a function of ground range. It can be seen the good agreement between the two simulators. The increasing difference that appears after approximately 80 s of flight is the result of a minor difference in the thrust correction method with altitude used in the two simulations. The results of flight speed have also a good agreement, as it can be seen in Fig. 9. In this case, the difference that appears at the end of the trajectory is due to the fact that ROSI simulates the payload reentry considering the drag force, while RTS considers a free flight during reentry. The difference in the results of altitude and speed can be better observed in Fig. 10, that shows this difference in relative terms.

One last result about VSB-30 concerns the roll speed. Figure 11 shows ROSI and RTS results, along with an actual flight data. It can be seen that ROSI and RTS differ after approximately 25 s of flight. This discrepancy is due to a simplification adopted in ROSI that does not consider the moment caused

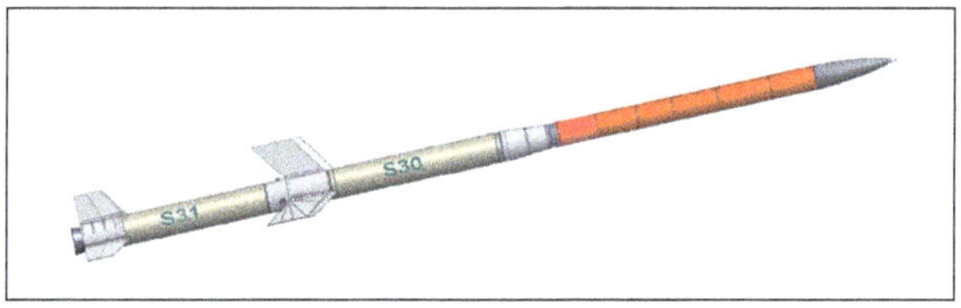

Figure 7. VSB-30 vehicle, adapted from Garcia *et al.* (2011).

by changes in the vehicle inertia. However, for rockets that reach high roll speed, like VSB-30, this moment cannot be neglected, as confirmed by the actual flight data presented in the figure. Is this case, this moment has the same order of magnitude than the aerodynamic moment, shown in Fig. 12.

The second scenario consisted of the VLM-1 simulation, the Brazilian microsatellite launcher, shown in Fig 13. VLM-1 is composed by three solid propellant stages (Agência Espacial Brasileira, 2012). The first and second stages are controlled by gimbaled nozzles, while the third one has no control system. The results of RTS are compared with those obtained with ASTOS®, a commercial simulation and optimization tool. ASTOS® performs a simulation considering three degrees-of-freedom of translation, while the vehicle rotation is considered to happen ideally. For this scenario, ASTOS® runtime was about 9 s and RTS runtime was about 180 s.

The trajectory of the mission SHEFEX III, a German suborbital mission, was simulated here. Figures 14 and 15 show the vehicle position and speed. It can be seen that the results of RTS and ASTOS® are in good agreement. Figure 16 shows the relative difference between the two simulations.

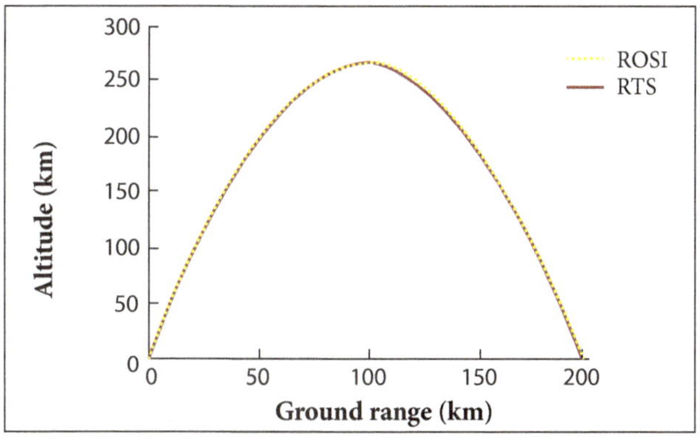

Figure 8. Altitude *versus* ground range of VSB-30.

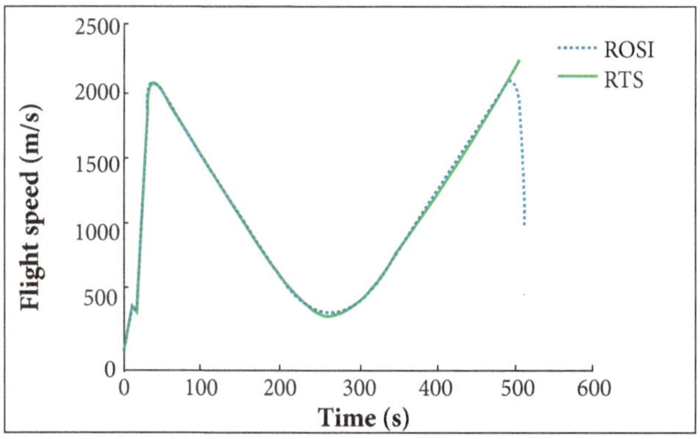

Figure 9. Flight speed *versus* time of VSB-30.

Despite the good agreement of the results just presented, the difference between a three and a six degrees-of-freedom simulation can be better observed when the parameters analyzed are directly related to the vehicle rotation dynamics.

For the VLM-1 trajectory simulation, it is necessary to provide the reference attitude that is expected for the vehicle to follow. Since the simulation performed with ASTOS® considers

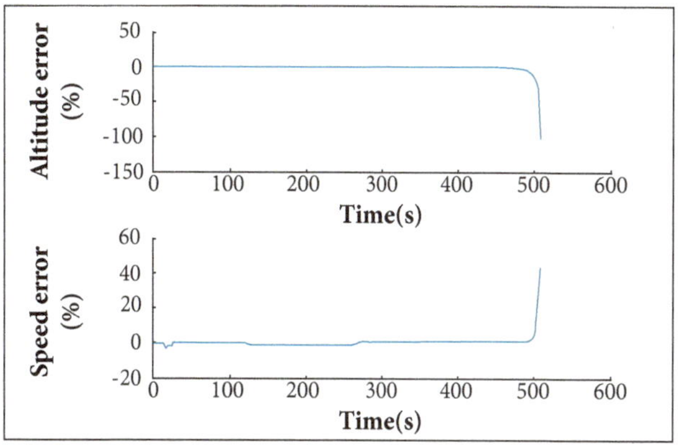

Figure 10. Altitude and flight speed relative error between RTS and ROSI.

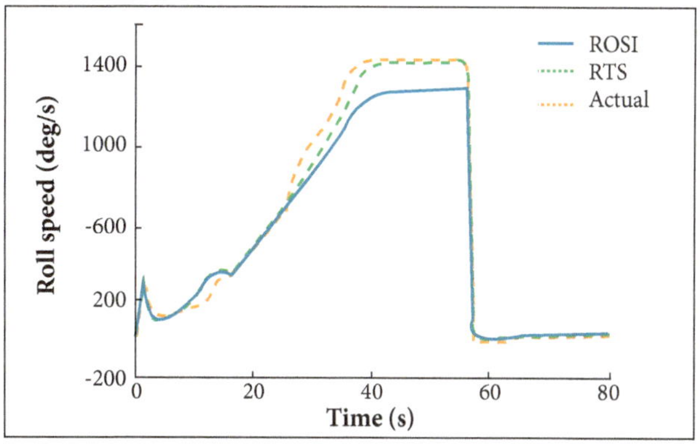

Figure 11. Roll speed of VSB-30.

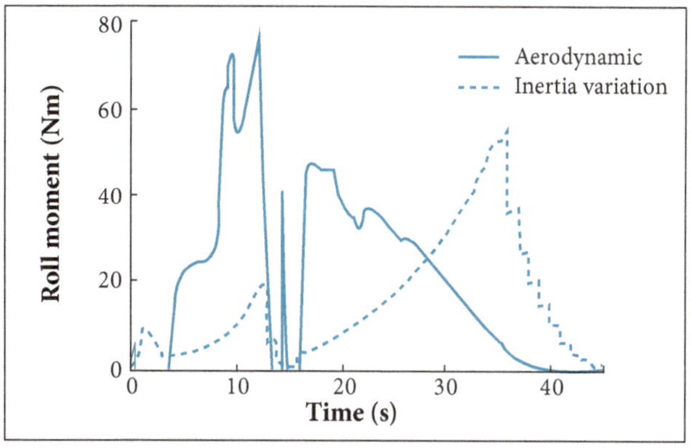

Figure 12. Comparison of roll moment due to aerodynamic and inertia variation effects.

an ideal rotation dynamics, the vehicle attitude at each instant will be exactly equal to the provided reference attitude. In RTS, on the other hand, the vehicle control system is responsible to drive the attitude to the reference trajectory. Figure 17 shows the pitch angle of the vehicle during the flight. Overall, both ASTOS® and RTS pitch attitude are in good agreement, meaning that the control system model used in RTS is adequate to follow the reference for the pitch angle. However, some differences can be seen in specific portions of the trajectory.

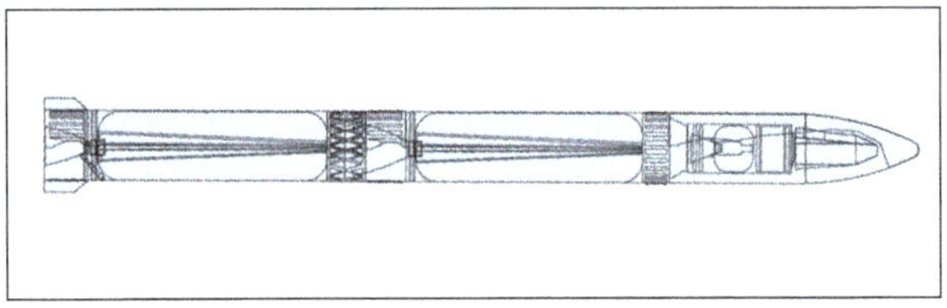

Figure 13. VLM-1 vehicle, adapted from Agência Espacial Brasileira (2012).

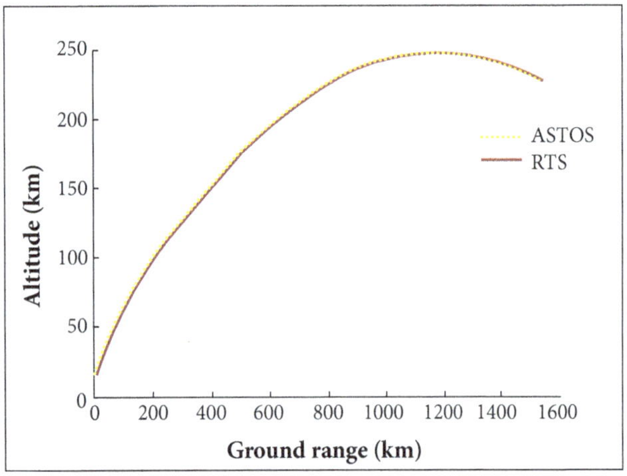

Figure 14. Altitude *versus* ground range of VLM-1.

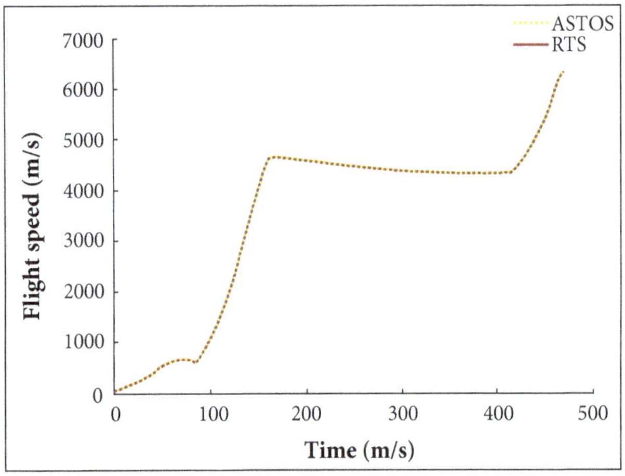

Figure 15. Flight speed of VLM-1.

Figure 18 shows the pitch angle at the beginning of flight. Due to its ideal rotation dynamics when simulated with ASTOS®, the vehicle is capable of instantaneous attitude changes, as indicated by the corners at the pitch curve provided by this tool. In RTS, since the control system is not ideal, the pitch angle slightly deviates from the pitch reference, mainly when the provided reference angle changes abruptly.

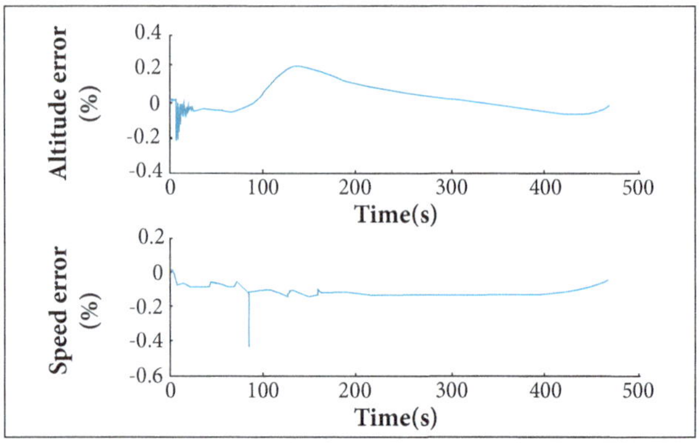

Figure 16. Altitude and flight speed relative error between RTS and ASTOS®.

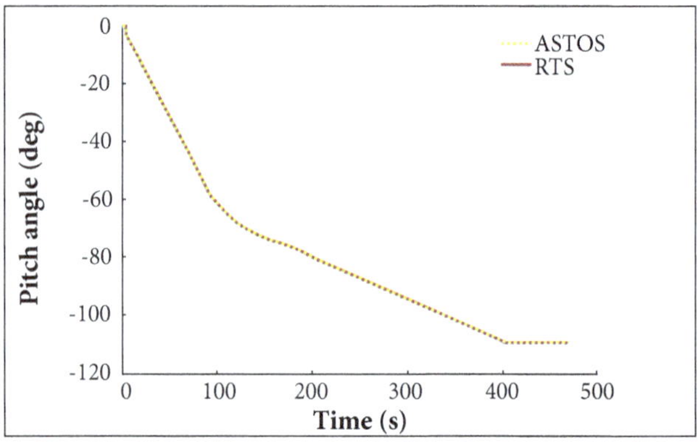

Figure 17. Pitch angle of VLM-1.

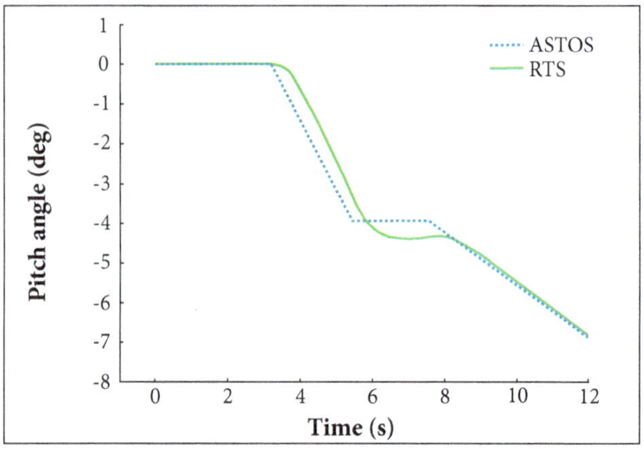

Figure 18. Pitch angle of VLM-1 at the beginning of flight.

CONCLUSION

This paper presented the development of RTS, a simulation tool suited to investigate the flight dynamics of launch vehicles. The tool was implemented considering requirements of flexibility, following a trend observed in research institutes and companies around the world. To fulfil this requirement, a modular programming strategy was used successfully. The models present in the library of RTS and also the tool's architecture have shown to be adequate to simulate the flight of different types of launch vehicles.

The test cases presented have shown a good agreement of RTS with other tools already used to simulate the flight dynamics of launch vehicles. The differences observed in the results of RTS with respect to ROSI and ASTOS® were explained by the differences in the used models. Notably, results concerning the roll speed indicated that the simplification adopted in ROSI, where the moment caused by the inertia variation is neglected, could compromise the simulation results.

ACKNOWLEDGEMENTS

The authors are thankful to INPE and IAE for the opportunity to develop this study.

REFERENCES

Agência Espacial Brasileira, 2012, "Programa Nacional de Atividades Espaciais, PNAE: 2012-2021", Ministério da Ciência, Tecnologia e Inovação, Agência Espacial Brasileira, Brasília, Brazil.

Albertson, C., Tartabini, P.V. and Pamadi, B.N., 2012, "End-to-End Simulation of Launch Vehicle Trajectories Including Stage Separation Dynamics", Proceedings of AIAA Atmospheric Flight Mechanics Conference, Minneapolis, USA.

Baldesi, G. and Toso, M., 2012, "European Space Agency's Launcher Multibody Dynamics Simulator Used for System and Subsystem Level Analyses", CEAS Space Journal, Vol. 3, No. 1-2, pp. 27-48. doi: 10.1007/s12567-011-0023-9

Betts, K.M., Rutherford, R.C., McDuffie, J., Johnson, M.D., Jackson, M. and Hall, C., 2007, "Time Domain Simulation of the NASA Crew Launch Vehicle", Proceedings of AIAA Modeling and Simulation Technologies Conference and Exhibit, Hilton Head, USA.

Cornelisse, J., Schöyer, H. and Wakker, K., 1979, "Rocket Propulsion and Spaceflight Dynamics", Pitman, London, United Kingdom.

Cremaschi, F., Huertas, I., Wiegand, A., Jung, W. and Scheuerpflug, F., 2010, "6-Dof Trajectory Simulation and Optimization for Sounding Rockets", Proceedings of International Conference on Astrodynamics Tools and Techniques, Madrid, Spain.

Falck, R. and Gefert, L., 2007, "Crew Exploration Vehicle Ascent Abort Trajectory Analysis and Optimization", Proceedings of AIAA Guidance, Navigation and Control Conference and Exhibit, Hilton Head, USA.

Garcia, A., Yamanaka, S.S.C., Barbosa, A.N., Bizarria, F.C.P., Jung, W. and Scheuerpflug, F., 2011, "VSB-30 Sounding Rocket: History of Flight Performance", Journal of Aerospace Technology and Management, Vol. 3, No. 3, pp. 325-330. doi: 10.5028/jatm.2011.03032211

Greensite, A., 1967, "Short Period Dynamics", NASA, Marshall Space Flight Center, Huntsville, USA.

Hughes, P.C., 2004, "Spacecraft Attitude Dynamics", Dover, Mineola, USA.

Ippolito, C.A. and Pritchett, A.R., 2000, "Software Architecture for a Reconfigurable Flight Simulator", Proceedings of AIAA Modeling and Simulation Technologies Conference, Denver, USA.

Lu, P. and Rao, P.P., 2004, "An Integrated Approach for Entry Mission Design and Flight Simulations", Proceedings of AIAA Aerospace Sciences Meeting and Exhibit, Reno, USA.

Mooij, E., 1997, "The Motion of a Vehicle in a Planetary Atmosphere", Delft University of Technology, Faculty of Aerospace Engineering, Delft, The Netherlands.

Sarma, I., Prasad, U. and Vathsal, S., 1978, "Computer Simulation Methods for Launch Vehicle Mission and Control Problems", Proceedings of the Indian Academy of Sciences Section C: Engineering Sciences, Vol. 1, No. 4, p. 423-440. doi: 10.1007/BF02842910

Steele, M.J., Mollaghasemi, M., Rabadi, G. and Cates, G., 2002, "Generic Simulation Models of Reusable Launch Vehicles", Proceedings of Winter Simulation Conference, San Diego, USA.

Zipfel, P., 2007, "Modeling and Simulation of Aerospace Vehicle Dynamics". 2nd edition, AIAA, Reston, USA.

14

Burning Rate Measurement of KNSu Propellant Obtained by Mechanical Press

Antônio Carlos Foltran[1], Diego Fernando Moro[1], Nicholas Dicati Pereira da Silva[1], Ana Eliza Gonçalves Ferreira[1], Luciano Kiyoshi Araki[1], Carlos Henrique Marchi[1]

ABSTRACT: The potassium nitrate (KNO_3)/sucrose ($C_{12}H_{22}O_{11}$) propellant, known as KNSu, is traditionally used in rocket studies by amateur groups. Performance tests of KNSu, whose composition is 65 wt% KNO_3 and 35 wt% $C_{12}H_{22}O_{11}$, cold-manufactured by mechanical press, are presented in this paper. The study determines the behaviour of density (ρ) and burning rate (r) as function of compression pressure (P_c) for the propellant grain manufacturing. Forty-four samples were prepared and tested, for seven different values of P_c. It was observed that ρ and r depend on P_c, according to quadratic polynomial functions, and the mass flux per unit area is a constant.

KEYWORDS: Potassium nitrate, Sucrose, KNSu, Experimental rocket motor, Model rocket, Solid propellant.

INTRODUCTION

The potassium nitrate (KNO_3)/sucrose ($C_{12}H_{22}O_{11}$) propellant, known as KNSu, is traditionally used in rocket studies by amateur groups. Both components are easy to buy, their combustion products are non-toxic and this propellant can produce relatively high specific impulse, with values higher than the ones obtained by black gunpowder commonly used in model rockets (Leslie and Yawn, 2002).

Traditional methods employed in the production of propellant grains involve fusion (recrystallization) process. In this case, both propellant components are grinded and mixed. After, the mixture is heated until the sucrose fusion around the nitrate grains, forming a doughy mixture which can be placed in molds with a proper propellant grain design, and is cooled in sequence (Vyverman, 1978; Nakka, 1984). This methodology produces high density propellant grains (Nakka, 1984); there are, however, risks associated to the propellant combustion by the heating process. Many amateur rocket builders, who use this methodology, consider it secure; nevertheless, it is indispensible the use of equipments that provide a slow and gradual heating process, as well as the control of temperature along the fusion process (Nakka, 1997).

In this paper, however, an alternative methodology for cold manufacturing was studied. It involves the grinding and mixture of components, as well as the preparation of propellant grain by compaction of components in a die with the use of a hydraulic press machine. Cold manufacturing, in different ways, was used by Marchi *et al.* (1990), Moro (2013) and Thompson (2014). According to Nakka (1997), heated propellant is inherently more hazardous than cool propellant

1. Universidade Federal do Paraná – Departamento de Engenharia Mecânica – Curitiba/PR – Brazil.

Author for correspondence: Luciano Kiyoshi Araki | Universidade Federal do Paraná – Depto. Engenharia Mecânica | Caixa Postal: 19.040 | CEP: 81.531-980 Curitiba/PR – Brazil | Email: lucaraki@ufpr.br for two reasons: (1) less energy is required to ignite it; and (2) burning rate of hot propellant is much greater, perhaps by a factor of five.

The aim of this paper was to provide the propellant grains production in a simple way, consisting only on grinding and mixing the components, in absence of a heating process. The used security recommendations were basically the ones relative to chemical products manipulation (Wallace, 1995), described by Nakka (1997). Furthermore, a study of how to determine the dependence of density on compression pressure was made, comparing different propellant batches.

METHODOLOGY

SAMPLE PREPARATION

The used propellant components for sample manufacture were potassium nitrate and refined sugar. Each component was milled in an individual mill with rotary blades. Each mill was carried up to 90% of maximum volume and used by 30 s.

After grinding, the components were stored in individual pots. The chemical composition adjustment was made by measuring the components' masses in a scale with 0.01 g resolution. The components were then placed in a third hermetic pot and each 100 g of propellant were manually shaken for 15 min.

Samples present a parallelepipedal shape, with approximately 60 mm x 20 mm in transversal section and height between 7.8 and 9.4 mm dependent on the applied compression pressure. All samples were manufactured with 16 g of powdered components. The experiment consists of nine tests. Each test was composed by five samples, except by the first one, which was made with four samples, totalizing 44 samples. Moreover, the propellants for these nine tests were prepared in batches (*i.e.* different mixture preparation operations): for tests 1 to 5, five different propellant batches were prepared; for tests 6 to 9, only one batch was made.

The applied forces on each sample, for a given test, were 2, 4, 6, 8 and 10 tons-force. Exceptions were: (1) the 1st test, for which one sample was compressed with 3 tons-force, two with 6 tons-force and one with 9 tons-force; and (2) the 6th test, for which all five samples were compressed using 6 tons-force. Because of this, seven nominal pressures were generated for different compression pressures, approximately, equal to 16, 24, 32, 48, 64, 72 and 80 MPa.

The following forces were neglected, since they are relatively small: friction forces of the hydraulic cylinder gasket; the deformation forces related to the return spring of the press hydraulic cylinder; and the friction forces between the punch and the die. Then, considering the pressurized system in equilibrium, the following relation can be applied:

$$P_c = p_2 = \frac{p_1 A_1}{A_2} = \frac{F_1}{A_2} \tag{1}$$

where:

p_1: pressure inside the hydraulic system; P_c or p_2: compression pressure; A_1 and A_2: areas of the hydraulic cylinder piston and the sample under compression pressure, respectively; F_1: force produced by the pressure of the hydraulic system acting on the piston area.

DENSITY EVALUATION

After the press process, the samples were removed from the die and cleaned with a cloth in order to remove small detached grains. For each sample, measurements consisted of: length (once), width (at three different positions along the sample length) and thickness (at six different positions along the sample length). These several measurements of width and thickness were made since there were small imperfections in both punch and die.

From the average of all dimension values, it was possible to evaluate the volume of each sample and to obtain the density as a ratio of the measured mass by volume. Since the KNSu samples are: (1) highly hygroscopic and (2) porous, it was preferable to evaluate the volume of each sample in a stereometric form in spite of immersing samples in a liquid, which is a common practice.

The obtained value for density could be compared to the theoretical density ρ_t, which considers the mass fraction of components and their respective densities as crystalline solids; in other words, it supposes a sample without any porosity, so it stands an upper limit to density to be obtained in tests:

$$\rho_t = \frac{1}{\frac{x_{SU}}{\rho_{SU}} + \frac{x_{KN}}{\rho_{KN}}} \cong 1888 \frac{kg}{m^3} \tag{2}$$

where:

ρ_{KN} and ρ_{SU}: density values of potassium nitrate (2,109 kg/m^3) and sucrose (1,581 kg/m^3), respectively; x_{KN} and x_{SU}: mass fractions of nitrate and sucrose (fixed as 0.65 and 0.35, in this work), in this order.

Atmospheric pressure variation was not significant for all tests, so it was irrelevant for analysis. Tests were conducted at atmospheric pressure specially by the fact that, under these conditions, burning rate presents the smallest possible values, what can be of useful for the design of small experimental rocket motors.

Otherwise, room temperature presented variation for all tests: temperature range of 20.0 to 28.0°C with average temperature of 25.4°C. However, for each test, samples were burnt basically at the same room temperature.

BURNING RATE MEASUREMENTS

All burn out process occurred in open and well ventilated conditions. A 30 fps video camera was placed orthogonally to the flame propagation direction, in order to minimize parallaxes error. A graduated scale was placed in parallel to each sample to enable the measurement of the burning rate at atmospheric conditions in Curitiba, Brazil, 916 m and 90.6 kPa.

All samples presented five faces isolated by a thick layer of glue, which allowed the flame propagation in only one surface. The chosen surface was the one with the smallest area, on which a thin layer of glue was placed. Black gunpowder was placed on it, as a way to guarantee a simultaneous burning ignition of all surface. For safety reasons, a fuse was used to ignite the powder.

The burning rate was obtained by measuring the time interval needed by the flame to cover the distance between two arbitrary points on each sample, namely: from 0 to 20 mm, from 20 to 40 mm, from 40 to 60 mm, from 10 to 30 mm, from 30 to 50 mm, from 0 to 30 mm, from 30 to 60 mm and from 0 to 60 mm. These sets of points were chosen to verify if the burning rate varies locally at the sample length.

According to Douglass *et al.* (1972) and Fordham (1980), most solid rocket propellants present a dependence of burning rate on pressure at combustion chamber, which can be described by:

$$r = a\,P_0^n \qquad (3)$$

where:

r: burning rate (mm/s); a: constant dependent on the propellant (mm/MPa·s); P_0: pressure at combustion chamber (in MPa); n: pressure coefficient (non-dimensional).

However, depending on the author, other sets of measurement units for both pressure and burning rate are employed (Vyverman, 1978; Nakka, 1984; Marchi *et al.*, 1990; Stancato and Miraglia, 1997; Stancato *et al.*, 2000). Nakka (1984, 1997) used Eq. 3 to determine, experimentally, the parameters a and n for KNSu propellant. In this work, only the burning rate for local atmospheric pressure (90.6 kPa) was evaluated.

RESULTS AND DISCUSSION

Figure 1 shows the burn process of a typical sample. At the two lower frames of Fig. 1, the burning surface is not clear because of the fact that traces of glue layer did not burn as easily as the propellant. Since the flame front is not completely homogeneous, it is difficult to decide exactly when it achieves the initial and final positions. Because of this, each measurement was repeated three times. This strategy allowed the evaluation of uncertainty ranges for the burning rate, based on t-Student distribution with 95% confidence level. The highest interval was observed for sample 6 and presented 0.20 mm/s of amplitude, *i.e.* 8.4% with respect to the average of sample 6.

For some samples, the burning rate was spoiled by the inhibitor inefficiency, which conducted to the burn process in faces other than the one activated by the black gunpowder. Sometimes the flame did not uniformly begin in all activated surface, as expected, or it presented some delay points. When

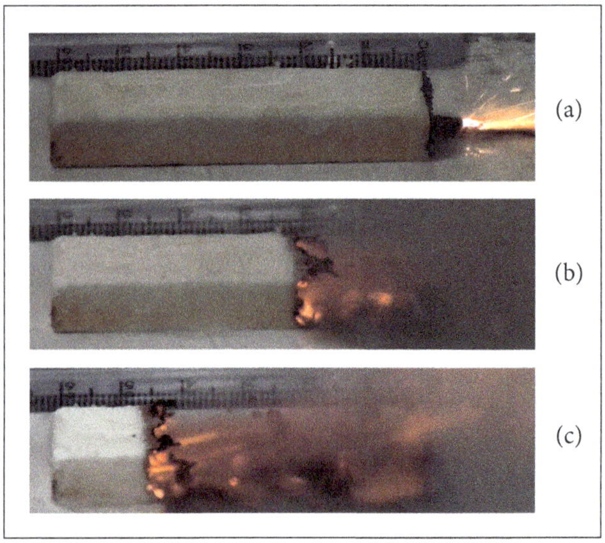

Figure 1. Stages of a sample burn: (a) burn ignition with a fuse; (b) and (c) flame front propagation.

these phenomena were observed, the flame front propagation was not parallel to the original surface activated with gunpowder. Because of this, from 44 samples, 44 density and only 33 burning rate measurements were obtained. Most problems were observed in initial tests, occasions in which the samples were not efficiently inhibited.

As expected, density values increase with higher compaction pressures. Nevertheless, they are lower than the ones obtained by Nakka (1997) and most ones obtained by Vyverman (1978), which used the fusion process. However, the highest values obtained in this work (at about 80 MPa) are compatible to the ones observed by Vyverman (1978). The observed values are higher than the ones obtained by using similar production processes, such as presented by Marchi *et al.* (1990) and one of the authors (Moro, 2013) for rocket motors made using PVC tubes (observing that, in this case, the burn process can occur at slightly higher pressures than the atmospheric one). Marchi *et al.* (1990) produced the propellant in a similar way to the present in this work, however, without using a hydraulic press machine; in this case, the propellant was mixed with acetone in order to obtain a kind of paste which was poured into the motor, manually pressed and then placed to dry.

From experimental results, three second-order polynomial functions could be adjusted to data: (1) correspondent to tests 1 to 5, *i.e.* data from five different batches — Eq. 4; (2) correspondent to tests 6 to 9, *i.e.* data from only one batch — Eq. 5; and (3) the general behaviour of all experimental data — Eq. 6. All polynomials describe the average density (ρ) as function of compaction pressure (P_c):

Tests 1 to 5:

$$\rho = -0.0356P_c^2 + 7.19P_c + 1311 \quad (4)$$

Tests 6 to 9:

$$\rho = -0.0469P_c^2 + 8.11P_c + 1288 \quad (5)$$

All tests:

$$\rho = -0.0396P_c^2 + 7.51P_c + 1302 \quad (6)$$

For these three equations, P_c and ρ are given, respectively, in MPa and kg/m^3.

Although the three curves given by Eqs. 4 to 6 are not exactly coincident, they present the same general behaviour. Experimental uncertainties for each group of tests were also evaluated and it was verified that differences among all results are not significant. For each group of batches (1 to 5 and 6 to 9), results are over and under the respective polynomial fit equation. Based on these facts, it was preferable to use the average behaviour of all tests, *i.e.* Eq. 6.

The behaviour of Eq. 6 can also be graphically seen in Fig. 2, with experimental results for all samples of all tests. Figure 2 also includes results from other authors and the theoretical density, described by Eq. 2.

Tests 6 to 9 were prepared from the same propellant batch; because of this, similar results for these tests were expected. However, these tests also presented a similar behaviour to the first five ones. Test 6, for which all samples were pressed with 48 MPa, produced samples with almost the same density values and, because of this, there are several overlying marks of these five samples in Fig. 2.

Values of all samples were grouped according to compaction pressure and, based on t-Student distribution with 95% confidence level, uncertainty intervals were obtained. They are shown in Fig. 2 and the highest value was observed for the 64 MPa group, for which the amplitude was 180 kg/m^3, *i.e.* 11% related to the average value obtained by Eq. 6. A similar analysis, made only for samples of test 6, presented an uncertainty interval of 21 kg/m^3, *i.e.* 1.3% related to the average value obtained by Eq. 6.

The burning rate was measured for different regions of samples and the results were analyzed. All samples exhibited small variations in local burning rate; however, none evident behaviour was observed and, because of this, the average burning rate observed from 0 to 60 mm was used as reference for all analyses of this paper.

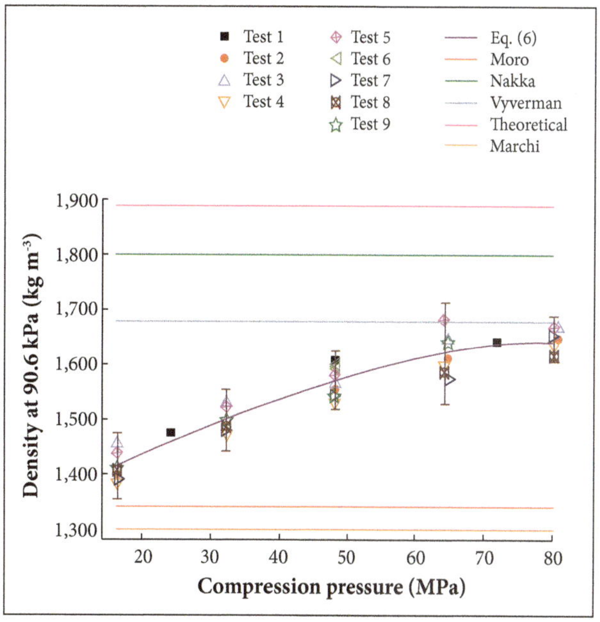

Figure 2. Density *versus* compression pressure.

From experimental data, a second-order polynomial function could be adjusted to evaluate the average burning rate (r) as function of compression pressure (P_c) for: (1) tests 1 to 5 — Eq. 7; (2) tests 6 to 9 — Eq. 8; and all experimental data — Eq. 9:

Tests 1 to 5:

$$r = 3.15\text{x}10^{-5}P_c^{\,2} - 9.14\text{x}10^{-3}P_c + 2.88 \quad (7)$$

Tests 6 to 9:

$$r = 4.18\text{x}10^{-5}P_c^{\,2} - 9.38\text{x}10^{-3}P_c + 2.75 \quad (8)$$

All tests:

$$r = 5.68\text{x}10^{-5}P_c^{\,2} - 1.13\text{x}10^{-2}P_c + 2.86 \quad (9)$$

where:

P_c and r are given, respectively, in MPa and mm/s.

Although the average burning rate equation correctly predicts all experimental data (including experimental uncertainties), the burning rates present a kind of dependence on the batch: tests conducted with samples from batches 1 to 5 presented higher values for burning rate when compared to tests conducted with the samples from the other batch. It is emphasized, however, that the average behaviour, represented by Eq. 9, predicts well all experimental data.

The confidence intervals shown in Fig. 3 were evaluated based on t-Student distribution with 95% confidence level. The amplitude of the major interval, for all tests, was observed for the sample group with 32 MPa compaction pressure. The obtained value was 0.55 mm/s, which corresponds to 21% related to the average of such group. The same analysis, however, considering only test 6 (48 MPa), presents an amplitude of 0.15 mm/s, *i.e.* 6.4% related to the average of test 6.

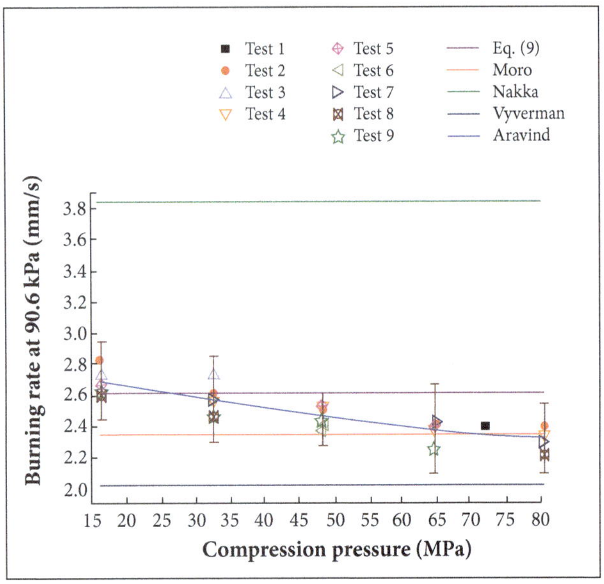

Figure 3. Burning rate *versus* compression pressure.

Figure 3 also includes results from other authors: Aravind *et al.* (2013), Moro (2013), Nakka (1984) and Vyverman (1978). As can be seen, the burning rate presents an opposite behaviour compared to density, since it decreases with higher values of compaction pressure. Since the sample preparation in this work takes a different process from Nakka and Vyverman ones, these authors did not study the influence of compression pressure on the burning rate. The observed experimental results, however, are slightly closer to Vyverman's ones than from Nakka's; they are also nearer Aravind's results, which also produced the propellant by a different process, than the one proposed in this work.

Since density and burning rate presents opposite behaviors in relation to compression pressure, it was analyzed how it affects the combustion gases mass flux, which is given by (Sutton and Biblarz, 2010):

$$\dot{m} = \rho\, r\, A_b \quad (10)$$

where:

A_b: burn surface area.

The result from burning rate times density can be identified as gases mass flux ($\dot{m}$) per unit burn surface (A_b), which is graphically presented in Fig. 4. In this figure, results of all samples that presented a regular burn process are presented. It was also observed that the gases mass flux per unit of burnt surface presents an average value of 3.83 kg/(m²s). It was verified,

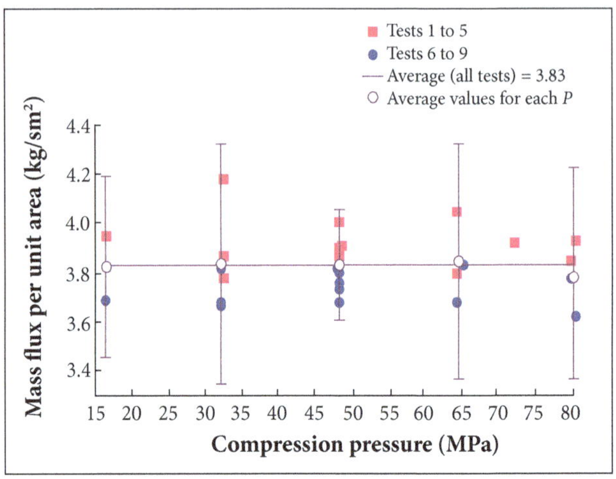

Figure 4. Mass flux per unit area *versus* compression pressure.

however, that depending on the batch, this average can slightly change: for tests 1 to 5, the average is about 3.93, while for tests 6 to 9, it reduces to 3.74.

Figure 4 also presents the confidence intervals based on t-Student distribution with 95% confidence level related to experimental data. The highest interval occurs for 32 MPa pressure and is equal to 0.98 kg/(m^{2}s), *i.e.* it varied 26% related to the average. In this case, in spite of the average of each group, the average of all samples was employed, since it is nearly independent of the compaction pressure. Even if the averages of different tests are different, considering the experimental uncertainties, the average of all tests correctly represents all data. The confidence intervals for all analyses are coincident and, because of this, one cannot affirm that the three averages are different.

A last important aspect to be mentioned is related to the solid and liquid residua accumulation on the surface where the samples were burnt. Apparently, non-gaseous phases, specially potassium carbonate (K_2CO_3), can adhere to the rocket nozzle. Figure 5 shows accumulated residua after the burn process of four samples. The quantity of residua was not measured, specially because it was strongly adhered to the plaque surface and a fraction of it was ejected far from the test place. The two samples in the upper portion of Fig. 5 seem to have generated more carbonate residua and the two lower portions, more carbon residua. It can indicate that the compounds mixture was not exactly homogenous in these samples.

Figure 5. Accumulated residua after the burn process of four samples.

CONCLUSION

In this paper, KNSu propellant was processed by cold manufacturing. Although the observed density values were, in general, lower than the ones obtained by fusion process, for the higher applied compression pressures (at about 80 MPa), results were similar to the ones obtained by Vyverman (1978). It corroborates the feasibility of KNSu cold manufacturing by mechanical press.

From experimental data, it was observed that density (ρ) increases with higher values of the propellant compression pressure (P_c) according to a second-order polynomial function. Moreover, the propellant burning rate (r) also depends on P_c according to a second-order polynomial function; however, r decreases with higher values of P_c.

It was also observed that the product ρ times r, which represents the mass flux per unit area, is nearly constant for samples obtained by the cold manufacturing.

ACKNOWLEDGEMENTS

The authors thank Prof. Juarez Souza de Oliveira and Technician Walderson Klitzke for the use of Ceramic Laboratory dependences at Universidade Federal do Paraná. The authors also acknowledge the financial support of Conselho Nacional de Desenvolvimento Científico e Tecnológico (CNPq), Agência Espacial Brasileira (AEB), by Uniespaço Program, Coordenação de Aperfeiçoamento de Pessoal de Nível Superior (CAPES) and the referees by their suggestions. The first four authors receive scholarships supported by CAPES. The last author receives a scholarship supported by CNPq.

REFERENCES

Aravind, M., Jerin, J., Rariprasad., V., Ganesh, B.K., Aswin, C. and Kumar, V.R.S., 2013, "In-house Manufacture of Variable Burning Rate Propellant Bangles for Nozzleless Propulsion", Proceedings of the International Conference on Aerospace, Mechanical, Automotive and Materials Engineering, Paris, France.

Douglass, H.W., Collins, J.H. and Keller, R.B., 1972, "Solid Propellant Design and Internal Ballistics", NASA, SP-8076.

Fordham, S., 1980, "High Explosives and Propellants", 2nd edition, Pergamon Press, Oxford, UK.

Marchi, C.H., Besozzi, B.M., Ciceroni, M. and Hering, S.F., 1990, "Experimental Verification of Nozzle and Propellant Influence on Netuno-R Model-Rocket Performance" (in Portuguese), Federal University of Santa Catarina, Report, Universidade Federal de Santa Catarina, Florianópolis, Brazil.

Moro, D.F., 2013, "Personal communication to C.H. Marchi", Curitiba.

Nakka, R.A., 1984, "Solid Propellant Rocket Motor Design and Testing", Bachelor of Science Thesis, University of Manitoba, Manitoba, Canada.

Nakka, R.A., 1997, "Richard Nakka's Experimental Rocketry Web Site", Retrieved in March 19, 2014, from http://www.nakka-rocketry.net/

Leslie, S. and Yawn, J., 2002, "Proposion for the Inclusion of KNO3/ Sugar Propellants in the TRA Experimental Rocketry Program", Retrieved in April 9, 2015, from http://www.aeroconsystems.com/motors/sugar_motor/SugarPro_Proposal.pdf

Stancato, F. and Miraglia, J., 1997, "Internal Ballistic Principles: a Methodology and an Experimental Didactic Experimentation", Proceedings of the VI International Mobility Technology Conference and Exhibit, São Paulo, Brazil.

Stancato, F., Racca, J.G.C., Castaldo, F., Carbonari, G., Souza, L.A. and Ballarotti, M.G., 2000, "A 3000 Ns Sugar-Sodium Nitrate Motor Development for Space Education", Proceedings of the 51st International Aeronautical Congress, Rio de Janeiro, Brazil.

Sutton, G.P. and Biblarz, O., 2010, "Rocket Propulsion Elements", John Wiley & Sons, New Jersey, USA.

Thompson, G., 2014, "How to Make Sugar Rockets", Retrieved in January 13, 2015, from https://www.youtube.com/watch?v=12fR9neVnS8

Vyverman, T., 1978, "The Potassium Nitrate — Sugar Propellant", Report, Belgium.

Wallace, W., 1995, "FMX — the Revised Black Book: a Guide to Field-Manufactured Explosives", Paladin Press, Boulder, USA.

15

Effect of the Fiber Orientation Relatively to the Plasma Flow Direction in the Ablation Process of a Carbon-Phenolic Composite

Sônia Fonseca Costa e Silva[1], Humberto Araújo Machado[1,2], Edison Bittencourt[3]

ABSTRACT: Composite materials are vastly employed in rocket nozzle extensions and as thermal protection shields in aerospace industry. In both cases, ablative and thermal properties play an important role for material choice and equipment sizing. These properties are strongly dependent on the way the pieces are manufactured. For the case of nozzle extensions, wrapping is the most common technique employed, which can be done by diverse ways. In this work, samples of the carbon reinforced carbon fiber (CRCF) were extracted from pre-forms obtained by biased and parallel tape wrapping processes with a resol-type phenolic resin used as polymer matrix, in order to study the influence of the fiber angle in the ablative behavior and thermal conductivity. Experimental results for both manufacturing techniques were obtained through ablation tests carried out in a plasma torch, were compared and had the different behaviors discussed. A computational simulation based in a two-front ablation model was also performed, in order to explain the influence of thermal conductivity on the different behaviors presented for biased fiber samples. Results show that the angle of the fibers relative to the heat flux direction is an important parameter to determine the ablative properties and the thermal performance of composite wrapped structures.

KEYWORDS: Ablation, Carbon-phenolic, Plasma torch, Computational simulation.

INTRODUCTION

The thermal resistance and ablation characteristics of composite materials are properties which play a very important role in the field of aerospace technology. When composite materials with excellent characteristics are exposed to extremely high temperatures, above 1,500°C, not only their surface and structure but also their intrinsic features can be protected from the decomposition, which can occur at these temperatures. The ablation of a material depends on its intrinsic conditions and also on extrinsic conditions such as thermal, chemical and mechanical factors according to the variables of theoretical or practical environments (Schmidt, 1971). Ablation is an endothermic and erosive process which involves complex physical and chemical reactions.

There are several types of materials used in these ablative thermal protection systems, as each material presents certain advantages and disadvantages in terms of its properties (specific mass, mechanical strength, melting temperature etc.) and environmental conditions that they are subjected to (heat flow, temperature, mechanical stress etc.).

Carbon-based materials have been considered to be promising candidate for severe environments with super high temperature and high pressure for their high temperature capability (over 2,000°C), high thermal shock resistance, good ablation resistance and low density (Kanevce *et al.*, 1999).

Phenolic resin reinforced with carbon fabric has been a popular ablative material for rocket nozzles. Phenolic resin is the preferred matrix material due to its ability to efficiently

1.Instituto de Aeronáutica e Espaço – São José dos Campos/SP – Brazil 2.Universidade Estadual do Rio de Janeiro – Resende/RJ – Brazil 3.Universidade Estadual de Campinas – Campinas/SP – Brazil.

Author for correspondence: Sonia Fonseca Costa e Silva | Instituto de Aeronáutica e Espaço | Praça Marechal Eduardo Gomes, 50 – Vila das Acácias | CEP: 12.228-904 São José dos Campos/SP – Brazil | Email: sonia_sfcs@hotmail.com form char during endothermic pyrolysis (Tick *et al.*, 1965). Carbon-fabric is preferred as reinforcement due to its ability to resist high-temperature erosive aerodynamic shear forces because of its high thermal stability, low thermal conductivity and higher inter-laminar shear strength (ILSS) in the composite form (Savage, 1993). However, there is always an urge to improve the ablation resistance of carbon-phenolic composites to further realize relatively thinner ablative composite structures to reduce the overall weight of aerospace systems.

Carbon fiber reinforced phenolic matrix composites (FRFC) have been used in the aerospace industry for the thermal-structural components. These materials can support aggressive thermal environments, such as in rocket nozzles throats, where a large amount of hot gases, due to the burning of propellant, pass through. These rocket nozzle extensions must submit, when in operation, low wear ablation on the inner walls. Furthermore, the wear should occur fairly homogeneous, *i.e.* distributed over the entire inner surface of the structure that has direct contact with the exhaust gases.

The carbon-phenolic ablative behavior should be influenced by the manufacturing process. The carbon-phenolic nozzle are commonly produced by a wrapping process, which consists in wrapping the carbon fiber tissue around a conic or bell shaped guide, in order to obtain the nozzle profile. The winding process with woven tape ("tape wrapping") basically consists on wrapping a tape over a steel mandrel. These strips are obtained by cutting fabric-based glass pre-impregnated with phenolic resin fiber, silica or carbon (Clements and Ward, 1966). The winding system is equipped with a compactor roller with pneumatic actuation, which compresses the "prepreg" tape as it is being deposited on the mandrel, as shown in Fig. 1.

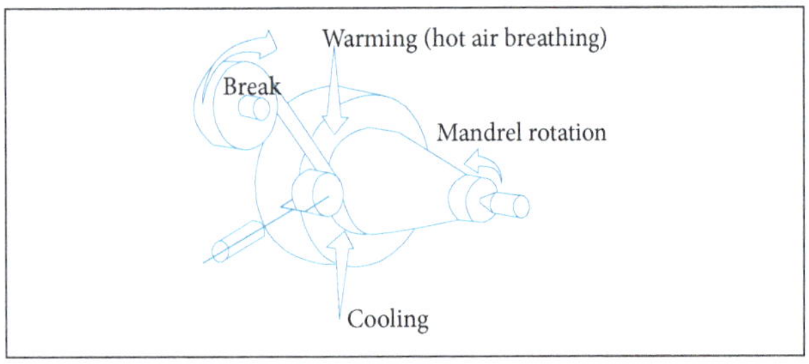

Figure 1. Scheme of a typical tape wrapping equipment.

The tape wrapping technique can be divided into:

- Parallel wrapping: consists on wrapping the tapes in the direction parallel to the main axis of the winding mandrel, as shown in Fig. 2. In this technique, the tapes used typically have fiber oriented towards 0° and 90° in a balanced manner, *i.e.* with an equal amount of filaments per unit length in the warp direction (the longitudinal direction of the fabric) and weft (transverse direction of the fabric), compared to the same principal directions (Clements and Ward, 1966). There is no pattern of weaving preferred for the tapes to be wounded in parallel. Figure 2 shows a plan view of the winding parallel deposition system showing the movement of the equipment's axes operating on a mandrel in the shape of a truncated cone.
- Biased wrapping: consists on wrapping the tapes in an angle (b) relative to the main axis of the winding mandrel direction (Clements and Ward, 1966). Compared to the parallel winding, this technique presents major practical difficulties. Most of the technological problems of this type of process arise from the fact that a single tape, at one point, has to go through different perimeters along the width, as evidenced in Fig. 3. For this to be possible in practice, the tapes to be used should allow a certain relative movement between the various fibers that constitute them. The fibers in this case should be directed to ± 45° regarding the same principal directions and, as in the previous case, these fibers should be distributed

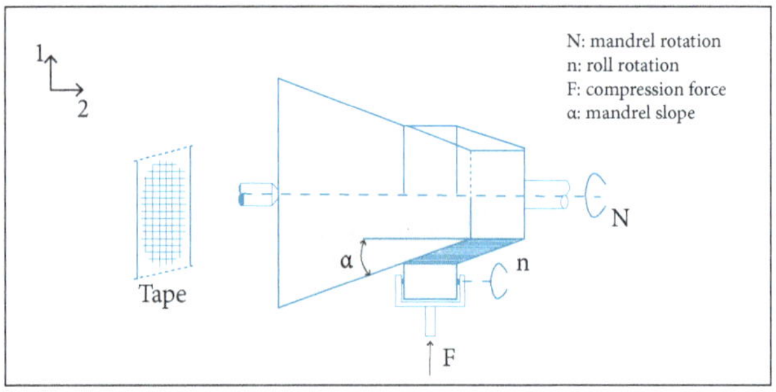

Figure 2. Parallel wrapping on conical mandrel.

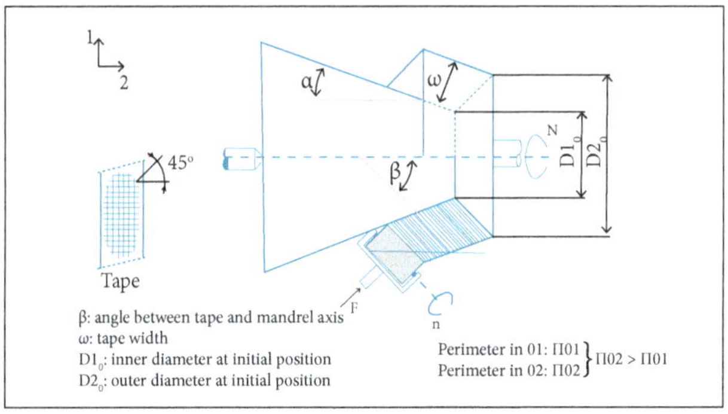

Figure 3. Biased wrapping on conical mandrel.

in a balanced way in the weft (direction 2) and warp directions (direction 1).

For optimum levels of quality and performance of ablation be achieved, it is necessary that the polymerization of the composite be carried out under high pressure. For the polymerization of ablative high-performance composites, such as those proposed for the structure of the divergent nozzle and components, it is strongly recommended that more specialized equipment be used like hydroclaves, where the pressurizing medium is water in its liquid state (Clements and Ward, 1966).

In this paper, thermal degradation experiments were carried out using high flow of plasma jet (Fujimoto *et al.*, 2003), in order to simulate the warmed gas flow through a rocket nozzle engine. The experimental results were obtained for three different angles of carbon fiber in test samples. A computational simulation is also performed, in order to compare the accuracy between the theoretical model for composite ablation and the empirical data.

EXPERIMENTAL PROCEDURE

MATERIALS

The carbon fiber used in the present work is T22R-ECHO with thickness of 0.45 mm, weave pattern 2 x 2 Twill, areal weight of 350 ± 35 g/m^2 and density of 1.55 g/cm^3. A phenolic resin resol-type matrix with viscosity of 1,000 cPs at 20°C has been used. The "prepreg" process was carried out in the Plastflow company facilities. After wetting, the fiber volume reached 55 – 60%.

The inclined and parallel wrapping process was performed on a property equipment of the Instituto de Aeronáutica e Espaço (IAE) of the Departamento de Ciência e Tecnologia Aeroespacial (DCTA), The polymerization cycle was carried out in an hydroclave at high pressure, which gives the perfect polymerization of the phenolic resin used and an ablative structure with high compaction and low void content (porosity). The cycle used for polymerization (curing) was: 1 h at 100°C and 3 h at 165°C with a pressure of 7.0 MPa.

After the wrapping and curing process, the samples were extracted in different positions, according to the manufacturing process. Both the position of extraction and the wrapping process play important roles in the thermal properties and ablative behavior of the composite, since they directly affect the fiber orientation in the sample (Fig. 4), where, for parallel wrapping, $\alpha = 22°$ and, for biased wrapping, $\beta = 20°$.

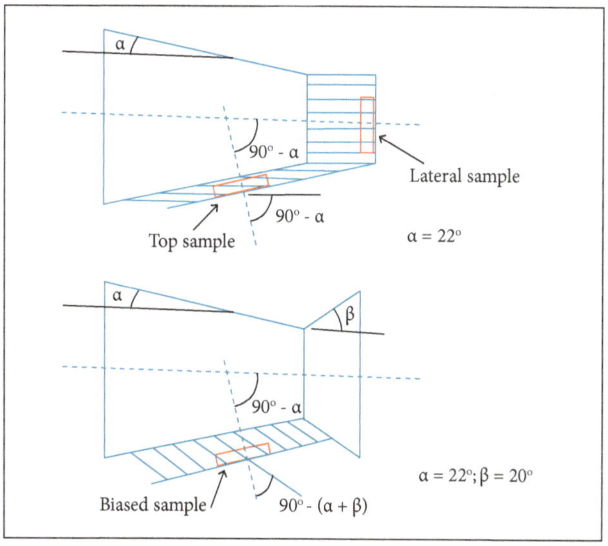

Figure 4. Positions for sample extraction.

ABLATION TEST

A plasma torch test was performed to investigate the ablation property of carbon-phenolic composites. The operation was carried out in atmospheric pressure. A direct current (DC) arc plasma system was designed for continuous working at a power up to 50 KW. The intensity of current was adjusted to 135 A with tension of 300 V. So that, the maximum achievable power obtained, due the power supply, is about 30 KW, which gives a plasma enthalpy of about 5.5 MJ/kg. The gas flow was maintained at 4.5 x 10^{-3} kg/s. The specimen was placed vertically to the flame direction in air. Three fiber orientations were assessed in relation to exposure to plasma jet: parallel fibers to the plasma flow (parallel wrapping - lateral sample - 0°); fibers at an angle of 68° in relation to the plasma flow (parallel wrapping - top sample); and fibers at an angle of 48° in relation to the plasma flow (biased sample), as shown in Fig. 5.

The ablation test was carried out in a time from 30 to 120 s and the distance between the nozzle tip of the plasma gun and the front surface of the specimen varied from 4 to 10 cm. The surface of the samples reached temperatures in the range from 1,200 to 1,900 K and it was measured by an optical pyrometer Mod. IR-AH 3SU Chino. Figure 6 shows the schematic illustration of the apparatus utilized for ablation test using an arc plasma torch.

The burn-through time was measured. The erosion rate was calculated by dividing the mass change before and after the test into a burn-through time for each specimen. One representative sample was used for each condition of testing. The specimens were cut in samples with diameter of 1.2 cm.

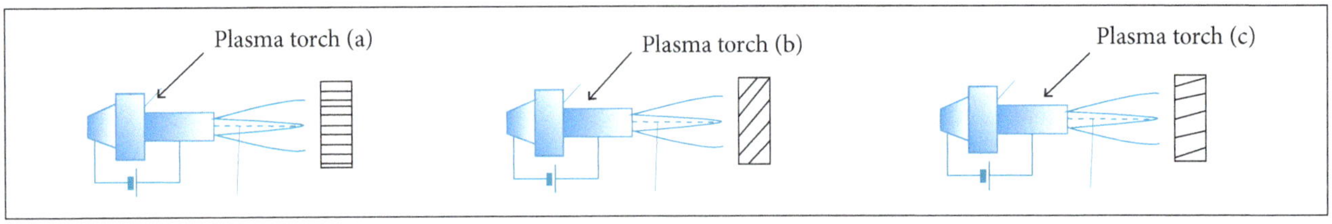

Figure 5. Orientation of the fibers in relation to the plasma flow: (a) parallel wrapping (lateral sample cut) - 0°, (b) parallel wrapping (top sample cut) - 68°, (c) biased wrapping (biased sample cut) - 48°.

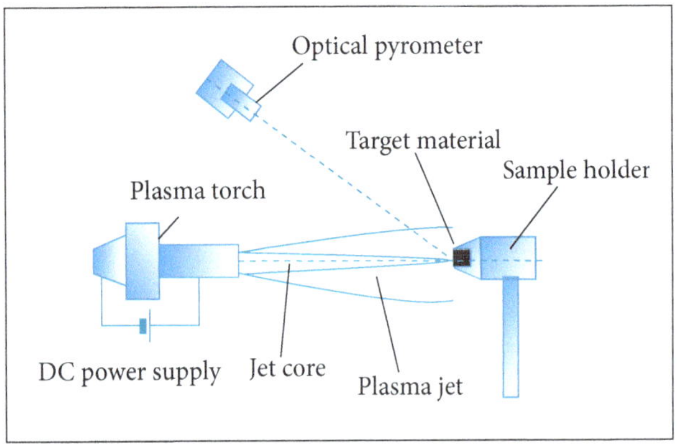

Figure 6. Scheme of an apparatus for ablation test using arc plasma torch.

COMPUTATIONAL SIMULATION

A computational code was employed to simulate a one-dimensional ablation process, alike that conducted experimentally, where material samples were exposed to a plasma jet and had their temperature and heat flux measured. In this study, the target was not a perfect match between experimental and theoretical data but the qualitative behavior of the physical approach, in order to verify its coherence between the experimental and numerical results.

The ablation model proposed (Machado, 2014) for a composite material will consider the presence of two layers: the virgin material and the char layer that appears after the pyrolysis front. Two moving fronts will be accounted: the pyrolysis and the ablation front of the char layer. Both will appear at constant and specific temperatures. The hypotheses considered to build the mathematical model are:

- Solid materials are isotropic with constant properties.
- The pyrolysis front is considered to have zero thickness.
- Char layer recession is supposed to occur due oxidation or sublimation at constant temperature.
- Melting layer is absent.
- The resulting gases fully react and perfectly mix with the external air layer, without influence over its properties.
- Air is considered a perfect gas.
- Radiation is absorbed or emitted by the external surface, but is not transmitted.

The moving boundary problem was solved by the interface tracking method, introduced by Unverdi and Tryggvason (1992) and employed by Juric (1996) in the solution of phase-change problems. In this method, a fixed uniform Eulerian grid is generated, where the conservation laws are applied over the complete domain. The interface acts as a Lagrangian referential, where a moving grid is applied. The instantaneous placement of the interface occurs through the constant remeshing of the moving grid, and each region of the domain is characterized by the indicator function, which identifies the properties of the wall and the air around it. The method can be extended for various interfaces, through the use of one particular indicator function for every phase within the domain of simulation (Machado, 2013). This method allows the representation of any geometry and also the characterization of every layer separately. It is accomplished without a high increase in the computational cost and does not need any pre-processing (construction of unstructured grid or coordinate transformation). In this work, this method is employed to estimate the ablative performance of the carbon-phenolic composite, considering a two-dimensional approach in both the heat conduction and the moving boundary problem. Both the set of equations and the numerical solution are described in detail in the work of Machado (2014).

RESULTS

The specimens submitted to ablation test were extracted to pre-form obtained by biased and parallel wrapping using a hole saw to remove small cylinders with a diameter of 12.5 mm. After cutting, each sample was identified and weighed on a precision balance (± 0.0001 g). A reweighing of the specimens was also taken after exposure to heat flow, to determine the rate of mass removed per unit of area and time of exposure to the plasma jet (kg/m^{2}s) and to estimate the experimental mass loss rate. According to the results, the mass loss of the samples increases with time of exposure to the plasma jet in both manufacturing processes, biased and parallel wrapping (side section and top cut), as shown in Figs. 7, 8 and 9. The samples obtained by parallel wrapping with the side cut exposure had higher removing mass in relation to exposed samples cut from the top and samples obtained by biased wrapping, due to the orientation of the fibers.

If the fibers are oriented parallel to the plasma flow, high mechanical strength enhancement of the orientation plane is produced and a minimum internal heat conduction is also achieved because heat transfer occurs almost completely through the resin and, with strict packaging of the fibers, it becomes more difficult. The resulting effect is a high non-uniform ablation and the tendency to mechanical erosion is increased by gases formed in the substrate material, which are trapped by closely packed fibers and yield pressure rising, resulting in a crack in the material. We observed that the mass loss is lower in the samples with a certain angle relative to the plasma flow (Figs. 10 and 11).

Fibers oriented perpendicular to the direction of flow offer greater resistance to shear flow and minimize the strengthening mechanical surface erosion, reducing ablation speed and maximizing the amount of heat ablation compared to fibers oriented in other directions, because of the greater length of

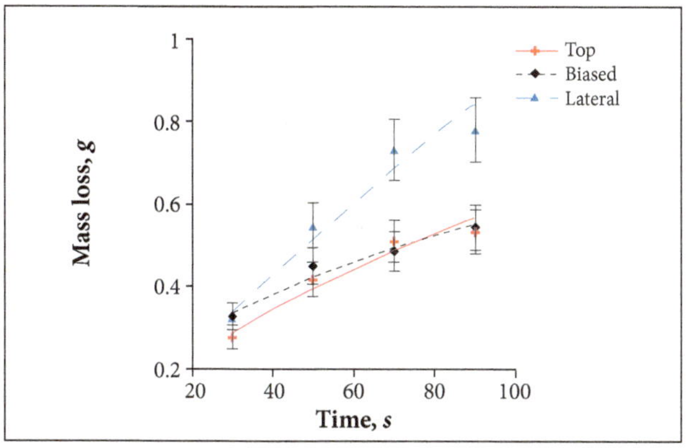

Figure 7. Variation of mass loss over time for a heat flux of 0.626 MW/m².

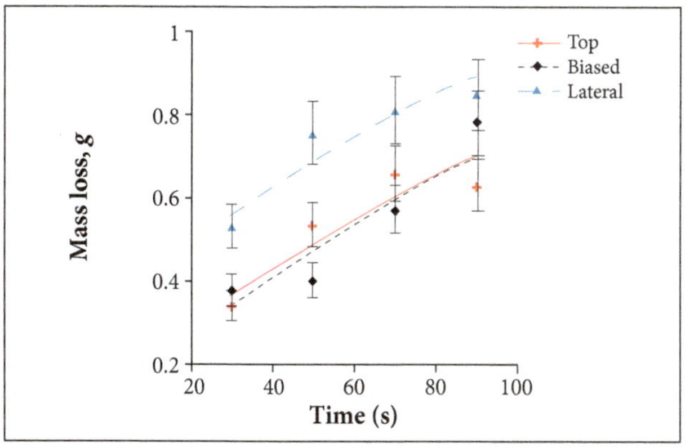

Figure 8. Variation of mass loss over time for a heat flux of 0.903 MW/m².

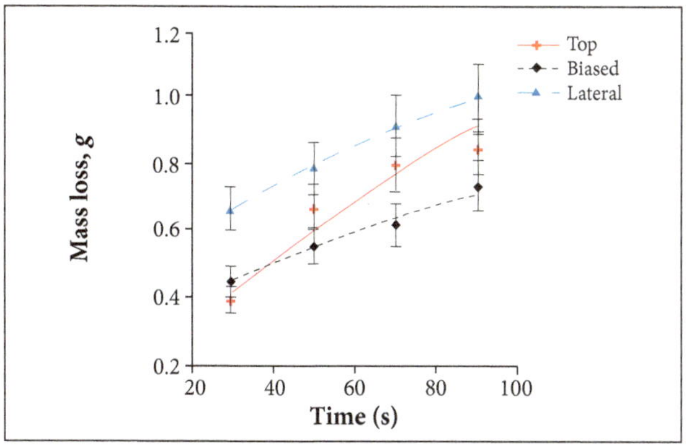

Figure 9. Variation of mass loss over time for a heat flux of 1.376 MW/m².

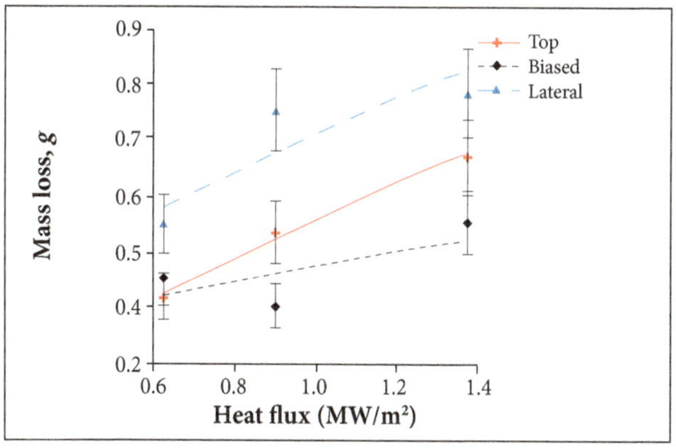

Figure 10. Results of the variation of total mass flow in the heat exposure time of 50 s.

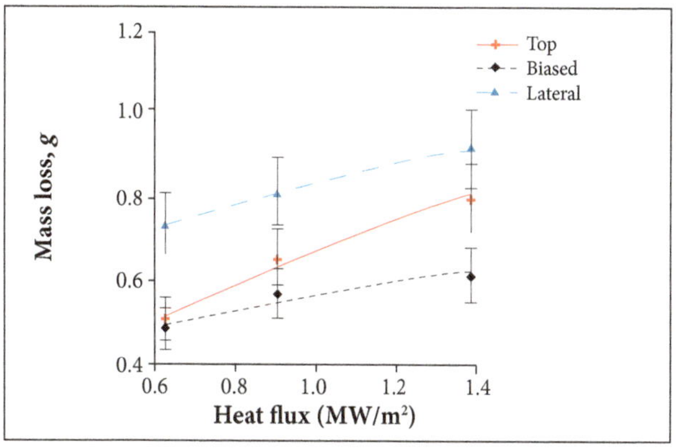

Figure 11. Results of the variation of total mass flow in the heat exposure time of 70 s.

continuous fiber supported by the phenolic resin, since it is not easily removed by the high-velocity gas flow, causing the material to be capable of presenting greater resistance to mechanical erosion. This erosion resistance regularly decreases as the fibers approach the parallel orientation. There are also basic limitations for this type of orientation. The mechanical properties are relatively low at the plane perpendicular to the fibers; this leads to occasional material removal and failure by stress during ablation. With the fibers in this orientation, the heat transfer rate of the substrate material can be appreciably higher because they provide a direct drive line to the inside, increasing the thickness of the material required to produce the desired insulation.

The mass loss rate was the parameter chosen in order to compare the experimental data and numerical results. The rate was obtained dividing the mass loss by the time of exposure. The same procedure was employed to obtain the mean mass loss rate in the simulations. However, the calculation of the mass loss rate is done instantaneously through the integration at each time step, while the experimental method shows the final result after the selected period.

The properties used in the simulation are presented in Table 1 (Machado, 2013). Although some of the thermophysical properties of the ablative materials were measured during the experiments, most of them had to be extracted from diverse literature sources. The values for thermal conductivity depend on the fiber orientation. The higher value (11 W/m^2°C) is considered when the fiber is normal to the surface were the heat flux is imposed (parallel to the heat flux direction) and the lower value (0.8 W/m^2°C) is considered when the fibers lie parallel to the surface and normal to the heat flux direction. Both cases were considered for the numerical simulation.

Table 1. Properties of the carbon-phenolic resin.

Property	Virgin	Char
k (W/m°C)	**0.8–11**[1]	1.73[2]
Cp (J/kg°C)	1,300[1]	1,800[3]
ρ (kg/m^3)	1,700[1]	1,184[3]
e	0.92[4]	0.7[3]
Heat of pyrolysis (MJ/kg)	2.33[5]	
Pyrolysis temperature (°C)	350[6]	
Heat of melting/sublimation (MJ/kg)		10.5[3]
Melting temperature (°C)		3,700[3]

[1]Barros (2008); [2]Sutton (1970); [3]Sawatimskii (2003); [4]Khlybov and Makhankov (2008); [5]Russel (1996); [6]Tick *et al.* (1965); Khlybov and Makhankov (2008).

The one-dimensional heat conduction problem was solved considering prescribed heat flux in the warmed face and adiabatic surface in the opposite side, as boundary conditions. The initial temperature was 27°C (300 K). The same heat fluxes measured in the samples — 0.464 MW/m^2, 0.626 MW/m^2, 0.903 MW/m^2, 1.376 MW/m^2 and 1.725 MW/m^2 — were imposed to the warmed face.

Table 2 shows the temperature of sample's surface in the side were the plasma jet was impinged (for the experiment) and the heat flux was imposed (simulation). Deviation of the numerical results relative to the average experimental data rises from -7.82 to 18.7%. Even considering the uncertainty of the experimental results, the simple approach used in the mathematical model (that neglects the influence of various physical processes) and the values of the properties used in the calculation (which were extracted from diverse sources and consequently are subjected to different conditions of evaluation) are considered to present physical coherence and both curves, numerical and experimental, have a similar behavior, as shown in Fig. 12.

Figures 13, 14 and 15 show the comparison among the experimental results of all samples for each value of thermal conductivity used in the computational simulation and the mass loss rate with the time. The experimental results present the points resulting of measurements and the respective fitting curves, according to the slope of the fiber. In all cases, the mass loss rate seems to be underestimated by numerical simulation. The numerical results show little influence of thermal conductivity variation. The results for both values used (0.8 to 11 W/m^2°C) are relatively close. In Fig. 13, the sample curves obtained by

Table 2. Surface temperature after 90 s warming.

Case	Heat flux (MW/m^2°C)				
	0.464	0.626	0.903	1.376	**1.725**
Simulation - k = 0.8 W/m^{2}K	1,113	1,342	1,624	1,942	2,097
Simulation - k = 11 W/m^{2}K	1,070	1,321	1,620	1,944	2,103
Simulation - average	1,091	1,331.5	1,622	1,943	**2,100**
Parallel wrapping - top sample	1,200	1,190	1,450	1,640	1,745
Parallel wrapping - lateral sample	1,167	1,275	1,375	1,667	1,792
Parallel wrapping - average	1,183.5	1,232.5	1,412.5	1,653.5	1,768.5
% deviation (relative to experiment)	-7.82	8.03	14.8	17.5	18.7

parallel wrapping (lateral and top sample) and biased wrapping produce the same slope compared with the numerical simulation, without crossing at any point. The crossing between the curves appears in all cases for numerical simulation, but only in the two last cases for the experimental results (Figs. 14 and 15). Although the first case does not present such crossing, these curves feature similar behavior of their numerical counterparts. In all cases, the lateral sample curves present a level of mass loss rate above the two other samples. It indicates that the influence of the fiber orientation and the thermal conductivity variation affects the ablation process.

When the fibers are disposed normally to the heat flux, conduction occurs mainly through the resin, and the result is a high and non-uniform ablation as described for each value of heat flux observed for the samples obtained by biased and top cut parallel wrapping. The samples where fibers are perpendicular or biased to the direction of gas flow provide greater resistance to cutting gas effort and minimize the mechanical erosion of the surface, reducing the ablation rate, as shown in Figs. 13, 14 and 15, which is confirmed by the proximity of both curves in all results.

The discrepancy between the experimental and numerical temperatures and mass loss rates can be the result of other processes. The catalycity (Carleton and Marinelli, 1992; Kovalev and Kolesnikov, 2005; Barbato *et al.*, 2000) effect of the surface over the recombination of species produced by decomposition due to high temperature can increase the heat flux. Since this is an exothermic reaction, such increasing may have resulted in higher heating rates and mass loss rates, which is not considered in the numerical simulation.

Resin pyrolysis yields gaseous products that percolate toward the heated surface and are injected into the boundary layer which changes its properties, usually resulting in a reduction in convective heating. In addition, chemical reactions between the surface material of the boundary layer and pyrolysis-resulting species can result in consumption of the leading surface material and surface recession. These reactions can be endothermic (vaporization, sublimation) or

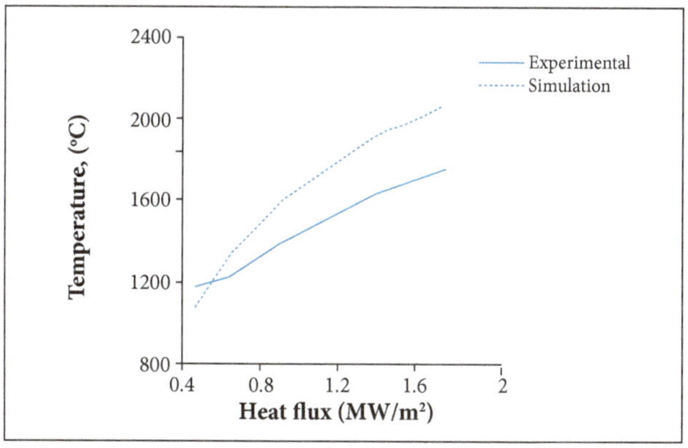

Figure 12. Temperature at t = 90 s in the sample's warmed surface, according to heat flux.

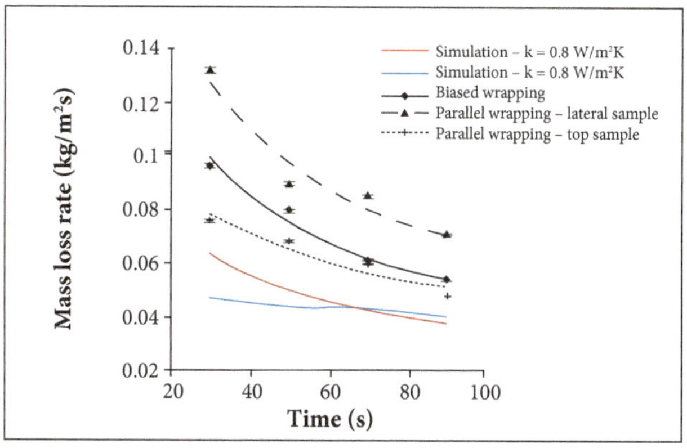

Figure 13. Mass loss rate with time: simulation and experimental results for q = 0.626 MW/m².

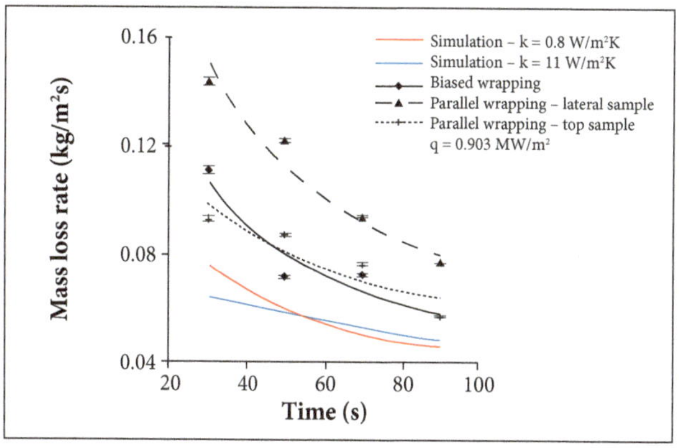

Figure 14. Mass loss rate with time: simulation and experimental results for q = 0.903 MW/m².

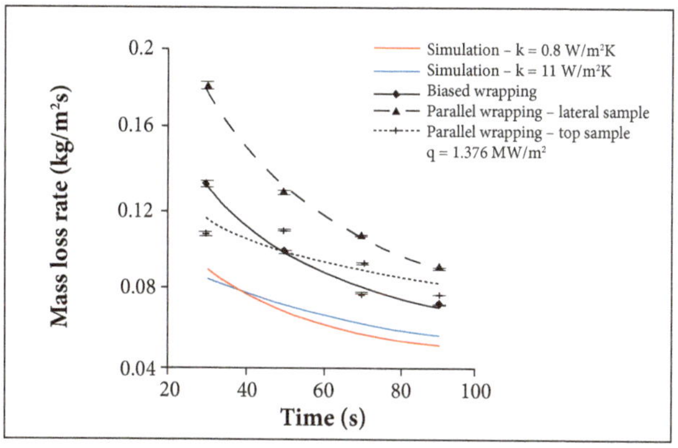

Figure 15. Mass loss rate with time: simulation and experimental results for q = 1.376 MW/m².

exothermic (oxidation) and will have an important impact on energy to the surface.

These effects may result in increasing mass loss rate and decreasing temperature, when compared with the mathematical model. Such effects are implicit in the physical parameters that were measured under specific conditions, which are not exactly applicable to the actual physical phenomenon and occur at high temperatures. It is expected that these effects occur during the tests in plasma torch because the temperatures in samples' surface can reach values between 3,600°C – 5,000°C.

The intersection between the curves for the mass loss rate with time among the biased and top parallel wrapping processes appear to be mainly due to the thermal conductivity variation with the angle of inclination of the fibers, since the two curves are far from the lateral parallel wrapping curve in which the fibers are parallel to the heat flow; consequently, ablation processes involved are different from those in which the fibers are inclined in relation to the flow. When the experimental results are compared with the numerical simulation for the two extreme values of thermal conductivity (0.8 W/m^{2}°C when the fiber is perpendicular to the flow and 11 W/m^{2}°C when the fiber is parallel), results present the same behavior. It is noted that, as the heat flux increases, the crossover point tends to the left, in the direction of convergence of the curves on the initial times. A possible explanation for this behavior, based on numerical results, is related to the difference in conductivity between the carbonized layer ("char") and the virgin material. The "char" works as an insulator, with a conductivity slightly above the lowest value considered for the virgin material (1.73 W/m^{2}°C) and reaches higher temperatures than this, because it is directly exposed to the plasma jet and its maximum surface temperature (melting point) can reach 3,700°C.

Figure 16 allows understanding the process. The first column shows the position of the pyrolysis front, which separates the virgin material and the char layer. In the case of lower conductivity, the heat flux incident on the surface takes longer to spread into the material and the pyrolysis temperature is reached earlier on a smaller thickness, starting the formation of char layer. In the case of higher conductivity, the heat spreads faster, yielding a lower temperature gradient, which delays the surface to achieve pyrolysis temperature. In both cases, when the pyrolysis temperature is reached, the velocity of recession of the pyrolysis front shoots to a peak and then reduces gradually, as seen in the right column of

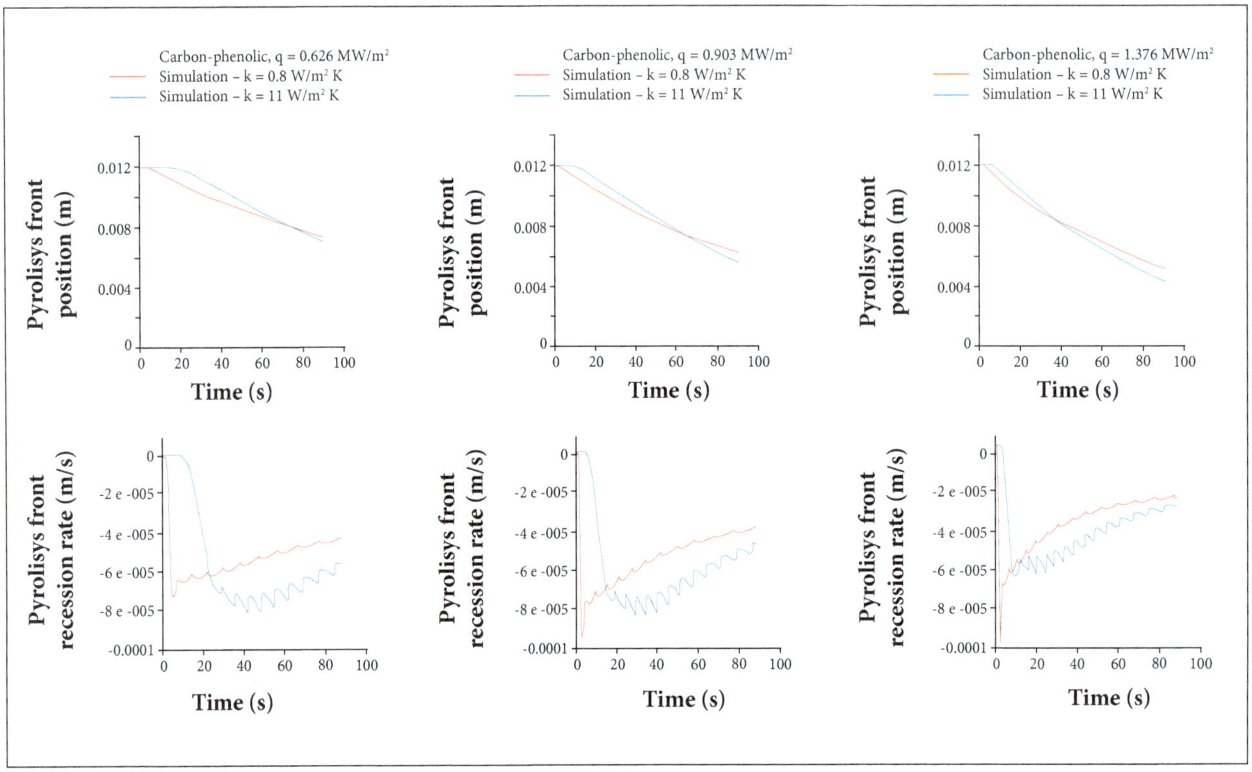

Figure 16. Effects of thermal conductivity difference between carbonized layer ("char") and the virgin material over the pyrolysis dynamics.

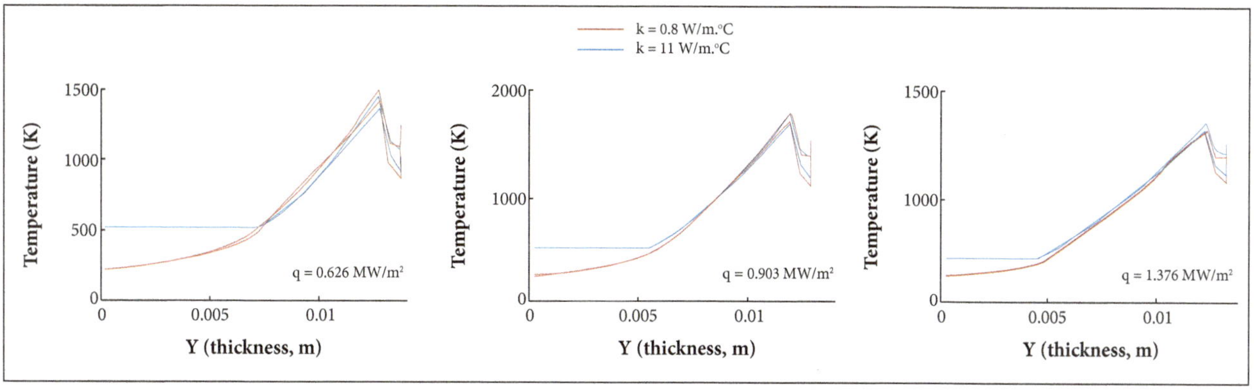

Figure 17. Temperature profile in the char and virgin layers of "char" after 90 s of jet exposure.

Fig. 16, with the resulting char layer working as a thermal insulator. In every case (high and low conductivity), the speed of pyrolysis fronts converge with the time and the difference between them reduces as heat flux increases.

Figure 17 shows results for temperature distribution after a period of 90 s. In the case of lower thermal conductivity, the highest temperature gradients are observed in the virgin material layer, which results in lower temperatures at the opposite surface to the heat flux. In the case of higher thermal conductivity, there is almost no temperature gradient in the virgin material layer, resulting in a higher temperature on the opposite side to the heat flux. This fact is relevant for a thermal protection system, since one of the objectives of the ablative heat shield is to ensure that temperatures remain below the operating limits. In both cases, the temperature profile of the char layer is quite similar.

CONCLUSION

In this paper, the effect of the fiber orientation in a carbon-phenolic ablator was studied. Experimental results were obtained from a test with a plasma jet emulating the gas flow within a rocket nozzle engine, using samples made by two different wrapping procedures and extracted from two different ways. The experimental data were compared with numerical results of computational simulation, obtained for the maximum and minimum values of thermal conductivity for the composite, according to the fiber orientation (perpendicular or parallel to the heat flux).

Comparison of surface temperature at 90 s of warming showed relatively good agreement between experiment and computational simulation, in spite of all uncertainties involved in both methods. The numerical results showed little effect of thermal conductivity over the mass loss rate. However, the experimental results showed a considerable difference between the top sample (fibers normal to heat flux) and the lateral and biased sample (fibers inclined to heat flux direction), which indicates that the fiber slope relative to the heat flux affects other physical processes involved in ablation, like pyrolysis gas releasing and surface catalycity effect.

Although the difference between numerical and experimental results and between the top samples and the lateral and biased samples, the order of magnitude and the curve behavior were considered similar, which attested the physical coherence of all results.

REFERENCES

Barbato, M., Reggiani, S., Bruno, C. and Muylaert, J., 2000, "Model for Heterogeneous Catalysis on Metal Surfaces with Applications to Hypersonic Flows", Journal of Thermophysics and Heat Transfer, Vol. 14, No. 3, pp. 412-420. doi: 10.2514/2.6539

Barros, E.A., 2008, "Thermal Plasma for Ablation in Materials Used in Heat Shield of Space Systems" (in Portuguese), Master's Dissertation, Instituto Tecnológico de Aeronáutica, São José dos Campos, Brazil.

Carleton, K.L. and Marinelli, W.J., 1992, "Spacecraft Thermal Energy Accommodation from Atomic Recombination", Journal of Thermophysics and Heat Transfer, Vol. 6, No. 4, pp. 650-655. doi: 10.2514/3.11547

Clements, H.R. and Ward, G.T., 1966, "Fabrication of Ablative Liners for Large Solid Booster Nozzles", Journal of Spacecraft and Rockets, Vol. 3, No. 4, p. 483-488. doi: 10.2514/3.28480

Fujimoto, K., Shioya, T. and Satoh, K., 2003, "Degradation of Carbon-Based Materials due to Impact of High-Energy Atomic Oxygen", International Journal of Impact Engineering, Vol. 28, No. 1, pp. 1-11. doi: 10.1016/S0734-743X(02)00057-X

Juric, D., 1996, "Computations of Phase Change", Ph.D. Thesis, University of Michigan, Ann Arbor, USA.

Kanevce, L.P., Kanevce, G.H. and Angelevski, Z.Z., 1999, "Comparison of Two Kinds of Experiments for Estimation of Thermal Properties of Ablative Composite", Proceedings of the 3° International Conference on Inverse Problems in Engineering, Port Ludlow, USA.

Khlybov, V.I. and Makhankov, S.A., 2008, "Experimental Studies to Determine Ablation Parameters and Emissivity of Thermal-Protective Materials", Technical Report, OAO-Makeyev GRT (Restricted).

Kovalev, V.L. and Kolesnikov, A.F., 2005, "Experimental and Theoretical Simulation of Heterogeneous Catalysis in Aerothermochemistry (a Review)", Fluid Dynamics, Vol. 40, No. 5, pp. 669-693. doi: 10.1007/s10697-005-0106-4

Machado, H.A., 2013, "Simulation of Ablation in a Sounding Rocket Thermal Protection System Via an Interface Tracking Method with Two Moving Fronts", Journal of Aerospace Technology and Management, Vol. 5, No. 4, pp. 409-420. doi: 10.5028/jatm.v5i4.262

Machado, H.A., 2014, "Simulation of Two Moving Fronts Ablation in a Composite Thermal Protection System Via an Interface Tracking Method", Proceedings of the 11° IAAA/ASME Joint Thermophysics and Heat Transfer Conference, Atlanta, USA.

Russel, G.W., 1996, "Evaluation of Thermal of an Epoxy Composite Using a Complex Modeling Technique", Proceedings of the AIAA, ASME and ASEE, 32° Joint Propulsion Conference and Exhibit, Lake Buena Vista, USA.

Savage, G., 1993, "Carbon-Carbon Composites", Chapman & Hall, London, 389p.

Savvatimskii, A.V., 2003, "Melting Point of Graphite and Liquid Carbon", Physics-Uspekhi, Vol. 46, No. 12, pp. 1295-1303. doi: 10.1070/PU2003v046n12ABEH001699

Schmidt, D.L., 1971, "Ablative Polymers in Aeroespace Technology", In: D'Alelio, G.F. and Parker, J.A. (Eds.), "Ablative Plastics", Marcel Dekker, New York, USA, pp. 1-2.

Sutton, K., 1970, "An Experimental Study of a Carbon-Phenolic Ablation Material", NASA Technical Note D-5930.

Tick, S.J., Huson, G.R. and Griese, R., 1965, "Design of Ablative Thrust Chambers and Their Materials", Journal of Spacecraft and Rockets, Vol. 2, No. 3, pp. 325-331. doi: 10.2514/3.28179

Unverdi, S.O. and Tryggvason, G., 1992, "A Front-Tracking Method for Viscous, Incompressible, Multi-fluid Flows", Journal of Computational Physics, Vol. 100, No. 1, pp. 25-37. doi: 10.1016/0021-9991(92)90307-K

Nonlinear Characteristics of Revolute Joints with Clearance

Liu Rong-qiang[1], Zhang Jing[1], Guo Hong-wei[1], Deng Zong-quan[1]

ABSTRACT: The nonlinear contact characteristic of revolute joint with clearance can degrade the performance of deployable structure. In this article, tensile and compressive tests are adopted to investigate the accuracies of the simplest conformal model and nonconformal model for contact, which are used to calculate the revolute joint deformation. In order to study the applicability of the two models, different clearances of joints are introduced in the tests. The results of the two contact models do not well agree with the experimental results. A model for the calculation of contact force and deformation of revolute joint, considering geometric constraints and cylinder contact characteristic, is presented. By comparison with the simplest conformal model and nonconformal model for contact, the proposed model is more accurate to calculate the nonlinear contact of revolute joint. The theoretical and experimental analysis of revolute joint with clearance is helpful to improve the reliability of deployable structure simulation.

KEYWORDS: Nonlinear stiffness, Revolute joint, Clearance, Conformal contact.

INTRODUCTION

The deployable structures have always been used to reduce the packaging volume of the spacecraft. Many appendage structures (Larsen *et al.*, 2009; Meguro *et al.*, 2009), such as communication antennas, have been successfully deployed in space. The truss elements of deployable structures are constructed by revolute joints. In order to achieve connection and large-scale rotation of truss elements, there are small clearances in the joints. It affects the performance of deployable structures and causes noise, wear, and vibrations (Ingham and Crawley, 2001; Schwab *et al.*, 2002; Parenti-Castelli and Venanzi, 2005; Flores *et al.*, 2006a,b, 2010; Qi *et al.*, 2010). The presence of clearance in the revolute joints often degrades the structure stiffness and has a significant effect on the dynamic behavior of deployable structures. Formulation for the contact of clearance joint is very important for researchers to study the nonlinear dynamics of structures and simulate the multibody system precisely (Liu *et al.*, 2006). So it is necessary to analyze and calculate the stiffness of clearance joint.

The heart of the revolute joint formulation is the intermittent and continuous contact simulation of joints. Various types of continuous models have been proposed by many researchers. There are two main types of contact models – Kelvin-Voigt viscoelastic model and Hertz contact model (Johnson, 1985), which are based on elasticity theory. The models have been applied in joint deformation calculation and impact simulation. Dubowsky and Freudenstein (1971) proposed deformation expression of a pin inside a cylinder. Energy dissipation characteristic of impact is simplified into the product of the rebound force and the coefficient of restitution (Bengisu *et al.*, 1986; Ravn, 1998). Rhee and Akay (1996) studied the

1. Harbin Institute of Technology – Harbin, China

Author for correspondence: Zhang Jing | School of Mechatronics Engineering | Harbin Institute of Technology | 92 West Dazhi Street | Nan Gang District | Post Code 150001 Harbin – China | Email: free1985216@163.com response of four-bar mechanism with clearance joint, including the impact force and friction. Tian *et al.*, (2009) deduced the lubricant force in lubricated joints based on Reynolds' equation. In these plane mechanism studies, the Hertz law expression about two spheres has been used to calculate the contact force in revolute joint (Lankarani and Nikravesh, 1990; Jia *et al.*, 2002; Khemili and Romdhane, 2008; Shi and Jin, 2008; Erkaya and Uzmay, 2012; Olyaei and Ghazavi, 2012). It is difficult to obtain accurate contact deformation and load of two cylinders by using sphere contact model and nonconformal contact model. So the rationality of the nonconformal contact model and conformal contact model has to be testified.

The main methods of building the contact model are the elastic theory and discrete method. Based on the contact force model with hysteresis damping given by Lankarani and Nikravesh (1990), Flores and Ambrosio (2004), and Flores *et al.*, (2006a,b) gave the normal contact model of planar revolute joints and compared different contact models. The contact area and load can be separated into its normal and tangential components (Abdo and Shamseldin, 2005). Discrete method is usually achieved by using finite element method (FEM), which has been used to give the force–displacement relation of sphere and the parameters influence on contact stress (Zhang and Vu-Quoc, 2001; Knight *et al.*, 2002). However, it will take a long time to simulate the contact of joint with FEM.

Because a simple and accurate calculation method has not yet been developed, the particular focus in this article lies on the evaluation and formulation of the relationship among contact force, depth, and width of revolute joint, which is based on joint experiment, geometric constraint, and elastic theory. The layout of this article is as follows: in Hertz model that has been widely used to calculate the revolute joint contact in mechanism and conformal model that has been most studied are presented. In revolute joint experiment is achieved. The results of the two models are compared with the experimental results when the clearance size changes. The new model of revolute joint with clearance is proposed, which is more precise than Hertz model and conformal model in a number of cases.

TWO EXISTING MODELS FOR REVOLUTE JOINT WITH CLEARANCE

Nonconformal contact model

The revolute joint is composed of clevis, tang, and pin, shown in Fig. 1a. Some amount of clearance always exists in revolute joints to allow relative rotation between the clevis and tang. The contact force and the deformation of revolute joint with clearance are usually calculated by Hertz model. In this section, the contact model of two cylinders is presented first, which is based on plane strain theory. Figure 1b depicts the contact of revolute joint, in which the radial clearance, ΔR, is defined as difference between the pin and tang radius, R_1 and R_2, respectively. It is assumed that the two contact bodies are isotropic materials. The Young's modulus and Poisson's ratio of tang and pin are represented by E_1, E_2, υ_1, and υ_2, respectively, a is the contact width and ε is the semiangle of contact corresponding to the contact relationship.

According to the Hertz laws for two-dimensional contact of cylindrical bodies (Johnson, 1985), two cylinder contact can be simplified from the three-dimensional contact into two-dimensional contact when their axes parallel. The relationship of contact force and width is obtained based on the constraints of contact. The sign of the curvature is related with the shape of the contact surface. When the contact surface is concave or saddle shaped, the radius curvature sign of the contact solid is negative. So the relative radius of the pin and the tang contact can be expressed as

$$R = \frac{R_1 R_2}{R_2 - R_1} \tag{1}$$

The equivalent Young's modulus can be written

$$\frac{1}{E} = \frac{1-u_1^2}{E_1} + \frac{1-u_2^2}{E_2} \tag{2}$$

The contact pressure distribution is given by

$$p(x) = \frac{2P}{\pi a^2}(a^2 - x^2)^{\frac{1}{2}} \tag{3}$$

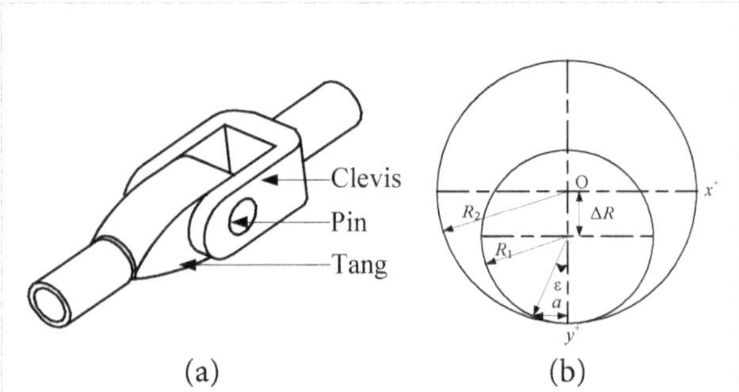

Figure 1. Model of revolute joints with clearance: (a) The revolute joint (Lake and Lee, 1996) and (b) contact model for revolute joint with clearance.

where P is the contact load under the per unit length of the contact cylinder.

The contact load P can be expressed as (Johnson, 1985)

$$P = \frac{\pi a^2 E}{4R} \quad (4)$$

Because a can be expressed by δ, as

$$a = \sqrt{\delta R} \quad (5)$$

where δ is the mutual approach of distant points on the pin and the hole of the tang.

So the contact load can be expressed by the contact depth

$$P = \frac{\pi}{4} E\delta \quad (6)$$

In Hertz laws (Johnson, 1985), three assumptions need to be satisfied: surfaces are continuous, nonconforming, and frictionless; the strain is small; each solid could be considered as an elastic half-space. However, the contact pair of revolute joints with clearance has conformal surfaces that cannot satisfy the assumption of Hertz theory. So the conformal model must be considered.

Conformal contact model

In the previous part, the nonconformal contact model is expressed to calculate the contact force and penetration depth. Because the clearance in revolute joint is small, the contact of revolute joint is conformal contact. The normal contact of conformal surfaces has also been considered by Steuermann and Persson (Johnson, 1985). The profiles are represented as a polynomial to achieve a required degree approximation. So the initial separation of the contact bodies' boundary can be expressed as

$$h = A_1x^2 + A_2x^4 + A_3x^6 + ... + A_nx^{2n} + ... \quad (7)$$

The external normal load and compression between centers of the contact bodies can be expressed as

$$P_n = \frac{4A_nEna^{2n+1}}{(2n+1)} \frac{2\cdot 4\cdots 2n}{1\cdot 3\cdots(2n-1)} \quad (8)$$

$$\delta = \frac{2\cdot 4\cdots 2n}{1\cdot 3\cdots(2n-1)} A_n \quad (9)$$

The second-order profiles assumed in this model is corresponding to n=1 in the Hertz theory. So the simplest initial separation in the revolute joint contact can be given by

$$h = A_1x^2 \quad (10)$$

The compression can be expressed as

$$\delta = 2A_1a^2 \quad (11)$$

The corresponding Hertz approximation can also be used to express the compression by

$$\delta = \left(\frac{1}{R_1} - \frac{1}{R_2}\right)a^2 \quad (12)$$

The total load can be written

$$P_n = \frac{4nEa\delta}{2n+1} \quad (13)$$

It is clear in Eq. 13 that the stiffness of the contact pair is increased when the number of the polynomial is improved. So the simplest conformal model can be helpful to evaluate the error of other number conformal model.

Based on Eqs. 12 and 13, the contact load can be written

$$P = \frac{4E}{3}\delta^{\frac{3}{2}}\left(\frac{1}{R_1} - \frac{1}{R_2}\right)^{-\frac{1}{2}} \quad (14)$$

REVOLUTE JOINT EXPERIMENT AND COMPARISON

The experiment, which is shown in Fig. 2, is set up to evaluate the accuracy of nonconformal model and conformal model in calculating the stiffness of revolute joint. The joint was fixed between the base and the moving beam. The moving beam moved in the vertical direction. So the load applied on the joint was also in the vertical direction. The displacement of the moving beam and force along the joint can be collected by encoder and sensor, which correspond to the variable δ and P. The Young's modulus of revolute joint is 2.06×10^5 MPa. The experiment introduces three joints with different radial clearances which are 0.067, 0.094, and 0.104 mm, respectively. The contact length is taken as one

(a) Revolute joint installation (b) Revolute joint

Figure 2. Revolute joint experiment.

unit length for the comparison of the experimental data with other model results, because the coaxially usually cannot be guaranteed in experiment. The comparisons of the joint deformation among nonconformal and conformal model results with experimental results demonstrate the suitability of the models, shown in Fig. 3.

As shown in Fig. 3, it is found that the results based on nonconformal model and conformal model are not close to the experimental results. The average errors of the two models are more than 50% and 30% separately. The nonconformal model results based on Hertz theory have large errors in small clearance and small errors in large clearance. However, conformal model is closer to the experiment results compared with Hertz model when the clearance is small. Because the Hertz model is base on nonconformal contact, it is more applicable to large clearance contact. The conformal model is more suitable for small clearance contact.

NONLINEAR CONTACT MODEL BASED ON THE GEOMETRIC CONSTRAINT

In the previous comparison, it is obvious that Hertz theory cannot solve the contact of revolute joint with clearance exactly, which is based on the assumption of nonconformal contact and the simplest conformal equation. So the dynamics simulation of multibody system cannot be performed conveniently and accurately. A new contact model is proposed in this section and is proved by comparing with above models.

Geometric constraints

To find the relationship between the contact depth and the width, the new model must satisfy the geometric constraint. Based on the model of Fig. 1, the contact deformation and other assistant parameters are shown in Fig. 4.

The coordinates of the points G_1 and G_2 in the border of the pin and hole can be written as:

$$\begin{cases} x_1 = R_1 \cdot sin\theta_1 \\ y_1 = R_1 \cdot cos\theta_1 \end{cases} \tag{15}$$

$$\begin{cases} x_2 = R_2 \cdot sin\theta_2 \\ y_2 = -\Delta R + R_2 \cdot cos\theta_2 \end{cases} \tag{16}$$

where θ_1 and θ_2 are the center angles of the G_1 and G_2.

Based on Eqs. 15 and 16, the distance of points having the same x coordinate separately in different circles can be written as

$$h = -\Delta R + R_2 \cdot cos\theta_2 - R_1 \cdot cos\theta_1 \tag{17}$$

Because the two points G_1 and G_2 have the same x coordinate, θ_1 and θ_2 satisfy

$$R_1 \cdot sin\theta_1 = R_2 \cdot sin\theta_2 \tag{18}$$

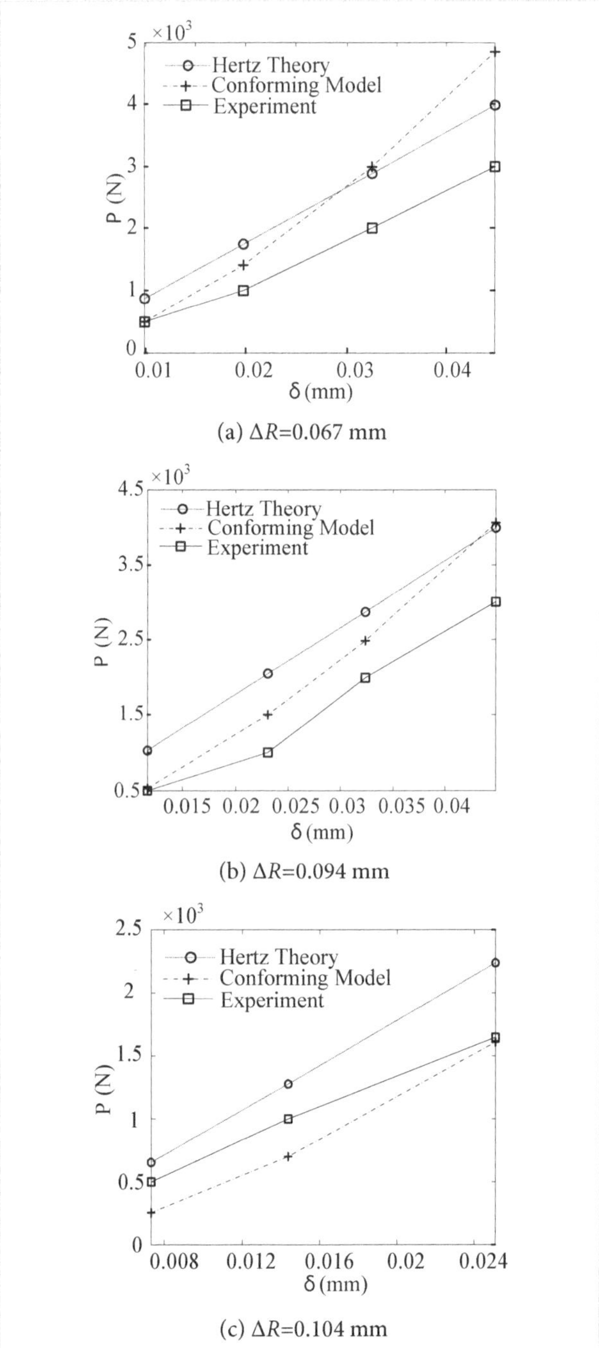

Figure 3. The results of the models and experiment.

Considering Eq. (18) and trigonometric relationship $sin^2\theta_2 + cos^2\theta_2 = 1$, h in Eq. (17) can be redefined into

$$h(\theta_1) = -\Delta R + R_2\sqrt{1-\left(\frac{R_1}{R_2}\cdot sin\theta_1\right)^2} - R_1\cdot cos\theta_1 \quad (19)$$

When θ_1 is equal to contact semiangle ε, $h(\theta_1)$ corresponds to the distance of the edge points H_1 and H_2. Due to the deformation of the hole of the tang and the pin, the total displacement of the tang is equal to δ and $h(\varepsilon)$, which can be found in Fig. 4. Based on Eq. 19, δ can be written as

$$\delta = -\Delta R + R_2\sqrt{1-\left(\frac{R_1}{R_2}\cdot sin\,\varepsilon\right)^2} - R_1\cdot cos\,\varepsilon \quad (20)$$

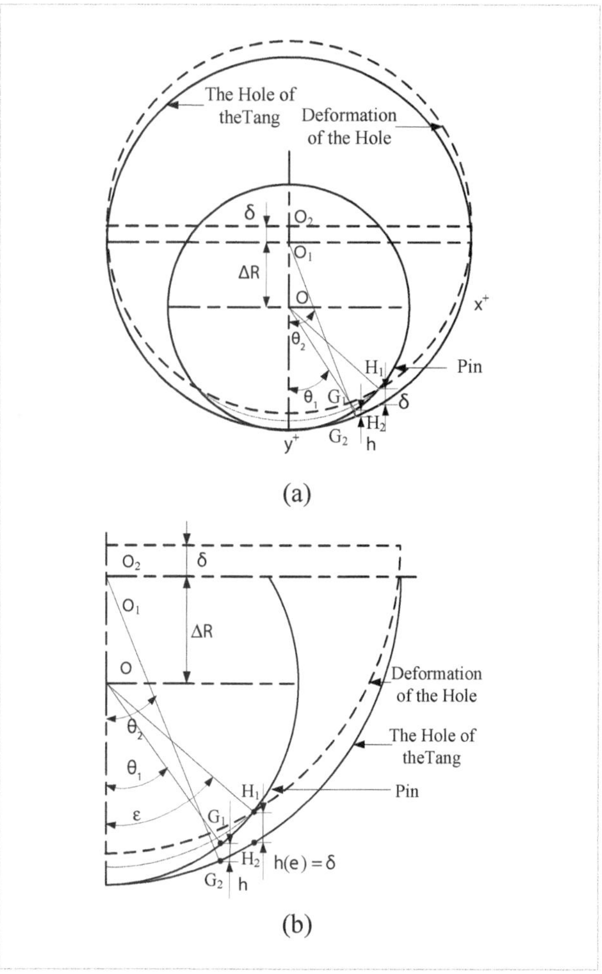

Figure 4. The geometry model of revolute joint with clearance: (a) integral geometry model and (b) A quarter of the geometry model.

Based on the relationship of contact width with pin radius and contact semiangle $a = R1.sin\varepsilon$ which is shown in Fig. 4, Eq. 20 can be expressed as

$$\delta = -\Delta R + \sqrt{R_2^2 - a^2} - \sqrt{R_1^2 - a^2} \quad (21)$$

Eq. 21 can also be expressed as

$$(\delta + \Delta R)\left(\sqrt{R_2^2 - a^2} + \sqrt{R_1^2 - a^2}\right) = R_2^2 - R_1^2 \quad (22)$$

So

$$\frac{R_2^2 - R_1^2}{(\delta + \Delta R)} = \sqrt{R_2^2 - a^2} + \sqrt{R_1^2 - a^2} \quad (23)$$

By adding the Eq. 23 to Eq. 21, $\sqrt{R_1^2 - a^2}$ can be eliminated. The contact width a can be given by

$$a = \sqrt{R_2^2 - \frac{1}{4}\left(\frac{R_2^2 - R_1^2}{\delta + \Delta R} + \delta + \Delta R\right)^2} \quad (24)$$

Approximate model for revolute joint with clearance

The difficulty of establishing the contact model of revolute joints with clearance is how to determine the distribution of contact pressure and stress in the contact body when the geometry constraint is satisfied. In this article, the research is limited to the case that deformations of contact bodies are elastic and there is no friction. Based on the assumptions that: the shape of the contact area of the pin and role satisfies the geometric constraint given by Eq. 24; based on the half-space theory, the inner stress distribution of contact bodies has been considered; the contact pressure is assumed as ellipse distribution of Hertz theory. The deformation of the pin can be given by (Johnson, 1985)

$$\delta_b = 2P\frac{1-\nu^2}{\pi E}\left(2ln\left(\frac{4R_1}{a}\right)\ 1\right) \quad (25)$$

Because the clevis and tang contact with pin at relative direction, the total deformation of contact pair of clevis and pin is

$$\delta = 2\delta_b \quad (26)$$

Substituting Eq. 21 into Eq. 26, the contact load on per unit length of cylinder can be expressed as

$$P = \frac{\pi E}{4\left(1-\nu^2\right)}\frac{-\Delta R + \sqrt{R_2^2 - a^2} - \sqrt{R_1^2 - a^2}}{\left(2ln(4R_1/a) - 1\right)} \quad (27)$$

The relationship of contact depth and force can be expressed by

$$P = \frac{\pi E \delta}{4\left(1-\nu^2\right)}\left(2ln\left(4R_1 / \sqrt{R_2^2-\frac{1}{4}\left(\frac{R_2^2-R_1^2}{\delta+\Delta R}+\delta+\Delta R\right)^2}\right)-1\right)^{-1} \quad (28)$$

Comparisons of three models and experiment for revolute joints with clearance

To verify the new contact model for revolute joint, the comparisons among the Hertz model, conformal model, and new model (experiment) are conducted. The results are presented in Table 1. The normal load P versus the normal displacement δ curve is shown in Fig. 5 which is obtained from the Hertz model given by Eq. 6, new model expressed by Eq. 28, simplest conformal model written as Eq. 14, and experimental data.

Observed from Fig. 5, it is found that the new contact model is closer to the experimental results than Hertz model and conformal model. The error of Hertz model decreases with the increase of the clearance. The errors of simplified conformal model and new model are small compared with Hertz model when the clearance is small. Because the contact of clearance revolute joint is close to conformal contact, the new model is more suitable for calculating the contact of revolute joint.

Figure 6 shows the penetration depth of simplest conformal and nonconformal contact when the radial clearance is 0.067 mm. It is clearly shown that the nonconformal and simplest conformal contact depth cannot be very large, because the contact depth has linear relationship with the square of contact width. The maximum displacement increases rapidly in the new model when the contact width is close to the radius of the surface. However, the penetration depth of conformal model and corresponding Hertz model always rise steadily with the contact width increase and is hard to reach a large value when the contact width approaches to the radius of the contact surface.

CONCLUSIONS AND FUTURE WORKS

This article investigates the conformal contact model of revolute joint with clearance. The revolute joint experiment is conducted to evaluate the nonconformal and the simplest polynomial fitting conformal model. By comparing the two models with experimental results, it

Table 1. The comparison of the Hertz and conformal model with experimental results.

Δ*R*(mm)	δ (×10⁻²mm)	Experiment	Hertz theory		Conformal model	
		p-value (N)	*p*-value (N)	Relative error (%)	*p*-value (N)	Relative error (%)
0.067	0.981	500	871	74.20	494	-1.20
	1.971	1000	1751	75.10	1410	41.00
	3.253	2000	2889	44.45	2988	49.40
	4.484	3000	3983	32.77	4836	61.20
0.094	1.156	500	1026	105.20	527	5.40
	2.308	1000	2054	105.40	1502	50.20
	3.232	2000	2871	43.55	2483	24.15
	4.484	3000	3983	32.77	4056	35.20
0.104	0.734	500	653	30.60	252	-49.60
	1.436	1000	1275	27.50	696	-30.40
	2.512	1640	2231	36.04	1607	-2.01

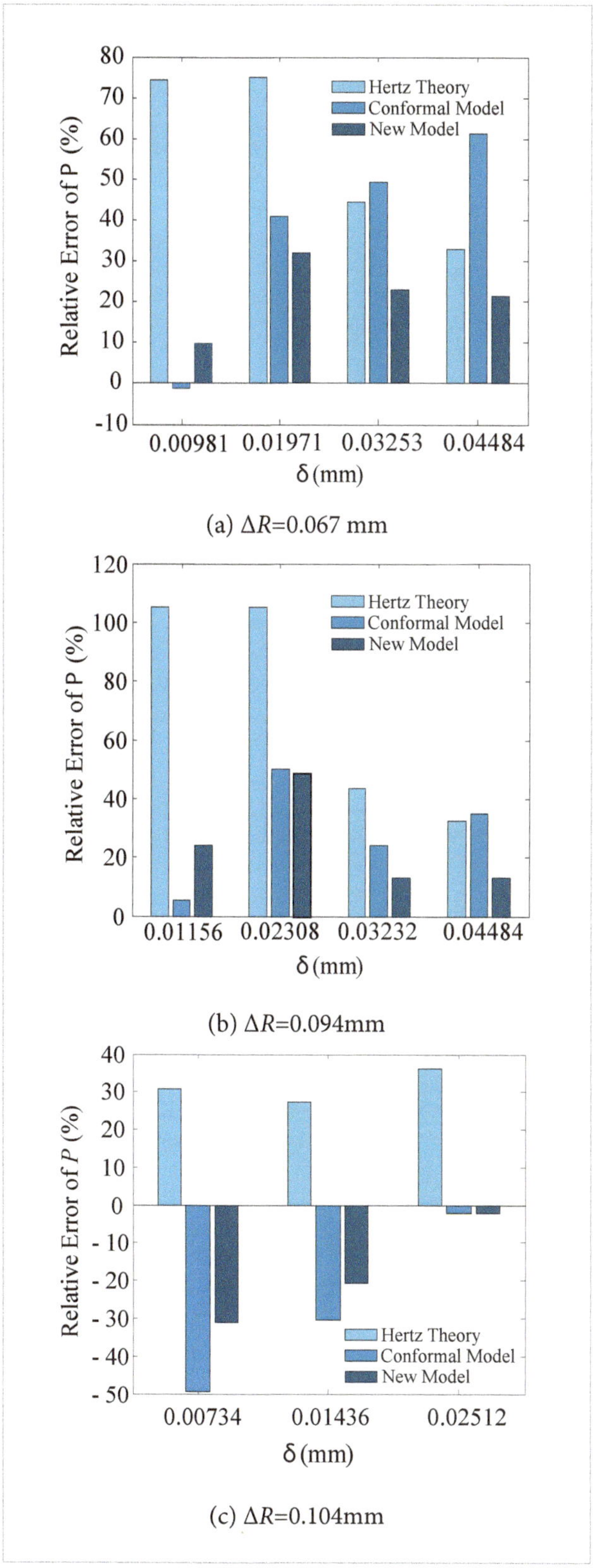

(a) ΔR=0.067 mm

(b) ΔR=0.094mm

(c) ΔR=0.104mm

Figure 5. Relation of the normal force and the total contact depth.

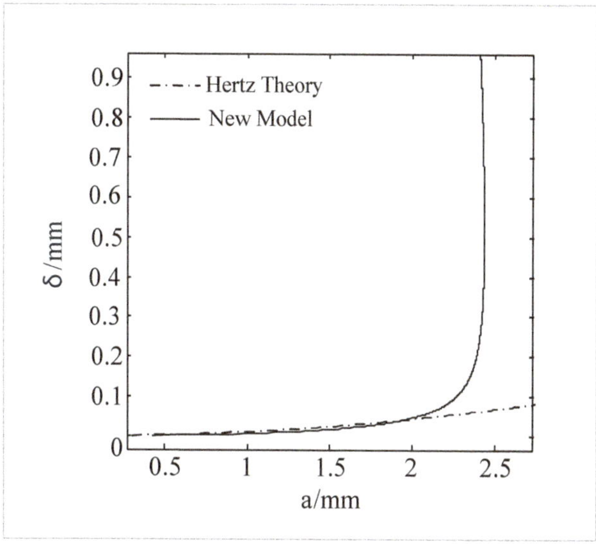

Figure 6. Relation of the contact width and the penetration depth.

is found that the Hertz model is just suitable to solve the revolute joint contact when the contact load is not very large. The stiffness calculated by the simplest conformal model is higher than the nonconformal model and it is more accurate when the clearance is not small. The new model is built to solve the conformal contact problem which is based on geometric restriction, the half space theory, pressure distribution of Hertz theory, elastic deformation, and frictionless. When the clearance is large, the errors of conformal model and new model are close to each other by comparing these models with experiment. The error of new model is the smallest when the clearance is not very large. And it has not changed rapidly with the increase of clearance. So the new model is more simple and reliable to calculate the contact of revolute joint.

Future work focuses on the dynamic characteristics of the revolute joints. The friction and damping will be considered. The corresponding dynamic experiment will be conducted to get the dynamic characteristic.

ACKNOWLEDGMENTS

The authors acknowledge the support from the Natural Science Foundation of China (Project No. 50935002&11002039).

REFERENCES

Abdo, J. and Shamseldin, E., 2005, "Modeling of Contact Area, Contact Force, and Contact Stiffness of Mechanical Systems with Friction", 2005 ASME International Mechanical Engineering Congress and Exposition, Orlando, pp. 1-8.

Bengisu, M.T., Hidayetoglu, T. and Akay A., 1986, "A Theoretical and Experimental Investigation of Contact Loss in the Clearances of a Four-Bar Mechanism", ASME, Vol. 108, pp. 237-244.

Dubowsky, S. and Freudenstein, F., 1971, "Dynamic analysis of mechanical systems with clearances, Part 1: Formulation of dynamic model", Journal of Engineering for Industry, Vol. 93, pp. 305-309.

Erkaya, S. and Uzmay, I., "Effects of balancing and link flexibility on dynamics of a planar mechanism having joint clearance", Sharif University of Technology Scientia Iranica, doi: 10.1016/j.scient.2012.04.011.

Flores, P. and Ambreósio, J., 2004, "Revolute joints with clearance in multibody systems", Computers and Structures, Vol. 82, pp. 1359-1369.

Flores, P., Ambrósio, J., Claro, J. C. P. and Lankarani, H. M, 2006a, "Dynamics of Multibody Systems with Spherical Clearance Joints", Transactions of the ASME, Vol. 1, pp. 240-247.

Flores, P., Ambrósio, J., Claro, J.C.P., Lankarani, H.M. and Koshy, C.S., 2006b, "A Study on Dynamics of Mechanical Systems Including Joints with Clearance and Lubrication", Mechanisms and Machine Theory, Vol. 41, pp. 247-261.

Flores, P., Leine, R. and Glocker. C., 2010, "Modeling and Analysis of Planar Rigid Multibody Systems with Translational Clearance Joints Based on the Non-smooth Dynamics Approach", Multibody System Dynamics, Vol. 23, pp. 165-190.

Ingham, M.D. and Crawley E.F., 2001, "Micro Dynamic Characterization of Modal Parameters for a Deployable Space Structure", AIAA, Vol. 39, pp. 331-338.

Jia, X., Jin, D., Ji, L. and Zhang, J., 2002, "Investigation on the Dynamic Performance of the Tripod-ball Sliding Joint with Clearance in a Crank-sliding Mechanism. Part 1. Theoretical and Experimental Results", Journal of Sound and Vibration, Vol. 252, pp. 919-933.

Johnson, K.L., 1985, "Contact Mechanics", Cambridge University Press, Cambridge, 99 p.

Khemili, I. and Romdhane, L., 2008, "Dynamic analysis of a flexible slider-crank mechanism with clearance", European Journal of Mechanics A/Solids, Vol. 27, pp. 882-898.

Knight, M.G., de Lacerda, L.A., Wrobel, L.C. and Henshall, J.L., 2002, "Parametric Study of the Contact Stresses Around Spherical and Cylindrical Inclusions", Computational Materials Science, Vol. 25, pp. 115-121.

Lake, M.S. and Lee, D., 1996, "A revolute joint with linear load-displacement response for precision deployable structures", The 37th AIAA/ASME/ASCE/AHS/ASC Structures, Structural Dynamics, and Materials Conference, Salt Lake City, pp. 1-14.

Lankarani, H.M. and Nikravesh, P.E., 1990, "A contact force model with hysteresis damping for impact analysis of multibody systems". Journal of Mechanical Design, Vol. 112, pp. 369-376.

Larsen, J.J., Jenkins, C., Banik, J. and Murphey, T., 2009, "Continuum Structural Representation of Flexure and Tension Stiffened 1D Spacecraft Architectures", 50th AIAA/ASME/ASCE/AHS/ASC Structures, Structural Dynamics, and Materials Conference, California, pp. 1-13.

Liu, C.S., Yang, L. and Zhang, K., 2006, "Normal Force-Displacement Relationship of Spherical Joints with Clearances", Transactions of the ASME, Vol. 1, pp. 160-167.

Meguro, A., Shintate, K., Usui, M and A. Tsujihata., 2009, "In-orbit Deployment Characteristics of Large Deployable Antenna Reflector Onboard Engineering Test Satellite VIII", Acta Astronautica, Vol. 65, pp. 1306-1316.

Olyaei, A.A. and Ghazavi, M.R., 2012, "Stabilizing slider-crank mechanism with clearance joints", Mechanism and Machine Theory, Vol. 53, pp. 17-29.

Parenti-Castelli, V. and Venanzi, S., 2005, "Clearance Influence Analysis on Mechanisms", Mechanism and Machine Theory, Vol. 40, pp. 1316-1329.

Qi Z.H., Xu, Y., Luo, X. and Yao, S., 2010, "Recursive Formulations for Multibody Systems with Frictional Joints Based on the Interaction Between Bodies", Multibody System Dynamics, Vol. 24, pp. 133-166.

Ravn, P., 1998, "Continuous Analysis Method for Planar Multibody Systems with Joint Clearance". Multibody System Dynamics, Vol. 2, pp. 1-24.

Rhee, J. and Akay, A., 1996, "Dynamic response of a revolute joint with clearance", Mechanisms and Machine Theory, Vol. 31, pp. 121-134.

Schwab, A.L., Meijaard, J.P. and Meijers, P., 2002, "A Comparison of Revolute Joint Clearance Models in the Dynamic Analysis of Rigid and Elastic Mechanical Systems", Mechanism and Machine Theory, Vol. 37, pp. 895-913.

Shi, B. and Jin, Y., 2008, "Dynamic analysis of the reheat-stop-valve mechanism with revolute joint in consideration of thermal effect", Mechanism and Machine Theory, Vol. 43, pp. 1625-1638.

Tian, Q., Zhang, Y., Chen, L. and Flores, P., 2009, "Dynamics of spatial flexible multibody systems with clearance and lubricated spherical joints", Computers and Structures, Vol. 87, pp. 913-929.

Zhang, X., and Vu-Quoc, L., 2001, "Modeling the Dependence of the Coefficient of Restitution on the Impact Velocity in Elasto-Plastic Collisions", International Journal of Impact Engineering, Vol. 27, pp. 317-341.

Aerospace Vehicle Development Applications of Atmospheric Thermodynamic Inputs

William Walton Vaughan[1], Dale Leroy Johnson[2]

ABSTRACT: Aerospace engineering has always had the need for various natural environment parameters to be used as inputs in research and engineering analyses used in the design and development of aircraft and launch/reentry vehicles. Although winds are indeed the main natural environment parameter used as inputs in vehicle design, the thermodynamic atmospheric parameters and models are also of great value and much needed as inputs. This paper will help the design engineer, chief engineer, or project manager understand the role that these thermodynamic parameters/models play.

KEYWORDS: Aerospace meteorology, Launch vehicle development, Mission operations, Atmospheric thermodynamic parameters/models.

INTRODUCTION

This paper is a follow-up to the article "Aerospace Meteorology: An Overview of Some Key Environmental Elements", published in the Journal of Aerospace and Technology Management (Vaughan and Johnson, 2013). It has presented an overview with background on how the natural environment plays a major role in the design and development of launch vehicles, with an emphasis on what is presented in the NASA technical report, "Terrestrial Environment (Climatic) Criteria Guidelines for Use in Aerospace Vehicle Development" (Johnson, 2008), as a help to the design engineer. This paper presents a few specific examples of how the atmospheric thermodynamic parameters (pressure, temperature and density) along with how thermodynamic model atmospheres play a significant role as input in engineering vehicle design and analysis.

Surface and upper air winds are usually the main natural environment driver used in launch vehicle design and development. A launch vehicle's flight control and structural systems are sensitive to extreme wind speed, turbulence, wind gust, and wind shear that may occur during launch and ascent (or lift-off) throughout the Earth's terrestrial atmosphere (from 0 to 90 km altitude).

However, the three thermodynamic parameters of the atmosphere (pressure, temperature and density) are also very important in aerospace vehicle planning, design, development, testing and launch/reentry. This paper presents a historical look at the use of some of these thermodynamic procedures/models. The NASA Terrestrial Environment (Climatic) Criteria Guidelines Technical Memorandum (Johnson, 2008); hereafter referred to

1.University of Alabama – Huntsville/AL – USA 2.NASA (retired) – USA
Author for correspondence: William W. Vaughan | 301 Sparkman Drive Northwest | 35899 Huntsville/AL – USA | Email: vaughan@nsstc.uah.edu

as TM (Johnson, 2008) has over the years been a major natural environments initial source document. It contains 13 key technical natural terrestrial environment parameters that can be used in the engineering design and development of launch/space vehicles. It presents the surface and in-flight thermodynamic parameters of the atmosphere in a statistical and modeling mode. The applicable model should be selected for design use based on the operational requirements for the aerospace vehicle. Mean and extreme values of these thermodynamic parameters can be used in application to many aerospace vehicle design and operational problems, such as (1) research planning and engineering design of remote Earth sensing systems, (2) vehicle design and development, and (3) vehicle trajectory analysis, dealing with vehicle thrust, dynamic pressure, aerodynamic drag, aerodynamic heating, vibration, structural and guidance limitations, and reentry analysis.

Model atmospheres have also been used and improved, starting with use of the U.S. Standard Atmosphere 1976 (Anon., 1976), and its predecessors, for any U.S. site in general. However, if sufficient surface and aloft atmospheric data (measurements) exist for the exact launch site location, a Range Reference Atmosphere (RRA) can then be developed specifically for that site. Finally a Global Reference Atmospheric Model (GRAM-2010) (Leslie and Justus, 2011), has been developed so that thermodynamic values *versus* altitude can be obtained for any site on planet Earth. A launch or a re-entry trajectory can also be run through GRAM.

Atmospheric density normally has been the main thermodynamic parameter affecting launch vehicle development for flight within the terrestrial environment. Initially, mean or median values of density (from standard or reference atmospheres for a specific site) were used as input for certain engineering calculations. Extremes of density, i.e. vertical profiles of maximum and minimum have also been used for all altitudes within the terrestrial atmosphere. However, this is very unrealistic in the real atmosphere as density is not an extreme at all terrestrial altitudes. Therefore, this brought about the construction of hot (summer) and cold (winter) atmosphere development for the various launch sites of interest to NASA, specific sites such as Kennedy Space Center (KCS)/FL, Vandenberg Air Force Base (VAFB)/CA, and Edwards Air Force Base (EAFB)/CA, etc. Finally, the NASA MSFC Global Reference Atmospheric Model (Leslie and Justus, 2011) provides in-flight atmospheric thermodynamic variables for all global geographical sites.

Another unique thermodynamic procedure developed over the years is also presented within this paper. The Buell statistical relationships (Buell, 1954; Buell, 1970) between the independent variables of atmospheric pressure (P), temperature (T), and density (ρ) has been developed. This method allows one to obtain simultaneous values of two thermodynamic variables at discrete altitude levels, given that the third variable is an extreme value. Whenever an extreme thermodynamic parameter like ρ is given, the associated P and T values can then be calculated from this statistically developed procedure (equations). Likewise, also for the other two atmospheric thermodynamic parameters, should they be extreme values.

Finally, if meteorological balloon and rocket data exist only up to about 55-km altitude, there is another statistical technique which extrapolates the thermodynamic data from the last level of measurement up to the 90 km terrestrial altitude level, with acceptable accuracy (Graves *et al.*, 1973). This extrapolation procedure is not presented within this paper, but can be obtained directly from Graves, et al.(1973).

The Standard Atmosphere section of TM (Johnson, 2008) presents a more detailed and complete overview of all atmospheric thermodynamic models and procedures than the ones presented here. Most of what is presented in this paper was taken from the Johnson (2008), and much of the description and text presented here has been taken directly from all the various references cited. Keep in mind that the TM (Johnson, 2008) is a terrestrial natural environment engineering applications document that has been maintained and updated by NASA for over the last fifty years, as it provides to the engineer or program/project manager all the various natural terrestrial environment statistics and models that can be used in the planning, design and development of aerospace vehicles and payloads.

What is presented within this paper is a discussion of (1) atmospheric density *versus* altitude, (2) the atmospheric thermodynamic parameters and how they were applied within the Space Shuttle Program design, (3) the Standard and Reference atmospheres, (4) typical hot and cold atmospheres, (5) the Earth Global Reference Atmospheric Model, and (6) the application of the Buell statistical relationships regarding an extreme thermodynamic parameter and the two associated parameters.

DISCUSSION

A general discussion of atmospheric density, atmospheric pressure, pressure decrease with altitude, atmospheric temperature, extreme cold compartment temperatures, and atmospheric density (surface and at altitude) variability is presented in detail in Johnson (2008). The other key or unique thermodynamic parametric relationships for this paper are discussed below.

IN-FLIGHT ATMOSPHERIC DENSITY AT ALTITUDE

The density of the atmosphere decreases rapidly with height, decreasing to one-half of the surface value at ~7 km altitude at mid latitudes. Density is also variable at a fixed altitude, with the greatest relative variability occurring at ~70 km altitude in the high northern latitudes (60°N). Other altitudes of maximum density variability occur around the surface and 16 km. Altitudes of minimum variability occur around 8, 24 and 90-km altitude. Figure 1 shows these altitude density tendencies for the Kennedy Space Center/FL area along with the extreme density envelope, and hot/cold density values as a percent of the Patrick Reference Atmosphere 1963 (Smith and Weidner, 1964). These extreme density values approximate the ±3 σ (corresponding to the normal distribution) density values. Figure 2 provides the associated virtual temperature profiles for the KSC extreme, and hot/cold, and PRA-63 Models.

Density varies with latitude in each hemisphere, with the mean annual density near the surface increasing toward the poles. In the region around 8-km altitude in the Northern Hemisphere; e.g. the density variation with latitude and season is small. Above 8-km to ~28-km, the mean annual density decreases toward the north. Mean monthly densities between 30-km and 90-km increase toward the north in July and toward the equator in January (Smith, 1964; Johnson, 2008). Drag on a reentering spacecraft, which is a direct function of atmospheric density at a given altitude for a specific vehicle, like the Space Shuttle, has varied up to 19% over a few seconds of flight time, resulting from "patchy" density variations (density "pot holes"). The designer must recognize that atmospheric density variations do occur, and they will highly influence engine performance, specific fuel consumption, drag, and flight control. GRAM-2010 (Leslie and Justus, 2011) has been designed to reproduce typical density variations that can be encountered along a given flight path and should be considered in vehicle design, both ascent and reentry.

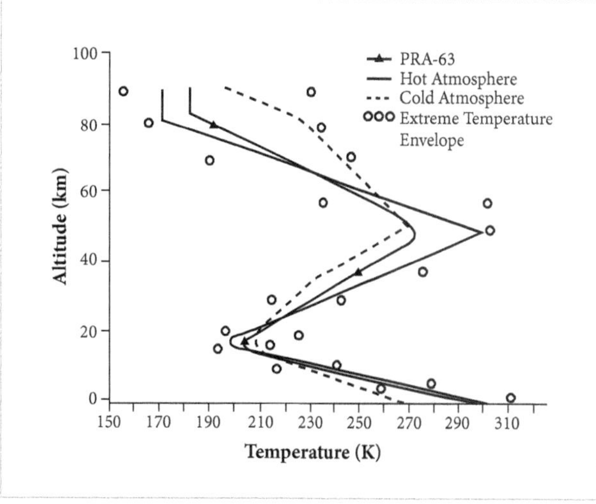

Figure 2. Extreme KSC virtual temperature profiles and the KSC hot & cold temperatures shown with the PRA-63 Temperature (Johnson, 2008).

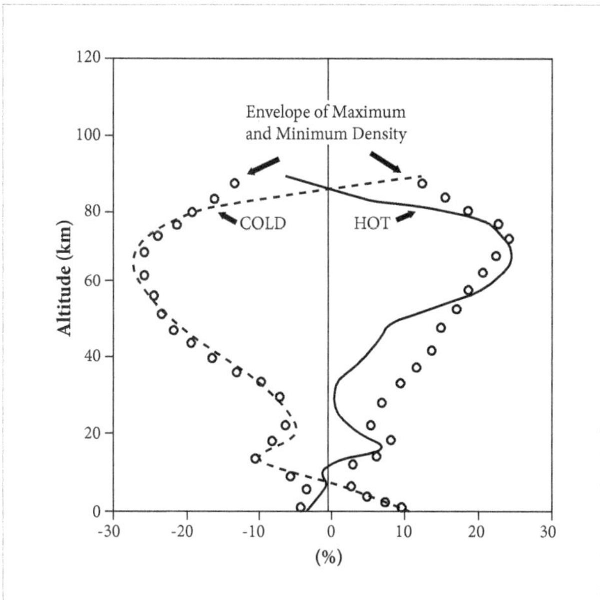

Figure 1. Relative deviations (%) of extreme KSC density profiles with respect to PRA-63 density. KSC hot & cold density also shown (Johnson, 2008).

THERMODYNAMIC PARAMETER REQUIREMENTS USED IN THE SPACE SHUTTLE PROGRAM

The Space Shuttle Requirements document, NASA NSTS 07700 Book 2, Volume X, Appendix 10.10 (Anon., 1999), contains very specific atmospheric wind and thermodynamic parametric requirements to use for the design and development of the Space Shuttle launch vehicle. Requirements are given with regard to the atmospheric thermodynamic parameters, as shown in the paragraph below, taken directly from cited document. Key models and purposes are presented. Most of these requirements refer back to the Terrestrial Environment

(Climatic) Criteria Guidelines document of 1973 (Daniels, 1973), which was base-lined early-on for the Space Shuttle program. Since the Earth GRAM was not developed when the Shuttle Requirements Document (Anon., 1999) was first assembled, it was not included initially.

NASA NSTS 07700 (Anon., 1999), contains the neutral atmosphere requirements. The following atmospheric models are base-lined for the Space Shuttle as indicated for the following engineering functions:

- NASA-GRAM: used in Vehicle Design (for Ascent/for Reentry, except at lower altitudes) along given flight path; used as subroutine in TRAJECTORY, ORBIT PROPAGATOR, or SIMULATIONS of IN-FLIGHT programs.
- NASA-Hot & Cold Atmospheres: used in ASCENT DESIGN for all ALTITUDES; used in REENTRY STUDIES from 30 km to SURFACE; used in DESIGN CALCULATIONS (aerodynamic heating, engine performance, and ferry operation); used for aerospace ferrying vehicles, in conjunction with hot or cold day design ambient air temperatures over runways, producing the extreme atmosphere resulting in the maximum vehicle design requirement.
- Cape Kennedy Reference Atmosphere (PRA-63): used as NOMINAL CRITERIA for SURFACE to ORBIT INSERTION, ABORT TRAJECTORY and ANALYSES for LAUNCHES; used as nominal criteria 30 km altitude to surface, for vehicle reentry; used for return to landing site (RTLS), and EXTERNAL TANK analysis and disposal design.
- Buell Extrapolation Technique: used for design analyses requiring a knowledge of the two atmospheric variables that are associated with a third extreme variable at discrete altitudes.
- U.S. Standard Atmosphere 1976: used as a STANDARD DAY for purposes of ENGINE RATINGS and COMPARISONS THEREOF (SEA level values); used in ORBITER ENTRY (90 km down to 30 km altitude); used for non-RTLS EXTERNAL TANK analysis and disposal design.

STANDARD ATMOSPHERE

A standard atmosphere is a vertical description of atmospheric temperature, pressure and density that is usually established by international agreement and taken to be representative of the Earth's atmosphere. The first standard atmospheres established by international agreement were developed in the 1920s, primarily for the purposes of pressure altimeter calibrations and aircraft performance calculations. Later, some countries, notably the United States, also developed and published standard atmospheres. The term standard atmosphere has, in recent years, also been used by national and international organizations to describe vertical descriptions of atmospheric trace constituents, the ionosphere, aerosols, ozone, atomic oxygen, winds, water vapor, planetary atmospheres, etc. (Vaughan, 2010).

The standard sea-level values of temperature, pressure, and density that have been used for decades are: temperature of 288.15 K, or 15°C; pressure of 1013.25 mbar or 760 mmHg; and density of 1225.00 g/m^3 (Anon., 1976).

The history of standard and reference atmospheres are presented and summarized in Vaughan (2010). Key atmospheric engineering models are given in Johnson (2008). The 1966 U.S. Standard Atmosphere Supplements (Anon., 1966) present different latitudinal atmospheres for the United States. It includes tables of temperature, pressure, density, sound speed, viscosity, and thermal conductivity for five northern latitudes (15, 30, 45, 60, 75 degrees), for summer and winter conditions.

REFERENCE ATMOSPHERES

The term reference atmosphere is used to identify vertical descriptions of the atmosphere for specific geographical locations or globally, such as the Range Reference Atmospheres (RRA) and the Earth-GRAM-2010 version 2 (Leslie and Justus, 2011). These RRA were developed by organizations for specific applications, especially as the aerospace industry began to mature after World War (Vaughan, 2010).

In design and preflight analysis of aerospace vehicles, average atmospheric models are used to represent the mean or median thermodynamic conditions with respect to altitude. For general worldwide design, the U.S. Standard Atmosphere, 1976 (Anon., 1976) has been used but site-specific atmosphere models are needed at each launch location. A group of 17 RRA from the RCC/MG as documented in Anon. (1984) have been prepared to represent the thermodynamic medians within the first 70-km altitude at various ranges and launch locations. So far, a total of over 29 different site RRA have been issued. Anon.(1966), Cole and Kantor (1978) – the supplemental atmospheres – together with GRAM (Leslie and Justus, 2011), are also useful in this regard. The NASA Marshall Earth – Global Reference Atmospheric Model – 2010 Version (Leslie and Justus, 2011) was constructed such that it provides a close approximation to the respective RRA. For the derivation of GRAM-2010, a total of 21 RRA from different sites have been utilized within GRAM-2010. These include the sites:

- Argentia, Newfoundland (St. Johns Airport)
- Ascension Island, Atlantic
- Barking Sands, Hawaii (Lihue)
- Cape Canaveral, Florida
- China Lake Naval Air Weapons Center, California
- Dugway Proving Ground (Salt Lake City), Utah
- Edwards Air Force Base, California
- Eglin AFB, Florida
- El Paso, Texas
- Fairbanks, Alaska
- Ft. Huachuca Elec Prvng Grnd (Tucson), Arizona
- Great Falls, Montana
- Kwajalein Missile Range, Pacific
- Nimes-Courbessac, France (STS TAL Site)
- Nellis AFB, Nevada (Mercury)
- Point Mugu Naval Air Weapons Center, California
- Taguac, Guam (Anderson AFB)
- Vandenberg AFB, California
- Wallops Island, Virginia (NASA)
- White Sands Missile Range, New Mexico
- Yuma Proving Ground, AZ (San Diego, California)

A major new feature of the GRAM-2010 (Leslie and Justus, 2011) is the optional ability to use data (in the form of vertical profiles) from a set of RRA as an alternate to the usual GRAM climatology. With this feature, it is possible, for example, to simulate a flight profile for an aerospace vehicle that takes off from the location of one RRA site, e.g. EAFB/CA using the range reference atmospheric data to smoothly transition into an atmosphere characterized by the GRAM climatology, then smoothly transition into an atmosphere characterized by a different RRA site, e.g. WSMR/NM, to be used as the landing site in the simulation. The user can also prepare data for any other site desired for use in this mode. The GRAM-2010 model will be discussed in more detail later in this paper.

The Patrick Reference Atmosphere (PRA-63) (Smith and Weidner, 1964) is a more extensive site specific annual reference atmosphere presenting data to 700-km altitude for KSC/FL. Because of the utility of this atmosphere, a simplified version is given extracted from Johnson (2008). Reference atmospheres are also available for VAFB (Carter and Brown, 1971; Johnson, 2008) and EAFB (Johnson, 1975; Johnson, 2008). These provide an annual reference atmosphere model to 700 km and have been designated as computer subroutines VRA-71 and ERA-75, respectively.

Currently, some of the most commonly used standard and reference atmospheres used in the U.S. include:

- COSPAR International Reference Atmosphere (CIRA), 1986
- ISO Standard Atmosphere, 1975
- NASA Earth Global Reference Atmosphere Model (GRAM), 2010
- NRL MSIS Reference Atmosphere, 2000
- RCC/MG Range Reference Atmospheres (RRA)
- U.S. Standard Atmosphere, 1976
- U.S. Standard Atmosphere Supplements, 1966

A detailed listing and description of many worldwide reference and standard atmospheric models is given in Vaughan (2010). In 1996, the American Institute of Aeronautics and Astronautics first published a *Guide to Reference and Standard Atmosphere Models* (Vaughan, 2010). This document has been updated since then and provides information on the principal features for over 70 global, regional, middle atmosphere, thermosphere, test ranges, Earth and planetary reference and standard atmospheric models.

EXTREME HOT AND COLD ATMOSPHERIC PROFILES FOR KSC, VAFB, AND EAFB

Johnson (2008), gives the two extreme density profiles that correspond to the summer (hot) and winter (cold) extreme atmospheres for KSC, VAFB and EAFB. See Johnson (1973) for VAFB and Johnson (1975) for EAFB, for detailed information pertaining to these extreme atmospheres. The associated values of temperature and pressure versus altitude are also tabulated. These extreme atmospheric profiles should be used in ascent design analyses at all altitudes. For reentry studies, they are to apply only from 30 km to the surface for vehicles to be used at KSC, VAFB, or EAFB. For those aerospace vehicles with ferrying capability, design calculations should use these extreme profiles in conjunction with the hot or cold day design ambient air temperatures. The extreme atmosphere producing the maximum vehicle design requirement should be utilized to determine the design.

The extreme envelopes of density deviations given in Fig. 1 imply that a typical individual extreme density profile may be represented by a similarly shaped profile, e.g. deviations of density are either all negative or all positive from sea level to 90-km altitude. However, examination of many individual density profiles shows that when large positive deviations of density occur at the surface, correspondingly large negative deviations will occur near 15-km altitude and above. Such a situation occurs during the winter season (cold atmosphere). The reverse is also true — density profiles with large negative deviations at lower levels will have correspondingly large positive deviations at higher levels. This situation occurs in the summer season (hot atmosphere) and is presented in Fig. 1.

The two extreme (hot & cold) KSC density profiles of Fig. 1 are shown as percent deviations from the Patrick Reference Atmosphere, 1963 density profile (Smith and Weidner, 1964). The two profiles obey the hydrostatic equation and the ideal gas law (i.e. the hypsometric relation). The extreme density profiles shown up to 30-km altitude were observed/measured in the atmosphere. The results shown above 30-km altitude are somewhat speculative because of the limited data from that region of the atmosphere. Quasi-isopycnic levels (levels of minimum density variation) are noted at approximately 8 and 86-km. Another level of minimum density variability is seen at ~24 km, and levels of maximum variability occur at zero, ~15 and ~68-km altitude. The associated extreme hot and cold virtual temperature profiles for KSC are given in Fig. 2. Temperatures below ~10-km altitude are virtual temperatures. Virtual temperature includes moisture to avoid computation of the specific gas constant for moist air (Johnson, 2008).

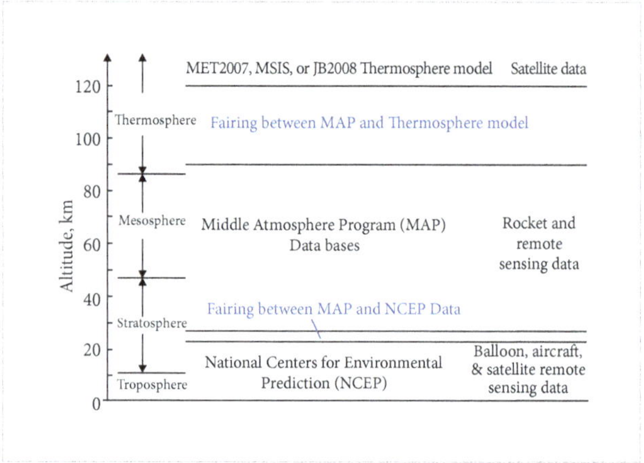

Figure 3. Summary of the three atmospheric regions in the GRAM-2010 program, sources for the model, and data on which the mean monthly GRAM-2010 values are based. See GRAM-2010 for more details (Leslie and Justus, 2011).

NASA-MSFC EARTH GRAM-2010

Reference or standard atmospheric models have long been used for design and mission planning of various aerospace systems. The NASA Marshall Space Flight Center Global Reference Atmospheric Model (Leslie and Justus, 2011) was developed in response to the need for a design reference atmosphere (using empirical data bases) that provides complete global geographical variability and complete altitude coverage (surface to orbital altitudes), as well as complete seasonal and monthly variability of the thermodynamic variables and wind components. Figure 3 provides a graphical summary of the data sources and height regions of GRAM-2010. In addition to providing the geographical, height, and monthly variation of the mean atmospheric state, it includes a perturbation model that has the ability to simulate spatial and temporal perturbations in these atmospheric parameters, if dispersions are desired (e.g., fluctuations due to turbulence and other atmospheric perturbation phenomena). When a large number of Monte Carlo type model profile dispersions are generated at any location, the mean and standard deviation of these data will match those of the observations. The model is statistically equivalent to available measurements. The $\pm 2\ \sigma$ envelopes encompasses ~95.45% of the observations, while $\pm 3\ \sigma$ encompasses ~99.73% of the observations (according to the normal distribution). This perturbation feature makes Earth GRAM especially useful for Monte Carlo dispersion analyses of guidance and control systems, thermal protection systems, and similar applications. Some of these applications have included operational support for Space Shuttle entry, flight simulation software for other vehicles, entry trajectory and landing dispersion analyses for the Stardust and Genesis missions, planning for aero-capture and aero-braking for Earth-return from lunar and Mars missions, 6 degree-of-freedom entry dispersion analysis for the Multiple Experiment Transporter to Earth Orbit and Return (METEOR) system, and the Crew Exploration Vehicle (CEV). GRAM can also input a spacecraft trajectory and output the various atmospheric parameters along this trajectory path.

The atmospheric model recommended for all reentry analyses, except lower altitudes specified in Johnson (2008), is GRAM-2010 (Leslie and Justus, 2011). GRAM is also suitable for use as a subroutine in a trajectory code or orbit propagator program or other programs used for simulations of in-flight or on-orbit atmospheric variability in density, temperature, or winds.

GRAM TRAJECTORY/RE-ENTRY EXAMPLE

Figure 4 presents a typical re-entry spacecraft trajectory plot of its latitude-longitude location with height and time as indicated. De-orbiting in from a 57° inclination orbit to land at EAFB, in January. Figure 5 shows this same typical trajectory passing through the GRAM-99 derived mean January cross-sectional map of density as a function of height (altitude) vs. longitude (shown as a ratio of U.S.76 Standard density). The density variability along the trajectory starts at ~20% higher density than the U.S.76 Standard, then goes into a region of ~25% lower than standard, and finally ~5% higher than standard before landing at Edwards. Figure 6 presents the same trajectory, and density results, as before; but now includes one Monte Carlo type density perturbation run along this trajectory. This time the trajectory is plotted showing the mean calculated GRAM-99 density, along with the ±2 σ densities, and the one perturbed density track, all plotted on a density ratio *versus* longitude graph. Note that the perturbed density does exceed the +-2 σ boundary from time to time, as ±2 σ represents approximately 95.45% of the observations, according to the normal distribution.

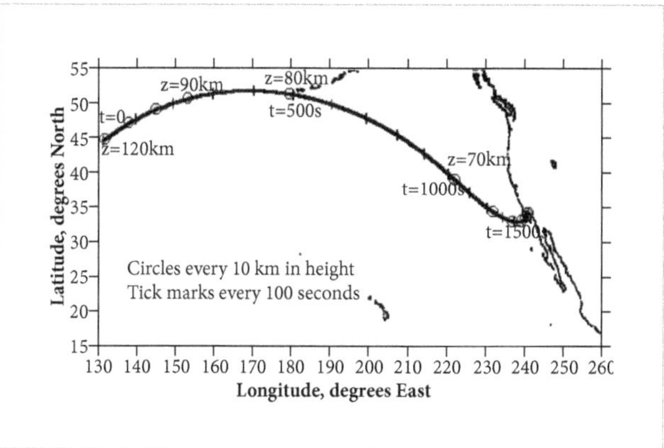

Figure 4. January Ground Track of a Typical Re-entry Trajectory (57° Inclination Orbit) Landing at Edwards AFB, CA).

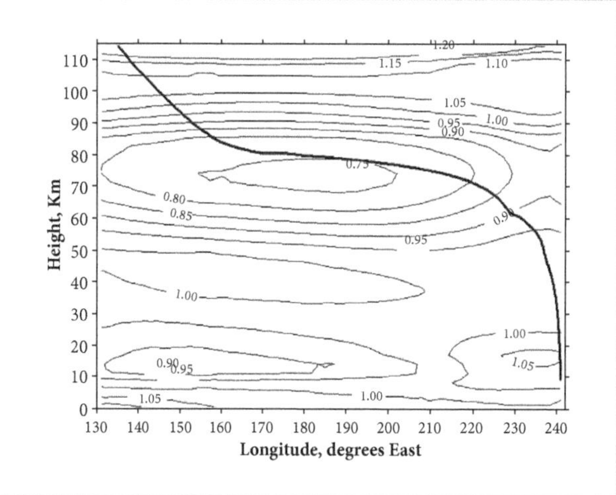

Figure 5. Typical trajectory through GRAM-99 Mean January Height *versus* Longitude Cross Section of Density (as ratio of US 76 Standard Density).

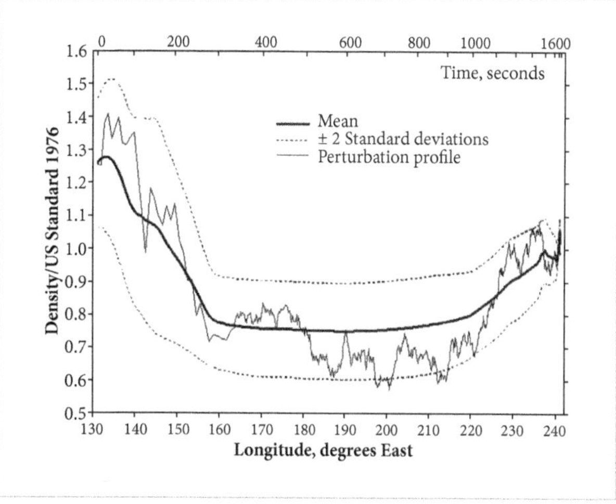

Figure 6. Typical trajectory through GRAM-99 Mean January Atmospheric Density with ±2 s Density Envelopes (95.45%) and one Monte-Carlo Density Perturbation Profile *versus* Longitude (as a Ratio of US76 Standard Density).

SIMULTANEOUS VALUES OF KSC TEMPERATURE, PRESSURE, AND DENSITY AT DISCRETE ALTITUDES

This section presents simultaneous values of atmospheric temperature, pressure, and density as guidelines for aerospace vehicle design considerations. The necessary assumptions and the lack of sufficient statistical data samples restrict the precision with which these data can currently be presented. The analysis is currently limited to KSC, FL only. The Buell (Buell, 1954; Buell, 1970) statistical relationships between atmospheric P, T, and ρ, when 1 is an extreme value, are presented here.

Method of Determining Simultaneous Values

An aerospace vehicle design problem that often arises in considering natural environmental data is stated by the following question: "How should the extremes (maxima or minima) of temperature, pressure, and density be combined (1) at discrete altitude levels and (2) versus altitude?" As an example, suppose one wants to know what temperature and pressure should be used simultaneously with a maximum density at a discrete altitude. From statistical principles set forth by Buell (1970), the solution results by allowing mean density plus three standard deviations to represent maximum density and using the coefficients of variations, correlations, and mean values as expressed in Eq. 1 below (Johnson, 2008):

$$\text{Maximum } \rho = \left(\bar{\rho} + 3\sigma_\rho\right) = \bar{\rho}\left(1 + 3\frac{\sigma_\rho}{\bar{\rho}}\right) = \bar{\rho}\left\{1 + 3\left[\underbrace{\left(\frac{\sigma_P}{\bar{P}}\right) r(P\rho)}_{A} - \underbrace{\left(\frac{\sigma_T}{\bar{T}}\right) r(\rho T)}_{B}\right]\right\}. \quad (1)$$

The associated values for pressure and temperature are the last two terms of Eq. 1, (A) and (B), multiplied by $\bar{P}$ and $\bar{T}$, respectively, and then this result is added to $\bar{P}$ and $\bar{T}$, respectively. Appropriate values of the thermodynamic correlation coefficients (r) and coefficients of variation (CV) are obtained from Johnson (2008).

In general, the three extreme ρ, P, and T equations of interest are:

$$\text{Extreme } \rho = \left(\bar{\rho} \pm M\sigma_\rho\right) = \bar{\rho}\left[1 \pm M\left(\frac{\sigma_\rho}{\bar{\rho}}\right)\right] = \bar{\rho}\left\{1 \pm M\left[\left(\frac{\sigma_P}{\bar{P}}\right) r(P\rho) - \left(\frac{\sigma_T}{\bar{T}}\right) r(\rho T)\right]\right\} \quad (2)$$

$$\text{Extreme } P = \left(\bar{P} \pm M\sigma_P\right) = \bar{P}\left[1 \pm M\left(\frac{\sigma_P}{\bar{P}}\right)\right] = \bar{P}\left\{1 \pm M\left[\left(\frac{\sigma_\rho}{\bar{\rho}}\right) r(P\rho) + \left(\frac{\sigma_T}{\bar{T}}\right) r(PT)\right]\right\} \quad (3)$$

$$\text{Extreme } T = \left(\bar{T} \pm M\sigma_T\right) = \bar{T}\left[1 \pm M\left(\frac{\sigma_T}{\bar{T}}\right)\right] = \bar{T}\left\{1 \pm M\left[\left(\frac{\sigma_P}{\bar{P}}\right) r(PT) - \left(\frac{\sigma_\rho}{\bar{\rho}}\right) r(\rho T)\right]\right\}, \quad (4)$$

where M denotes the multiplication factor to give the desired deviation. The values of M for the normal distribution and the associated percentile levels are shown in Johnson (2008). The two associated atmospheric parameters that deal with a third extreme parameter are listed in more detail in Table 1.

It must be emphasized that this procedure is to be used at discrete altitudes only, and holds for a specific site (KSC FL). Whenever extreme vertical profiles of pressure, temperature, and density are required for engineering application, the use of these correlated variables at discrete altitudes is not satisfactory. In TM (Johnson, 2008) there is a section (hot and cold atmospheres) that deals directly with this problem, since profiles of only extreme values of pressure, temperature, or density at every level from zero to 90-km altitude is unrealistic in the real atmosphere. GRAM can also be used to represent by Monte Carlo process the respective range of thermodynamic parameters for a given geological site.

Hot Surface/Cold Tropopause Temperature Example

The following equations from Table 1 will be used, for an Extreme Cold 18-Km Summer time Temperature example at KSC:

Table 1. Associated Parameters for Extreme Density, Temperature, and Pressure (Johnson, 2008).

	For Extreme Density	For Extreme Temperature	For Extreme Pressure
$P_{\text{assoc.}}$	$\bar{P}\left[1 \pm \left\{M\left(\frac{\sigma_P}{\bar{P}}\right) r(P\rho)\right\}\right]$	$\bar{P}\left[1 \pm \left\{M\left(\frac{\sigma_P}{\bar{P}}\right) r(PT)\right\}\right]$	
$T_{\text{assoc.}}$	$\bar{T}\left[1 \pm \left\{M\left(\frac{\sigma_T}{\bar{T}}\right) r(\rho T)\right\}\right]$		$\bar{T}\left[1 \pm \left\{M\left(\frac{\sigma_T}{\bar{T}}\right) r(PT)\right\}\right]$
$\rho_{\text{assoc.}}$		$\bar{\rho}\left[1 \pm \left\{M\left(\frac{\sigma_\rho}{\bar{\rho}}\right) r(\rho T)\right\}\right]$	$\bar{\rho}\left[1 \pm \left\{M\left(\frac{\sigma_\rho}{\bar{\rho}}\right) r(P\rho)\right\}\right]$

Use + sign when extreme parameter is maximum and use – sign when extreme parameter is minimum. M is the 1, 2, or 3 sigma (normal) for the extreme parameter in question.

$$\rho_{assoc.} = \bar{\rho}\left[1 \pm \left\{M\left(\frac{\sigma_\rho}{\bar{\rho}}\right) r(\rho T)\right\}\right] \quad (5)$$

and

$$P_{assoc.} = \bar{P}\left[1 \pm \left\{M\left(\frac{\sigma_P}{\bar{P}}\right) r(PT)\right\}\right] \quad (6)$$

And, for Extreme High Summer time Density at 18 km:

$$T_{assoc.} = \bar{T}\left[1 \pm \left\{M\left(\frac{\sigma_T}{\bar{T}}\right) r(\rho T)\right\}\right] \quad (7)$$

The following example is shown for the KSC 18-km altitude level during a hot summer day when the surface value of temperature is high (hot) and surface density low. Whereas the value of temperature at the18-km (tropopause) altitude level is very low (cold) while the atmospheric density is very high. To compute the Eqs. 5 and 6 values for the associated density and pressure for this KSC extreme case of low (cold) temperature at 18-km altitude, one just needs to insert the thermodynamic CV and r values from Table 2 into Eqs. 5 (and 6) of Table 1. The associated value of 18-km density then turns out to be:

$\rho_{assoc.}$ = 0.132392 [1 -{ 3 (0.0275) (-0.7904)}] or
$\rho_{assoc.}$ = 0.141025 kg/m^3

Which gives a higher density value than the annual PRA-63 density value does at 18-km. The percentage is 6.5% greater than the PRA-63 18-km density value (from Table 3).

Table 2. KSC atmospheric thermodynamic coefficients of variation (CV) and correlation coefficients (r) between *P*, *T*, and ρ, at 18-km altitude (from Johnson, 2008).

CV (%)		
$[\sigma_\rho/\bar{\rho}]$	$[\sigma_P/\bar{P}]$	$[\sigma_T/\bar{T}]$
2.75	1.75	1.70
[r]		
$r(P\rho)$	$r(PT)$	$r(\rho T)$
0.8036	-0.2706	-0.7904

CV = σ/mean. All parameters are unit less.

Atmospheric deviations as a percentage from the annual PRA-63 were computed by the following Eq. 8 (only density shown):

Table 3. Patrick Reference Atmosphere 1963 (PRA-63) Annual Thermodynamic Parameters of T, P, and ρ at 18 km altitude (from Johnson, 2008).

Altitude (km)	Mean Temperature (K)	Mean Pressure (mb)	Mean Density (kg/m³)
0	296.68	1017.01	1.18355
18	205.30	78.0974	0.132392

$$\%\Delta\rho = \frac{\rho_{\text{max or min}} - \rho_{PRA63}}{\rho_{PRA63}} \text{ x } 100 \quad (8)$$

Next, the associated value for pressure at 18 km, from Eq. 6, is calculated as follows:

$P_{assoc.}$=78.0974 [1-{ 3 (0.0175) (-0.2706)}] or
P_{assoc}=79.20689 mb

Which is 1.4% greater than the PRA-63 18-km pressure value at 18 km altitude.

Next, we should compute the calculated low temperature value at 18 km altitude, given a high value maximum density condition at that same altitude level. So, using Eq. 7, for a given high density extreme, we can compute the associated value of temperature as follows:

T_{assoc}=205.3 [1+{ 3 (0.0170) (-0.7904)}] or
T_{assoc}=197.02 K

Which is -4.0% of the PRA-63 temperature value at 18-km altitude.

Finally, we should determine if these calculated values of associated density, pressure, and temperature, given an extreme parametric condition at 18-km altitude (i.e. extreme low temperature, extreme high density), are close to reality or not. We have obtained a few extremely hot summer day vertical sounding measurement taken at KSC. One sounding in particular (June 16, 1958) is very extreme at the surface (hot temperature) and at 18-km altitude (cold temperature), as presented in Table 4. By the way, this sounding was one of the key soundings used in the derivation of the KSC Hot atmosphere.

Comparing the Buell calculated values of associated pressure, temperature and density at altitude of 18 km with the measured observations (Radiosondes-Raob) values gives good results as shown in Table 5. The three parametric value percent differences are all within 0.8% of the measured Raob values. This is to be expected due to the fact that the inter- and intra-level correlations are very high at and between the surface level and the tropopause level for temperature and density.

Table 4. Actual KSC Atmospheric Raob Sounding Measurement for 1958-06-16-00Z.

Altitude (km)	Temperature (K)	Pressure (mb)	Density (kg/m³)
Sfc.	307.46	1010.00	1.1366
18	195.5	79.7707	0.1422

SUMMARY REMARKS

This paper presents some of the function and use that atmospheric thermodynamic parameters of pressure, temperature and density have had in the design and development of aerospace launch and reentry vehicles for flight through the terrestrial atmosphere. The application and use of these atmospheric models, or statistical values, will help the program/project managers and engineers with useful atmospheric inputs for engineering design, usage and trade studies. Most of the information presented here came from (Johnson and Vaughan, 2012) and especially (Johnson, 2008).

Table 5. Calculated *versus* KSC Measured Values of T, P, and ρ at 18 km altitude.

Calculations	Temperature	Pressure	Density
PRA-63 values	205.30 K	78.09740 mb	0.132392 kg/m³
Buell Computed (% of PRA-63)	197.02 K	79.20689 mb	0.141025 kg/m³
	-4.0 %	**+1.4 %**	**+6.5 %**
Raob Measured (% of PRA-63)	195.50 K	79.77020 mb	0.142200 kg/m³
	-4.8 %	**+2.1 %**	**+7.4%**
KSC-Hot values (% of PRA-63)	200.0 K	81.36950 mb	0.141732 kg/m³
	-2.6 %	**+4.2 %**	**+ 7.1 %**
(% of Computed from Measured)	**+0.78 %**	**-0.71 %**	**-0.83 %**

REFERENCES

Anon., 1966, "U.S. Standard Atmosphere Supplements, 1966", U.S. Government Printing Office, Washington, DC, 1966, pp. 3-36.

Anon., 1976, "U.S. Standard Atmosphere, 1976", U.S. Government Printing Office, Washington, DC, October 1976.

Anon., 1984, "Range Reference Atmosphere Documents" published by Secretariat, Range Commanders Council (RCC), Meteorology Group (MG), White Sands Missile Range, NM.

Anon., 1999, "NASA NSTS 07700 Natural Environment Space Shuttle Flight and Ground System Specification, Book 2, Vol. 10, Apx. 10.10".

Buell, C.E., 1954, "Some Relations Among Atmospheric Statistics", Journal of Atmospheric Sciences, Vol. 11, No. 3, pp. 238-244.

Buell, C.E., 1970, "Statistical Relations in a Perfect Gas", Journal of Applied Meteorology, Vol. 9, No. 5, pp. 729-731.

Carter, E.A. and Brown, S.C., 1971, "A Reference Atmosphere for Vandenberg AFB, California, Annual (1971 Version)," NASA/TM X-64590, NASA Marshall Space Flight Center, AL.

Cole, A.E. and Kantor, A.J., 1978, "Air Force Reference Atmospheres", AFGL-TR-78-0051, Air Force Surveys in Geophysics, No. 382.

Daniels, G.E., 1973, "Terrestrial Environment (Climatic) Criteria Guidelines for Use in Aerospace Vehicle Development, 1973 Revision", NASA/TMX-64757.

Graves, M.E., Lou, Y.S. and Miller, A.H., 1973, "Specification of Mesospheric Density, Pressure, and Temperature by Extrapolation", NASA/CR-2223, March 1973. Retrieved in January 23, 2013,from http://ntrs.nasa.gov/archive/nasa/casi.ntrs.nasa.gov/19730012637_1973012637.pdf

Johnson, D.L., 1973, "Hot and Cold Atmospheres for Vandenberg AFB, California (1973 Version)", NASA TM-64756, NASA Marshall Space Flight Center, AL.

Johnson, D.L., 1975, "Hot, Cold, and Annual Reference Atmospheres for Edwards Air Force Base, California (1975 Version)", NASA/TM X-64970, NASA Marshall Space Flight Center, AL.

Johnson, D.L., 2008, "NASA/TM-2008-215633 Terrestrial Environment (Climatic) Criteria Guidelines for Use in Aerospace Vehicle Development, 2008 Revision", Dec. 2008. Retrieved in January 23, 2013, from http://ntrs.nasa.gov/archive/nasa/casi.ntrs.nasa.gov/20090022159_2009021428.pdf

Johnson, D.L. and Vaughan, W.W., 2012, "How Atmospheric Thermodynamic Parameters and Model Atmospheres Have Been Used to Help Engineering in Aerospace Launch Vehicle Design & Development", 50th AIAA 2012 Aerospace Sciences Meeting, Nashville, TN.

Leslie, F.W. and Justus, C.G., 2011, "The NASA Marshall Space Flight Center Earth Global Reference Atmospheric Model-2010 Version 2", NASA/TM—2011-216467.

Smith, J.W., 1964, "Density Variations and Isopycnic Layers", Journal of Applied Meteorology, Vol. 3, No. 3, pp. 290-298.

Smith, O.E. and Weidner, D.K., 1964, "A Reference Atmosphere for Patrick AFB, Florida, Annual (1963 Revision)", NASA/TM X-53139, NASA Marshall Space Flight Center, AL.

Vaughan, W.W., 2010, "Guide to Reference and Standard Atmospheric Models", AIAA G-003C-2010, American Institute of Aeronautics and Astronautics, Reston, VA.

Vaughan, W.W. and Johnson, D.L., 2013, "Aerospace Meteorology: An Overview of Some Key Environmental Elements", Journal of Aerospace Technology and Management, Vol. 5, No. 1, pp. 7-14. doi: 10.5028/jatm.v5i1.188.

18

Design, Fabrication and Flight Demonstration of a Remotely Controlled Airship for Snow Scientists

Rajkumar Sureshchandra Pant[1]

ABSTRACT: A remotely controlled airship was designed, fabricated and demonstrated within a tight timespan of under a month after receiving the go-ahead. The main design requirement for this airship was to be able to operate from a Helipad located at an altitude of 6,572 feet AMSL under ISA+20 deg.C. The images of the terrain below were recorded during the flight and transmitted in real-time to a ground based system using an onboard telemetry system. The paper describes the methodology followed for sizing of the envelope and key components of the airship, and the procedure followed for in-house fabrication and testing. The major issues that cropped up during the operation of the airship in harsh weather conditions of rain and mild snow, as well as at night, are also highlighted. The demonstration established the efficacy of remotely controlled airships for aerial photography and data collection by snow scientists.

KEYWORDS: Lighter-Than-Air systems, Airships, Blimps.

INTRODUCTION

Aerial surveillance usually requires monitoring the activities of people and/events over a designated area and long period of time from an airborne platform. A monitoring process such as this would essentially be based on steady visuals from the area of interest. The design features and attributes that an aerial surveillance airborne platform must possess should help accomplish this mission completely and efficiently. For instance, such a system should have the ability to take-off and land vertically or from limited areas, since such airborne platforms may need to operate from remote locations, some of which might lack large flat open areas to get airborne and/or land.

The onboard surveillance equipment suite has to steadily focus on the desired area on the ground, from a particular altitude. The quality of the surveillance is directly related to the level of stability that the platform can offer. Hence the airborne platform must be able to hover at an altitude and stay put without causing serious fluctuations to the dynamic surveillance data collection. The platform should be considerably inert to disturbances like crosswind, gust, wavering of the payload surveillance equipment due to maneuvering and induced vibrations by the platform itself.

The platform should also require minimal expertise for operation, and the cyclic procedure from storage to deployment should be prompt and easy. Another important attribute is low cost of manufacture and operation of the system, not to mention the strict environmental regulations that need to be adhered to in this age of rapid degradation of our planet. Hence an airship is arguably the most suitable and preferred prospect

1.Indian Institute of Technology Bombay – Mumbai/Maharashtra – India
Author for correspondence: Rajkumar Sureschandra Pant | Adi Shankaracharya Marg, Powai | 400076 Mumbai/Maharashtra – India | Email: rkpant@aero.iitb.ac.in

for an aerial surveillance system, taking into account all the above mentioned desirable attributes of such a system.

Airships are aerodynamically shaped bodies filled with a "Lighter-Than-Air" (LTA) gas that displaces the ambient air, which results in a net upward force due to buoyancy. This is called static (or buoyant) lift, since it is due to the inherent buoyant force, which is generated even when airship is stationary. In addition, an airship also can generate dynamic lift due to the action of aerodynamic forces acting on it as it moves through the air, just like an airplane. Once airborne, airships can perform much like helicopters, remaining nearly geo-stationary for extended periods of time, albeit with much lesser fuel consumption, and noise and vibration levels. A remotely controlled (RC) airship is perhaps much more suitable than a remotely controlled aircraft for aerial surveillance due its long endurance loiter and lower fuel consumption.

Design, fabrication and flight testing of prototypes of several LTA systems for various scientific applications has been carried out at IIT Bombay since 2002 (Gawale and Pant, 2002). Some recent projects include indoor remotely controlled airships for neural network control hardware implementation (Sangole *et al.*, 2006), aerostats as platforms for low-cost re-locatable wireless communications systems (Gawande *et al.*, 2007), outdoor remotely controlled airships for product promotion (Gawale *et al.*, 2008), aerial river ferry using superheated steam-filled balloons (Banerjee *et al.*, 2008), and aerostats for snow cover evaluation (Bhandari and Raina, 2009). The following sections highlight the design process and fabrication.

DESIGN PROCESS

An RC airship can broadly be divided into the five main components, viz., Envelope, Stabilizer and Fins (empennage), Gondola, Propulsion system and Remote Control system. The conceptual design and sizing of the airship is based on the methodology developed by Pant (2008), utilizing the aerostatic and aerodynamic analysis described in Cheeseman (1999). This methodology arrives at the baseline specifications of a manned airship with inputs on the user specified performance and operational requirements and has been suitably adapted by Gawale *et al.* (2008) for the design and sizing of RC airships. The scheme of arriving at baseline specifications of an airship to meet the user-specified operating and performance requirements is shown in Fig. 1.

DESIGN REQUIREMENTS

The key requirement of the RC airship was the ability to deploy the airship from a Helipad located at an altitude of 6,572 ft. AMSL, while carrying a payload camera that could take high resolution photographs of the terrain below and transmit them in-real time. As we shall observe later, the airship had to operate in very cold weather at a high altitude and the envelope material selection and fabrication process had to conform to such harsh environmental conditions. In addition, the airship specifications included limitations on its length and volume. As mentioned earlier, adequate stability of the airship during flight would obviously generate clear photographs for surveillance and this requires the airship to be controlled effectively. The stabilizers and fins were also designed keeping this requirement in mind. Of all these, the prime concern was of the propulsion system which needed to operate at altitudes where density of the air is lower than at sea level. Table 1 lists some of these key design requirements for the airship.

Table 1. Airship design requirements.

Parameters	Value
Payload	3 kg
Operating altitude	200 ft AGL
Base station altitude	6,572 ft AMSL
Design wind speed	15 m/s
Off Standard temperature	20°C
Operational time	15 days

Apart from the above mentioned requirements, which describe a more involved design process, more often than not we also have the constraint on lowering the cost of manufacturing and operation to a level acceptable to user.

The following sections briefly explain the design of the important components of the RC airship.

ENVELOPE DESIGN

The envelope is the most crucial component of the airship and its design demands to optimally integrate the aspects of aerodynamics, stability and payload. Based on the constraints of length and payload, the GNVR shape was chosen for the envelope profile. The GNVR shape shown in Fig. 2 consists of three standard

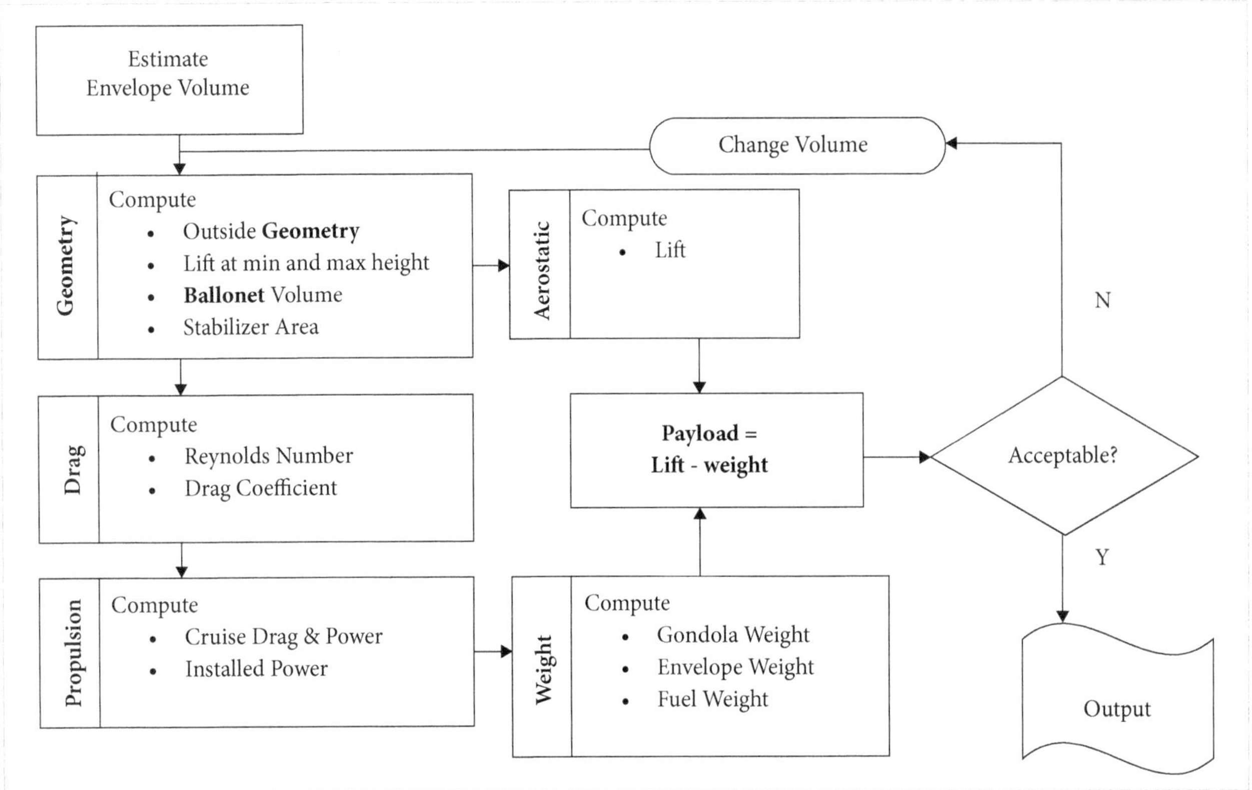

Figure 1. Scheme for determination of baseline specifications of an airship.

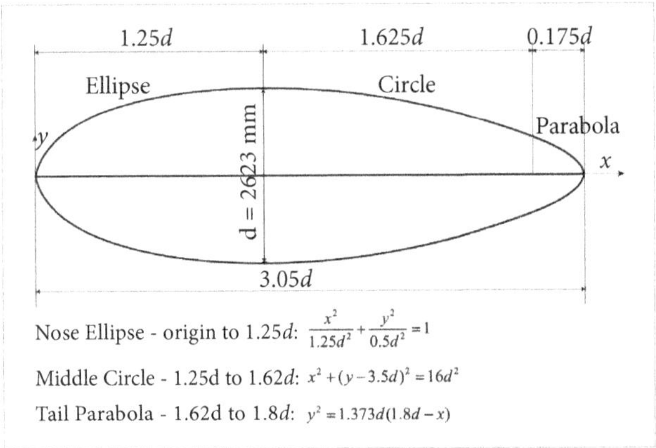

Figure 2. Geometrical Construct of GNVR profile.

sections, and its entire geometry is analytically parameterized in terms of its max diameter *d*. The length and volume of the airship GNVR envelope shape are 8 m and 26.64 m^3. Experimental (Narayana and Srilatha, 2000) and numerical (Sundaram, 2000) studies have shown that this shape is very well suited for tethered aerostats operating at an altitude up to 1 km.

To optimize the fabric thickness, the total load acting on the envelope was calculated as the summation of circumferential and longitudinal loads respectively as given by the equation:

$$L_{tot} = P_{\text{int}} \cdot r \pm \frac{M}{\pi r^2} \tag{1}$$

where M is the bending moment and P_{int} is the internal static pressure.

The pressure distribution on the outer envelope surface has been calculated to identify the points with maximum differential pressure. The total differential pressure was estimated using the equation given below:

$$P_{tot} = P_{aero} + P_{\text{int}} + \rho_{he} gh - \rho_{air} gh \tag{2}$$

where P_{aero} is the dynamic pressure, P_{int} is the internal static pressure and h is the radius of the envelope.

The center of pressure of this section is located at 33% of its length, which is also the point at which maximum differential pressure occurs. The flow over the hull is assumed to be turbulent and the drag is estimated in terms of drag coefficient C_{DV}, maxitmum velocity, volume and the reference area from the formula

$$D = \frac{1}{2} C_{DV} \rho V^2 S \tag{3}$$

The envelope material has to be appropriately chosen to ensure adequate strength, durability, as well as low weight. A qualitative comparison of the properties of various fabrics that are used for load bearing was carried out and a light weight, yellow coloured PVC fabric was used, keeping in mind the weight, cost and availability constraints. The key properties of this material are listed in Table 2 (Gupta and Malik, 2002).

Table 2. Properties of envelope material used for airship.

Property	Value
H_2 permeability @ 25 cm H_2O column	3.75 liters/m²/day
Break strength along warp direction	1.68 kg/cm
Break strength along weft direction	1.76 kg/cm
Specific weight	0.14 kg/m²
Thickness	0.11 mm

STABILIZER AND RUDDER

The stabilizer and rudder have been designed to have a cruciform shape on the envelope and the procedure for sizing has been adopted from Gawale and Pant (2002) by scaling down the dimensions of the stabilizer and rudder of existing airships. However, care has been taken to ensure that the scaling is done using data of airships which have envelopes of similar shapes and fineness ratios. A schematic view of the stabilizer for the airship is shown in Fig. 3 and its relevant dimensions are tabulated in Table 3.

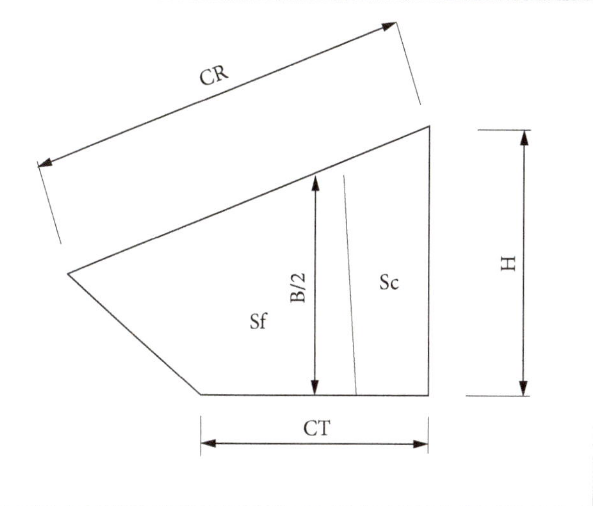

Figure 3. Schematic of the stabilizer & rudder.

Table 3. Dimensions of the stabilizers.

Parameter	Fixed surface	Moving surface
Surface area	0.61 m² (S_f)	0.13 m² (S_c)
C_T (tip chord)	0.86 m	0.17 m
C_R (root chord)	1.45 m	0.30 m
H (height)	0.86 m	
$B/2$ (mean half span)	0.69 m	

PROPULSION SYSTEM

A detailed overview of the power plant system, design issues in engine sizing and selection, advances in engine technology, various concepts for thrust vectoring and a methodology for sizing and selection of design features of an airship engine has been outlined by Gawale and Pant (2004). The specifications of the engine used for this airship have been listed in Table 4.

Table 4. Specifications of the engine.

Type	0.15 LA-S (OSMG1415)
Displacement	2.49 cc
Bore	15.2 mm
Stroke	13.7 mm
RPM	2,500 to 18,000
Power output	0.41 BHP @ 17,000 rpm
Mass	0.13 kg

STABILITY ANALYSIS

For a fully stable flight, the center of gravity (CG) of the airship should be exactly at the same location as that of the center of buoyancy (CB). In order to ensure this, a detailed mass breakup and CG analysis was carried out (Table 5). As shown in Table 5, the airship CG and CB were estimated to be located 3.79 and 3.64 m from the nose, respectively. This small difference in the CG and CB location helped to create a slightly unstable airship which would be easily maneuverable.

Table 5. Component mass and center of gravity distribution.

Component	Weight (kg)	CG location from nose (m)
Envelope	12.53	3.74
Fin velcro strip + hooks	0.59	6.60
Fins	1.32	6.97
Batons	0.70	0.50
Gondola	3.00	3.01
Airship assembly	18.14	3.82
Ballast	3.37	3.64
Total weight	21.51	3.79

FABRICATION

This section briefly outlines the fabrication methods of the components of the airship.

ENVELOPE

The envelope is usually constructed by joining smaller sections called petals, as seen in Fig. 4, which are cut out from the fabric sheets using a shape template. The shape template consists of the cross-sectional shape of the envelope generated using the coordinates of the GNVR profile. The envelope petals were then seamed together, as shown in Fig. 5, depending on their orientations using radio frequency (RF) heat welding, thereby forming the whole envelope.

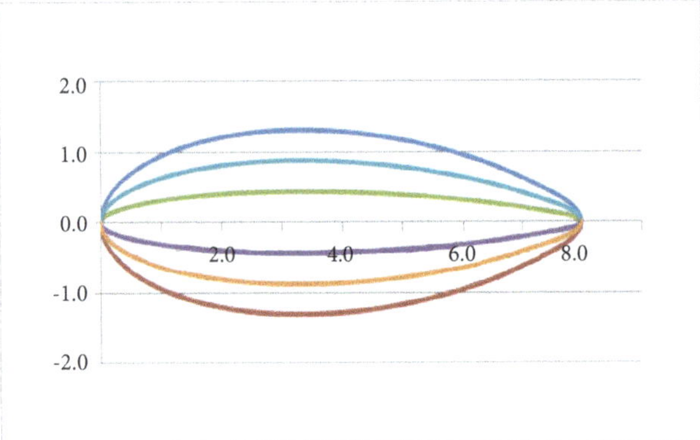

Figure 4. Envelope profile.

Figure 5. Radio frequency welding of petals.

GONDOLA

The gondola was sized to accommodate the receiver, battery package, fuel tank, engine and payload. The gondola mounting slots were welded below the envelope and the gondola has been attached directly to the envelope using contoured aluminum rods going into these slots. It was carefully set up upon determining its location and mass, to have as stable an airship as possible.

STABILIZER AND RUDDER

The fabrication of stabilizer was done using flat sheet of high density foam (expanded polystyrene). Balsa wood was attached at edges to impart stiffness and avoid local damage to the foam due to impact loads. Servo controllers with boosters were fitted to each rudder in such a way that all control surfaces are free to rotate around 30° about the hinge. In order to attach the stabilizer surfaces, a velcro sheet was pasted on the envelope. A photograph of the ventral stabilizer and rudder is shown in Fig. 6.

Figure 6. Stabilizer and Rudder.

FLIGHT DEMONSTRATION

Some of the major concerns dealt before the flight testing was of transportation and leak testing of the envelope. The envelope underwent a leak detection test, where light sources illuminated the holes, if any. Subsequent to the gas leak testing was the complete integration of all the sub-assemblies and the inflation of the envelope with Helium. The CAD model of the airship along with the fins and a side view of airship are shown adjacently in Fig. 7. It was observed that an additional ballast weight had to be added to the gondola in order to attain stability.

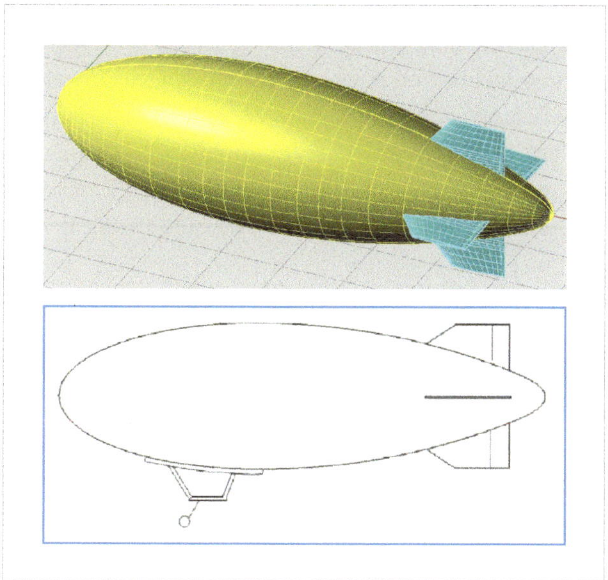

Figure 7. CAD model and airship side view.

The airship was successfully flown during three different weather conditions, viz., dry day, mild rain and mild snowfall. The low density of air at such a high altitude could always resulted in reduced performance of a propeller engine and a small engine set up can amplify this problem. Despite this, the flight performance was quite satisfactory with comfortable take off and climbing flight. The airship was also maneuvered to undergo both level and climbing turning flight, including a touch-down and take-off, which demonstrated its control effectiveness. The completely fabricated airship is shown in Fig. 8. Figure 9 shows some of the images of the airship during its flight.

Figure 8. Airship upon complete fabrication.

The airship carried an operating payload which consisted of an autofocus Internet Protocol (IP) camera, which took live footage during its flight. The IP camera was capable of continuously taking photographs at the rate of around one photograph per second and the transmitted sequence of pictures were then combined in a video by playing them at a faster rate. Figure 10 shows the photograph of IP camera mounted on the front of the gondola.

Figure 11 shows two photographs of the aerial footage taken by the onboard IP camera. Figure 12 shows a sequence of four images taken during flight when the LTA gas storage facility (pointed by the red arrow) was being monitored. The onboard camera was capable of procuring clear and crisp images of the people on the ground and layout of the area.

Figure 9. Airship during flight.

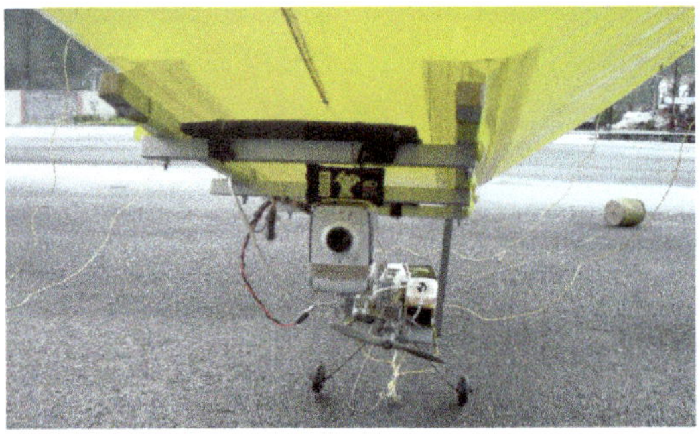

Figure 10. IP camera.

Figure 11. Two aerial photographs taken by onboard IP camera.

CONCLUSIONS

The design methodology adopted for the design of a RC airship not only covers the basic aspects of airship design but also governs the various options that can be utilized depending on the user's constraints. The fabrication, system integration and operation of the RC airship have stablished the tremendous potential that airships possess to be a highly viable airborne platform for aerial surveillance. The images obtained are steady, clear and can be used to monitor all activities at a given location on the ground. The manufacturing costs and operational issues of such an effective aerial surveillance system have also been found to be much lower than for fixed wing aircraft. The demonstration established the efficacy of remotely controlled airships for aerial photography and data collection by snow scientists.

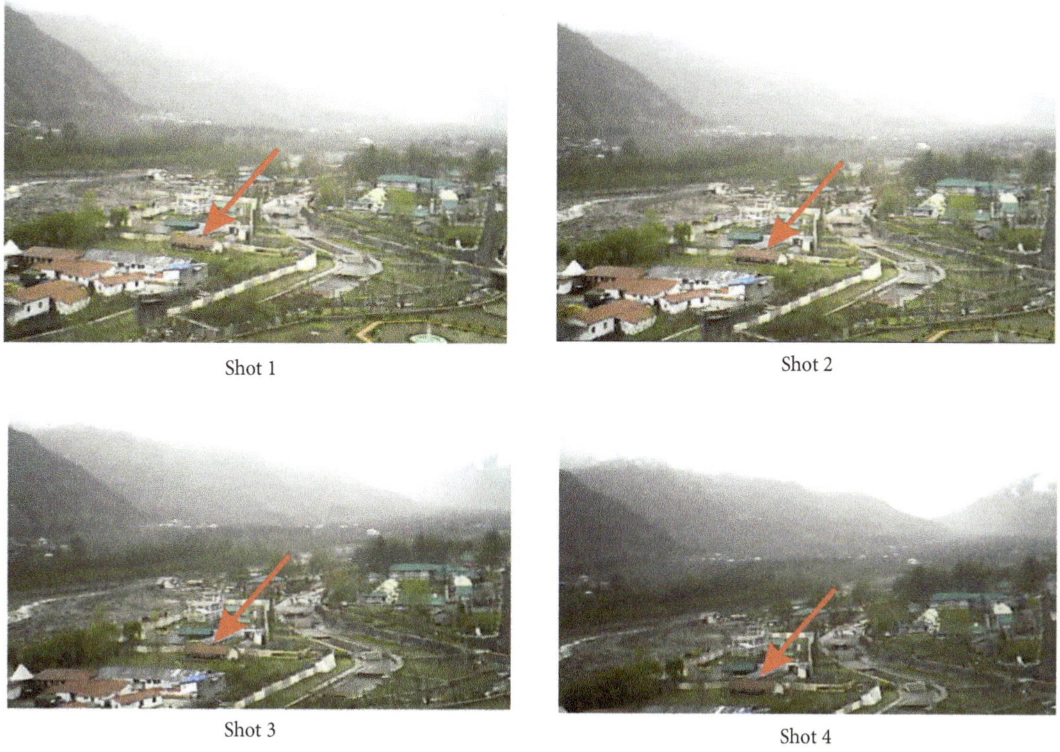

Figure 12. Sequence of photographs taken by the onboard camera.

ACKNOWLEDGMENTS

The author would like to thank Dr. R. N. Sarwade, Director, and Mr. Jimmy Kansal, Deputy Director of Snow and Avalanches Study Establishment, DRDO, for inviting the team from LTA systems Lab. of IIT Bombay to carry out the demonstration flight of this airship, and express his gratitude towards the following members Mr. Vishal Chaugule, Mr. Kaviresh Bhandari, Mr. Maxwell Rodrigues, Mr. Kushal Moolchandani, Mr. Nilesh Dhanve and Mr. V. D. Patil, who participated in this effort. The help and assistance from Mr. Y. P. Jahagirdar, Director of Research Institute for Model Aeronautics, Ahmednagar, and his team in system integration and flight demonstration is also gratefully acknowledged.

REFERENCES

Banerjee, S., Raina, A.A. and Pant, R.S., 2008, "Low Cost Trans-river Aerial Ferry", Proceedings of AIAA's 8th Aviation Technology Integration and Operations (ATIO) Conference, Anchorage, Alaska, USA, AIAA-2008-8851.

Bhandari, K.M. and Raina A.A., 2009, "Conceptual Design Of A High Altitude Aerostat For Study Of Snow Pattern", Proceedings of International Symposium on Snow & Avalanches (ISSA-09), SASE, Manali, India.

Cheeseman, I., 1999, "Aerodynamics", In: Khoury, G.A. and Gillet, J.D., Airship Technology, Cambridge Aerospace Series 10, Cambridge University Press, New York, pp. 25-38.

Gawale, A.C. and Pant, R.S., 2002, "Design, Fabrication and Flight Testing of Remotely Controlled Airships", Proceedings of National Conference on LTA Technologies, Aerial Delivery R&D Establishment, Agra, India.

Gawale, A. and Pant, R., 2004, "Design Studies of a Powerplant System of Non-Rigid Airship", Proceedings of the 5th International Convention of the Airship Association, Oxford, England.

Gawande, V.N., Raina A.A., Bilaye, P., Pant, R.S. and Desai, U.B., 2007, "Design and Fabrication of an Aerostat for Wireless Communication in Remote Areas", Proceedings of AIAA's 17th Aviation, Technology, Integration, and Operations (ATIO) Conference and 17th Lighter-Than-Air Systems Technology Conference, Belfast, Northern Ireland, UK, AIAA-2007-7832.

Gawale, A., Raina, A.A., Pant, R.S. and Jahagirdar, Y.P., 2008, "Design Fabrication and Operation of Low Cost Remotely Controlled

Airships in India", Proceedings of AIAA's 18th Aviation Technology Integration and Operations (ATIO) Conference, Anchorage, Alaska, USA, AIAA-2008-8853.

Gupta, S. and Malik, S., 2002, "Envelope Details for Demo Airship", PADD Project Report, Aerial Delivery R&D Establishment.

Narayana C.L. and Srilatha, K.R., 2000, "Analysis of aerostat configurations by panel methods", BLISS Project Document CF 0010, National Aerospace Laboratories, Bangalore, India.

Pant, R.S., 2008, "Methodology for Determination of Baseline Specifications of a Non-rigid Airship", Technical Note, AIAA Journal of Aircraft, Vol. 45, No. 6, pp. 2177-2182.

Sangole, R., Agashe, S., Palshikar, K. and Nakanekar, G., 2006, "Design, Fabrication and Testing of Remotely Controlled Indoor Airship employed for Neural Networks Hardware Experimentation", Undergraduate Project Thesis, Mechanical Engineering Department, PVG College of Engineering, University of Pune, India.

Sundaram, S., 2000, "Wind Tunnel tests on 1:7 and 1:28 scale Aerostat Models", Experimental Aerodynamics Divisions, National Aerospace Laboratories, Bangalore, India.

Estimation of Pico-Satellite Attitude Dynamics and External Torques via Unscented Kalman Filter

Halil Ersin Söken[1], Chingiz Hajiyev[2]

ABSTRACT: In this study, an Unscented Kalman Filter (UKF) algorithm is designed for estimating the attitude of a pico-satellite and the in-orbit external disturbance torques. The estimation vector is formed by the satellite's attitude, angular rates, and the unknown constant components of the external disturbance torques acting on the satellite. The gravity gradient torque, residual magnetic moment, sun radiation pressure and aerodynamic drag are all included in the estimated external disturbance torque vector. The satellite has magnetometers and gyros onboard as the attitude sensors. Because of the inherent nonlinear dynamics and the nonlinear measurement model, the UKF, which is a nonlinear version of the Kalman Filter, is selected as the filter algorithm. Performance of the proposed algorithm is demonstrated via simulations for a cube pico-satellite and the results are analyzed for different scenarios.

KEYWORDS: Pico-satellite, Attitude estimation, Unscented Kalman filter, Disturbance torques.

INTRODUCTION

Although there are numerous researches on cubesats, and this number is increasing day by day, the investigations are still far from being concluded. The Cubesat is basically a cubic pico-satellite which has a volume of 1 liter and mass of no more than 1.3 kg. These types of satellites are the outcomes of a search for lighter, smaller and cheaper spacecrafts, and recently, they have mostly been considered as a part of research projects of organizations like universities (Toorian *et al.*, 2008).

The biggest difficulty of cubesat applications is the limitations on the size and mass of the satellite. All the subsystems must be designed regarding these limitations and, unquestionably, a trade-off between the performance and applicability might be required. Specifically for the attitude determination and control system (ADCS), we cannot use highly accurate sensors and actuators onboard a cubesat, since they are usually heavy and large. The attitude of the satellite must be determined and controlled using miniaturized sensors and actuators, which are usually coarser and less accurate (Candini *et al.*, 2012). In this case, one technique to increase the system performance is to properly select the onboard running ADCS algorithm.

The Extended Kalman Filter (EKF) is a nonlinear Kalman Filter (KF), which has been widely used for satellite attitude estimation (Psiaki *et al.*, 1990; Lefferts *et al.*, 1982). On the other hand, the EKF has some disadvantages, especially for the highly nonlinear systems. Generally, this is caused by the mandatory linearization phase of the EKF procedure and so the Jacobians derived with that purpose. For most of the applications, the generation of the Jacobians is time consuming, difficult and

1.Graduate University for Advanced Studies – Sagamihara/Kanagawa – Japan 2.Istanbul Technical University – Istanbul – Turquey
Author for correspondence: Chingiz Hajiyev | Aeronautics and Astronautics Faculty | Istanbul Technical University | Maslak, 34469 Istanbul – Turkey | Email: cingiz@itu.edu.tr

prone to human errors (Julier and Uhlmann, 2004; Sekhavat *et al.*, 2007). Nonetheless, the linearization brings about an unstable filter performance when the time steps for the update is not sufficiently small and so the estimation or identification procedure fails as the filter diverges (Julier *et al.*, 1995). Per contra, small time steps increases the computational burden because of the larger number of Jacobian calculations. As a result of these facts, the EKF may be efficient only if the system is almost linear on the timescale of update intervals (Julier and Uhlmann, 2004).

A relatively new Kalman filtering technique, which does not have the shortcomings of the EKF for the nonlinear systems, is the Unscented Kalman Filter (UKF). The UKF generalizes the Kalman filter for both the linear and nonlinear systems and, in case of nonlinear dynamics, it provides relatively more accurate estimation results than other known observer design methodologies such as the EKF. The essence of the UKF is the fact that the approximation of a nonlinear distribution is easier than the approximation of a nonlinear function or transformation (Julier *et al.*, 2000). The UKF introduces sigma points for catching higher order statistics of the system. It satisfies both better estimation accuracy and convergence characteristics by securing higher order information (Sekhavat et al., 2007). Furthermore, the UKF is more robust against the initial estimation errors, so even in case of inaccurate *a priori* knowledge about the initial condition of the states, it performs well. The only disadvantage of the UKF is the increase in the computational burden when compared to the EKF. Yet, the computational load depends on the type of application and there are several researches aiming at making the UKF computationally more efficient (Li *et al.*, 2013).

There are many documented studies for the UKF as an estimation algorithm in astronautics. In (Crassidis and Markley, 2003) it is used as a state estimator, while both the states and the parameters of the satellite are estimated by the UKF in Dyke *et al.* (2004), Sekhavat *et al.* (2007) and Sekhavat *et al.* (2009). Moreover, in Vinther *et al.* (2011) the UKF is preferred as a part of the inexpensive cubesat attitude estimation method, where also the magnetometer biases are estimated and in Inamori *et al.* (2009) it is used for in-orbit magnetic disturbance estimation and compensation. In Soken and Hajiyev (2011) and Soken and Sakai (2011), both the magnetometer and the gyro biases are estimated as well as the attitude of the satellite by using the UKF. In Soken and Sakai (2013), two different UKFs are run onboard a nanosatellite for attitude estimation, sensor calibration and residual magnetic moment compensation. In Oliveria *et al.* (2014), the UKF is used for attitude estimation of a nanosatellite and the results are compared with the EKF. In Li *et al.* (2013), an adaptive version of the UKF is implemented as a part of the ADCS of a nanosatellite. Lastly, in Cornejo *et al.* (2010), the UKF is used for state estimation and fault detection purposes within the nanosatellite attitude control system.

In this study, an UKF algorithm is designed for estimating a pico-satellite's attitude and the in-orbit external disturbance torques acting on the pico-satellite. Gravity-gradient torque, sun pressure, aerodynamic drag and residual magnetic moment are included in the estimated disturbance torque vector. The satellite, for which the algorithm is proposed, has magnetometers and gyros onboard as attitude sensors. The main contribution of this study is to show that it is possible to estimate the unknown external torques as well as the attitude dynamics parameters of the satellite via a simple UKF based algorithm. In addition to the investigations in (Soken and Hajiyev, 2009), a more detailed discussion is given for the torque estimation with an extended simulation scenario. Moreover the case for uncertainty in the satellite's moments of inertia is covered.

The paper proceeds as follows: the mathematical model of the pico-satellite attitude dynamics is presented in the following section. In "The sensors measurement models" the sensors' measurement models are given. "Unscented Kalman filter (UKF) for the estimation of attitude dynamics and external torques" contains information about the UKF algorithm for the attitude state and unknown external torque estimation. In "Simulations", the performance of the proposed algorithm is demonstrated via simulations. And in the last section of this study, there is a brief summary of the obtained results and the conclusion.

PICO-SATELLITE ATTITUDE DYNAMICS MATHEMATICAL MODEL

Although the Euler angles may have singularity in some certain cases as an attitude kinematics representation technique, they are physically more significant and easier to interpret than the quaternions. Moreover, when the quaternions are used for kinematics representation, implementing the UKF becomes more difficult because of the quaternion norm constraint (Crassidis and Markley, 2003). Hence, in this study, we preferred to use the Euler angles for simplicity and presumably to reduce the computational load.

When we use the Euler angles the mathematical model of the satellite can be expressed with a 9 dimensional system vector.

The Euler angles representing the attitude of the satellite (φ is the roll angle about x axis; θ is the pitch angle about y axis; ψ is the yaw angle about z axis), the body angular rates with respect to the inertial axis frame and the constant components of the external disturbance torques form the state vector:

$$\boldsymbol{x} = \begin{bmatrix} \varphi & \theta & \psi & \omega_x & \omega_y & \omega_z & N_x & N_y & N_z \end{bmatrix}^T, \tag{1}$$

where

$$\boldsymbol{\omega}_{BI} = \begin{bmatrix} \omega_x & \omega_y & \omega_z \end{bmatrix}^T, \tag{2}$$

$$\boldsymbol{N}_d = \begin{bmatrix} N_x & N_y & N_z \end{bmatrix}^T. \tag{3}$$

ω is the angular velocity vector of the body frame with respect to the inertial frame and N_d is the external disturbance torques vector. Since we only estimate the constant components of the torques in this study:

$$\frac{d\boldsymbol{N}_d}{dt} = 0. \tag{4}$$

The dynamic equations of the satellite can be derived by the use of the angular momentum conservation law (Wertz, 1998):

$$J\frac{\boldsymbol{\omega}_{BI}}{dt} = \boldsymbol{N}_d + \boldsymbol{N}_c - \boldsymbol{\omega}_{BI} \times \left(J\boldsymbol{\omega}_{BI}\right), \tag{5}$$

where J is the inertia matrix, consisting of the main moments of inertia as $J=diag(J_x,J_y,J_z)$ and N_c is the applied control torque. In this study it is assumed that there is no attitude controller so $N_c=0$

Kinematic equations of the pico-satellite in terms of Euler angles can be given as:

$$\begin{bmatrix} \dot{\varphi} \\ \dot{\theta} \\ \dot{\psi} \end{bmatrix} = \begin{bmatrix} 1 & s(\varphi)t(\theta) & c(\varphi)t(\theta) \\ 0 & c(\varphi) & -s(\varphi) \\ 0 & s(\varphi)/c(\theta) & c(\varphi)/c(\theta) \end{bmatrix} \begin{bmatrix} p \\ q \\ r \end{bmatrix}. \tag{6}$$

Here, $c(.)$, $s(.)$ and $t(.)$ are the cosine, sine and tangent functions, respectively. Besides, p, q and r are the components of the ω_{BR} vector, which indicates the angular velocity of the body frame with respect to the reference frame (orbit frame). ω_{BI} and ω_{BR} can be related via:

$$\boldsymbol{\omega}_{BR} = \boldsymbol{\omega}_{BI} + A\begin{bmatrix} 0 & -\omega_0 & 0 \end{bmatrix}^T. \tag{7}$$

where ω_0 denotes the angular velocity of the orbit with respect to the inertial frame, found as $\omega_0 = \left(\mu / r_0^3\right)^{1/2}$. A represents the direction cosine matrix (Wertz, 1998):

$$A = \begin{bmatrix} c(\theta)c(\psi) & c(\theta)s(\psi) & -s(\theta) \\ -c(\varphi)s(\psi)+s(\varphi)s(\theta)c(\psi) & c(\varphi)c(\psi)+s(\varphi)s(\theta)s(\psi) & s(\varphi)c(\theta) \\ s(\varphi)s(\psi)+c(\varphi)s(\theta)c(\psi) & -s(\varphi)c(\psi)+c(\varphi)s(\theta)s(\psi) & c(\varphi)c(\theta) \end{bmatrix}. \tag{8}$$

THE SENSORS' MEASUREMENT MODELS

The investigated cubic pico-satellite has two types of sensors onboard: the magnetometers, which measure the strength of the Earth's magnetic field; and the gyros, which provide the angular rates of the satellite with respect to the inertial frame.

In this section, measurement models of these two sensors are presented.

THE MAGNETOMETER MEASUREMENT MODEL

The Earth magnetic field vector components can be modeled in the orbit frame as a function of time (Sekhavat *et al.*, 2007):

$$H_1(t) = \frac{M_e}{r_0^3}\left\{\cos(\omega_0 t)\left[\cos(\varepsilon)\sin(i) - \sin(\varepsilon)\cos(i) \cos(\omega_e t)\right] - \sin(\omega_0 t)\sin(\varepsilon)\sin(\omega_e t)\right\} \tag{9}$$

$$H_2(t) = -\frac{M_e}{r_0^3}\left[\cos(\varepsilon)\cos(i) + \sin(\varepsilon)\sin(i)\cos(\omega_e t)\right], \tag{10}$$

$$H_3(t) = \frac{2M_e}{r_0^3}\left\{\sin(\omega_0 t)\left[\cos(\varepsilon)\sin(i) - \sin(\varepsilon) \cos(i)\cos(\omega_e t)\right] - 2\cos(\omega_0 t)\sin(\varepsilon)\sin(\omega_e t)\right\}. \tag{11}$$

Here, M_e is the magnetic dipole moment of the Earth (M_e = 7.943 x 10^{15} $Wb.m$); μ is the the Earth Gravitational constant (μ = 3.98601 x 10^{14} m^3/s^2); i is the orbit inclination; ε is the magnetic dipole tilt (ε = 11.7^0); ω_e is the spin rate of the Earth (ω_e = 7.29 x 10^{-5} rad/s); r_0 is the distance between the centre of mass of the satellite and the Earth.

Three onboard magnetometers of the pico-satellite measures the components of the magnetic field vector in the body frame. Therefore, for the measurement model, which characterizes the measurements in the body frame, the magnetic field vector terms must be transformed by the use of the direction cosine matrix A. The overall measurement model may be given as:

$$\begin{bmatrix} H_x(\phi,\theta,\psi,t) \\ H_y(\phi,\theta,\psi,t) \\ H_z(\phi,\theta,\psi,t) \end{bmatrix} = A \begin{bmatrix} H_1(t) \\ H_2(t) \\ H_3(t) \end{bmatrix} + \eta_1, \quad (12)$$

where, $H_1(t)$, $H_2(t)$ and $H_3(t)$ represent Earth's magnetic field vector components in the orbit frame as a function of time, and H_x (ϕ, θ, ψ, t) H_y (ϕ, θ, ψ, t) and H_x (ϕ, θ, ψ, t) show the measured Earth magnetic field vector components in the body frame as a function of time and varying attitude. Furthermore, η_1 is the zero average Gaussian white noise with the characteristic of

$$E\left[\eta_{1k}\eta_{1j}^T\right] = I_{3x3}\sigma_m^2\delta_{kj}. \quad (13)$$

Here I_{3x3} is the identity matrix with the dimension of 3 x 3, σ_m is the standard deviation of each magnetometer error and δ_{kj} is the Kronecker delta.

In this study, we assume that the magnetometers are calibrated using either an on-ground or in-orbit technique so no bias terms are included in Eq.(12).

THE GYRO MEASUREMENT MODEL

As aforementioned, the other attitude sensor onboard the satellite is the gyro. Widely used model for the gyro measurements is as follows:

$$\boldsymbol{\omega}_{BI,meas} = \boldsymbol{\omega}_{BI} + \eta_2. \quad (14)$$

where, $\omega_{BI,meas}$ is the measured angular rates of the satellite, and η_2 is the zero mean Gaussian white noise with the characteristic of:

$$E\left[\eta_{2k}\eta_{2j}^T\right] = I_{3x3}\sigma_v^2\delta_{kj}, \quad (15)$$

Here, σ_v is the standard deviation of each rate gyro random error. Same as the magnetometers, the gyros are assumed to be calibrated.

UNSCENTED KALMAN FILTER (UKF) FOR THE ESTIMATION OF ATTITUDE DYNAMICS AND EXTERNAL TORQUES

The essence of the UKF is the unscented transform, a deterministic sampling technique, that we use for obtaining a minimal set of sample points (or sigma points) from the a priori average and covariance of the states. Then, these sigma points go through nonlinear transformation. The posterior average and the covariance are determined using the transformed sigma points (Julier *et al.*, 1995; Crassidis and Markley, 2003).

The UKF is derived for discrete-time nonlinear equations, so the system model is given by;

$$\boldsymbol{x}_{k+1} = f(\boldsymbol{x}_k, k) + \boldsymbol{w}_k \quad (16)$$

$$\tilde{\boldsymbol{y}}_k = h(\boldsymbol{x}_k, k) + \boldsymbol{v}_k \quad (17)$$

Here x_k is the state vector and $\tilde{\boldsymbol{y}}_k$ is the measurement vector. Moreover w_k and v_k are the process and measurement error noises, which are assumed to be Gaussian white noise processes with the covariances of Q (k) and R (k) respectively.

The UKF is based on the determination of $2n$+1 sigma points with an average of x ($k|k$) and a covariance of P ($k|k$). For an n dimensional state vector, these sigma points are obtained by:

$$\boldsymbol{x}_0(k|k) = \hat{\boldsymbol{x}}(k|k) \quad (18)$$

$$\boldsymbol{x}_l(k|k) = \hat{\boldsymbol{x}}(k|k) + \left(\sqrt{(n+\kappa)P(k|k)}\right)_l \qquad l=1\ldots n \quad (19)$$

$$\boldsymbol{x}_{l+n}(k|k) = \hat{\boldsymbol{x}}(k|k) - \left(\sqrt{(n+\kappa)P(k|k)}\right)_l, \qquad l=1\ldots n \quad (20)$$

where, $x_0(k|k)$, $x_l(k|k)$ and $x_{l+n}(k|k)$ are sigma points, n is the state number and k is the scaling parameter which is used for fine tuning, and the heuristic for choosing this parameter is n+k=3 (Julier and Uhlmann, 2004). $\left(\sqrt{(n+\kappa)P(k|k)}\right)_l$ corresponds to the lth column of the indicated matrix and l is given as l=1... n.

The next step of the UKF process is transforming each sigma point by the use of system dynamics:

$$\boldsymbol{x}_l(k+1|k) = f\left[\boldsymbol{x}_l(k|k), k\right]. \qquad l=0\ldots 2n \quad (21)$$

Then, these transformed values are utilized for gaining the predicted average and covariance (Crassidis and Markley, 2003; Soken and Hajiyev, 2011)

$$\hat{\boldsymbol{x}}(k+1|k) = \frac{1}{n+\kappa}\left\{\kappa\boldsymbol{x}_0(k+1|k) + \frac{1}{2}\sum_{l=1}^{2n}\boldsymbol{x}_l(k+1|k)\right\}, \quad (22)$$

$$P(k+1|k)=\frac{1}{n+\kappa}\Big\{\kappa\big[x_0(k+1|k)-\hat{x}(k+1|k)\big]\big[x_0(k+1|k)-\hat{x}(k+1|k)\big]^T + \frac{1}{2}\sum_{l=1}^{2n}\big[x_l(k+1|k)-\hat{x}(k+1|k)\big]\big[x_l(k+1|k)-\hat{x}(k+1|k)\big]^T\Big\}+Q(k). \quad (23)$$

Here, $\hat{x}(k+1|k)$ is the predicted average and $P(k+1|k)$ is the predicted covariance. Furthermore, the predicted observation vector is:

$$\hat{y}(k+1|k)=\frac{1}{n+\kappa}\left\{\kappa y_0(k+1|k)+\frac{1}{2}\sum_{l=1}^{2n} y_l(k+1|k)\right\}, \quad (24)$$

where

$$y_l(k+1|k)=h\big[x_l(k+1|k),k\big]. \qquad l=0\dots 2n \quad (25)$$

After that, the observation covariance matrix is determined as:

$$P_{yy}(k+1|k)=\frac{1}{n+\kappa}\Big\{\kappa\big[y_0(k+1|k)-\hat{y}(k+1|k)\big]\big[y_0(k+1|k)-\hat{y}(k+1|k)\big]^T + \frac{1}{2}\sum_{l=1}^{2n}\big[y_l(k+1|k)-\hat{y}(k+1|k)\big]\big[y_l(k+1|k)-\hat{y}(k+1|k)\big]^T\Big\}, \quad (26)$$

where the innovation covariance is

$$P_{vv}(k+1|k)=P_{yy}(k+1|k)+R(k+1) \quad (27)$$

Here $R(k+1)$ is the measurement noise covariance matrix. On the other hand, the cross correlation matrix can be obtained as:

$$P_{xy}(k+1|k)=\frac{1}{n+\kappa}\Big\{\kappa\big[x_0(k+1|k)-\hat{x}(k+1|k)\big]\big[y_0(k+1|k)-\hat{y}(k+1|k)\big]^T + \frac{1}{2}\sum_{l=1}^{2n}\big[x_l(k+1|k)-\hat{x}(k+1|k)\big]\big[y_l(k+1|k)-\hat{y}(k+1|k)\big]^T\Big\}. \quad (28)$$

The following part is the update phase of the UKF algorithm. At that phase, first by using measurements, $y(k+1)$, the innovation sequence is found as:

$$e(k+1)=y(k+1)-\hat{y}(k+1|k), \quad (29)$$

and then the Kalman gain is computed via the equation:

$$K(k+1)=P_{xy}(k+1|k)P_{vv}^{-1}(k+1|k). \quad (30)$$

At last, the updated states and covariance matrix are determined by:

$$\hat{x}(k+1|k+1)=\hat{x}(k+1|k)+K(k+1)e(k+1), \quad (31)$$

$$P(k+1|k+1)=P(k+1|k)-K(k+1)P_{vv}(k+1|k)K^T(k+1). \quad (32)$$

Here, $\hat{x}(k+1|k+1)$ is the estimated state vector, and $P(k+1|k+1)$ is the estimated covariance matrix.

SIMULATIONS

The proposed UKF based estimation method is tested in this section. The simulations are performed for 40,000 s, which is the period for almost 7 orbits of the satellite. The orbit of the cubesat is a circular orbit with an altitude of r=550km and an inclination of i=97°. The inertia matrix of the satellite is $J=diag(2.1\text{x}10^{-3},2.0\text{x}10^{-3},1.9\text{x}10^{-3})kg.m^2$, which corresponds to a 10cm cubic satellite with an approximate mass of 1.2kg, as mentioned.

The magnetometer sensor noise is characterized by zero mean Gaussian white noise, with a standard deviation of σ_m=300 nT. The rate gyro random error is taken as $\sigma_{gv}=0.274\,[\text{deg}/\sqrt{h}]$. As the filter parameter for the UKF, κ is selected as κ=-3, which is different than the suggested heuristic.

The initial attitude errors for all the simulations are set to few degrees. The initial estimates for the angular rates and the external disturbance torques are all zero. Besides, the initial value of the covariance matrix is taken as $P_0=10^{-10}$ for all the estimated states, while the process noise covariance matrix is selected as $Q=10^{-29}$ for the attitude and $Q=10^{-20}$ for the rest of the states. The R matrix is composed of the sensor noise covariances for the magnetometers and gyros, which means $9\text{x}10^{-12}\ T^2$ as 3 diagonal components corresponding to magnetometer measurements and $64\text{ x }10^{-10}$ rad^2/s^2 as 3 diagonal elements corresponding to the gyro measurements. The constant torque term used for modeling purposes in the simulations (the values which will be estimated) is $Nd=[5\ -3\ \ 4]^T\text{ x }10^{-1}\ \mu Nm$.

The algorithm was tested in two cases. We are running a dynamics based model for the estimations (in other words, since we are using the satellite dynamics in addition to the kinematics) so there might be uncertainties, mainly caused by the mismatch between the real dynamics and the model used in the filter. The only source for such uncertainty in our simulations might be the inertia terms, which are not exactly known. Hence the two scenarios for the simulations are the cases with and without the uncertainty in the inertia terms.

In the first scenario, there is no uncertainty in inertia terms. In Fig.1, an Euler angle estimation example is given. The top plot compares the estimation with the actual value, and the lower one presents the estimation error. As seen, the UKF accurately estimates the attitude, and after almost 1.5 orbits (≈11000s.) the estimation error falls below ±1deg. If the attitude determination accuracy requirement for such pico-satellite without attitude control is considered (±1deg of determination accuracy may be accepted as sufficient), then we may state that the UKF algorithm works well and provides sufficient accuracy for pico-satellite missions. Besides, as presented in Fig. 2, the UKF also has a good estimation characteristic for the angular rates.

As to the main purpose of the proposed algorithm, the Fig. 3 presents an example for the estimation of the external disturbance torques. As clearly seen, the UKF converges to the real values after 10,000th s, and gives sufficiently accurate estimations for the torque terms. Therefore, we can say that the given UKF algorithm estimates all of the states accurately. The estimation results for the rest of the parameters are similar and can be seen in Figs. 4-6. The key contribution of the method is proving that such estimation can be achieved by a simple algorithm that is computationally nondemanding and efficient enough for cubesat applications.

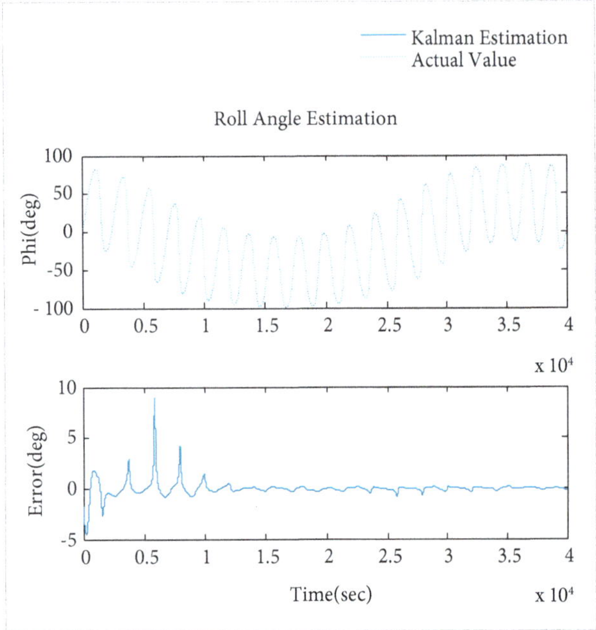

Figure 1. Roll angle estimation.

The second simulation is performed for a scenario where we do not know the inertia of the satellite accurately, and there are 5% and 10% uncertainty in inertia terms, respectively. For understanding sake, the results are given in table with comparison with the results obtained for the

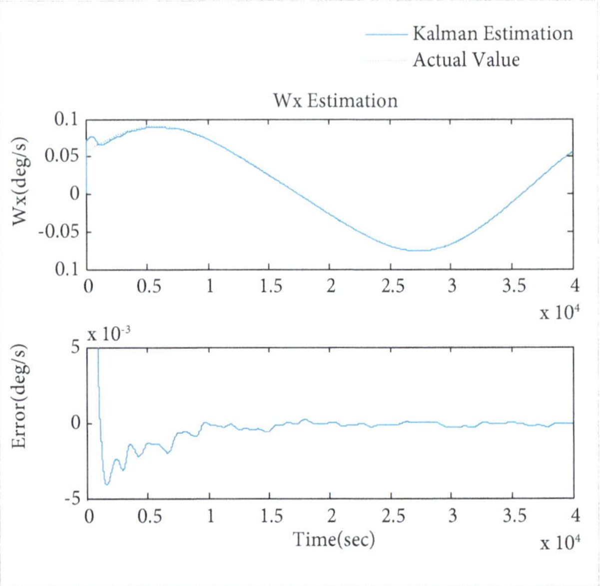

Figure 2. Estimation of the angular velocity about "x" axis.

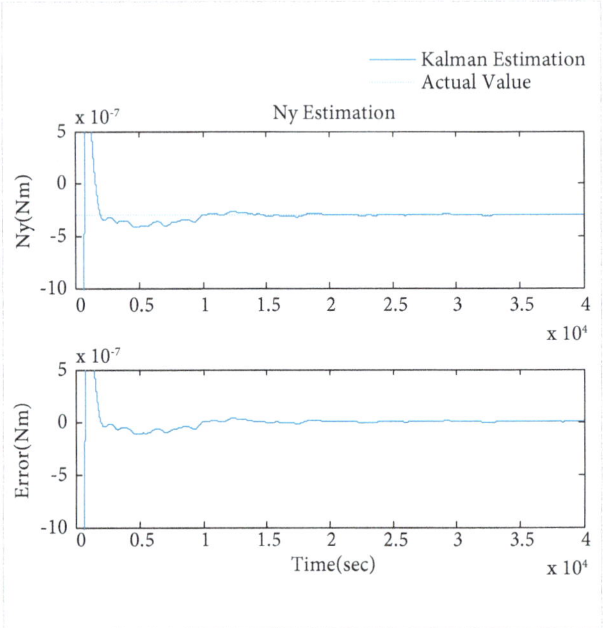

Figure 3. Estimation of the constant external torque about "y" axis.

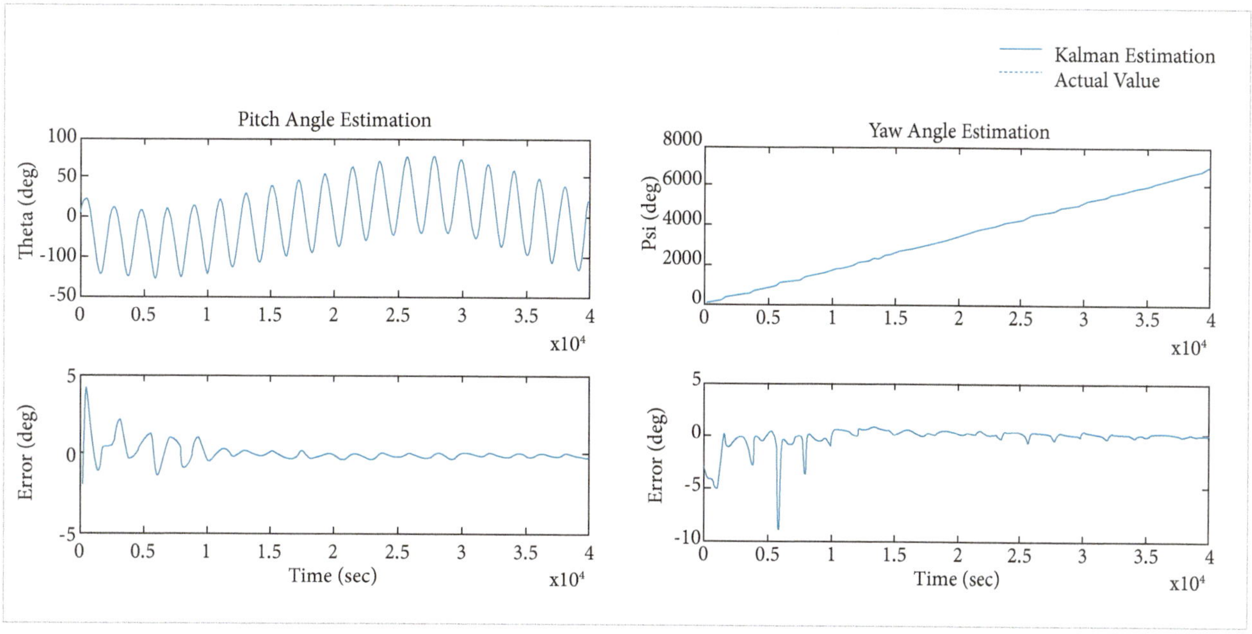

Figure 4. Estimation of the pitch and yaw angles.

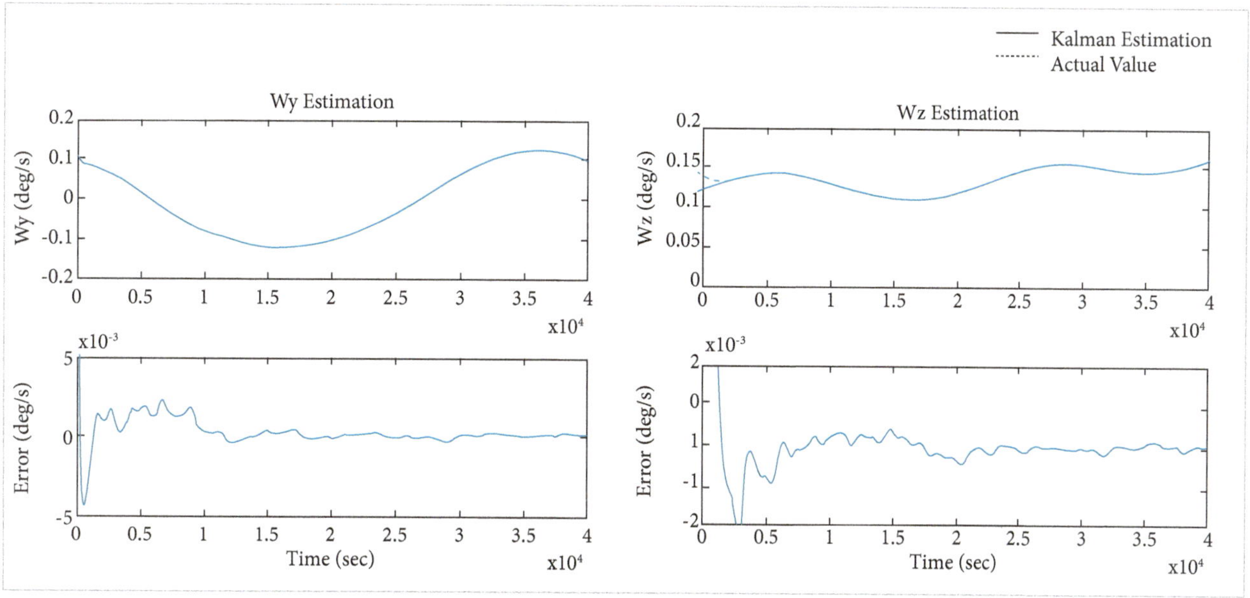

Figure 5. Estimation of the angular velocities about "y" and "z" axes.

first scenario, beforehand. In Table 1, root mean square errors are tabulated for 10,000 s. between 20,000th and 30,000th seconds such that;

$$RMSE_\gamma = \sqrt{\frac{1}{10000}\sum_{k=20001}^{30000}\left[\boldsymbol{x}_\gamma(k) - \hat{\boldsymbol{x}}_\gamma(k)\right]^2} \qquad \gamma = 1\ldots n \tag{33}$$

As it can be seen, even for a small uncertainty in inertia terms of the satellite, the estimation error increases for all the estimated states. Yet, for the simulation case with 5% uncertainty, estimation error is still within the acceptable bounds for cubesat applications, regarding that the attitude estimation accuracy is better than ±1deg. However, if the level of uncertainty is 10%,

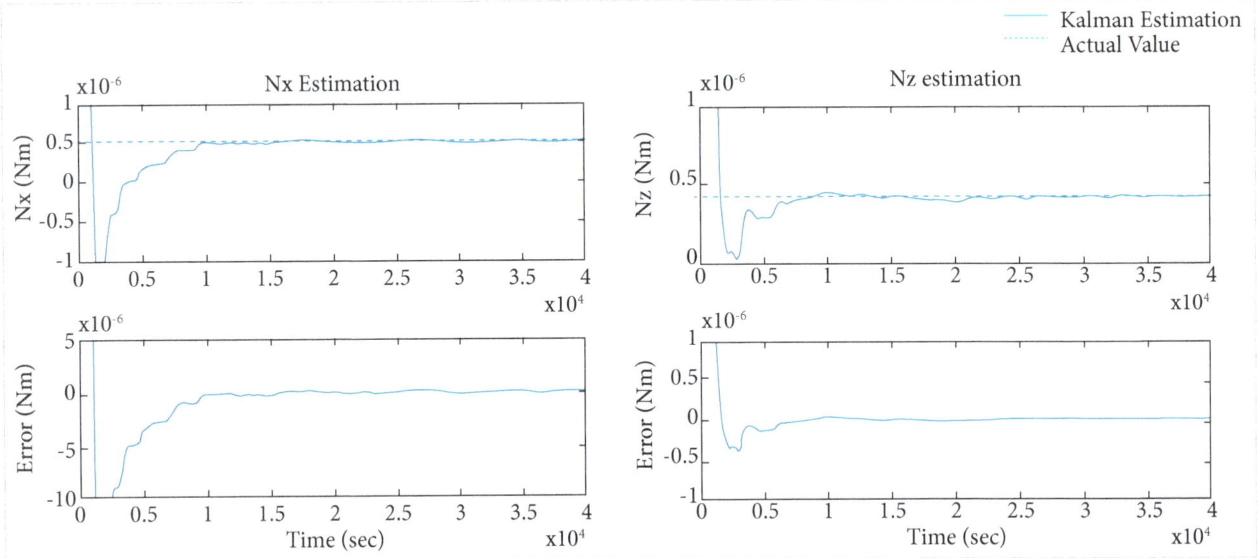

Figure 6. Estimation of the constant external torques about "x" and "z" axes.

Table 1. Root mean square error for the cases with and without uncertainty in the inertia terms of the satellite.

	Without Uncertainty	With 5% Uncertainty	With 10% Uncertainty
$\varphi(^0)$	0.1302	0.2719	3.7194
$\theta(^0)$	0.0905	0.2487	1.4792
$\psi(^0)$	0.3100	0.3868	2.8478
$\omega_x(^0/s)$	9.8034e-5	5.5424e-4	0.0041
$\omega_y(^0/s)$	1.3251e-4	3.4723e-4	0.0068
$\omega_z(^0/s)$	1.0042e-4	3.2553e-4	0.0056
N_x (Nm)	6.8174e-9	1.6826e-7	2.9664e-7
N_y (Nm)	4.8459e-9	1.5259e-8	8.7539e-7
N_z (Nm)	6.7456e-9	5.4487e-8	7.1730e-7

then it is not possible to satisfy these requirements and the attitude estimation error becomes more than ±1deg. Further examinations show that if there is even more uncertainty in the inertia terms of the satellite, the filter may diverge in long term.

CONCLUSION

In this study, an Unscented Kalman Filter (UKF) algorithm is designed for estimating a pico-satellite's attitude and the in-orbit external disturbance torques acting on the pico-satellite. The gravity-gradient torque, sun pressure, aerodynamic drag and residual magnetic moment are included in the estimated disturbance torque vector. The satellite, for which the algorithm is proposed, has magnetometers and gyros onboard as attitude sensors. The main contribution of this study is to show that it is possible to estimate the unknown external torques as well as the attitude dynamics parameters of the satellite via a simple UKF based algorithm. The results show that by using the presented algorithm it is possible to estimate the attitude of the cubesat with an accuracy better than ±1deg, which is accurate enough for such satellites. Moreover the case for uncertainty in the satellite's moments of inertia is addressed with additional simulations. It is shown that if the uncertainty is less than 5%, the requirements for the attitude estimation performance are still satisfied although the estimation error increases.

REFERENCES

Candini, G.P., Piergentili, F. and Santoni, F., 2012, "Miniaturized Attitude Control System for Nanosatellites", Acta Astronautica, Vol. 81, pp.325-334. doi: 10.1016/j.actaastro.2012.07.027.

Cornejo, N.E., Amini, R. and Gaydadjiev, G., 2010, "Model-based Fault Detection for the DELFI-N3XT Attitude Determination System", Proceedings of 2010 IEEE Aerospace Conference, Montana, USA, pp. 1-8. doi: 10.1109/AERO.2010.5446801.

Crassidis, J.L. and Markley, F.L., 2003, "Unscented Filtering for Spacecraft Attitude Estimation", Journal of Guidance, Control, and Dynamics, Vol. 26, pp. 536-542. doi: 10.2514/2.5102.

Dyke, M.C., Schwartz, J.L. and Hall, C.D., 2004, "Unscented Kalman Filtering for Spacecraft Attitude State and Paremeter Estimation", Proceedings of the AAD/AIAA Space Flight Mechanics Conference, No. AAS 04-115, Hawaii, USA.

Inamori, T., Nakasuka, S. and Sako, N., 2009, "In-orbit Magnetic Disturbance Estimation and Compansation Using UKF in Nano-Satellite Mission", Proceedings of AIAA Guidance, Navigation, and Control Conference, Chicago, USA.

Julier, S.J. and Uhlmann, J.K., 2004, "Unscented Filtering and Nonlinear Estimation", Proccedings of IEEE, Vol. 92, No. 3, pp. 401-422. doi: 10.1109/JPROC.2003.823141.

Julier, S.J., Uhlmann, J.K. and Durrant-Whyte, H.F., 1995, "A New Approach for Filtering Nonlinear Systems", Proceedings of American Control Conference, Vol.3, pp. 1628-1632. doi: 10.1109/ACC.1995.529783.

Julier,S., Uhlmann, J. and Durrant-Whyte, H.F., 2000, "A New Method for the Nonlinear Transformation of Means and Covariances in Filters and Estimators", IEEE Transactions on Automatic Control, Vol.45, No. 3, pp. 477-482. doi: 10.1109/9.847726 .

Lefferts, E.J., Markley, F.L. and Shuster, M.D., 1982, "Kalman Filtering for Spacecraft Attitude Estimation", Journal of Guidance, Control, and Dynamics, Vol.5, pp. 417-429. doi: 10.2514/3.56190.

Li, J., Post, M.A. and Lee, R., 2013, "A Novel Adaptive Unscented Kalman Filter Attitude Estimation and Control Systems for 3U Nanosatellite", Proceedings of European Control Conference, Zurich, Switzerland, pp.2128-2133.

Oliveira, G.F., Nehme, P.H.D. and Cappelletti, C., 2014, "Analysis and Simulation of Attitude Determination and Control for the Serpens Nanosatellite," Proceedings of 2nd IAA Conference on Dynamics and Control of Space Systems, Rome, Italy, pp.1-20.

Psiaki, M.L., Martel, F. and Pal, P.K., 1990, "Three-axis Attitude Determination via Kalman Filtering of Magnetometer Data", Journal of Guidance, Control, and Dynamics, Vol.13, pp. 506-514. doi: 10.2514/3.25364.

Sekhavat, P., Gong, Q. and Ross, I.M., 2007, "NPSAT I parameter Estimation Using Unscented Kalman Filter", Proceedings of 2007 American Control Conference, New York, pp. 4445-4451.

Sekhavat, P., Karpenko, M. and Ross, I.M., 2009, "UKF-Based Spacecraft Parameter Estimation Using Optimal Excitation", Proceedings of AIAA Guidance, Navigation and Control Conference, Chicago, USA.

Soken, H.E. and Hajiyev, C., 2009, "UKF for Identification of the Pico-Satellite Attitude Dynamics Parameters and External Torques on IMU and Magnetometer Measurements", Proceedings of 4th International Conference on Recent Advances in Space Technologies, Istanbul, Turkey, pp. 541 – 546. doi: 10.1109/RAST.2009.5158254

Soken, H.E. and Hajiyev, C., 2011, "In-flight Calibration of Pico Satellite Attitude Sensors Via Unscented Kalman Filter", Gyroscopy and Navigation, Vol.2, No.3, pp.156-163. doi: 10.1134/S2075108711030114.

Soken, H.E and Sakai, S., 2011, "UKF Based On-Orbit Gyro and Magnetometer Bias Estimation as a Part of the Attitude Determination Procedure for a Small Satellite", Proceedings of 11th International Conference on Control, Automation and Systems, Seoul, Korea, pp. 1891-1896.

Soken, H.E. and Sakai, S., 2013, "A Study on an Accurate yet Simple Attitude Estimation Scheme for Nanosatellites", Proceedings of 5th Nano-satellite Symposium, Tokyo, Japan, pp.1-21.

Toorian, A. Diaz, K. and Lee, S., 2008, "The Cubesat Approach to Space Access", Proceedings of 2008 IEEE Aerosacpe Conference, Montana, USA, pp. 1-14. doi: 10.1109/AERO.2008.4526293.

Vinther, K., Jensen, K.F., Larsen, J.A. and Wisniewski, R., 2011, "Inexpensive Cubesat Attitude Estimation Using Quaternions and Unscented Kalman filtering", Automatic Control in Aerospace, Vol. 4, No. 1.

Wertz, J.R., 1998, "Spacecraft Attitude Determination and Control", Kluwer Academic Publishers, Dordrecht, Holland.

The Update of an Aerodynamic Wind-Tunnel for Aeroacoustics Testing

Leandro Dantas Santana[1], Micael Carmo[2], Fernando Martini Catalano[1]

ABSTRACT: This paper describes the update and characterization of a previously pure aerodynamics wind-tunnel into a facility able to simultaneously execute aerodynamics and aeroacoustics testing. It is demonstrated that the application of high-performance acoustic materials on strategic positions of the wind-tunnel circuit and punctual actions can substantially reduce the background-noise level. This paper shows efficient measures which resulted to broadband-noise reduction of up to 5 dB and practically complete removal of spectral tones. In addition, it is demonstrated that the applied acoustic treatment reduced the turbulence level, measured at the test-section at maximum operational velocity, from the previous 0.25% level to 0.21%. As a minor penalty, the acoustic treatment reduced the flow velocity in 2% for the same electric-power input. Finally, the work described in this paper resulted on a wind-tunnel with good flow quality and capacity for aeroacoustics testing.

KEYWORDS: Aeroacoustic testing, Wind-tunnel noise, Wind-tunnel acoustic treatment.

INTRODUCTION

The recent air transportation growth raised concern to authorities and civil institutions regarding its environmental consequences, among them, the noise. In order to regulate this important environmental impact, authorities established noise restrictions to aircraft certification and operations. To comply with these increasingly stringent regulations, the aeronautic industry has been, for decades, developing means to reduce the engine noise. With the current state of the art technology, the aircraft engine reached the same noise level as the airframe. Therefore, regarding noise, the engine optimization is no longer the exclusive preoccupation of the aeronautic industry, now, aeroacoustic development of airframe components also challenges the minds of engineers and aeronautical researchers.

This paper is result of a work-package from the Silent Aircraft project – sponsored by *Empresa Brasileira de Aeronáutica* (EMBRAER) and *Fundação de Amparo à Pesquisa do Estado de São Paulo* (FAPESP) with objective to develop critical knowledge on aeroacoustic testing and simulations.

In an international context of building new wind-tunnels for aeroacoustics testing such as the works of Kim *et al.* (2001), Wickern and Lindener (2000) and Sarradj *et al.* (2009), or for the adaptation of previously pure aerodynamic wind-tunnel into an aeroacoustic facilities (Remillieux *et al.*, 2008; Künstner *et al.*, 1995), it was decided that upgrading a previously pure aerodynamic facility was the optimal solution for the project requirements and constraints. The chosen wind-tunnel was the closed circuit wind-tunnel LAE-1, situated in the Aerodynamic Laboratory at *Escola de Engenharia de São Carlos* from the *Universidade de São Paulo,* Brazil. The work subject of this paper was developed in the period between October 2008 and March 2010.

1. Escola de Engenharia de São Carlos – São Carlos/SP – Brazil 2. Empresa Brasileira de Aeronáutica – São José dos Campos/SP – Brazil

Author for correspondence: Leandro Dantas Santana | Escola de Engenharia de São Carlos | Avenida Trabalhador São Carlense, 400 – Centro | CEP: 13566-590 São Carlos/SP – Brazil | Email: leandro_dantas@yahoo.com.br

In order to define minimum required levels for noise reduction, it was decided to set, as a target experiment, the measurement of a NACA-0012 trailing-edge airfoil noise at velocity of 30 m/s. The beamforming technique was used for noise sources identification and possible quantification (Sijtsma and Holthusen, 1999; Brooks and Humphreys, 2006; Dougherty, 2005). Wind-tunnel background-noise is, in this paper, defined as the noise generated by the wind-tunnel running in stable regime and empty testing chamber.

The LAE-1 closed circuit wind-tunnel was designed originally as a 3/8 scale facility prototype of an automotive wind-tunnel to be built in the future. The wind-tunnel was originally built in the period between 1997 and 2002, having as predominant material the naval plywood. Due to the Brazilian automotive industry difficulties in the early 2000s, and the national aeronautic industry rebirth, this previously automotive wind-tunnel, became a multi-task facility with instrumentation mainly focused on aeronautical testing, able to perform a diverse range of industrial and academic tests. Figure 1 presents the drawings of the LAE-1 closed circuit wind-tunnel.

The wind-tunnel test section dimensions are 3.00 m long, 1.30 m high and 1.70 m wide. The maximum design flow speed

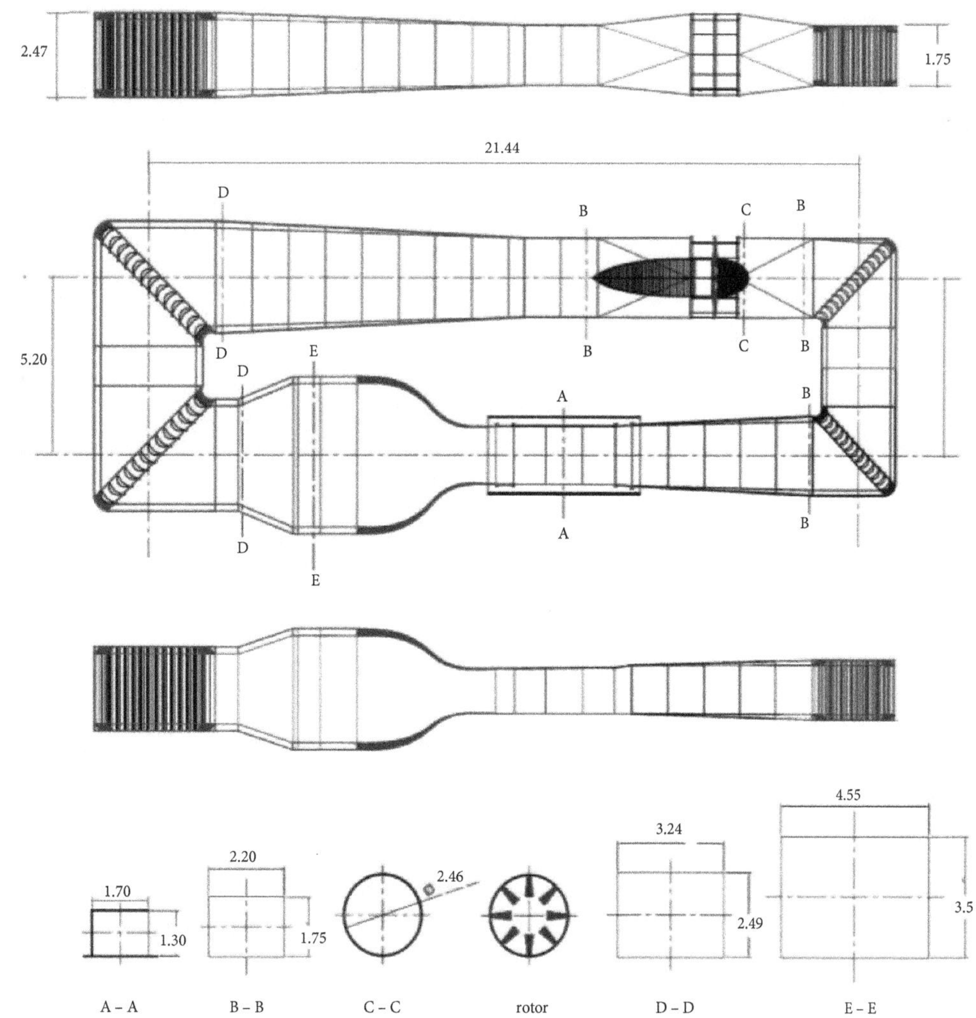

Figure 1. Plan view of the LAE-1 wind-tunnel.

is 50 m/s, with turbulence level of 0.25%, nowadays, due to operational safety and components endurance, the maximum speed is limited to 45 m/s. Its 110-HP-electric motor drives an eight-blade fan, with seven straighteners located downstream the fan. In the flow stabilization section, there are two mesh screens with 54% porosity, followed by a 1:8 contraction cone designed using two 3rd order polynomials, joined at 45% from the inflection point. The relative low turbulence level, considering the installation of only two screens and no honeycomb, is attributed to the care taken during the design and construction of low angle diffusers, low-drag corner-vanes and high-efficiency rotor blades, optimized using a combination of Computation Fluid Dynamics (CFD) and semi-empirical techniques (Catalano, 2001).

THE BACKGROUND NOISE REDUCTION PROCESS

The LAE-1 background noise reduction campaign was developed in three phases: a first phase, where conceptual solutions, either innovative or inspired on literature, were multidisciplinary and evaluated as to their efficiency, feasibility and budgetary constraints. This first phase was followed by two steps of effective implementation. In the first step, the effect of pure wall treatment was evaluated, whose results indicated the need of a second step of acoustic treatment, where, in order to achieve the noise reduction target, the addition of an acoustic baffle and fan treatment was implemented.

PRELIMINARY CONCEPTS FOR THE PROJECT

In order to establish the noise reduction target, it was defined an objective experiment to be done after the wind-tunnel adaptation. The requirements for this experiment were its representativeness of future tests campaign and the detailed availability of results in the literature to be used as a validation data, the self-noise measurement of a NACA-0012 airfoil, carried-out by Brooks *et al.* (1989), was chosen as an ideal experiment to fulfill those requirements .

An initial wind-tunnel noise assessment was conducted in order to determine the most critical noise condition. From this evaluation, it was found that the flow speed of 31 m/s is the most critical condition, due to the existence of fan excited structural resonance. For this reason, this speed was set as reference for the measurements shown in this paper. Figure 2 shows a comparison of the LAE-1 baseline background noise level and the NACA-0012 airfoil noise.

At Fig. 2, the NACA-0012 self-noise was calculated using semi-empirical methods described by Brooks *et al.* (1989). For these calculations, it was considered a 0.20 m chord airfoil spanning the wind-tunnel test section, subjected to a 31 m/s flow speed and an angle of attack of two degrees.

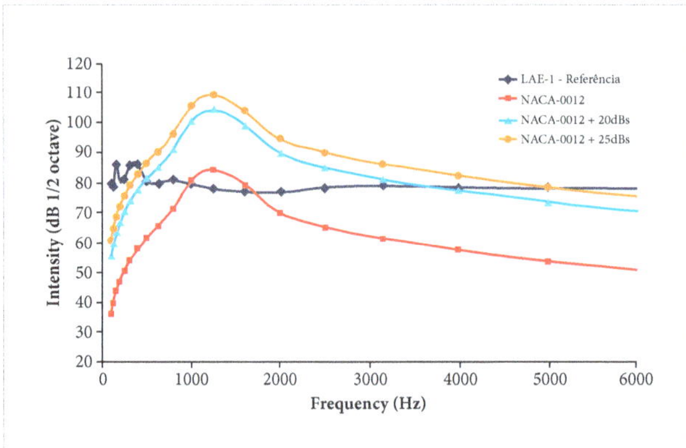

Figure 2. NACA-0012 noise compared with the wind-tunnel background noise.

From Fig. 2, it is noticed that, without post-processing for noise source separation, it is only possible to measure three octave band frequencies, localized from 1 kHz to 1.6 kHz. This clearly justifies the need of wind-tunnel background noise reduction in conjunction with the use of a microphone array to increase the measurable bandwidth. For this paper, the post-processing technique considered is the traditional delay-and-sum beamforming. For a microphone array, it is a rule of thumb used to compute the maximum signal to noise ratio (DdB), measurable by an array with M microphones as proposed by Shin *et al.* (2007):

$$\Delta dB = 10\log_{10} M \tag{1}$$

For the current project, it was defined to use an array with M=106 microphones, consequently, it is concluded that it is possible to distinguish a source with an intensity of 20.25 dB less than the background noise by exclusively using regular post-processing techniques. If advanced decorrelation techniques of beamforming measurements are considered, it is possible to subtract 5 dB more in relation to the noise level estimated on Eq. 1 (Shin *et al.*, 2007). Finally, it can be concluded that with the gain given by the instrumentation and signal processing, it is possible to identify sources 25 dB less intense than the wind-tunnel background noise level. Adopting these hypotheses, it can be concluded that it is possible to measure a NACA-0012 airfoil noise sources for all the region of the orange line of Fig. 2. In addition to that, after a wind-tunnel background noise reduction of 5 dB, it would be possible to measure the full range of frequencies of interest, from 500 to 8.000 Hz, for the NACA-0012 airfoil.

THE FIRST WIND-TUNNEL NOISE REDUCTION PHASE

The first phase of acoustic treatment consisted of applying melamine foam on the selected section walls. Several regions of the wind-tunnel were considered for the foam application. Sections with adverse pressure gradient were discarded due to the boundary layer separation risk induced by increasing of the roughness, possible imperfections on the foam superficial finishing, or steps between the foam plates, which may occur on installation or during the operational life of the tunnel. The settling chamber was also discarded due to the risk of negative effects on the flow stabilization. Therefore, the remainder regions A and B from Fig. 3 were chosen as the best candidates for foam application.

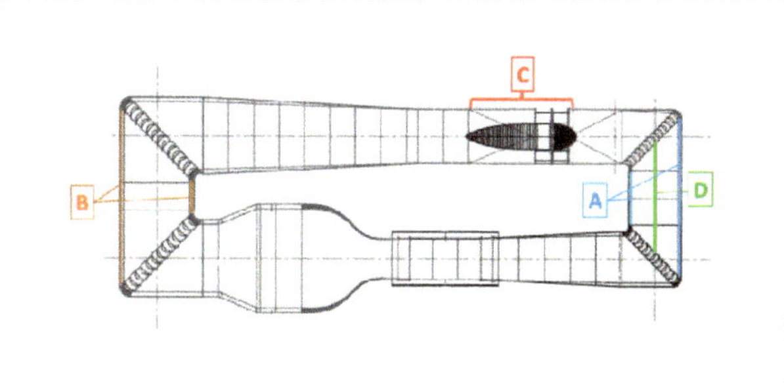

Figure 3. Wind-tunnel regions that received acoustics treatments.

The next step was to define the thickness of the noise absorbent layer. It was performed an analysis of sensitivity for reducing the transversal area, due to the foam thickness on the testing chamber flow velocity. Calculations, made with the help of the wind-tunnel design model, showed that the section reduction on the region of the walls A (see Fig. 3) leads to a very sensible impact on the wind-tunnel flow velocity. The main reason for this is its comparatively high flow speed and, therefore a significant pressure loss. The reduction of the cross-section of the region of the walls B (see Fig. 3) has a minor effect on the wind-tunnel flow velocity, once this area has a lower flow speed and bigger cross-sectional area. To achieve the acceptable margin of 1.5% of flow velocity reduction, on the testing chamber, it was decided to apply a 2 cm thick foam on walls A and 5 cm thick foam on walls B.

THE SECOND WIND-TUNNEL NOISE REDUCTION PHASE

In order to increase the noise reduction achieved so far, the second and deeper noise reduction task was conducted.

To reduce the noise at lower frequencies, an acoustic baffle (element D from Fig. 3) was installed between the walls A, dividing the inner sections from the first and second corner vanes. With the installation of this baffle, it was expected a flow speed reduction of no more than 2%, referring to the wind-tunnel baseline conditions. The acoustic baffle construction scheme is shown in Fig.4.

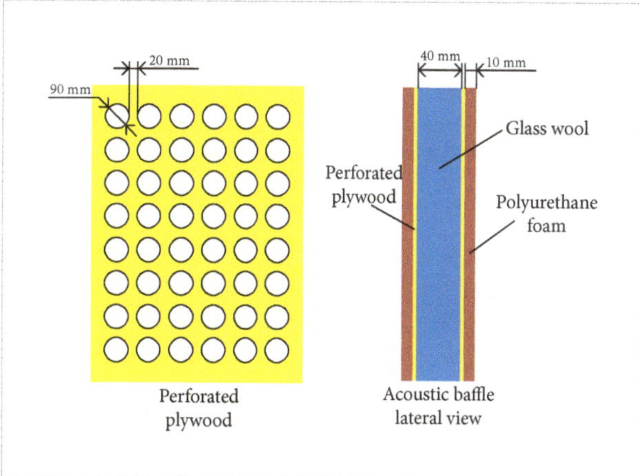

Figure 4. Acoustic baffle construction scheme.

For the design of the baffle, it was considered the feasibility, efficiency and budgetary constraints. This baffle was composed of a sandwich assembly filled with glass wool. The perforated plywood open area was optimized in order to generate a maximum open area to the mechanical strength relationship. The perforated plate was covered with polyurethane foam to assure surface smoothness.

A second action was taken based on the opportunity to improve the wind-tunnel fan performance and, simultaneously, reduce its noise. Due to circularity defects of the metal shield that involves the fan, the tip to wall space varies from 1.5 cm to 3 mm. This space is known to generate a tip vortex and non-uniform loading to the fan blades, which is a noise source

(Camargo *et al.*, 2007). To avoid this problem, polyurethane foam was used as filling material in the regions where the gap was greater than 3 mm. The foam leading edge and trailing edge were shaped to guarantee smooth geometric transition. Figure 5 (a) presents the region with greater gap and (b) shows the same region, after the tip treatment, with polyurethane foam.

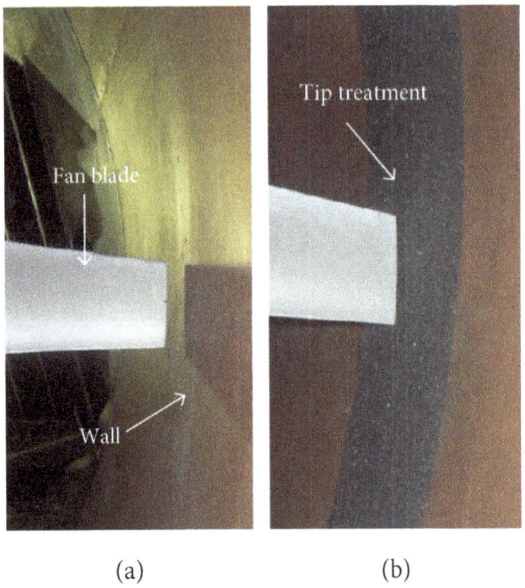

Figure 5. Region with higher fan tip to wall gap before (a) and after (b) the treatment.

The baseline wind-tunnel noise spectra, measured in the testing chamber, showed the existence of a very intense tone localized at the frequency of 5 kHz and its multiples. The investigation of the origin of this tone and harmonics showed that they were generated by the motor inverter, and they were related to the motor rotational speed controller. The inverter set-up allows changing its carrier frequency to a limit of 10 kHz. Considering that higher frequencies are better damped by the medium, the highest achievable frequency was adopted as the carrier frequency for the inverter.

WIND-TUNNEL BACKGROUND NOISE: RESULTS AND DISCUSSION

Typically the main wind-tunnel noise sources are the walls laminar/turbulent boundary layer, corner vanes, fan trailing edge and the electric motor that drives the fan. This last noise source is mainly related with the speed controller that excites the electrical coil inside the motor producing noise.

It is quite difficult to identify and treat each noise source separately; instead it is more practical and efficient to line the noise propagation path in order to isolate a region of interest, and consequently this approach is adopted in this work.

For low flow speeds, it can be considered that the main wind-tunnel noise source is the sum of the fan and the turbulent boundary layer. Considering this, it was decided to line the region downstream and upstream of the fan, in order to treat both directions of the propagation path of the noise generated by the fan. The upstream distance from the fan to the testing chamber is smaller than the downstream one; therefore, it was preferred to add, as much as possible, acoustic treatment on the region comprised by the walls A (see Fig. 3). In the upstream direction, the settling chamber screens can be considered as a sudden increase of acoustic resistance, which tends to reflect the acoustics waves back, attenuating the noise generated by the fan propagated upstream. To better identify and differentiate the noise sources generated by the fan from the one generated by the flow, the noise spectra were analyzed on three different flow speeds: 15 m/s, 31 m/s and 37 m/s. The results are presented on Figs. 6, 7 and 8. For low speed (15 m/s), it is expected the fan noise to predominate, while for flow speeds up to 37 m/s the noise from turbulent boundary layer will be the main source. The velocity of 31 m/s is a special point of interest due to the occurrence of a structural resonance induced by the fan rotation.

Analyzing the first phase of noise reductions, regarding the Figs. 7, 8 and 9, it is noticed that for all flow speeds the foam treatment was effective on reducing the noise at frequencies from

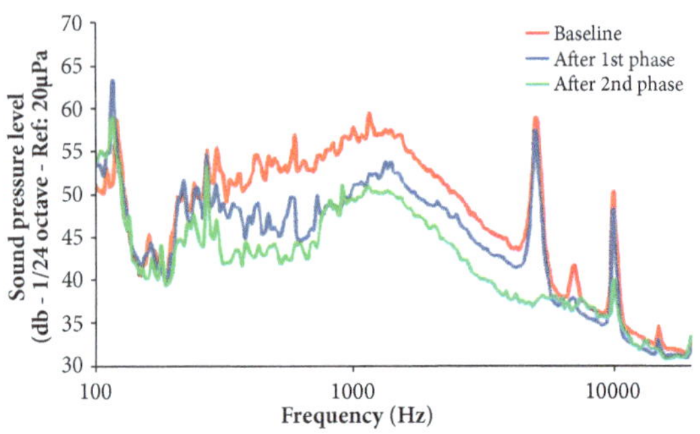

Figure 6. Wind-tunnel noise spectra measured on the testing chamber for a flow velocity of 15 m/s.

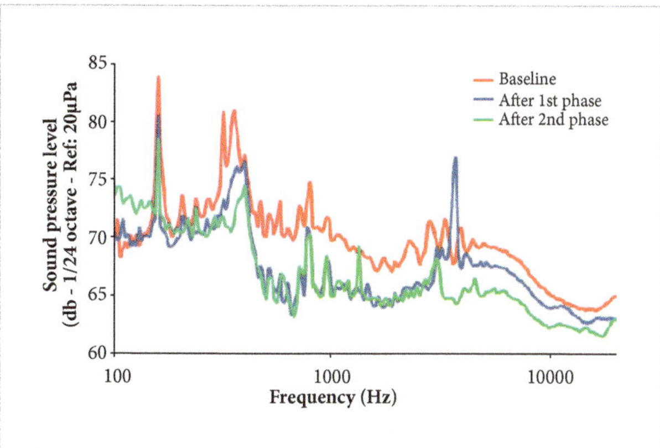

Figure 7. Wind-tunnel noise spectra measured on the testing chamber for a flow velocity of 31 m/s.

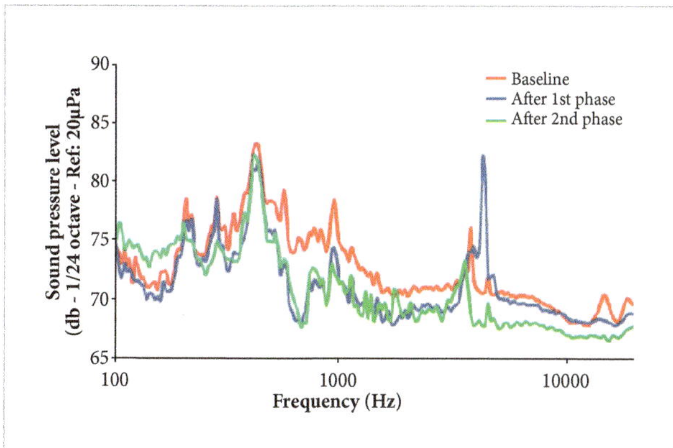

Figure 8. Wind-tunnel noise spectra measured on the testing chamber for a flow velocity of 37 m/s.

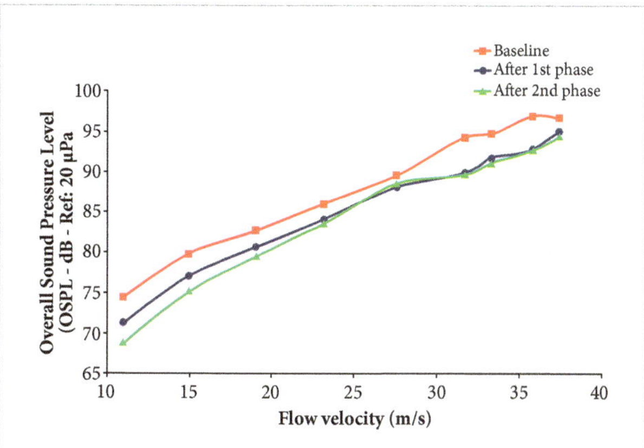

Figure 9. Overall Sound Pressure Level (OSPL) variation with the wind-tunnel flow speed.

300 Hz to 4 kHz. From Figs. 6 to 8, it is seen that the first phase of noise reduction was effective on absorbing the noise generated by the fan. For lower flow speeds, the noise reduction reached 5 dB, for the frequency of 1 kHz. Opposite to that, when the flow speed increases to a higher speed (e.g. 37 m/s), the boundary layer noise becomes a significant noise source, and its reduction reaches 2 dB for the same frequency. Finally, it can be noticed that the acoustic treatment was very effective in reducing the fan noise at low flow speeds. As a side effect, it can be noticed a noise increase of up to 2 dB at high frequencies and high flow velocities.

Observing specially Figs. 8 and 9, it is noticed that the final noise treatment phase resulted on undesired noise increment at frequencies below 200 Hz. Even thought all noise level increase is undesired, this side-effect was considered acceptable because it will not affect future noise measurement since this frequency is far below the one of interest for future tests.

An important result in Fig. 6 is the vanishing of the peak at 5 kHz after the second phase of the acoustic treatment. This can be explained by the parameter modification of the inverter that controls the fan rotation. Since this carrier frequency was moved to 10 kHz, this tone almost disappeared from the spectrum.

It can be concluded that the first acoustic treatment phase acted mainly at low frequencies while the second one was more effective at high frequencies.

Figure 9 presents the Overall Sound Pressure Level (OSPL) variation as a function of the flow speed. From Fig. 9, it is remarkable that the first phase of noise treatment reduced the OSPL in an average value of 4 dB for all flow speeds. Opposed to that, the second acoustic treatment phase reduced the OSPL just for flow velocities below 20 m/s, and had no noticeable effect at high flow speeds.

THE ACOUSTICS TREATMENT EFFECT ON THE FLOW

The acoustic treatment effect on the flow is verified by the values of the turbulence intensity and the flow velocity. The turbulence was measured with the help of a DANTEC constant temperature hot-wire anemometer system. In this system, a probe model 55P01 was rigidly fixed in the center of the wind-tunnel testing chamber. This probe was calibrated using a DANTEC system described by Bruun (2002).

The hot-wire anemometer data was acquired using a sampling ratio of 6400 Hz during 5 seconds. The measured turbulence raw data was processed with in-house software that calculates the turbulence intensity and auto-power-spectra. A parametric study showed that the calculation of the wind-tunnel turbulence, free of structural vibration and electrical noise influence, requires the use of a band-pass filter, which filters frequencies outside the 3 Hz and 1.000 Hz range.

Figure 10 shows the evolution of the turbulence intensity with the velocity measured in the testing chamber.

Figure 10 shows that the turbulence grows with velocity up to, approximately, 19 m/s, reaching a constant level, for velocities up to 30 m/s, and, consecutively, decays for higher velocities. This behavior is present in all phases of the acoustics treatment. The explanation to this behavior is based on the fact that this wind-tunnel was designed for a 50 m/s flow, meaning that, as the velocity increases, the flow passes from a laminar/low-turbulence to an "on design" condition, where the constant pitch blade reaches its optimum point. Since, for high velocities, the turbulence eddies, with largest length scales, have their characteristic length reduced, when compared to low velocities, which propitiates a more effective viscous dissipation. As a macro-effect of this, there is a measurable decreasing in the working section turbulence level as shown in Fig. 10.

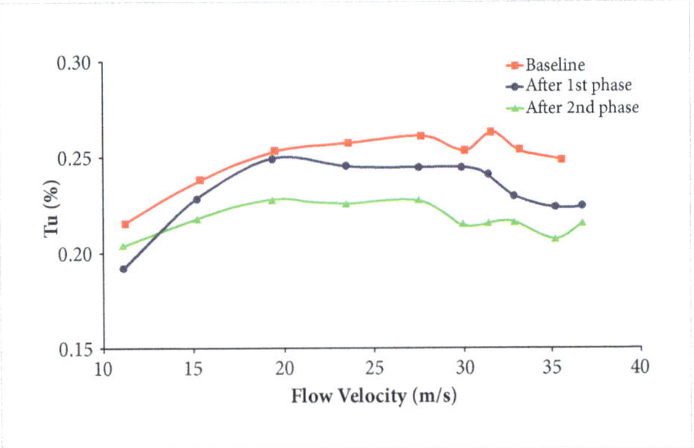

Figure 10. Wind-tunnel turbulence measured on the testing chamber.

A second point, coming from Fig. 10 analysis, is the fact that the acoustic treatment resulted to a positive effect on the wind-tunnel turbulence intensity levels. This is explained by the reduction of the acoustic noise, known as an excitation source on the non-linear instabilities mechanisms that lead the flow to turbulence. In addition to this is the fact that, for high Reynolds numbers, the statistics of small eddies is universally and uniquely determined by the viscosity and the rate of energy dissipation, not related to the noise, in such way that the turbulence generation has been decreased but the dissipation rate has been kept constant.

The turbulence spectrum is presented in Figs. 11 and 12.

It is desirable, for good quality wind-tunnels, that its turbulence spectrum is free of tones. It is noticed from Figs. 11 and 12 that the acoustic treatment does not introduce any effect on the turbulence spectra.

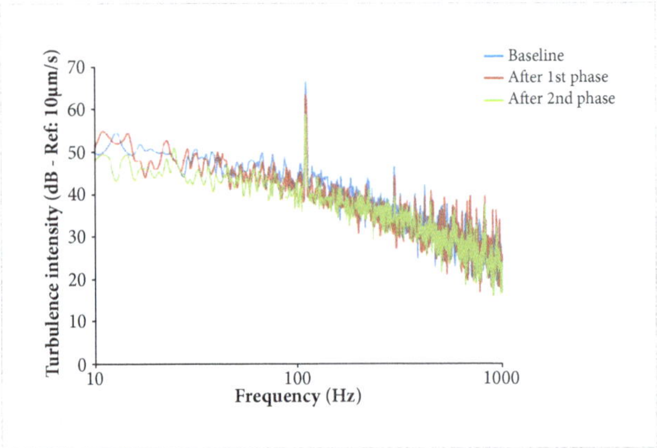

Figure 11. Wind-tunnel turbulence spectrum measured after background noise reduction process for a flow speed of 20 m/s.

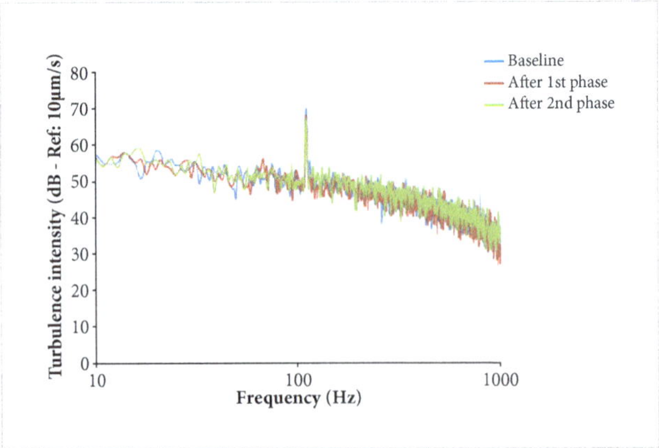

Figure 12. Wind-tunnel turbulence spectrum measured after background noise reduction process for a flow speed of 31 m/s.

The last analysis is the influence of the acoustic treatment to the flow velocity, for the same input power. The electric power input for the motor that drives the wind-tunnel fan, will be calculated based on the manufacturer information that a constant torque is kept with the rotational speed variation. Consequently, the flow velocity is a direct function of the pressure losses, and it is possible to compare the flow velocity reduction after the noise treatment phases. As previously mentioned, the foam installation on the walls increased its local velocity due to the cross-section reduction. Therefore, a slightly increase to the pressure loss, at the treated sections, are expected with the related decrease of the test section velocity. Figure 13 shows the acoustic treatment phases effect on the test section velocity.

Analyzing Fig. 13 results, it is seen that the flow velocity reduction was in agreement with the expected within the two phases of noise treatment. The second acoustic treatment phase caused the most important flow speed reduction. This fact can, mainly, be attributed to the hydraulic diameter reduction, caused by the baffle installation. Considering that the pressure loss is proportional to the inverse of the hydraulic diameter, reducing the hydraulic diameter in 2 times implies on a sectional pressure loss of 8 times. Since section B pressure loss is small,

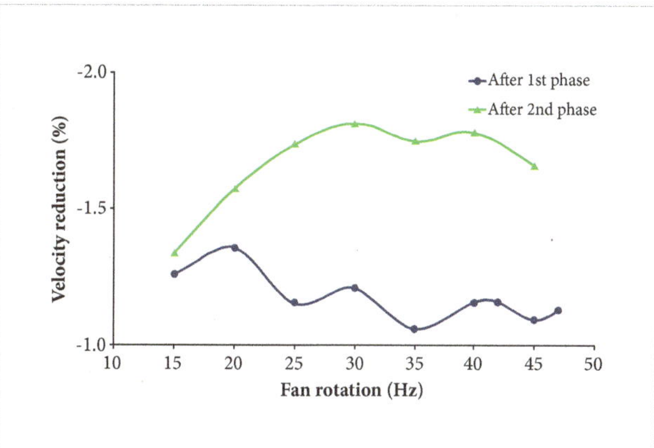

Figure 13. Wind-tunnel flow velocity reduction due the acoustic treatment.

compared to the rest of the wind-tunnel, a minor effect on the total wind-tunnel velocity reduction is expected.

CONCLUSIONS

The present paper shows the process of background noise reduction for the LAE-1 wind-tunnel. This process was divided into two main phases, where, in a first phase, the effect acoustic treatment on strategic parts of the wind-tunnel circuit was evaluated. Results showed that this treatment was more effective at frequencies ranging from 400 to 4 kHz. A second phase, complementary to the first, acted mainly on frequencies above 4 kHz. These two noise treatment phases resulted on a wind-tunnel background noise reduction of up to 5 dB, and a desirable turbulence level reduction from the original 0.25% to 0.21%, with no addition of tones on the turbulence spectra. The minor drawback of this work was the reduction of 2% in the flow speed, considering a constant electric power input.

ACKNOWLEDGEMENTS

The authors acknowledge FAPESP and EMBRAER for the financial support for this project.

REFERENCES

Brooks, T.F. and Humphreys, W.M., 2006, "A Deconvolution Approach for the Mapping of Acoustics Sources (DAMAS) Determined from Phased Microphone Arrays", Journal of Sound and Vibration, Vol. 294, pp. 856-879.

Brooks, T.F., Pope, D.S. and Marcolini, M.A., 1989, "Airfoil Self-Noise and Prediction", NASA Reference Publication 1218.

Bruun, H.H., 2002, "Hot-Wire Anemometry – Principles and Signal Analysis", 2nd Edition, Oxford Science Publications.

Camargo, H., Remillieux, M., Burdisso, R., Crede, E. and Devenport, W., 2007, "The Virginia Tech Stability Wind Tunnel from an aerodynamic into an aeroacoustic facility", 19th International Congress on Acoustics, Madrid.

Catalano, F.M., 2001, "The new closed circuit wind tunnel of the aircraft laboratory of University of São Paulo", 16th Brazilian Congress of Mechanical Engineering, Vol. 6, pp. 306-312.

Dougherty, R.P., 2005, "Extension of DAMAS and benefits and limitations of deconvolution in beamforming", 11th AIAA/CEAS Aeroacoustics Conference.

Kim M., Lee J., Kee J. and Chang J., 2001, "Hyundai Full Scale Aero-acoustic Wind Tunnel", SAE Technical Papers, 2001-01-0629.

Künstner R., Potthoff J. and Essers U., 1995, "The Aero-Acoustic Wind Tunnel of Stuttgart University", SAE Technical Papers, 950625.

Remillieux M.C., Crede E.D., Camargo H.E., Burdisso R.A., Devenport W.J. and Rasnick M., Van Seeters P., Chou A., 2008, "Calibration and Demonstration of the New Virginia Tech Anechoic Wind Tunnel", 14th AIAA/CEAS Aeroacoustics Conference, AIAA 2008-2911.

Sarradj E., Fritzsche C., Geyer T. and Giesler J., 2009, "Acoustic and aerodynamic design and characterization of a small-scale aeroacoustic wind tunnel", Applied Acoustics, Vol. 70, pp. 1073–1080.

Shin, H.C., Grahan, W.R., Sijtsma, P., Andreou, C. and Faszer, A.C., 2007, "Implementation of a Phased Microphone Array in a Closed-Section Wind Tunnel", AIAA Journal, Vol. 45, No. 12.

Sijtsma, P. and Holthusen, H., 1999, "Source Location by Phased Array Measurements in Closed Wind Tunnel Test Sections", AIAA-99-1814.

Wickern G. and Lindener N., 2000, "The Audi Aeroacoustic Wind Tunnel: Final Design and First Operational Experience", SAE Technical Papers, 2000-01-0868.

21

Numerical Simulation on the Radar Cross Section of Variable-Sweep Wing Aircraft

Kuizhi Yue[1,2], Bing Hu[1], Weigang Guo[1], Yunliang Wang[2]

ABSTRACT: The influence of variable-sweep wing on the aircraft's radar cross section (RCS) characteristics has been studied to reduce the aircraft's RCS as well as its detection probability by the hostile radar. With the help of CATIA, a 3-D digital model of the variable-sweep wing aircraft is built to generate a series of digital grids. Using MATLAB, a numerical simulation on the RCS of variable-sweep wing aircraft is conducted based on physical optics (PO) method and equivalent currents method (ECM). The results of mathematical statistics and comparative analysis show that: (i) the RCS peak value in the head direction of the aircraft decreases non-linearly with the sweep angle of the wing's leading edge; (ii) the azimuth angle corresponding to one of the peak values of the aircraft's RCS is equal to the leading edge's sweep angle; (iii) when the leading edge's sweep angle is 33°, the arithmetic average value of the RCS values in the head direction of the aircraft is 0.644% of the average value when the sweep angle is 0°; (iv) the larger the sweep angle is, the lower the probability that the aircraft is detected.

KEYWORDS: Aircraft conceptual design, Radar cross section, Variable-sweep wing, Stealth performance, Numerical simulation.

INTRODUCTION

Stealth aircraft or low-observability aircraft is a very important developing direction for modern combat aircraft. In view of this trend, the main military powers all over the world are conducting studies on the stealth aircraft's designing, manufacturing, arming and application. Typical stealth aircraft developed by the U.S. include F-22, B-2, F-35, X-45 and X-47 (Nangia and Palmer, 2005; Vogel, 2005). These aircraft have become or will become the backbone elements of the U.S. Air Force and U.S. Navy. Russia is also developing its Su-47, MG-1.44, T-50, and other stealth combat aircraft, which are likely to become the future main force of the Russia Air Force. China's aviation industry has also taken part in this competition and has achieved considerable success in developing its stealth combat aircraft. Variable-sweep wing aircraft have a glorious history. Classical aircraft such as the U.S. F-14, F-111, have participated in many wars. They are once the real backbone elements of the U.S. Navy and Air Force. In Russia, MG-23, Su-24, Tu-160, and other aircraft are still performing a variety of combat missions. This paper conducts feasibility studies on the stealth design of the variable-sweep wing aircraft. The RCS features are explored to provide reference and theoretical support for the aircraft's conceptual design and stealth optimization and thus provide technical reserves for the development of new stealth aircraft.

Researchers have performed lots of work on designing stealth aircraft for many years and have achieved quite many results. The different shapes of the stealth aircraft, including the aircraft's head, fuselage, wings, inlets and other strong scattering parts of electromagnetic waves, have been studied (Wood and Bauer, 2001). Stealth materials are

1.Beijing University of Aeronautics and Astronautics – School of Aeronautic Science and Engineering – Beijing 100191 – China. 2.Naval Aeronautical and Astronautical University – Department of Airborne Vehicle Engineering – Yantai 264001 – China.

Author for correspondence: Yue Kuizhi | Beijing University of Aeronautics and Astronautics – School of Aeronautic Science and Engineering | Beijing 100191 – China
Email: yuekuizhi_2000@sohu.com

analyzed and the key technologies of radar stealth, infrared stealth, visible and acoustic stealth are explored as well. Moreover, multidisciplinary optimization algorithms on pneumaticity, stealth and structure are investigated. Bai and Liu (2007) have studied the parametric modeling approach for fuselage sections. Taking into account the aircraft's stealth performance, their approach generates the body shape by adjusting the body parameters. Tom and Alfred (2010) have studied the relationship between the aircraft's configuration and its stealth performance when the resistance is the lowest. In Bao and Wang (2012), the characteristic parameters of flying wing's stealth performance and aerodynamic performance are investigated, and a wind tunnel test of the flying wing is conducted. Ji *et al.* (2009) have studied the impact of the relative bending of the vertically- and horizontally-bended inlet on the inlet's microwave scattering characteristics. Huang and Liu (2008) have studied the microwave scattering characteristics of the aircraft's surface cracks by RCS measuring. He *et al.* (2010) have studied the influence of inhomogeneous plasma on the attenuation of planar electromagnetic waves. Sun and Zhang (2008) have summarized the characteristics of F/A-22 and F-35 stealth fighter jets and have proposed a conceptual solution to reduce the antenna aperture size along with the antenna aperture characteristic signal, adopting the technology of low intercept probability. Lu and Wang (2009) have studied the infrared radiation model of the aircraft's surface, analyzed the aircraft's infrared characteristics, and summarized the infrared stealth reduction strategies. Hu and Yu (2011) have studied the applications of multidisciplinary optimization to aircraft's conceptual design, where the aircraft's aerodynamic performance, stealth performance, structure design, general layout and weight constraints are all taken into consideration.

Generally speaking, a series of issues about the stealth aircraft have been studied deeply; however, there are few research reports about the influence of angle changes of the variable-sweep wing aircraft on its stealth characteristics. The qualitative and quantitative relationship between the wing's leading edge sweep angle of the variable-sweep wing aircraft and the aircraft's stealth performance still remains unclear. This paper focuses on the stealth characteristics of the variable-sweep wing aircraft. Physical Optics (PO) method and the Equivalent Currents Method (ECM) are used. A numerical simulation is conducted to obtain an analytical report of the RCS characteristics, so as to provide reference for the designing of stealth aircraft.

THEORETICAL BASIS

The theoretical basis of this study consists of two parts: RCS prediction methods and radar detection probability model.

RCS PREDICTION METHODS

A series of numerical simulations of RCS characteristics of variable-sweep wing aircraft are conducted in this paper. Scattering surface elements are calculated using PO method. Edge diffraction is calculated by ECM. In PO method, it is assumed that the surface current on the incident point of the wave is equal to the surface current when the incident wave is just on the tangency plane of the incident point.

The formula of PO method is as follows (Yue *et al.*, 2014b):

$$\sqrt{\sigma_{po}} = \begin{cases} \dfrac{\hat{\boldsymbol{n}} \cdot (\hat{\boldsymbol{e}}_r \times \hat{\boldsymbol{h}}_i) e^{j\frac{2\pi}{\lambda} \boldsymbol{r}_o \cdot (\hat{s} - \hat{i})}}{\sqrt{\pi} |\boldsymbol{p}|^2} \displaystyle\sum_{m=1}^{3} \xi_m & \boldsymbol{p} \neq 0 \\ \dfrac{j2\sqrt{\pi}}{\lambda} \hat{\boldsymbol{n}} \cdot (\hat{\boldsymbol{e}}_r \times \hat{\boldsymbol{h}}_i) e^{j\frac{2\pi}{\lambda} \boldsymbol{r}_o \cdot (\hat{s} - \hat{i})} A & \boldsymbol{p} = 0 \end{cases} \tag{1}$$

$$\xi_m = \boldsymbol{p} \cdot \boldsymbol{L}_m e^{jk\boldsymbol{r}_m \cdot (\hat{s} - \hat{i})} \frac{\sin\left(\frac{\pi}{\lambda} \boldsymbol{L}_m \cdot (\hat{\boldsymbol{s}} - \hat{\boldsymbol{i}})\right)}{\frac{\pi}{\lambda} \boldsymbol{L}_m \cdot (\hat{\boldsymbol{s}} - \hat{\boldsymbol{i}})} \tag{2}$$

where:

$\sqrt{\sigma_{po}}$: RCS of a single surface element, in m^2; $\hat{\boldsymbol{n}}$: outward normal unit vector of the surface element; $\hat{\boldsymbol{e}}_r$: direction of electric field on the receiving antenna; $\hat{\boldsymbol{h}}_i$: direction of the incident wave's magnetic field; " . " is the dot product and "×" is the cross product; j: imaginary unit, and $j^2 = 1$; λ: incident wavelength, in m; $\boldsymbol{r}_0$: reference point on the surface element; $\hat{\boldsymbol{s}}$: scattering wave's direction; $\boldsymbol{i}$: incident wave's direction; $\boldsymbol{p} = \hat{\boldsymbol{n}} \times (\hat{\boldsymbol{s}} - \hat{\boldsymbol{i}})$; A: area of the surface element, in m^2; L_m: magnitude and direction of the mth edge; k: wave number, $2\pi/\lambda$; $\boldsymbol{r}_m$: vector from the reference point $\boldsymbol{r}_0$ to the midpoint of the mth edge.

The formula of ECM is as follows:

$$\sqrt{\sigma_{ecm}} = \frac{1}{\sqrt{\pi} \sin\theta} [(\boldsymbol{E}_0^i \cdot \boldsymbol{t}) f \boldsymbol{s} \times (\boldsymbol{s} \times \boldsymbol{t}) - Z_0 (\boldsymbol{H}_0^i \cdot \boldsymbol{t}) g \boldsymbol{s} \times \boldsymbol{t}] \cdot e^{-j2k\boldsymbol{r}_t \cdot \boldsymbol{s}} \frac{\sin(k\boldsymbol{l} \cdot \boldsymbol{s})}{k\boldsymbol{l} \cdot \boldsymbol{s}} \tag{3}$$

where:

t: mandatory edge unit vector's direction; i: incident wave; E_0^i: strength of the incident electric field; f and g: Yoffie Rousseff diffraction coefficients; s: scattering wave's direction; Z_0: wave impedance in vacuum; H_0^i: strength of the incident magnetic field; r_t: position vector of the middle of the edge; l: vector of the edge; θ: angle between i and t. Other symbols' definition can be found in Yue *et al.* (2014a, 2014c, 2014d, 2015).

The superposition formula of the RCS of variable-sweep wing aircraft is as follows:

$$\sigma = \left| \sum_{i=1}^{n} (\sqrt{\sigma_{po}})_i + \sum_{i=1}^{n} (\sqrt{\sigma_{ecm}})_i \right|^2 \tag{4}$$

The arithmetic average RCS of the aircraft is shown as follows:

$$\bar{\sigma}^{\beta}_{n\sim N} = \frac{1}{N-n+1} \sum_{\varphi=n}^{N} \sigma^{\beta}_{\varphi} \tag{5}$$

The unit-conversion formula of the aircraft's RCS is shown as follows:

$$\sigma_{\mathrm{dBm}^2} = 10\lg\sigma \tag{6}$$

where:

σ: RCS of the variable-sweep wing aircraft, in m^2; σ^{β}_{φ}: RCS in m^2 when the pitching angle of the incident wave is β and the azimuth angle of the aircraft is φ; $\bar{\sigma}^{\beta}_{n\sim N}$: average value of RCS when the pitching angle of the incident wave is β and the azimuth angle φ ranges from n to N; σ_{dBm^2}: RCS of the variable-sweep wing aircraft, in dBm^2.

RADAR DETECTION PROBABILITY MODEL

The probability that the aircraft is detected by a pulse Doppler radar can be estimated as follows Yue *et al.* (2010):

$$P_{Dk} = \exp\left(-\frac{9.5\sqrt{\Omega}(4\pi)^3 k_b T_0 B_n F_n L R^4}{\sqrt{\Delta\theta_\alpha f_r} P_t G^2 \lambda^2 E \sigma_{\varphi\gamma\beta}}\right) \tag{7}$$

where:

P_{Dk}: detection probability of the target; $\Delta\theta_a$: horizontal width of the radar lobe, in degrees; f_r: radar's pulse repetition frequency (PRF), in Hz; Ω: radar's scanning angular velocity, in degrees/s; P_t: radar transmitting power, in Watts (W); G: antenna gain, in dB; E: pulse compression ratio of the pulse compression radar; k_b: Boltzmann constant, and its value is 1.38×10^{-23}Ws/K; T_0: standard room temperature, and its value is 290K; B_n: noise bandwidth in the receiver, in Hz; F_n: noise coefficient in the receiver, in dB; L: total system losses, in dB; R: detection distance, in m; $\sigma_{\varphi\gamma\beta}$: RCS of the aircraft when the azimuth angle is φ, the rolling angle is γ, and the pitching angle is β, in m^2.

In order to demonstrate the influence of the varying sweep angle on the detection probability, we choose a typical ground-to-air searching situation for simplicity. In this situation, the aircraft is assumed to fly right towards the searching radar, and only the pitching angle varies with the detection distance. The pitching angle of the incident wave relative to the aircraft can be calculated as follows:

$$\beta = \beta_f - \frac{180}{\pi}\arcsin\left(\frac{h_d - h_m}{R}\right) \tag{8}$$

$$l \le \left\{ \sqrt{(h_m + R_d)^2 - R_d^2} + \sqrt{(h_d + R_d)^2 - R_d^2} \right\} \tag{9}$$

where:

β: pitching angle of the incident wave relative to the aircraft, in degrees. β is negative when the incident wave is below the aircraft; β_f: pitching angle of the aircraft relative to the ground, in degrees. It is negative when the aircraft's nose is upward; h_d: vertical distance between the aircraft and the sea level, in km; h_m: vertical distance between the radar and the sea level, in km; R_d: radius of the Earth and it is 6,371 km in general.

ALGORITHM VERIFICATION IN MICROWAVE ANECHOIC CHAMBER

In order to verify the effectiveness of the numerical simulation method that combines the PO method and the ECM, we have conducted a verification experiment in a microwave anechoic chamber.

The experiment procedure is as follows:

- Use CATIA to build a 3-D digital model of the aircraft, as shown in Fig. 1a.
- Use a 3-D printer to make a 1:36 scaled wax model, as shown in Fig. 1b.
- Transform the wax model into a sand model cavity and caste aluminum in the sand mold cavity to get a 3-D model in the state of cast-aluminum model, as shown in Fig. 1c.
- Conduct the RCS measurement experiment on the cast-aluminum model in the microwave anechoic chamber, as shown in Fig. 1d.

The initial conditions of numerical simulation are: the incident wave is at X band and of horizontal polarization. The pitching angle of the incident wave is 0°. The azimuth angle of the aircraft is between 0° ~ 180°. Use PO method and ECM to do the numerical simulation on the 3-D digital prototype shown in Fig. 1a, and the RCS characteristics curve can be obtained as shown in Fig. 2.

According to the similarity principle, the incident wavelength should be 0.83 mm when conducting the RCS measurement experiment in the microwave anechoic chamber. And the incident wave should also be of horizontal polarization. The measurement results are shown in Fig. 2.

From the comparison between the numerical simulation and experimental results, we can see that the two RCS characteristic curves basically match each other. After verifying the scientific character and accuracy of the numerical simulation method, we can use this method to analyze the RCS characteristics of the aircraft.

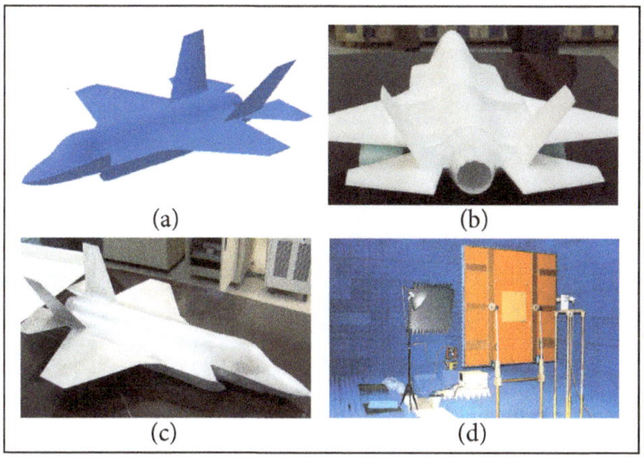

Figure 1. Microwave chamber experiment. (a) 3-D digital model; (b) Wax model; (c) Cast-aluminum model; (d) Microwave chamber.

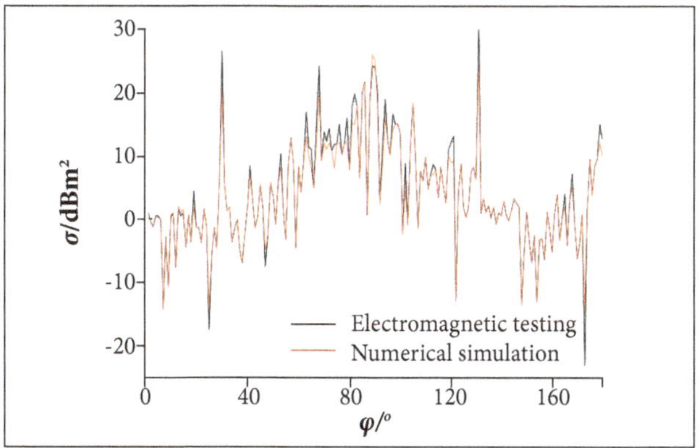

Figure 2. Comparison of the aircraft's RCS between numerical simulation and experiment results.

NUMERICAL SIMULATION OF AIRCRAFT'S RCS

Numerical simulation of aircraft's RCS consists of four components: CATIA modeling of the variable-sweep wing aircraft, numerical simulation of the aircraft's RCS, analysis of the impact of wing leading edge's sweep angle on aircraft's RCS, and the probability of being detected by a searching radar.

CATIA MODELING OF THE VARIABLE-SWEEP WING AIRCRAFT

The variable-sweep wing aircraft scheme in this paper has adopted features as follows: single-seat, twin-engine, twin vertical tails extraversion, variable-sweep wing aerodynamic layout, the band edge of the wing and blended wing body. Double S-shaped bend inlets are adopted on both sides of the fuselage, and three bomb bays are buried into the fuselage. Figure 3 shows the layout of the aircraft.

The CATIA model of the variable-sweep wing aircraft uses parametric designing method to design the wing leading edge's sweep angle, which is defined as χ (Fig. 4). Other basic parameters of the aircraft remain unchanged, in which the length of the aircraft is 22 m.

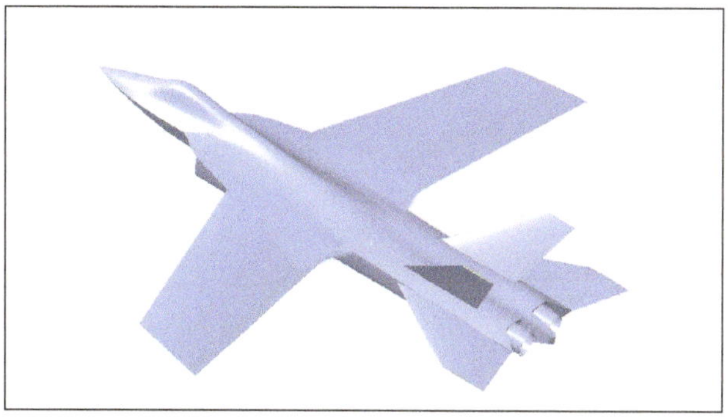

Figure 3. CATIA model of variable-sweep wing aircraft.

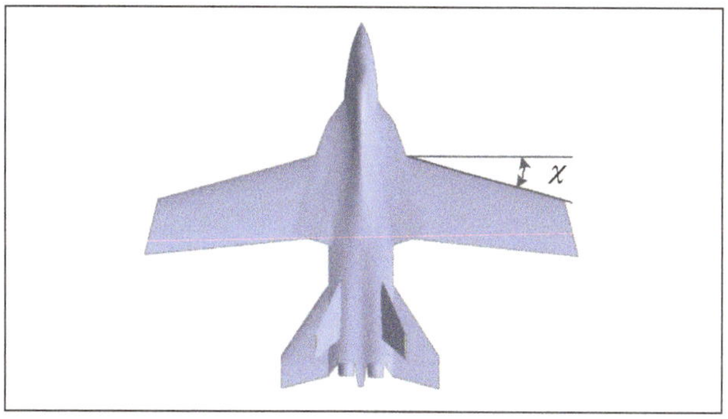

Figure 4. Schematic diagram of wing leading edge's sweep angle.

NUMERICAL SIMULATION OF THE AIRCRAFT'S RCS

CATIA is used in this study to establish a 3-D digital model for the variable-sweep wing aircraft. Analysis manager module is used to generate the 3-D grids (Fig. 5). Based on PO method and ECM, we import the aircraft 3-D grids to the RCS calculating program written in Matlab language. Under the conditions that the radar incident wave is at X band and of horizontal polarization, the pitching angle of the incident wave is −5°, 0° and +5°, and the azimuth angle of the aircraft is between 0° ~ 360°. In this section, we conduct a numerical simulation of the RCS of the variable-sweep wing aircraft and then obtain the RCS characteristic curve of the variable-sweep wing aircraft when χ = 15° (Fig. 6).

In the polar coordinates of the aircraft's RCS characteristic, the azimuth angle 0° corresponds to the head direction of the aircraft and the azimuth angle 180° corresponds to the tail direction of the aircraft. And when the azimuth angle is 90°, it corresponds to the right side of the aircraft.

From the data shown in Fig. 6, we can obtain some conclusions here: when χ = 15° and the pitching angle of the incident wave is 0°, the arithmetic average value of the aircraft's RCS between ± 10° is 3.59 dBm^2, the arithmetic average value of the aircraft's RCS between ± 30° is 20.43 dBm^2, and the RCS peak value in the head direction of the aircraft appears symmetrically and its value is 35.29 dBm^2, appearing at the azimuth angles of +15° and −15°.

Figure 5. 3-D grids of aircraft when χ = 15°.

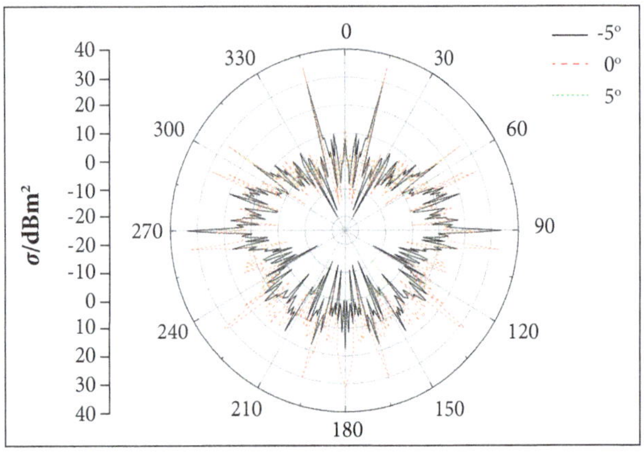

Figure 6. RCS characteristics of the aircraft when χ = 15°.

When the pitching angle of the incident wave is −5°, the arithmetic average value of the aircraft's RCS between ± 10° is 4.50 dBm^2, the arithmetic average value of the aircraft's RCS between ± 30° is 14.51 dBm^2, and the RCS peak value in the head direction of the aircraft appears symmetrically and its value is 29.07 dBm^2, appearing at the azimuth angles of +15° and −15°.

When the pitching angle of the incident wave is +5°, the arithmetic average value of the aircraft's RCS between ± 10° is 1.58 dBm^2, the arithmetic average value of the aircraft's RCS between ± 30° is 13.89 dBm^2, and the RCS peak value in the head direction of the aircraft appears symmetrically and its value is 28.56 dBm^2, appearing at the azimuth angles of +15° and −15°.

We can thus draw a qualitative conclusion that the azimuth angle of the RCS peak value of the variable-sweep wing aircraft is equal to the sweep angle χ.

ANALYSIS OF THE IMPACT OF THE AIRCRAFT'S SWEEP ANGLE ON ITS RCS CHARACTERISTICS

This section is an in-depth study of the qualitative conclusions obtained from the previous section. A series of numerical simulations of the RCS of the variable-sweep wing aircraft is conducted and the equivalence between the aircraft RCS peak azimuth angle and χ is then analyzed.

The purpose of calculating the average value of the aircraft's RCS is to analyze the influence of the variation of the sweep angle on the aircraft's RCS characteristics, from the perspective of mathematical statistics and the perspective of aircraft conceptual design and stealth design.

The leading edge's sweep angle χ varies from 0° to 66°, at an interval of 3°, and consequently 23 kinds of 3-D digital prototypes of the variable-sweep wing aircraft can be obtained. When calculating each aircraft prototype's RCS, the grids covering the whole aircraft are firstly generated, the radar incident wave is set to be at X band and of the horizontal polarization, and the pitching angle is 0°; 23 different characteristic curves of the aircraft's RCS can then be obtained. In this series of numerical simulations, the variable-sweep wing aircraft's grids are shown in Fig. 7, where χ is equal to 39°, 54° and 66°, respectively, and the RCS characteristic curves for these three different variable-sweep wing aircraft are shown in Fig. 8.

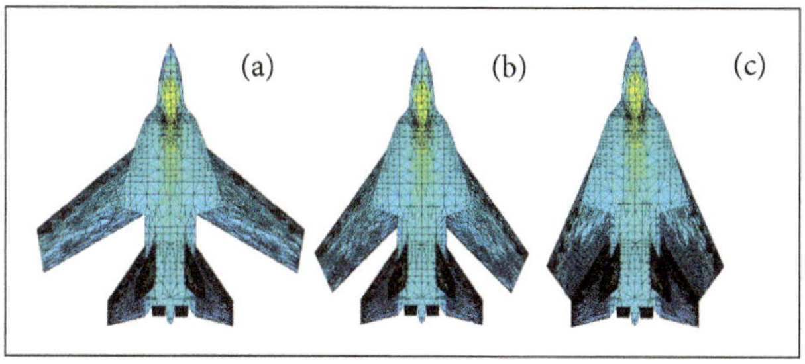

Figure 7. Trellis diagram in the type for χ of aircraft. (a) χ = 39°; (b) χ = 54°; (c) χ = 66°.

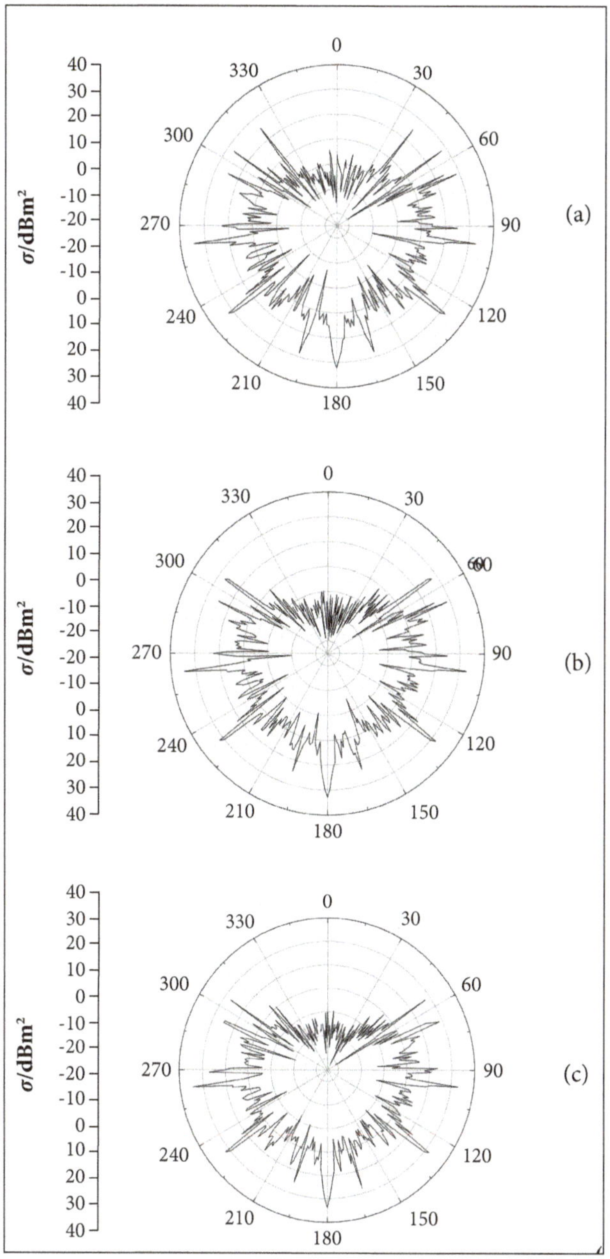

Figure 8. RCS characteristics in the type for χ of aircraft. (a) χ = 39°; (b) χ = 54°; χ = 66°.

From the data shown in Figs. 8a to 8c, we can obtain some conclusions here: (1) when the wing leading edge's sweep angle is 39°, the RCS arithmetic average value between ± 10° of the aircraft is –1.02 dBm^2, and the RCS arithmetic average value between ± 30° of the aircraft is –1.61 dBm^2. The RCS peak value, appearing symmetrically in the aircraft's head direction, is 24.85 dBm^2 and appears at both +39° and –39° azimuth angles. This proves that, in the head direction of the aircraft, the azimuth angle corresponding to the RCS peak value is equal to the wing leading edge's sweep angle; (2) when the wing leading edge's sweep angle is 54°, the RCS arithmetic average value between ± 10° of the aircraft is –5.901 dBm^2, and the RCS arithmetic average value between ± 30° of the aircraft is –5.59 dBm^2. The RCS peak, appearing symmetrically in the head direction of the aircraft, is 26.21 dBm^2 and appears at both +54° and –54° azimuth angles. This proves that, in the head direction of the aircraft, the azimuth angle corresponding to the RCS peak value is equal to the wing leading edge's sweep angle; (3) when the wing leading edge's sweep angle is 66°, the RCS arithmetic average value between ± 10° of the aircraft is –4.68 dBm^2, and the RCS arithmetic average value between ± 30° of the aircraft is –6.198 dBm^2. The RCS peak value appears twice symmetrically in the head direction of the aircraft, and the peak values are 27.72 and 25.1 dBm^2, appearing at both ± 55° and ± 66° azimuth angles. This proves that, in the head direction of the aircraft, the azimuth angle corresponding to the RCS peak value is equal to the wing leading edge's sweep angle. For another RCS peak, the aircraft forward azimuth is equal to the horizontal tail leading edge sweep angle, which is ± 55°.

Based on the mathematical statistics and comparative analysis of all the 23 aircraft RCS characteristic curves obtained from the simulations, the relationship curve between the RCS peak value and the sweep angle χ is shown in Fig. 9, and the relationship curve between the corresponding azimuth angle of the aircraft RCS forward peak value and the sweep angle χ is shown in Fig. 10, the relationship curves between the aircraft RCS forward arithmetic average value and sweep angle χ are shown in Fig. 11.

It can be seen from Fig. 9 that, when the wing leading edge's sweep angle χ is 0°, its aircraft's RCS forward peak value is 36.08 dBm^2. When χ = 18°, its aircraft's RCS peak value is 28.19 dBm^2, and, after unit conversion, we know that the aircraft's RCS peak value when χ = 18° is 16.25% of that when χ = 0°. When χ = 33°, its aircraft's RCS peak value is 26.19 dBm^2, and the converting results show that the aircraft RCS peak value when χ = 33° is 10.25% of that when χ = 0°. When χ = 42°, its aircraft's RCS peak value is 29.85 dBm^2, and the converting results show that, when $\chi = 42°$, the aircraft's RCS peak value is 23.82% of that when $\chi = 0°$. When $\chi = 54°$, its aircraft's RCS peak value is 26.21 dBm^2, and the converting results show that the aircraft's RCS peak value for $\chi = 54°$ is 10.30% of that when $\chi = 0°$. When $\chi = 66°$, its aircraft's RCS peak value is 25.10 dBm^2, and the converting results show that the aircraft's RCS peak value when $\chi = 66°$ is 7.97% of that when $\chi = 0°$. It can be seen that, when the sweep angle of the aircraft increases, the peak value of the RCS decreases non-linearly.

Figure 10 shows that the azimuth angle corresponding to one of the RCS peak value is equal to the aircraft's sweep angle.

From Fig. 11, we can see that:

- The arithmetic average value of the aircraft's RCS between ± 10° is $\bar{\sigma}_t^{\pm10°}$ and, when $\chi = 0°$, $\bar{\sigma}_t^{\pm10°} = 25.67$ dBm^2; when $\chi = 33°$, $\bar{\sigma}_t^{\pm10°} = 0$ dBm^2 and, after unit conversion, we know that when $\chi = 33°$, $\bar{\sigma}_t^{\pm10°}$ is 0.271% of that when $\chi = 0°$. When $\chi = 42°$, $\bar{\sigma}_t^{\pm10°} = -4.97$ dBm^2, and when $\chi = 42°$, $\bar{\sigma}_t^{\pm10°}$ is 0.087% of that when $\chi = 0°$.
- The arithmetic average value of the aircraft's RCS between ± 30° is $\bar{\sigma}_t^{\pm30°}$ and, when $\chi = 0°$, $\bar{\sigma}_t^{\pm30°} = 21.21$ dBm^2; when $\chi = 33°$, $\bar{\sigma}_t^{\pm30°} = -0.7$ dBm^2 and, after unit conversion, we know that, when $\chi = 33°$, $\bar{\sigma}_t^{\pm30°}$ is 0.644% of that when $\chi = 0°$; when $\chi = 42°$, $\bar{\sigma}_t^{\pm30°} = -2.67$ dBm^2 and, when $\chi = 42°$, $\bar{\sigma}_t^{\pm30°}$ is 0.409% of that when $\chi = 0°$.
- The arithmetic average value of the aircraft's RCS between ± 45° is $\bar{\sigma}_t^{\pm45°}$ and, when $\chi = 0°$, $\bar{\sigma}_t^{\pm45°} = 19.53$ dBm^2; when $\chi = 33°$, $\bar{\sigma}_t^{\pm45°} = 10.07$ dBm^2 and, when $\chi = 33°$, $\bar{\sigma}_t^{\pm45°}$ is 11.324% of that when $\chi = 0°$. When $\chi = 42°$, $\bar{\sigma}_t^{\pm45°} = 13.39$ dBm^2 and, when $\chi = 42°$, $\bar{\sigma}_t^{\pm45°}$ is 24.32% of that when $\chi = 0°$.

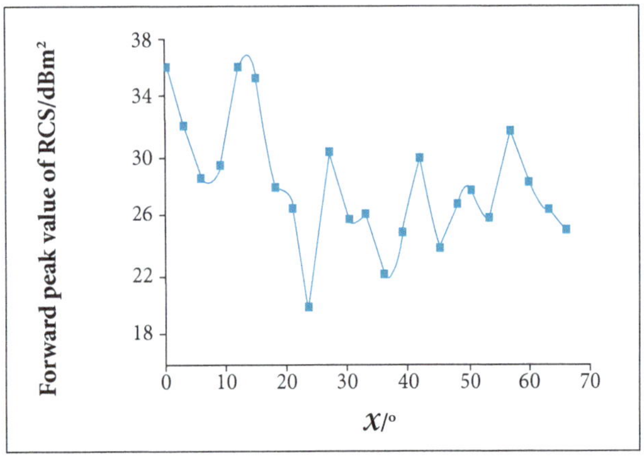

Figure 9. Relation between aircraft's RCS forward peak value and χ.

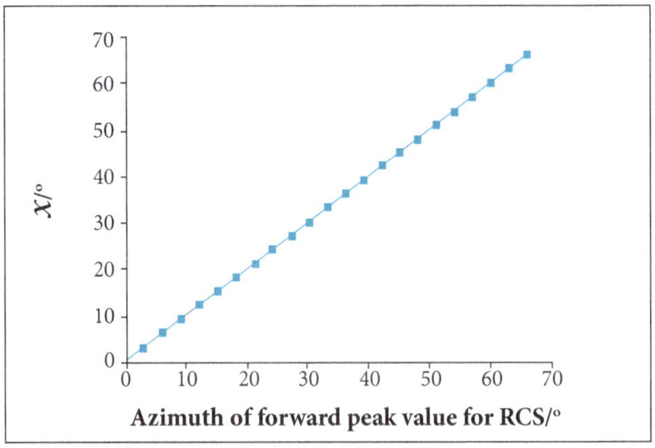

Figure 10. Relation between the corresponding azimuth angle of aircraft's RCS forward peak value and χ.

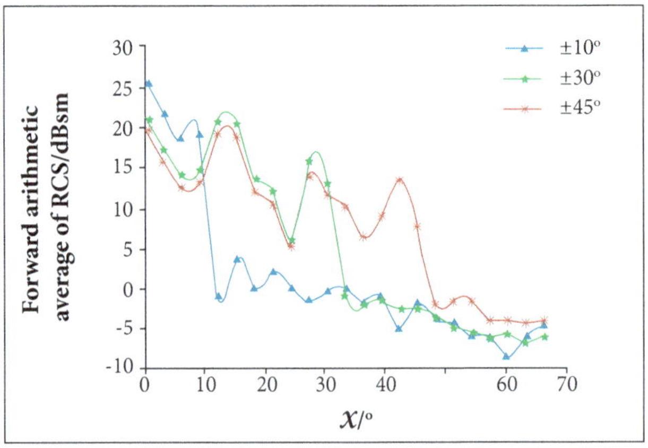

Figure 11. Relation between aircraft's RCS forward arithmetic average value and χ.

CALCULATING THE RADAR DETECTION PROBABILITY

The purpose of studying the RCS characteristics of the variable-sweep wing aircraft is to reduce the detection probability by the hostile radar. In this section, we study the relationship between the detection probability and the characteristics of the target's RCS and the detection distance.

In a typical ground-to-air searching situation, as mentioned in section Radar Detection Probability Model, the rolling and azimuth angles of the incident wave relative to the aircraft are both 0°, and β_f, h_d, h_m of Eqs. 8 and 9 are set to be −2.5°, 8 km, and 3 km, respectively. Figure 12 shows the relationship between the pitching angle of the incident wave and the detection distance.

As shown in Fig. 12, β decreases from −5° to below −30° when the detection distance decreases from 200 to 0 km. The corresponding RCS values of the aircraft under these attitudes can then be calculated using the PO method and

ECM. The incident wave is at X band and of horizontal polarization. Figure 13 shows the varying RCS value of the aircraft when it is flying towards the searching radar from 200 km away. Three types of lines indicate three different sweep angles, which are $\chi = 15°$, $39°$ and $54°$, respectively.

Figure 13 shows that, when the aircraft is flying towards the radar from 200 km away, its RCS value varies between about −20 and 10 dBm^2.

According to the RCS values shown in Fig. 13, the detection probability can be calculated using Eq. 7. The initial parameters of the pulse Doppler radar are set as follows: $\Delta\theta_a = 7°$, $f_r = 300$ Hz, $\Omega = 30°/s$, $P_t = 10$ MW, $G = 36$ dB, $\lambda = 0.03$ m, $E = 900$, $B_n = 0.077$ MHz, $F_n = 55$ dB, $L = 20$ dB. RCS values of Fig. 13 are expressed in dBm^2, when applied to Eq. 7, they should be converted to m^2, with the help of Eq. 6. Figure 14 shows the calculation results of the aircraft's detection probability.

As shown in Fig. 14, when the aircraft is flying right towards the radar from 200 km away, the detection probabilities are very low when R is beyond about 120 km, for all 3 types of variable-sweep wing aircrafts. When the distance decreases to within 100 km, the aircraft with a sweep angle of $\chi = 54°$ performs best and the aircraft with a sweep angle of $\chi = 15°$ performs worst. When the distance decreases to within 40 km, all 3 types of variable-sweep wing aircrafts perform almost the same.

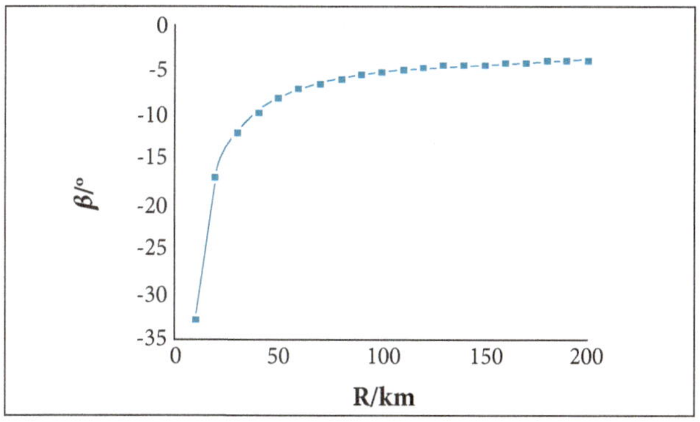

Figure 12. The pitching angle of the incident wave against the detection distance.

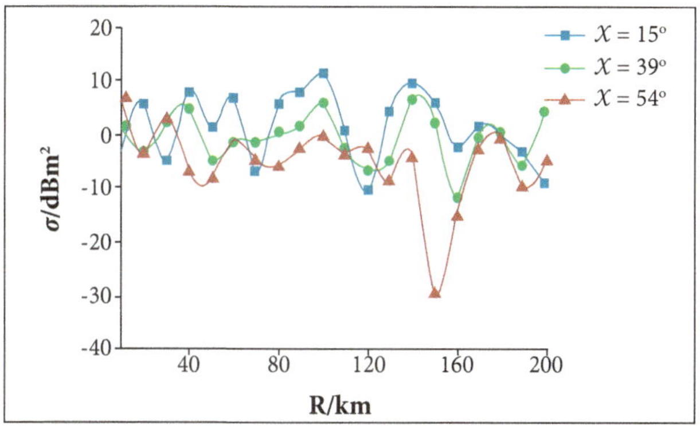

Figure 13. The RCS of variable-sweep wing aircraft against detection distance.

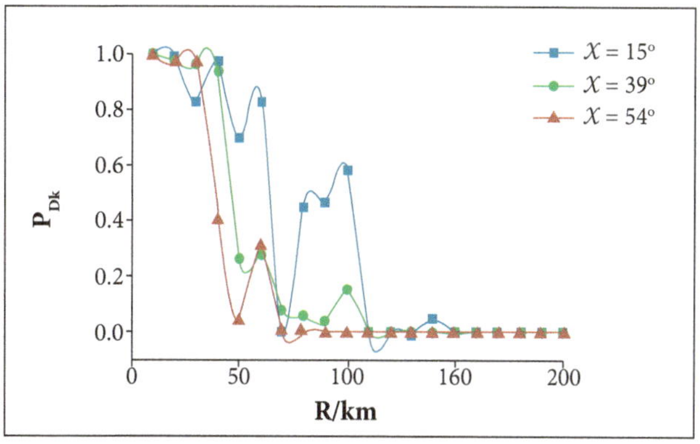

Figure 14. Detection probability of the aircraft when it is flying from 200 km away.

CONCLUSION

This paper utilizes the physical optics and the equivalent currents methods to study the RCS characteristics of the variable-sweep wing aircraft. Based on the numerical simulations of the aircraft, the following conclusions can be obtained:

The aircraft's RCS forward peak value decreases non-linearly with the sweep angle χ.

The sweep angle of the variable-sweep wing aircraft and the azimuth angle corresponding to one of the RCS peak values are identical.

When $\chi = 33°$, $\bar{\sigma}_t^{\pm30°} = -0.7$ dBm^2. After unit conversion, $\bar{\sigma}_t^{\pm30°}$ of the aircraft when $\chi = 33°$ is 0.644% of that when $\chi = 0°$.

The larger the sweep angle of the variable-sweep wing aircraft is, the lower the aircraft's detection probability will be.

We hope that the conclusions of this paper provide some reference and technical support for stealth aircraft's demonstration and designing.

ACKNOWLEDGEMENT

This work was supported by the Natural Science Foundation of China (51375490).

REFERENCES

Bai, Z.D. and Liu, H., 2007, "Parametric Modeling and Optimization of Observability Fuselage in Aircraft Conceptual Design", Journal of Beijing University of Aeronautics and Astronautics, Vol. 33, No. 12, pp. 1391-1394.

Bao, J.B. and Wang, G.L., 2012, "Optimization and Experimental Verification for Aerodynamic Scheme of Flying-Wing", Journal of Beijing University of Aeronautics and Astronautics, Vol. 38, No. 2, pp. 180-184.

He, X., Chen, J.P. and Ni, X.W., 2010, "Attenuation of Planar Electromagnetic Waves by Inhomogeneous Plasma", High Power Laser and Particle Beams, Vol. 22, No. 9, pp. 2115-2118.

Hu, T.Y. and Yu, X.Q., 2011, "Preliminary Design of Unconventional Configuration Aircraft Using Multidisciplinary Design Optimization", Acta Aeronautica et Astronautica Sinica, Vol. 32, No. 1, pp. 117-127.

Huang, P.L. and Liu, Z.H., 2008, "Research on Electromagnetic Scattering Characteristics of Slits on Aircraft", Acta Aeronautica et Astronautica Sinica, Vol. 29, No. 3, pp. 675-680.

Ji, J.Z., Wu, Z. and Liu, Z.H., 2009, "Research on the Curvature Parameter in the S-Shaped Inlet's Stealth Design", Journal of Xidian University, Vol. 36, No. 4, pp. 746-750.

Lu, J.W. and Wang, Qi., 2009, "Aircraft-Skin Infrared Radiation Characteristics Modeling and Analysis", Chinese Journal of Aeronautics, Vol. 22, No. 5, pp. 493-497.

Nangia, R.K. and Palmer, M.E., 2005, "A Comparative Study of UCAV Type Wing Planforms–Aero Performance and Stability Considerations", AIAA-2005-5078.

Sun, C. and Zhang, P., 2008, "LO Requirements and Solutions of Avionics/RF System for Advanced Aircraft", Acta Aeronautica et Astronautica Sinica, Vol. 29, No. 6, pp. 1472-1481.

Tom, R.B. and Alfred, G.S., 2010, "Optimization of Aircraft Configuration for Minimum Drag", AIAA-2010-3000.

Vogel, B.S., 2005, "Boeing X-45 UCAVs Complete Simulated Mission", Janes Defence Industry, UK.

Wood, R.M. and Bauer, S.X.S., 2001, "Flying Wings/Flying Fuselage", AIAA-2001-0311.

Yue, K.Z., Hou, Z.Q. and Yang, Y.K., 2010, "Markov Model of Efficiency of Volleying Anti-Ship Missiles Piercing into Ship-Bone Antiaircraft Missiles", Journal of System Simulation, Vol. 22, No. 6, pp. 1472-1475.

Yue, K.Z., Liu W.L., Li G.X., Ji J.Z. and Yu D.Z., 2015, "Numerical Simulation of RCS for Carrier Electronic Warfare Airplanes", Chinese Journal of Aeronautics, Vol. 28, No. 2, pp. 545-555.

Yue, K.Z., Gao, Y., Li, G.X. and Yu, D.Z., 2014a, "Conceptual Design and RCS Performance Research of Shipborne Early Warning Aircraft", Journal of Systems Engineering and Electronics, Vol. 25, No. 6, pp. 968-976.

Yue, K.Z., Sun, C. and Ji, J.Z., 2014b, "Numerical Simulation on the Stealth Characteristics of Twin-Vertical-Tails for Fighter", Journal of Beijing University of Aeronautics and Astronautics, Vol. 40, No. 2, pp. 160-165.

Yue, K.Z., Sun, C., Liu, H. and Su, M., 2014c, "Numerical Simulation on the RCS of Combat Aircraft for Mounted Missile", Systems Engineering and Electronics, Vol. 36, No. 1, pp. 62-67.

Yue, K.Z., Tian, Y.F., Liu, H. and Han, W., 2014d, "Conceptual Design and RCS Property Research of Three-Surface Strike Fighter", International Journal of Aeronautical and Space Sciences, Vol. 15, No. 3, pp. 309-319.

Optimization of Airfoils for Minimum Pitching Moment and Compressibility Drag Coefficients

Hamid Farrokhfal[1], Ahmad Reza Pishevar[2]

ABSTRACT: This paper concerns a numerical optimization method for designing airfoils based on adjoint method. The goal of present work is to reduce the compressibility drag or pitching moment of transonic airfoils without compromising on the lift coefficient. A new cost function based on this requirement is defined and the corresponding adjoint equations are discussed in details. At the end, by demonstrating some numerical results, we show that this technique is capable of converging to the optimum design point corresponding to the initial geometry of the airfoil.

KEYWORDS: Optimization, Pitching moment, Transonic airfoils, Adjoint equations, Euler equations.

INTRODUCTION

Improvement of aerodynamic performance of helicopter blade designs has been one of the most important research areas in rotorcraft aerodynamics. Low-pitching-moment airfoils found application primarily as helicopter rotor blades, but some attention has been given to the advantages of low pitching-moment sections for a "span-loader" vehicle. For such applications, a symmetric airfoil could conceivably be employed, but cambered airfoils can offer significant advantages (Barger, 1975). Therefore, designing cambered airfoils with low pitching-moment is attractive in helicopter blade design to achieve small control forces in rotor controls, and this task can be conducted via optimizing procedures.

Also, shock waves in transonic flow not only increase the wave drag but also cause unfavourable flutter or buffet to fixed wings. Helicopter rotor blades in forward flight are also frequently exposed to strong unsteady shock waves at the blade tip region. These shock waves increase the required torque, and become a source of undesirable noise and vibration. Thus, elimination or possible reduction of the strength of these shock waves would be desirable for the enhancement of the performance of helicopters and airplanes.

In order to improve airfoil performance under different flight conditions and to make the performance insensitive to off-design condition at the same time, Yu *et al.* (2010) have developed a multi-objective optimization approach considering robust design and applied it to airfoil tailoring. They used non-uniform rational B-spline (NURBS) representation, control points and related weights around airfoil as design variables. They could

1.Malek-Ashtar University of Technology – Isfahan – Iran 2.Isfahan University of Technology – Isfahan – Iran.

Author for correspondence: H. Farrokhfal | Department of Mechanical & Aerospace Engineering | Malek-Ashtar University of Technology – 83145-115 | Shahinshahr – Isfahan – Iran | Email: Farrokhfal@mut-es.ac.ir obtain a set of non-dominated airfoil solutions with robustness, by adopting multi-objective genetic algorithm that is based on non-dominated sorting. They showed that the multi-objective robust design method makes the airfoil performance robust under different off-design conditions.

In the gradient-based optimization design, the gradients of cost/constraint functionals with respect to design variables are important information in the design processes (Xie, 2002). A traditional approach to calculate gradients is the finite difference method which involves finite differencing performance functionals computed from the solutions to the governing equations with perturbed parameter values. The Finite difference method (FDM) with complex variables was explored by Anderson *et al.* (2001) to improve its accuracy. Even though it is very easy to implement FDM in program coding, its prohibitive computational costs (many times solving governing equations) motivate other time-efficient methods to calculate reduced gradients.

A way of performing an efficient aerodynamic shape optimization is to regard the design problem as a control problem in which the control is the shape of the boundary. This approach to optimal aerodynamic design was introduced by Jameson (1988, 1990) who examined the design problem for compressible flow with shock waves, and devised adjoint equations to determine the gradient for both potential flow and also flows governed by the Euler equations (Jameson, 1995; Reuther and Jameson, 1994).

Also, particular interest has been given to adjoint methods (Reuther *et al.*, 2001) in which the gradient, regarding an arbitrarily large number of parameters, can be calculated with roughly the same computational cost as two flow solutions. Once the gradient has been calculated, a descent method can be used to determine a shape change which will make an improvement in the design.

Based on the idea of adjoint method, an optimum aerodynamic design technique is presented by Ying *et al.* (2011), which can be applied to the optimum problems with a large number of design variables. The key of this method lies in that the optimization process is regarded as an unsteady evolution, i.e., the optimization is executed, simultaneously with solving the unsteady flow governing equations and adjoint equations. During the unsteady evolution, the airfoil surface is moving with time, and the dynamic grid technique is introduced in the grid generation to improve the computational efficiency (Ying *et al.*, 2011).

Nadarajah and Tatossian (2008) also have presented an adjoint method for the multi-objective aerodynamic shape optimization of unsteady viscous flows. Their method is beneficial when both the pitching angle and free stream Mach number are sinusoidally varied. They showed that the multi-objective cost function was able to preserve both the lift and pitching moment while simultaneously decreasing the overall drag (Nadarajah and Tatossian, 2008).

Gradient-based techniques are efficient direct aerodynamic shape optimization (ASO) methods for both incompressible and compressible flows, and many of them use continuous adjoint methods.

More recently, the introduction of surrogate-based optimization (SBO) methods to ASO have been successful in reducing the total computational cost. The overall objective of using SBO methods is to reduce the number of evaluations of the high-fidelity models, and thereby making the optimization process more efficient. Leifsson and Koziel (2010) introduced a computational design methodology which exploits surrogates constructed using low-fidelity flow analysis models and shape-preserving response prediction technique and demonstrated that their approach allows a rapid design improvement of airfoils at a very low computational cost corresponding to a few evaluations of the high-fidelity model (Leifsson and Koziel, 2010).

Then they replaced the direct optimization of an accurate, high fidelity airfoil model by an iterative re-optimization of a corrected low-fidelity model (Leifsson *et al.*, 2011). Their low-fidelity model is based on the same governing fluid flow equations as the high-fidelity one, but uses coarser discretization and related convergence criteria. With these techniques, they succeeded to improve the overall robustness of the optimization algorithm.

ASO problems in compressible viscous flow were also performed using simultaneous pseudo-time stepping (Hazra *et al.*, 2005). Hazra *et al.* (2005) could optimize airfoil surface by use of a preconditioner for convergence acceleration which stems from the reduced sequential quadratic programming (SQP) methods.

This paper describes the implementation of adjoint technique for designing transonic airfoils with minimum compressible drag or pitching moment coefficients while preserving the initial lift coefficient. For this purpose, the Euler and adjoint equations are solved sequentially on a two-block structured grid about the section. The finite-volume scheme of Jameson *et al.* (1981) is used to discrete governing equations. This scheme employs a cell centered discretization technique. A multistage time-stepping algorithm is used to advance the solution in time. A key feature in this scheme is applying the dissipative terms at the end of each time step for solving the flow and adjoint equations. The magnitude of the dissipative terms is adapted to the local properties of the flow by means of a Jameson sensor (Nadarajah and Jameson, 2001), based on the local pressure gradient. Acceleration techniques are applied to obtain faster steady-state convergence. These methods include local time stepping, variable coefficient implicit residual smoothing, and multi grid technique.

GOVERNING EQUATIONS

Let (u,v) denote the velocity components in the Cartesian coordinate system (x,y). Compressible Euler equations can be formulated as:

$$\frac{\partial W}{\partial t}+\frac{\partial f_x}{\partial x}+\frac{\partial f_y}{\partial y}=0 \tag{1}$$

Where:

$$W=\begin{bmatrix}\rho\\ \rho u\\ \rho v\\ \rho e\end{bmatrix},\ f_x=\begin{bmatrix}\rho u\\ \rho u^2+p\\ \rho uv\\ \rho uh\end{bmatrix},\ f_y=\begin{bmatrix}\rho v\\ \rho vu\\ \rho v^2+p\\ \rho vh\end{bmatrix} \tag{2}$$

Defining the flux vecto r as $f = f_x\hat{i} + f_y\hat{j}$, where $(\hat{i},\hat{j})$ are the unit vectors in the (x, y) coordinate system, Eq. 1 can be written as:

$$\frac{\partial W}{\partial t}+\nabla . f=0 \tag{3}$$

Equation 3 can be solved by the finite-volume method described in the proceeding section.

NUMERICAL METHOD

In order to apply the finite-volume method, Eq. 3 is written in the integral form as:

$$\frac{\partial}{\partial t}\int_V \boldsymbol{W}\,dV+\iint_{\partial S}\hat{n}.\boldsymbol{f}\,ds+\iint_{\partial S}\boldsymbol{u}_\Omega.\hat{n}\,\boldsymbol{W}ds=\int \boldsymbol{S}\,dV \tag{4}$$

The computational domain is meshed by a multi-block structured quadrilateral grid. Applying Eq. 4 to a typical (i,j) cell, for a cell center algorithm we obtain a system of ordinary differential equations as:

$$\frac{d}{dt}\left(\forall_{ij}W_{ij}\right)+\mathsf{F}_{ij}=0 \tag{5}$$

Where $\forall_{ij}$ is the cell area and F_{ij} represents the net convectional fluxes out of each cell.

The convective fluxes can be determined by a second order central scheme. However, the resultant scheme is not dissipative, and therefore, undamped oscillations at odd and even mesh points can be developed during the computation. To suppress the tendency for odd-even decoupling and to prevent the appearance of oscillations in regions containing severe pressure gradients near shock waves and stagnation points, the finite-volume scheme is modified by the addition of artificial dissipative terms as below:

$$\frac{d}{dt}\left(\forall_{ij}W_{ij}\right)+\mathsf{F}_{ij}-D_{ij}=0 \tag{6}$$

Where D_{ij} denotes the dissipative fluxes. Jameson has established that an effective form of dissipative terms for flows with discontinuities is a blend of second and fourth differences with coefficients which depend on the local pressure gradient (Jameson *et al.*, 1981). Equation 6 can be integrated in time by a multistage Runge-Kutta scheme.

BOUNDARY CONDITIONS

Special care must be given to the boundary conditions when solving Euler equations on a multi-block grid. When using more than a single block for a two-dimensional body, the attachment boundary condition should be used for the neighboring cells at the block interfaces. This can be done by considering a layer of ghost cells around each block. The boundary condition at the airfoil surface is the tangency condition of flow, that is:

$$\vec{V}.\hat{n} = 0 \tag{7}$$

Where $\vec{V}$ is the velocity vector and $\hat{n}$ is the unit vector normal to the airfoil surface.

Far-field boundary condition is also applied at the outer surface of the domain. The treatment of the far-field boundary condition is based on the concept of Riemann invariants for a one-dimensional flow normal to the boundary (Jameson and Baker, 1983).

DERIVATION OF THE INVISCID ADJOINT TERMS

The main objective of the present study is to achieve a modified airfoil which minimizes the inviscid compressibility drag without compromising the pitching moment under the constraint of maintaining the desired lift level. For this purpose, using the Lagrange multiplier, the cost function is defined as:

$$I = k_1\, Cd + k_2\, Cm^2 + \sigma\left(Cl - Cl_d\right) \tag{8}$$

Here k_1 and k_2 are the weight coefficients by which the airfoil shape can be modified for a particular application. When k_1 is chosen to be much larger than k_2, a reduction in the drag coefficient is the main priority of the designer while using a larger k_2 leads to an airfoil with smaller pitching moment coefficient. In this equation Cl_d indicates the desired value of Cl. The variation of cost function in relation to the angle of attack and other parameters can be written as:

$$\frac{\partial I}{\partial \alpha} = k_1 \frac{\partial Cd}{\partial \alpha}|_G + k_2 Cm \frac{\partial Cm}{\partial \alpha}|_G + \sigma \frac{\partial Cl}{\partial \alpha}|_G = 0 \tag{9}$$

$$\frac{\partial I}{\partial G}|_{Cl} = k_1 \frac{\partial Cd}{\partial G}|_\alpha + k_2 Cm \frac{\partial Cm}{\partial G}|_\alpha + \sigma \frac{\partial Cl}{\partial G}|_\alpha$$

Where $|_{(.)}$ denotes the partial derivative at the frozen quantity (.) and G denotes the flow or design variables. From Eq. 9, the Lagrange multiplier σ can be obtained as:

$$\sigma = -k_1 \frac{\partial Cd/\partial\alpha|_G}{\partial Cl/\partial\alpha|_G} - k_2 \frac{\partial Cm/\partial\alpha|_G}{\partial Cl/\partial\alpha|_G} \tag{10}$$

Substituting into Eq. 9 yields:

$$\frac{\partial I}{\partial G}|_{Cl} = k_1\left(\frac{\partial Cd}{\partial G}|_\alpha + \Omega_1 \frac{\partial Cl}{\partial G}|_\alpha\right) + k_2 Cm\left(\frac{\partial Cm}{\partial G}|_\alpha + \Omega_2 \frac{\partial Cl}{\partial G}|_\alpha\right) \tag{11}$$

Where: $\Omega_1 = -\dfrac{\partial Cd/\partial\alpha|_G}{\partial Cl/\partial\alpha|_G}$, $\Omega_2 = -\dfrac{\partial Cm/\partial\alpha|_G}{\partial Cl/\partial\alpha|_G}$

The airfoil lift, drag and pitching moment coefficients can also be written as:

$$Cl = -\int_B C_p S_{22}\, Cos\,\alpha + \int_B C_p S_{21}\, Sin\,\alpha$$

$$Cd = -\int_B C_p S_{21}\, Cos\,\alpha - \int_B C_p S_{22}\, Sin\,\alpha \tag{12}$$

$$Cm = -\frac{1}{c}\int_B C_p\left(x - x_{ref}\right) S_{22} - \frac{1}{c}\int_B C_p\, y\, S_{21}$$

Where x_{ref} is the reference point for computing the pitching moment and S_{21} and S_{22} are the components of unit vector normal to the airfoil surface.

Taking variation from the steady state flow equations in the computational domain leads to:

$$\frac{\partial}{\partial \xi_i}\delta F_i = 0 \tag{13}$$

Where: $\delta F_i = C_i\,\delta w + \delta S_{ij} f_j$, $C_i = S_{ij} A_j$, $A_j = \dfrac{\partial f_j}{\partial w}$

Multiplying Eq. 13 by a vector of co-state variables ψ and integrating over the domain,

$$\int_D \psi^T \left(\frac{\partial}{\partial \xi_k} \delta F_k \right) dD_\xi = 0 \qquad (14)$$

Finally, this equation can be integrated by parts to give:

$$\int_D \left(\frac{\partial \psi^T}{\partial \xi_k} \delta F_k \right) dD_\xi = \int_B \psi^T \left(\delta F_k . n_k \right) dE \qquad (15)$$

Thus, the variation in the cost function can be written as:

$$\begin{aligned}\delta I = k_1 \{ &- \int_B \delta C_p S_{21} \, Cos\,\alpha - \int_B C_p \delta S_{21} \, Cos\,\alpha - \\ &\int \delta C_p S_{22} \, Sin\,\alpha - \int C_p \, \delta S_{22} \, Sin\,\alpha \} + k_1 \Omega_1 \\ &\{ - \int \delta C_p S_{22} \, Cos\,\alpha - \int C_p \delta S_{22} \, Cos\,\alpha + \\ &\int_B \delta C_p S_{21} \, Sin\,\alpha + \int_B C_p \, \delta S_{21} \, Sin\,\alpha + \frac{k_2}{c} Cm \int_B [\delta C_p \\ &(x - x_{ref}) S_{22} + C_p \delta x \, S_{22} + C_p \left(x - x_{ref} \right) \delta S_{22} - \\ &\delta C_p y \, S_{21} + k_2 \Omega_2 Cm \{ \int_B [\delta C_p Sin\,\alpha \, S_{21} + \\ &C_p Sin\,\alpha \, \delta S_{21} - \delta C_p \, Cos\,\alpha \, S_{22} - C_p \, Cos\,\alpha \, \delta S_{22}]\} \\ &+ \int_B \psi^T \left(\delta F_k . n_k \right) dB - \int_D \frac{\partial \psi^T}{\partial \xi_k} \{ C_k \, \delta w + \delta S_{kl} f_l \} dD_\xi \end{aligned} \qquad (16)$$

and n_1=0 on boundary B.

Eliminating the term which contains δw from the above integrals, the adjoint equation can be obtained as:

$$\frac{\partial \psi}{\partial t} - C_k^T \frac{\partial \psi}{\partial \xi_k} = 0 \qquad (17)$$

Because n_1=0 on airfoil surface, the only remaining component of F_k is:

$$F_2 = S_{21} f_1 + S_{22} f \qquad (18)$$

Where from the tangent boundary condition of flow on the airfoil surface, the variation of F_2 becomes:

$$\delta F_2 = \begin{bmatrix} 0 \\ \delta S_{21} \\ \delta S_{22} \\ 0 \end{bmatrix} p + \begin{bmatrix} 0 \\ S_{21} \\ S_{22} \\ 0 \end{bmatrix} \delta p \qquad (19)$$

If the coordinate transformation is such that δS is negligible in the far field, then the boundary condition of adjoint Eq. 17 can be obtained as:

$$\begin{aligned} &S_{21}\psi_2 + S_{22}\psi_3 + \frac{k_1}{q_\infty} \{ (S_{21} Cos\,\alpha + S_{22} Sin\,\alpha) \\ &+ \Omega_1 \left(S_{22} Cos\,\alpha - S_{21} Sin\,\alpha \right) + \frac{k_2}{q_\infty c} Cm \\ &\{ \frac{1}{c} \left[y \, S_{21} - \left(x - x_{ref} \right) S_{22} \right] + \Omega_2 \} (S_{22} Cos\,\alpha - \\ &S_{21} Sin\,\alpha) = 0 \end{aligned} \qquad (20)$$

Where: $q_\infty = \frac{\gamma}{2} M_\infty^{\ 2} P_\infty$. Finally we have:

$$\begin{aligned} \delta I = -k_1 \{ &\int_B C_p \delta S_{21} \, Cos\,\alpha + \int_B C_p \, \delta S_{22} \, Sin\,\alpha + \Omega_1 \\ &\int_B C_p \delta S_{22} \, Cos\,\alpha - \Omega_1 \int_B C_p \, \delta S_{21} \, Sin\,\alpha \} + k_2 Cm \\ &+ k_2 Cm \int_B [C_p \delta x \, S_{22} + C_p \left(x - x_{ref} \right) \delta S_{22} - \\ &C_p \, \delta y \, S_{21} - C_p y \, \delta S_{21}] k_2 Cm \int_B [C_p Sin\,\alpha \, \delta S_{21} - \\ &C_p \, Cos\,\alpha \, \delta S_{22}] + \int_B C_p \left(\delta S_{21} \psi_2 + \delta S_{22} \psi_3 \right) dB - \\ &\int_D \frac{\partial \psi^t}{\partial \xi_k} \delta S_{kl} f_l dD_\xi \end{aligned} \qquad (21)$$

And the gradients can then be defined with respect to the design variables x_i as:

$$G(x_i) = \frac{\delta I}{\delta x_i} \qquad (22)$$

NUMERICAL METHOD FOR ADJOINT EQUATIONS

The semi discretized adjoint equations of Eq. 17 is obtained using the same approach used for the flow equations:

$$\forall_{ij} \frac{\partial \psi_{ij}}{\partial t} - R_{ij} = 0$$

$$R_{ij} = \frac{1}{2}\left\{C_{1\ ij}^{T}\left(\psi_{i+1,j} - \psi_{i-1,j}\right) + C_{2\ ijk}^{T}\left(\psi_{i,j+1} - \psi_{i,j-1}\right)\right\} - \{d_{i+\frac{1}{2},j} - d_{i-\frac{1}{2},j} + d_{i,j+1/2} - d_{i,j-1/2}\} \tag{23}$$

Introducing the artificial dissipation term to the Eq. 23 is crucially important to avoid discontinuities and to keep vector variable ψ differentiable. The five stage modified Runge-Kutta time stepping scheme is used to march the adjoint equations to the steady-state limit. Further details of the procedure of obtaining Eqs. 17 to 23 can be found in Jameson (1990, 1995).

OPTIMIZATION ALGORITHM

A simple optimization algorithm can be established by using a line search method. These methods require an algorithm to choose a direction p and search along this direction from the current design variables to obtain a new set of variables for the cost function value. Once the direction is chosen, then a step length λ is multiplied to the search direction to advance the optimization to the next iteration. A simple optimization algorithm can be defined by setting the search direction l=-G, to the negative of the gradient at every iteration:

$$l = -G \tag{24}$$

With a line search method, the step size λ is chosen so that the maximum reduction of the objective function I(X) is attained. The search procedure used in this work is a descent method, in which small steps are taken in a direction defined by the smoothed gradient. X represents the design variable, and G=∇I the gradient. Instead of making the step:

$$\delta X = \lambda\, l = -\lambda\, G \tag{25}$$

We replace the gradient ∇I by a smoothed value $\overline{\nabla I}$ to guarantee the generation of a sequence of smooth shapes. When smoothing is performed in the x direction, the smoothed gradient $\overline{\nabla I}$ may be calculated from a discrete approximation to:

$$\overline{\nabla I} - \frac{\partial}{\partial x}\varepsilon\frac{\partial}{\partial x}\overline{\nabla I} = \nabla I \tag{26}$$

where ε is the smoothing parameter.

Finding the new position of the boundary mesh points is also necessary, in order to estimate the variation of internal nodes accordingly. Jameson (1986, 1990) introduced a grid perturbation method that modifies the current location of the grid points based on perturbations at the geometry surface. This allows the variation of the grid point location in the equation for gradient evaluation, to be substituted with the variation of the surface points. The Optimization procedure can finally be summarized as follows:

a. Solve the flow equations for ρ, u_1, u_2, p and e.

b. Solve the adjoint equations for ψ subject to appropriate boundary conditions.

c. Evaluate the gradients $G=\frac{\delta I}{\delta X_s}$ and calculate the corresponding smoothed gradient $\overline{G}$;

d. Update the shape based on the direction of steepest descent $\delta X_s = -\lambda\, \nabla I$;

e. Solve the adjoint equations again for determining the gradients $G=\frac{\delta Cl}{\delta X_s}$ and consequently for updating angle of attack by using the equation:

$$\frac{\partial Cl}{\partial \alpha}\Big|_X \delta\alpha + \frac{\partial Cl}{\partial X}\Big|_\alpha \delta X_s = 0$$

or:

$$\delta\alpha = -\frac{\frac{\partial Cl}{\partial X}\big|_\alpha}{\frac{\partial Cl}{\partial \alpha}\big|_X}\delta X_s \tag{27}$$

where δX_s is the perturbed coordinates of nodes on the airfoil surface, obtained in step (d); and

f. Return to step (a) until convergence is reached.

RESULTS

In this section, we first validate our flow solver code by simulating several benchmark problems for the transonic inviscid flow over airfoils. For this purpose, transonic flow over NACA 0012 and RAE 2822 airfoils are considered. The validation is performed by comparing the obtained results to the previously published results by Pulliam *et al.* (1983) and Cook *et al.* (1979). In all cases, the far boundaries are located 5 chords away from the center of the airfoil. The NACA airfoil is set to three flow conditions: I) M_∞=0.63, α=2.0°; II) M_∞=0.8, α=0.0°; III) M_∞=0.8, α=1.25° and the free stream Mach number and the angle of attack for the RAE airfoil is IV)M_∞=0.67, α=1.06°.

The Courant number number was taken as 4.0 and five orders of magnitude reduction in the density residual is considered as the convergence criteria. In all cases, the convergence criterion is met in less than 1,500 time steps, as shown in Fig. 1.

The computational grid consists of 128 × 33 nodes stretched near the leading and trailing edges. The grid is also clustered towards the airfoil surface. Computations were performed on a personal computer with a dual core CPU for which the computational time for the aforementioned cases is less than 2 minutes.

The pressure coefficients C_p on the upper and lower surface of the airfoils are shown in Figs. 2 to 5 and compared with the other computational results. In these Figures, X has been normalized and denotes the distance from the airfoil centre. The agreement between the Euler calculation and the other reference data is obvious. Figures 6 to 8 display the pressure contour plots for case III and IV. The smoothness of rapid expansion near the nose region on the upper surface is notable. The results prove that the region near stagnation point is calculated without any excessive numerical difficulty.

This problem is very important for the shape optimization procedure and can deteriorate the smoothness of surface. In all cases, the shock location is also computed accurately with a resolution of 2 to 3 grid points. Case III is a more difficult one, where a strong shock forms on the upper surface and a weak shock on the lower one. The grid is clustered near the shock position on the upper-lower surfaces of the airfoil for this case.

The performance of the optimization algorithm for the cost function is examined and the effects of weight coefficients k_1 and k_2 are explored. As our first test case, we set k_1=1.0 and k_2=0.0, therefore, reducing the problem to a pressure drag minimization one. NACA 0012 airfoil is considered to be the initial geometry and the flow variables are set to M_∞=0.75 and angle of attack α=3.0°.

The lift and drag coefficients for the initial geometry are calculated as *Cl*=0.656 and *Cd*=0.0345. For the initial geometry, a strong shock is developed on the upper surface of the airfoil. The vertical coordinates of the boundary grid points are considered as the design variables and the aim of optimization process is to reach a new airfoil geometry

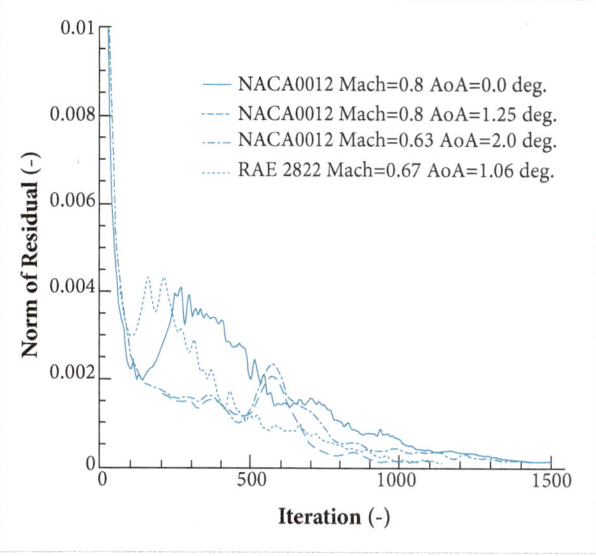

Figure 1. Convergence History of solutions.

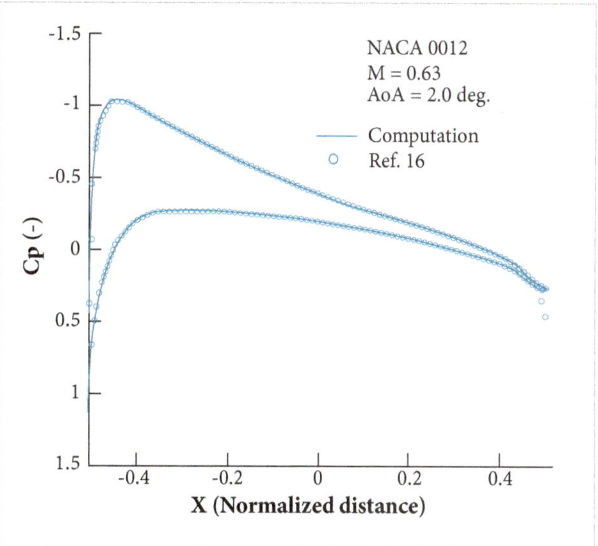

Figure 2. Comparison of pressure distribution for the case I.

for which the drag coefficient is minimum while the lift coefficient is maintained at the same value of Cl=0.656. An implicit smoothing technique is also used to smooth out the gradients before modifying the location of each point on the airfoil surface.

The convergence is quickly attained by this optimizing technique and only after four design cycles the drag coefficient is reduced to one third of its initial value under the fixed lift coefficient constraint $Cl = 0.656$. Comparisons of the pressure distribution for the new geometry in this figure and for the initial NACA 0012 indicate that the upper surface nodes are modified in a way that the local Mach number and the strength of the shock are attenuated on this surface. This, in turn, may lead to a lower wave drag but decreases the lift coefficient as well. As a result, the fall in the lift coefficient is compensated by modifying the shape of the trailing edge, as shown in Fig. 9 to recover the pressure difference near this region.

Figure 9 illustrates that the final design shaped is achieved after 50 designed cycles. The upper surface shock disappeared from the pressure distribution results and the drag coefficient is reduced to Cd=0.0448 , i.e., one eighth of its initial value.

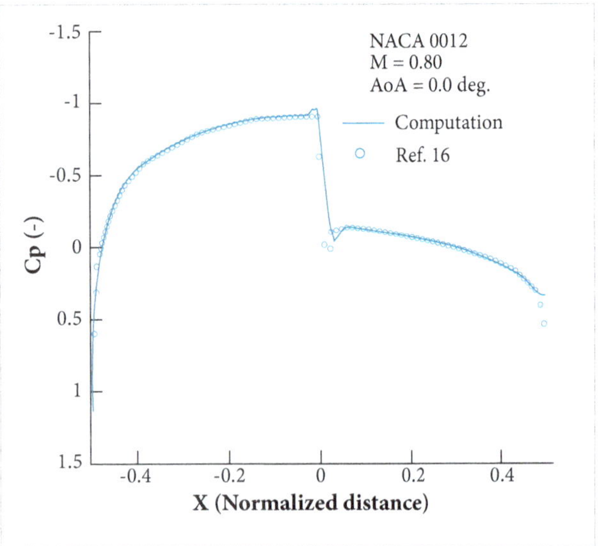

Figure 3. Comparison of pressure distribution for the case II.

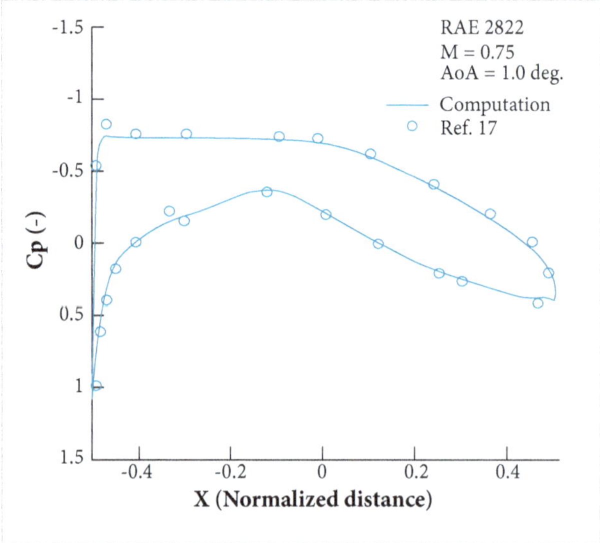

Figure 5. Comparison of pressure distribution for the case IV.

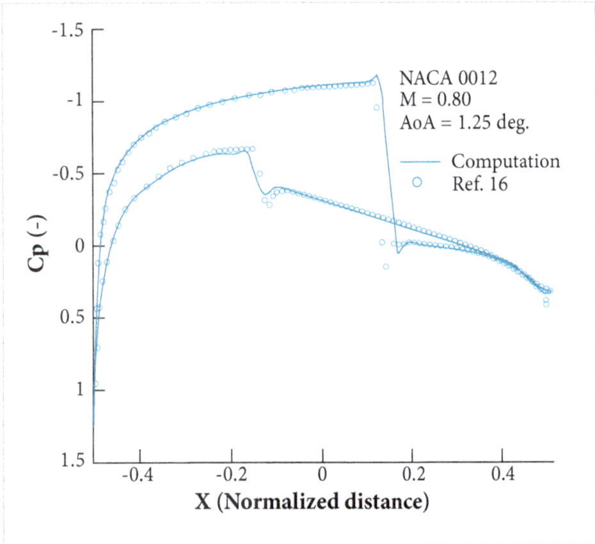

Figure 4. Comparison of pressure distribution for the case III.

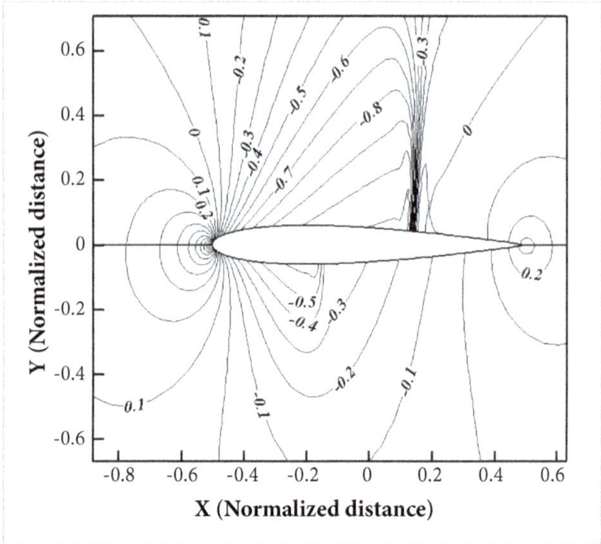

Figure 6. Pressure contour plots for the case III.

As can be seen in this Figure, the optimal shape is no longer symmetric, and a cusped shape is developed at the trailing edge.

The convergence history of the gradient norm and the drag coefficient are shown in Figs. 10 to 11. These Figures indicate that the convergence rate of the procedure is fast and that the optimization is accomplished mostly in the early cycles.

It is also required to show that the final result depends on the initial geometry. In order to do so, we repeat the computations for a five digit NACA 23012 airfoil as the initial geometry and change the angle of attack to reach the same lift coefficient as in the previous example.

To achieve this goal, the angle of attack must be changed to α=1.24°, as shown in Fig. 12. The initial drag coefficient is also calculated as Cd = 0.0433. Figure 13 illustrates that the optimization procedure converges to different solution for the final geometry and the pressure distributions and, therefore, the optimizing procedure is dependent to the initial

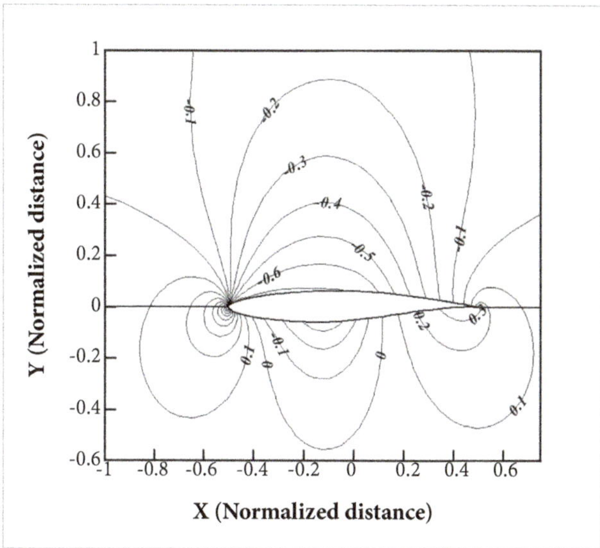

Figure 7. Pressure contour plots for the case IV.

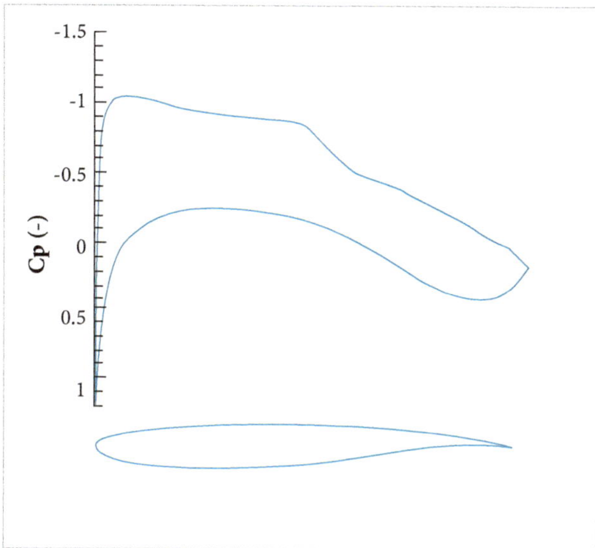

Figure 9. Inviscid drag minimization of NACA 0012, M=0.75, After 50 design iterations, Cl=0.656, Cd=0.0448, α=1.45°.

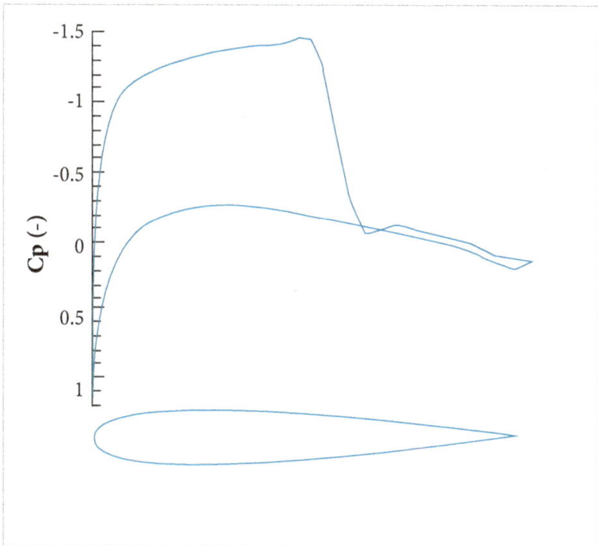

Figure 8. Inviscid drag minimization of NACA 0012, N=0.75, Cl=0.656, Initial Solution, Cd=0.0345, Cm=-0.0335, α=3.00°.

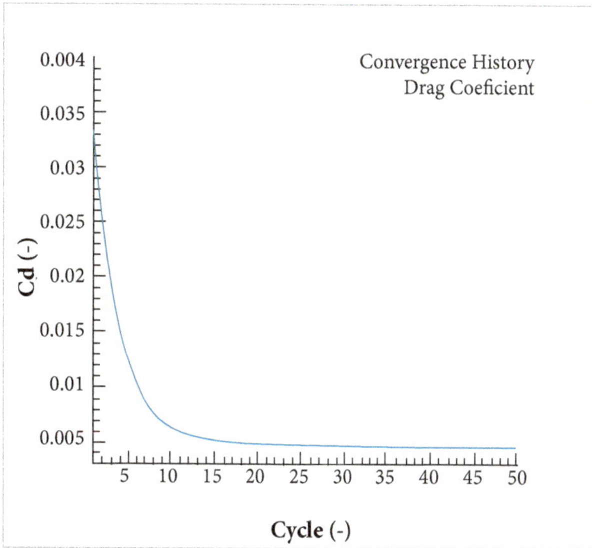

Figure 10. Convergence history of drag minimization of NACA 0012 airfoil.

conditions. We also notice that in this drag minimization problem, the pitching moment of the optimal shape is unfavorably increased from *Cm*=-0.0558 to *Cm*=0.154.

As our second example, the other extreme case of k_1=0.0 and k_2=1.0 is considered, i.e., the pitching moment coefficient is minimized under the constant lift constraint and without restricting the drag coefficient. The initial geometry and flow conditions are as in the previous example. As shown in Fig. 14, there is a 94% reduction in the pitching moment for the optimal shape, while the lift coefficient remains almost constant. The drag coefficient has also shown a 23% increase.

To minimize the drag coefficient under constant lift and pitching moment coefficient, the weight constants are set to k_1=0.50 and k_2=0.50.

Figure 15 indicates the obtained results for the same initial airfoil geometry and the flow conditions. The obtained results revealed that the drag coefficient is reduced by 76% while the lift and pitching moment coefficients are remained constant. This is the optimal shape for an airfoil that can reduce the compressible

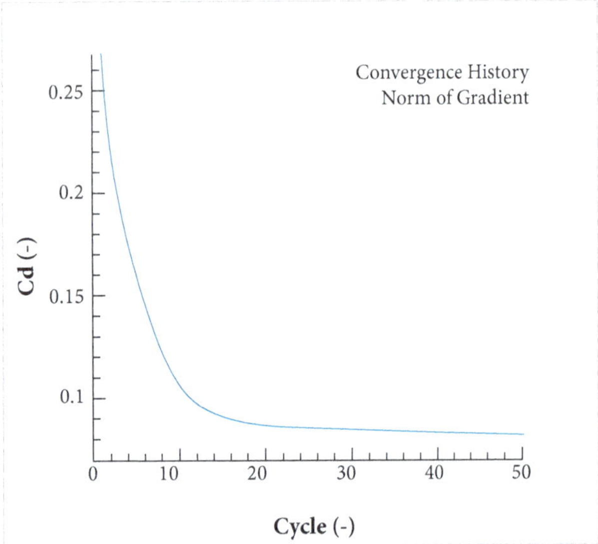

Figure 11. Decrement of gradient norm of drag minimization of NACA 0012 airfoil.

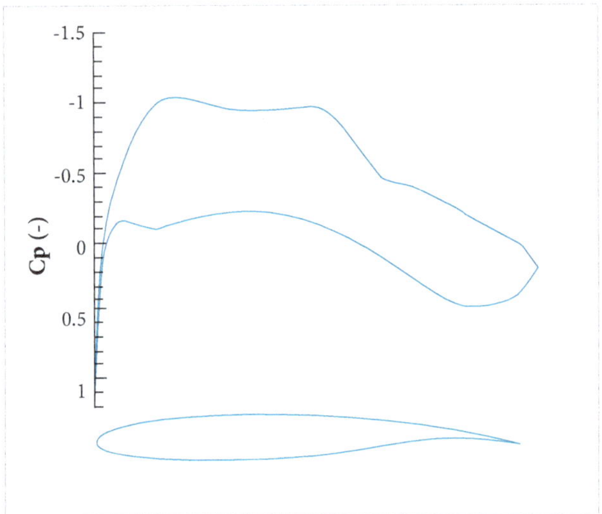

Figure 12. Inviscid drag minimization of NACA 23012 *M*=0.75, *Cl*=0.660, Initial solution, *Cd*=0.0433, *Cm*=-0.0558, α=1.24°.

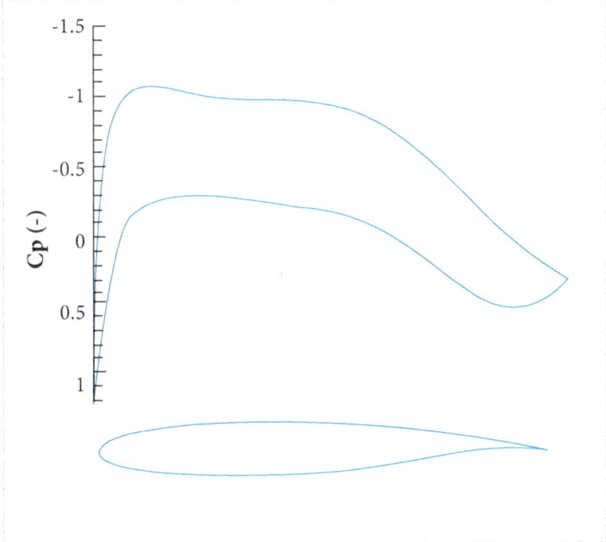

Figure 13. Inviscid drag minimization of NACA 23012, *M*=0.75, Final Solution, *Cl*=0.660 *Cd*=0.0436, α=-0.17°.

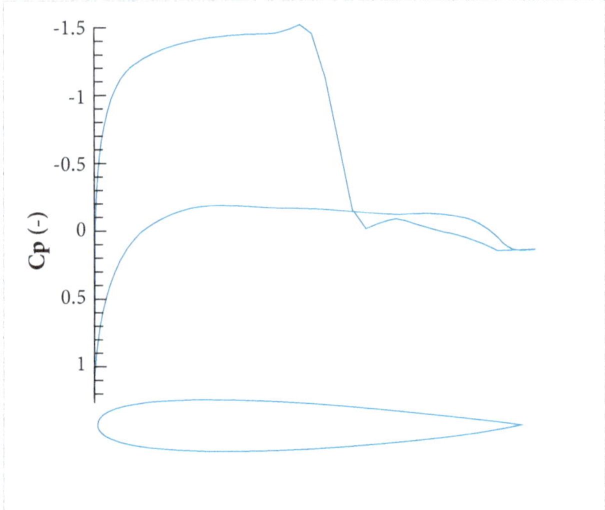

Figure 14. Pitching moment minimization of NACA 0012, *M*=0.75, After 36 design iterations, *Cl*=0.656, *Cd*=0.0424, Cm=-0.0002, α=3.71°.

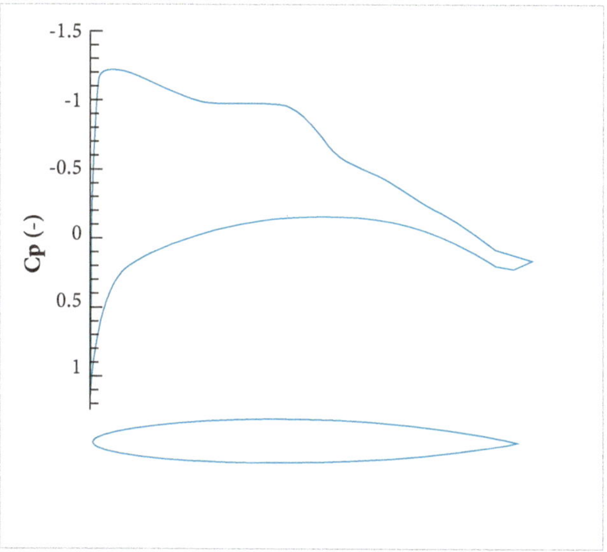

Figure 15. Optimization of NACA 0012, k_1=0.50, k_2=0.50,M=0.75, After 32 design iterations, Cl=0.656, Cd=0.0059, Cm=-0.0338, α=2.65°.

pressure drag at a given required lift without compromising the pitching moment coefficient.

The test cases above show that the proposed cost function for the ASO is very efficient for designing airfoil with specific requirements in subsonic and transonic flows.

CONCLUSION

An efficient methodology for ASO of airfoils in transonic or high subsonic regime was described. In this approach, the Euler equations are solved for a symmetric and asymmetric airfoil with a multi-block method. The results obtained from this approach for the pressure distribution on the airfoil surface were validated and it was shown that the aerodynamic coefficients are estimated with a rather good accuracy. Then the prepared solver is employed to optimize the airfoils based on the adjoint method for minimum pressure drag, while maintaining the lift and pitching moment coefficients.

Then, a new cost function was introduced for the problem of reducing the pitching moment and drag coefficients under the constant lift constraint. The procedure was first tested for the extreme case of reducing pitching moment under the constant lift constraint without restricting the drag coefficient which was led to a final shape with higher wave drag coefficient. However, by proper selection of cost function weight coefficients, it is possible to reduce the wave drag under constant pitching moment and lift coefficients. In this case, the final solution is an airfoil with a sharp trailing edge and a narrow leading edge.

Drag minimization of airfoils, in the case of constant lift coefficient, has led to a cusped shape of trailing edge and relatively flat upper surface in both symmetric and cambered airfoils; but the final solution is different for two cases, especially for the shape of leading edge of the airfoils.

These results indicate that the optimization procedure based on the adjoint method depends on the initial shape of airfoil. In other words, this technique can converge to a local extremum and not a global one.

The main advantage of this method is that satisfactory results can be achieved by minimum computational efforts, particularly for the grid generation problem and the required central processing unit of computer (CPU) time. This feature becomes important where a low fidelity accurate case such as rotor blade flow solver is required for primarily design or optimization purposes. However, the proposed method ignores important phenomena such as the viscous flow or shock boundary layer interactions.

REFERENCES

Anderson, W.K., Newman, J.C., Whitfield, D.L. and Nielsen, E.J., 2001, "Sensitivity analysis for Navier-Stokes equations on unstructured meshes using complex variables", AIAA Journal, Vol. 39, No. 1, pp. 56-63. doi:10.2514/2.1270.

Barger, R.L., 1975, "Procedures for the design of low-pitching-moment airfoils", NASA TN D-7982.

Cook, P.H., Mc Donald, M.A. and Firmin, M.C.P., 1979, "Aerofoil RAE 2822 pressure distributions, and boundary layer and wake measurement", AGARD Advisory Report No. 138.

Jameson, A., 1986, "Multigrid algorithms for compressible flow calculations", Proceedings of the 2nd European Conference on Multigrid Methods II, Lecture Notes in Mathematics, No.1228, pp. 166–201.

Jameson, A., 1988, "Aerodynamic design via control theory", Journal of Scientific Computing, Vol. 3, No. 3, pp. 233-260.

Jameson, A., 1990, "Automatic design of transonic airfoils to reduce the shock induced pressure drag", In Proceedings of the 31st Israel Annual Conference on Aviation and Aeronautics, pp. 5-17.

Jameson, A., 1995, "Optimum aerodynamic design using CFD and control theory", AIAA Paper 95-1729-CP.

Jameson, A. and Baker, T.J., 1983, "Solution of Euler equations for complex configurations", AIAA, pp. 83-1929.

Jameson, A., Schmidt, W. and Turkel, E., 1981, "Numerical solution of the Euler equations by finite volume methods using Runge-Kutta time-stepping schemes", AIAA, pp. 81-1259.

Hazra, S.B., Schulz, V., Brezillon, J. and , Gauger, N., 2005, "Aerodynamic shape optimization using simultaneous pseudo-timestepping", Journal of Computational Physics, Vol. 204, No. 1, pp. 46-64. doi:10.1016/j.jcp.2004.10.007.

Leifsson, L., Koziel, S. and Ogurtsov, S., 2011, "Inverse design of transonic airfoils using variable-resolution modeling and pressure distribution alignment", Procedia Computer Science, Vol. 4, pp. 1234-1243. doi:10.1016/j.procs.

2011.04.133.Leifsson, L. and Koziel, S., 2010, "Multi-fidelity design optimization of transonic airfoils using shape-preserving response prediction", Procedia Computer Science, Vol.1, No. 1, pp. 1311-1320. doi:10.1016/j.procs.2010.04.146.

Nadarajah, S. and Tatossian, C., 2008, "Multi-objective aerodynamic shape optimization for unsteady viscous flows", Journal of Optimization and Engineering, Vol. 11, No. 1, pp. 67-106. doi: 10.1007/s11081-008-9036-4.

Nadarajah, S. and Jameson, A., 2001, "Studies of the continuous and discrete adjoint approaches to viscous automatic aerodynamic shape optimization", AIAA 2001-2530, AIAA 15th. Computational Fluid Dynamic Conference, Anaheim, Ca, June 11-14.

Pulliam, T.H., Jespersen, D.C. and Childs, R.E., 1983, "An enhanced version of an implicit code for the Euler equations", AIAA-83-0344. January 10-13, Reno, Nevada.

Reuther, J.J., Jameson, A., Alonso, J.J., Rimlinger, M.J. and Saunders, D., 2001, "Constrained multipoint aerodynamic shape optimization using an adjoint formulation and parallel computers", parts 1 and 2, Journal of Aircraft, Vol. 36, No. 1, pp. 51-74.

Reuther, J. and Jameson, A., 1994, "Control based airfoil design using the Euler equations", AIAA paper 94-4272-CP.

Xie, L., 2002, "Gradient-based optimum aerodynamic design using adjoint methods".

Ying, G.Y., Feng, H. and Yu, S.M., 2011, "Aerodynamic airfoil design using the Euler equations based on the dynamic evolution method and the control theory", Science China Physics, Mechanics and Astronomy, Vol. 54, No. 4, pp. 697-702. doi:10.1007/s11433-011-4287-z.

Yu, L., Quan, C.X., Neng, L.Z. and Wu, X.J., 2010, "Multi-objective robust airfoil optimization based on non-uniform rational B-spline (NURBS) representation", Science China Physics, Technological Sciences, Vol. 53, No. 10, pp. 2708-2717. doi: 10.1007/s11431-010-4075-4.

A New Coupled Free Wake-CFD Method for Calculation of Helicopter Rotor Flow-Field in Hover

Hamid Farrokhfal[1], Ahmad Reza Pishevar[2]

ABSTRACT: This paper concerns a new coupled free wake-CFD method for proper calculation of aerodynamic loads on a two bladed helicopter rotor in hovering flight. Loading is computed by solving the three dimensional Euler equations in a rotating coordinate system. However, since direct simulation of the tip vortices and wake requires a very fine grid, the rotor's wake effects are modeled by a free wake approach and included into the CFD calculation by a transpiration boundary condition at the rotor surface. An influence coefficient solution method is used to find the rotor's wake shape, being steady in a rotating frame. Euler equations are also considered in the form of absolute flow variables and solved by a multi grid Jameson's finite-volume method. The accuracy of the proposed method is illustrated by comparing numerical results to the available experimental results for the pressure distribution on a blade of a model helicopter rotor at different tip Mach numbers.

KEYWORDS: Free wake, Euler equations.

INTRODUCTION

The flow field around rotors is inherently complex. In the vicinity of the blade tip, it is highly three dimensional and vortical and it is characterized by non-uniform inflow with blade-vortex interactions and transonic conditions. In hover, the complex' three dimensional rotor wake has a dominant influence on the blade loading (Jenny *et al.*, 1968; Clark and Langrebe, 1971; Landgrebe, 1972).

A proper evaluation of the rotor aerodynamic loading has an important role in the analysis and design of helicopter rotor blades. There are two general approaches for simulating the flow field around a rotor and determining its loads. In the first approach, the inviscid or viscous set of governing equations is solved by a proper Computational Fluid Dynamic (CFD) method (Roberts and Murman, 1985; Chen and McCroskey, 1988; Kramer *et al.*, 1988). Obviously, in this approach, the vortex system near the blade tips and the wake region below the rotor disk must be captured accurately for realistic pressure and load estimations. The computational cost is extremely increased for this approach (Roberts and Murman, 1985; Chen and McCroskey, 1988; Kramer *et al.*, 1988; Ramachandran *et al.*, 1989; Lee and Kwon, 2006; Chen *et al.*, 1987). In the second approach, the tip vortices and the helical wake system are not solved by the CFD calculation and instead, the influence of strong tip vortices and the geometry of the wake are included by a vortex-wake model.

Solution scheme which use the later idea are often grouped under methods using wake models encompassing the potential flow (Strawn and Caradonna, 1987; Strawn and

1. Malek-Ashtar University of Technology – Isfahan – Iran 2. Isfahan University of Technology – Isfahan – Iran.

Author for correspondence: H. Farrokhfal | Department of Mechanical & Aerospace Engineering | Malek-Ashtar University of Technology – 83145-115 | Shahinshahr – Isfahan – Iran | Email: farrokhfal@mut-es.ac.ir

Tung, 1986; Chang and Tung, 1985; Egolf and Sparks, 1987), the Euler (Sankar *et al.,* 1986, Agarwal and Deese, 1987; Chang and Tung, 1987) and the Navier-Stockes methods (Agarwal and Deese, 1988; Narramore and Vermeland, 1989; Wake and Sankar, 1989; Srinivasan and MacCroskey, 1988). Several approaches have been applied to model the vortical field of rotors. The most famous of these models is the prescribed wake model, which uses experimental data to drive empirical formulas relating the rotor blade geometric parameters and thrust coefficient (C_T) to the vortex' wake geometry (Kocurek and Tangler, 1976; Landgrebe, 1972; Landgrebe and Cheney, 1972; Landgrebe *et al.,* 1977). These schemes do not correctly model the flow characteristics and their reliance on experimental data causes to not appropriately predict blade loading for a new rotor configuration. Free wake models have been developed in an attempt to overcome these limitations by simulating in details the actual shape and motion of the wake (Crimi, 1965; Landgrebe, 1969; Clark and Leiper, 1970; Sadler, 1971; Johnson, 1980; Johnson, 1981; Summa and Clark, 1979; Summa and Maskew, 1981; Saberi, 1983; Rosen and Graber, 1988). This can be done by finding the geometry of the wake and the strength of tip and sheet vortices, which are modeled by a finite number of vortex line segments. Then, the effect of the vortex-wake is brought into CFD calculation by modifying either the boundary conditions at the blade surface (Sankar *et al.,* 1986, Agarwal and Deese, 1987; Chang and Tung, 1987) or explicitly modifying the blade angle of attack (Wake and Sankar, 1989; Srinivasan and McCroskey, 1988).

There are some advantages for calculating the rotor aerodynamics loads by a coupled wake-CFD method when compared to the direct method. The first benefit is that accurate enough results can be obtained with significant reduction in the computational expense. This is a key factor when a low fidelity approach is required for design or multidisciplinary optimization purposes. The second benefit is to relieve the difficulties arisen in meshing the 3-D domains around the rotor. Since the tip region is not computed directly in the CFD calculation, the 3-D grid can be obtained from an assembly of two-dimensional spanwise sectional grids (Caradonna *et al.*, 1984).

The shape of the rotor wake has an intense effect on the aerodynamic performance of a helicopter rotor in hover flight. The present research explains a new interactive wake-CFD solver for calculating the inviscid transonic flow on a two-bladed rotor in hover flight. In this method, the CFD flow solver is coupled with a free wake model by using the surface transpiration approach, i.e., effects of rotor vortex-wake are included into CFD calculation by using the estimated downwash velocity to modify boundary condition along the blades.

In most of the previously reported works, the induced velocity is computed prior to the CFD analysis by a separate program such as Hover, B-TRIM of McDonnell Douglas Helicopter Company (Agarwal and Deese, 1987), or Host (Benoit *et al.*, 2000). In this work, however, the induced downwash of rotor blades are computed along with the CFD solution and one can interact with it through an iterative procedure. The model is based on a lifting line theory, which provides estimation for induced downwash along the blade. In this method, the blade is substituted by a row of square vortex panels in radial direction and, at some location of blade span, a finite number of spiral vortices are shed downwards the disc rotor. Then, the positions of helices are corrected by induced velocities, which are computed by employing the Biot-Savart law and the use of the Basic Curved Vortex Element (BCVE) or Self-Induction Vortex Element (SIVE) elements (Quackenbush *et al.*, 1989). The procedure continues until the shape of the wake converges to a steady state and therefore, the self-induced velocities become from zero up to a tolerance.

Then, the transonic flow field about the rotor is calculated by the solution of the three-dimensional Euler equations on a multi-block structured grid about the blades. However, as mentioned earlier, the three-dimensional grid is constructed from an assembly of 2D grids about the blade sections which is extended only to the blade tip and consists of two blocks that are symmetric in relation to the planform. In this study the finite-volume scheme of Jameson, Schmidt, and Turkel (Jameson *et al.*, 1981) is used to discrete governing equations. The method is reformulated for a rotating coordinate system which is attached to the rotor blades, but equations are finally rewritten in terms of absolute-flow variables. This approach has been used previously by Holmes and Tong (1985) in order to develop an Euler solver for some applications in turbomachinery.

The iterative procedure starts by determining the spanwise circulation and the shape of the wake from the blade geometry. Then the flowfield around the blade is computed by the CFD solver, using the calculated downwash velocity along the blade as a boundary condition. Now, a new bound circulation can be obtained from the lift of each section and therefore, it is possible to improve the wake model accordingly. The procedure is repeated until the blade bound circulation or downwash velocity remains unchanged in two subsequent iterations.

The paper is organized as follows: the methodology used for modeling the rotor wake and tip vortices is explained. Governing equations for CFD calculations are presented and the numerical procedure for discretizing these equations is described. The necessary boundary conditions and the way through which the rotor wake model interacts with the CFD solver are explained. Finally, some illustrative numerical results are presented. Calculations are performed for the flow field of a model helicopter rotor in hover at various collective pitch angles. Comparisons with the experimental data of Caradonna and Tung (1981) are also presented.

ROTOR WAKE MODELING

The rotor wake model is based on the proposition that the free wake hover problem possesses a self-preserving state solution when viewed in a rotating frame that is fixed to blades (Bliss *et al.*, 1985). The fundamental idea in the model is that at each point on the rotor wake, when viewed in coordinates rotating with the blade, the velocity contributions due to all effects must be such that there is no convection to change the wake shape. Any velocity component in a plane, normal to the vortex filament curve, will convect the vortex to a new position, thereby changing the filament shape (Quackenbush *et al.*, 1989).

Applying the idea, two features of the wake are properly determined: the contraction of the helical wake as it extends in the axial direction below the rotor, and the interaction of the tip vortex of two blades. For modeling the wake, the two curved vortex elements introduced in Bliss *et al.* (1985) and Baskin *et al.* (1967) are used. These elements include the BCVE and SIVE. The BCVE is used in order to compute the velocity induced at any point in the flow field, not on the element itself, and the SIVE is used to evaluate the velocity induced by a vortex filament on itself, including the effect of distributed vorticity in the bent vortex core. Figure 1 shows two types of vortex elements on a typical helical vortex filament. The BCVE is a parabolic arc element positioned between control points whereas the SIVE is a circular arc element passing through each three adjacent control points.

The solution procedure involves a way of reducing the cross flow velocity to zero at every collocation point. Figure 2 shows the idea. In order to achieve this goal, for a given arrangement of the free wake collocation points, the net velocity in each cross flow plane is computed first. Next, the effect of small displacements of each collocation point on the velocity at every other collocation point, and the displaced point itself, are determined. These effects are known as influence coefficient.

In practice, the solution must be obtained numerically with the wake modeled by using discrete elements. The blade is represented by a set of vortex quadrilaterals in a way that

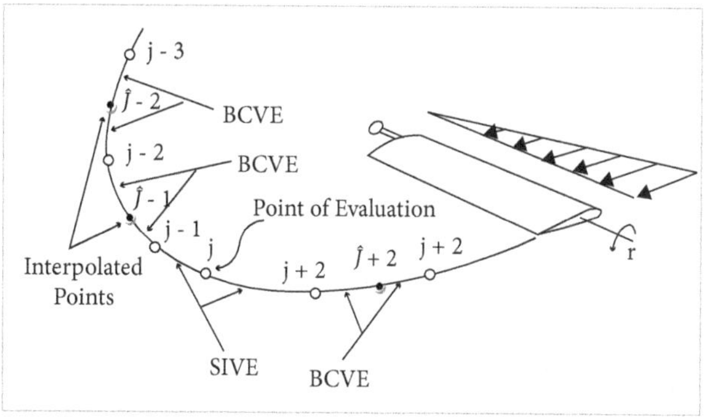

Figure 1. BCVE and SIVE elements for discretizing wake vortex filaments.

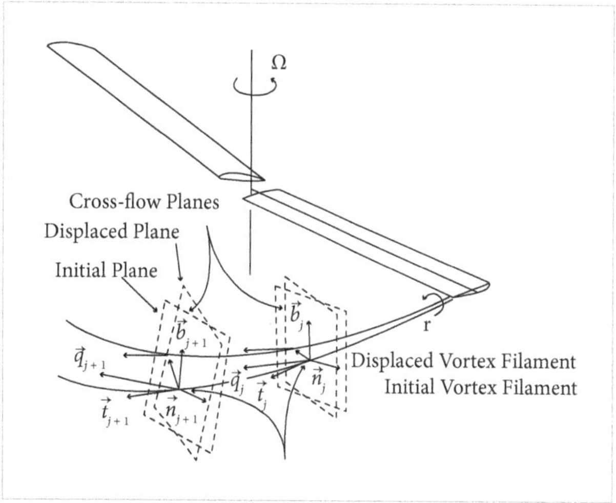

Figure 2. Method of displacing the vortex filament to eliminate cross flow.

their forward edges are located at the quarter chord line of the blade and the distance of their aft edges from the blade trailing edge is also one-fourth of the local chord. The vortex filaments of sheet vortex and the tip vortex are shed from the points on the aft legs of these panels, and the panels including the shed points are in the form of truncated vortices, as shown in Fig. 3.

Figure 4 shows a schematic of a typical load distribution on a rotor blade. It is assumed that the corresponding wake consists of four filaments, whose strengths are defined as shown in the figure. The locations of the release point are also computed from the Batchelor principle (Batchelor, 1967) as:

$$z_v^{\,2} = \frac{1}{\Gamma_2 - \Gamma_1}\int_{\Gamma_1}^{\Gamma_2} z^2\, d\Gamma \tag{1}$$

where z_v is the distance of the release point of the vortex filament from the center of rotation.

The velocity contribution of each individual vortex element is determined analytically using the Biot-Savart law. The total velocity vector at a collocation point includes the local self-induction effect (SIVE), the contributions from the rest of the wake (BCVE's), the velocity induced by the bound circulation on the rotor blades, and the swirl velocity component associated with the rotating coordinate system.

The influence matrix relates cross flow velocity perturbations to small displacements for a given wake configuration. This matrix can be used to predict the set of collocation point displacements, which will eliminate the cross flow velocity. If the free wake location is described in terms of the positions of N collocation points, and the normal and binormal cross flow velocity components at the ith point are denoted by q_1^n and q_2^n, respectively; then the N-dimensional vectors based on these velocity components are defined as:

$$\vec{V}_n = \left[q_1^n, q_2^n, \ldots, q_N^n\right] \tag{2}$$

The free wake solution is achieved when a set of collocation point locations is found, such as $\vec{V}_n = \vec{V}_b = 0$.

The amplitudes of small displacements on the ith collocation point in the local normal and binormal directions are denoted by Δn_1 and Δb_1. From these, changes of displacement vectors are defined as:

$$\begin{aligned} \vec{\Delta n} &= \left[\Delta n_1, \Delta n_2, \ldots, \Delta n_N\right] \\ \vec{\Delta b} &= \left[\Delta b_1, \Delta b_2, \ldots, \Delta b_N\right] \end{aligned} \tag{3}$$

and the normal and binormal velocity component changes, associated with the above displacement vectors, are:

$$\begin{aligned} \Delta\vec{V}_n &= \left[\Delta q_1^n, \Delta q_2^n, \ldots, \Delta q_N^n\right] \\ \Delta\vec{V}_b &= \left[\Delta q_1^b, \Delta q_2^b, \ldots, \Delta q_N^b\right] \end{aligned} \tag{4}$$

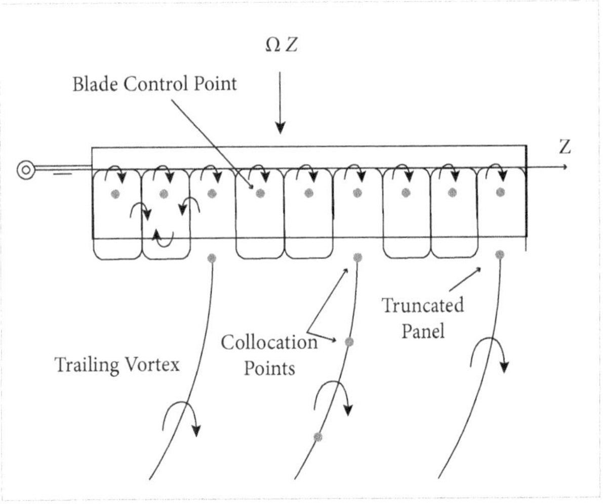

Figure 3. Schematic of the quadrilateral vortex panels of the blade.

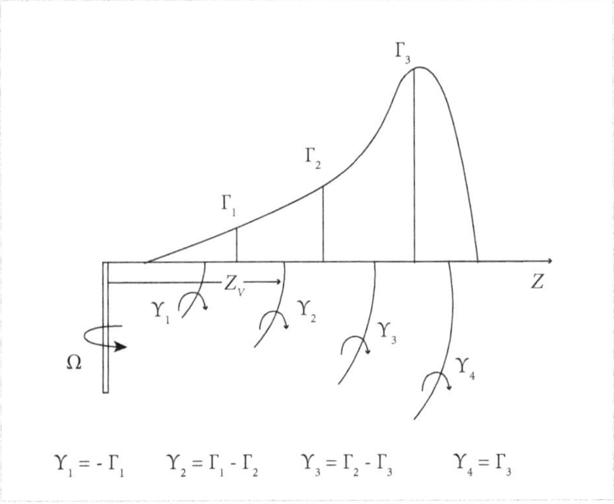

Figure 4. Defining the strength of trailing vortex filaments from the bound circulation.

The velocity change vectors can be related to the change of displacement vectors through the influence coefficient sub matrices, namely:

$$\begin{aligned} \tilde{Q}_{nn}\ \Delta\vec{n} + \tilde{Q}_{nb}\Delta\vec{b} = \Delta\vec{V}_n \\ \tilde{Q}_{bn}\ \Delta\vec{n} + \tilde{Q}_{bb}\Delta\vec{b} = \Delta\vec{V}_b \end{aligned} \quad (5)$$

For instance, the sub matrix $\tilde{Q}_{nb}$ gives the changes in normal velocity due to binormal displacements. These matrices must be evaluated numerically. To obtain zero net cross flow velocity, the collocation point displacements must be chosen so that the velocity changes cancel the net cross flow velocity, namely, $\Delta\vec{V}_n = -\vec{V}_n$ and $\Delta\vec{V}_b = -\vec{V}_b$, which means solving the following matrix equation for $\vec{\Delta n}$ and $\vec{\Delta b}$:

$$\begin{bmatrix} \tilde{Q}_{nn} & \tilde{Q}_{nb} \\ \tilde{Q}_{bn} & \tilde{Q}_{bb} \end{bmatrix} \begin{bmatrix} \vec{\Delta n} \\ \vec{\Delta b} \end{bmatrix} = \begin{bmatrix} \Delta\vec{V}_n \\ \Delta\vec{V}_b \end{bmatrix} = \begin{bmatrix} -\vec{V}_n \\ -\vec{V}_b \end{bmatrix} \quad (6)$$

Since the wake problem is actually nonlinear, this process must be repeated several times.

A free wake analysis involves only a finite number of turns of free vortex. Beyond the free wake region is an adaptive mid wake which completes the contraction process. The final part is also a semi-infinite helical wake having constant radius and constant pitch based on the momentum theory. The location and geometry of the adaptive mid wake depend on the position of the last points in the near section of the wake. In other words, the last points of near wake section are the beginning points of the mid-section, and after the mid-section the wake model is continued with the far section as shown in Fig. 5.

From the conservation of mass flow within the contracting mid wake and by applying the momentum theory, it is possible to drive the boundary shape of the mid wake (Bliss *et al.*, 1985) as:

$$r_b(y) = r_f + \left[r_b(y_0) - r_f \right] \exp\left[-\frac{(y-y_0)}{r_f} \right] \quad (7)$$

where y_0 and $r_b(y_0)$ are the axial and radial locations of the last free wake points, and r_f is the final contraction radius. In hover flight, the radial contraction exponent is given by:

$$\frac{y-y_0}{r_f} = \sqrt{\frac{C_T}{2}}\,(\psi-\psi_0) \quad (8)$$

where ψ_0 is the azimuth location of the last free wake point.

When calculating two-bladed rotors, the solution is symmetric from one blade to the next blade. This symmetry allows the induced effect of the second blade on the primary wake to be expressed in terms of the corresponding effect of the primary wake on the second blade.

Once all the different effects have been computed and summed, the resultant velocity vectors at each collocation point are expressed in the blade fixed reference frame. These resultant velocity vectors are solved into the local normal and binormal directions in order to obtain the cross flow velocity components and their changes, associated with point displacements. The velocity changes are normalized by the unit displacement magnitudes to obtain the influence coefficients. These coefficients, as described earlier, are used to predict how the collocation points should be displaced in order to reduce the velocity components in the cross flow planes. The obtained result at this stage is not the final solution due to the nonlinear nature of the problem. Therefore, starting from an initial wake shape, the solution procedure is repeated with a proper relaxation factor in quasi linear steps to improve the convergence rate (Bliss *et al.*, 1984).

The Initial wake shape can be estimated from a contracted wake model with a contraction ratio of $\frac{R_\infty}{R} = 0.85$, where R_∞ is the radius of the contracted far section of the wake.

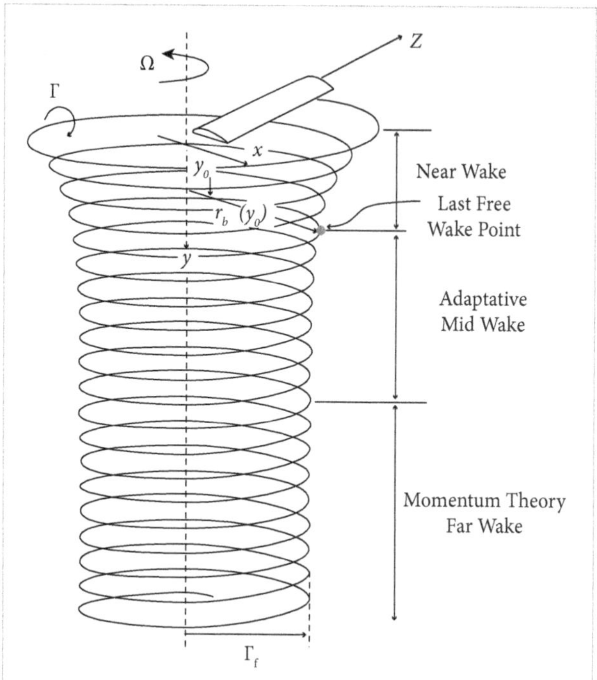

Figure 5. Three part wake model for the hover analysis.

The circulation distribution is also obtained from an approach explained by Baskin *et al.* (1967); in other words, equating the 2-D lift coefficient of blade section to the lift coefficient obtained from the Kutta-Joukowski theorem. As a result, a system of linear algebraic equation is obtained by defining the sought circulation. By knowing the circulation, the convergent shape of the wake is computed as explained in the previous section. Finally, at the end, the distribution of the downwash on blade surface is computed. This downwash is then fed to the CFD solver to include the compressibility effect and to correct the circulation. The circulation at this stage is computed from the integration of the pressure field around the blade sections and by using the Kutta-Joukowski theorem. With this new distribution of the circulation, the free wake solver provides a new downwash distribution and this process is iterated until a convergent solution for the downwash is attained.

GOVERNING EQUATIONS

Let (u,v,w) denote the absolute velocity components in the rotating Cartesian coordinate system (x,y,z), as shown in Fig. 6. Compressible Euler equations can be formulated as (Agarwal and Deese, 1987):

$$\frac{\partial W}{\partial t}+\frac{\partial F_x}{\partial x}+\frac{\partial F_y}{\partial y}+\frac{\partial F_z}{\partial z}+u_\Omega\frac{\partial W}{\partial x}+w_\Omega\frac{\partial W}{\partial z}=S \tag{9}$$

where:

$$W=\begin{bmatrix}\rho\\ \rho u\\ \rho v\\ \rho w\\ \rho e\end{bmatrix},\quad F_x=\begin{bmatrix}\rho u\\ \rho u^2+p\\ \rho uv\\ \rho uw\\ \rho uh\end{bmatrix},\quad F_y=\begin{bmatrix}\rho v\\ \rho vu\\ \rho v^2+p\\ \rho vw\\ \rho vh\end{bmatrix},\quad F_z=\begin{bmatrix}\rho w\\ \rho wu\\ \rho wv\\ \rho w^2+p\\ \rho wh\end{bmatrix},\quad S=\begin{bmatrix}0\\ \rho\Omega w\\ 0\\ -\rho\Omega u\\ 0\end{bmatrix} \tag{10}$$

and (u_Ω, v_Ω, w_Ω) denotes the rotational velocity components:

$$u_\Omega=\Omega z\ ,\ v_\Omega=0\ ,\ w_\Omega=-\Omega x \tag{11}$$

Defining the flux vector and rotational speed vector as:

$$F=F_x\,\hat{i}+F_y\,\hat{j}+F_z\,\hat{k}$$

and:

$$V_\Omega=u_\Omega\,\hat{i}+v_\Omega\,\hat{j}+w_\Omega\,\hat{k}$$

where ($\hat{i}$, $\hat{j}$, $\hat{k}$) are the unit vectors in the (x, y, z) coordinate system, Eq. (9) can be written as:

$$\frac{\partial W}{\partial t}+\nabla.F+V_\Omega.\nabla W=S \tag{12}$$

Equation (12) can be solved by the finite-volume method described in the next section.

NUMERICAL METHOD

In order to apply the finite-volume method, Eq. (12) is written in its integral form as:

$$\frac{\partial}{\partial t}\iiint_V W\,dV+\iint_{\partial S}\hat{n}\,F ds+\iint_{\partial S}u_\Omega.\,\hat{n}\,W\,ds=\iiint_V S\,dV \tag{13}$$

The computational domain is meshed by a multiblock structured hexahedral grid. Applying Eq. (13) to a typical (i,j,k) cell, for a cell center algorithm, we obtain a system of ordinary differential equations as:

$$\frac{d}{dt}\left(\forall_{ijk}W_{ijk}\right)+P_{ijk}+Q_{ijk}+R_{ijk}=\forall_{ijk}S_{ijk} \tag{14}$$

where $\forall_{ijk}$ is the cell volume, P_{ijk} represents the net convectional flux out of the cell, and Q_{ijk} and R_{ijk} are rotational fluxes out of each cell. We rewrite Eq. (14) as:

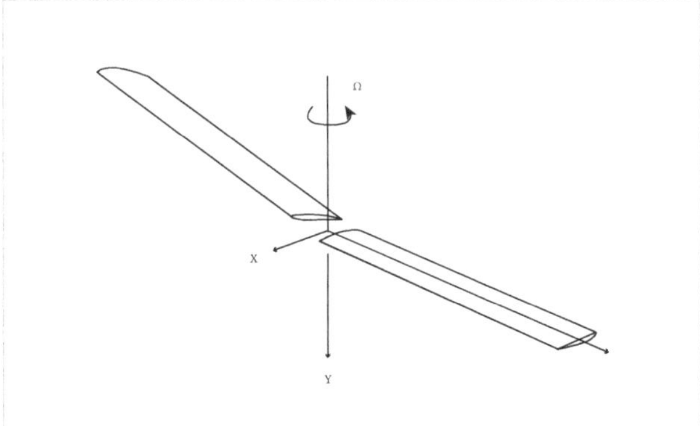

Figure 6. Rotating coordinate system.

$$\frac{d}{dt}\left(\forall_{ijk} W_{ijk}\right) + F_{ijk} = 0 \tag{15}$$

The convective fluxes can be determined by a second order central scheme. However, the resultant scheme is not dissipative, and therefore, undamped oscillations at odd and even mesh points can be developed during the computation. In order to suppress the tendency for odd-even decoupling, and to prevent the appearance of oscillations in regions containing severe pressure gradients near shock waves and stagnation points, the finite-volume scheme is modified by the addition of artificial dissipative terms as below:

$$\frac{d}{dt}\left(\forall_{ijk} W_{ijk}\right) + F_{ijk} - D_{ijk} = 0 \tag{16}$$

where D_{ijk} denotes the dissipative fluxes. Jameson *et al.* (1981) have established that an effective form of dissipative terms for flows with discontinuities is a blend of second and fourth differences, with coefficients which depend on the local pressure gradient. Therefore, dissipative terms are written as follows:

$$D_{ijk} = \left(D_x + D_y + D_z\right) W_{ijk} \tag{17}$$

Where:

$$D_x W_{i,j,k} = d_{i+\frac{1}{2},j,k} - d_{i-\frac{1}{2},j,k} \tag{18}$$

and $d_{i+\frac{1}{2},j,k}$ is defined as:

$$d_{i+\frac{1}{2},j,k} = \frac{\forall_{i+\frac{1}{2},j,k}}{\left(\Delta t_\ell\right)_{i+\frac{1}{2},j,k}} \times \left[ὸ^{(2)}_{i+\frac{1}{2},j,k} \Delta_x W_{i,j,k} - ὸ^{(4)}_{i+\frac{1}{2},j,k} \Delta_x^3 W_{i-1,j,k} \right] \tag{19}$$

Δ_x denotes the forward difference operator $\Delta_x = (W_{i+1,j,k} - W_{i,j,k})$ and $ὸ^{(2)}$ and $ὸ^{(4)}$ are adaptive coefficients defined as:

$$\upsilon_{ijk} = \left| \frac{p_{i+1,j,k} - 2p_{ijk} + p_{i+1,j,k}}{p_{i+1,j,k} + 2p_{ijk} + p_{i+1,j,k}} \right| \tag{20}$$

$$ὸ^{(2)}_{i+\frac{1}{2},j,k} = \kappa^{(2)} \max\left(\upsilon_{i-2,j,k}, \upsilon_{i-1,j,k}, \upsilon_{i,j,k}, \upsilon_{i+12,j,k}\right) \tag{21}$$

$$ὸ^{(4)}_{i+\frac{1}{2},j,k} = \max\left(0, \left| \kappa^{(4)} - ὸ^{(2)}_{i+\frac{1}{2},j,k} \right.\right) \tag{22}$$

Typical values of the constants $\kappa^{(2)}$ and $\kappa^{(4)}$ are $\kappa^{(2)} = 0.5$, $\kappa^{(4)} = 1/64$. Also, $\left(\Delta t_\ell\right)_{i,j,k}$ is defined as:

$$\left(\Delta t_\ell\right)_{i,j,k} = \frac{CFL}{\left(\lambda_\xi + \lambda_\eta + \lambda_\zeta\right)_{ijk}} \tag{23}$$

where CFL is the courant number and (λ_ξ, λ_η, λ_ζ) are the spectral radii for each cell:

$$\lambda_\xi = |U| + a\sqrt{\xi_x^2 + \xi_y^2 + \xi_z^2}$$
$$\lambda_\eta = |V| + a\sqrt{\eta_x^2 + \eta_y^2 + \eta_z^2} \tag{24}$$
$$\lambda_\zeta = |W| + a\sqrt{\zeta_x^2 + \zeta_y^2 + \zeta_z^2}$$

a is the speed of sound and (U, V, W) are the contavariant velocities:

$$U = \xi_x u_r + \xi_y v_r + \xi_z w_r$$
$$V = \eta_x u_r + \eta_y v_r + \eta_z w_r \tag{25}$$
$$W = \zeta_x u_r + \zeta_y v_r + \zeta_z w_r$$

(U_r, V_r, W_r) are the components of relative velocity obtained from:

$$\vec{V}_{r\Omega} = \vec{V} + \vec{V} \tag{26}$$

(ξ_x, ξ_y, ξ_z),(η_x, η_y, η_z) and (ζ_x, ζ_y, ζ_z) are the metrics of transformation for computational coordinates (ξ, η, ζ). Other dissipation terms, $D_y W_{y,j,k}$ and $D_z W_{y,j,k}$, in Eq. (17), are calculated in an analogous manner.

Equation (16) can be integrated in time by a multistage Runge-Kutta scheme. The conserved variables are updated to the new time level n+1 as:

$$W^{(n+1,0)} = W^{(n)}$$
$$W^{(n+1,1)} = W^{(n)} - \alpha_1 \frac{\Delta t}{\forall}\left[F\left(W^{(0)}\right) - D(W^{(0)})\right]$$
$$W^{(n+1,k)} = W^{(n)} - \alpha_k \frac{\Delta t}{\forall}\left[F\left(W^{(k-1)}\right) - D\left(W^{(k-1)}\right)\right] \tag{27}$$
$$W^{(n+1)} = W^{(n+1,m)}$$

where $\alpha_m = 1$, and $D(w^{(k)})=\beta^{(k)} D(W^{(n+1,k)})+(1-\beta^{(k)})D(W^{(k-1)})$. For a fifth-stage scheme the coefficients are given as:

$$\alpha_1 = \frac{1}{4}, \alpha_2 = \frac{1}{6}, \alpha_3 = \frac{3}{8}, \alpha_4 = \frac{1}{2}, \alpha_5 = 1$$
$$\beta_1 = 1, \beta_2 = 0, \beta_3 = 0.56, \beta_4 = 0, \beta_5 = 0.44 \quad (28)$$

In this work a local time stepping and residual smoothing procedure is also used in order to improve the convergence rate.

BOUNDARY CONDITIONS

Special care must be given to the boundary conditions when solving Euler equations for rotor blades of a helicopter by a coupled free wake-CFD method. The computational domain, as shown in Fig. 7, is a cylinder with radius of 1.5 times the blade span. A far-field boundary condition is applied at the outer surface of the cylinder and also at the blade tip plane. In addition to that, at the root of the blade, the symmetry boundary condition $\frac{\partial W}{\partial z} = 0$ is applied where z axis is along the blade span. At the solid surface of the blade, the transpiration condition is applied to take into account the effects of the wake in the CFD calculations. In this method, the usual velocity to the blade surface is calculated so that it exactly cancels out the effect of the wake normal downwash to the surface.

The boundary condition at the blade surface must be modified in order to include the wake effects. For invicid calculations, this becomes:

$$\vec{V}.\hat{n} = -\vec{V}_{DW}.\hat{n} \quad (29)$$

where $\vec{V}_{DW}$ is the wake induced velocity and $\hat{n}$ is the normal unit vector of the blade surface. Thus, the induced velocity is used to modify the conventional tangency of velocity vector at the solid surface. Also, the fluxes of density and the enthalpy at the solid surface are modified accordingly to meet the new boundary condition.

It is interesting to note that, for a computational region covering only a portion of the rotor disk, the effect caused by the other rotor blades and the flow outside the finite computational region are incorporated into the calculation by employing the above transpiration velocity technique if the effects of other blades are included in the induced downwash velocity.

The treatment of the far-field boundary condition is based on the concept of Riemann invariants for a one-dimensional flow normal to the boundary (Jameson and Baker, 1983). Let subscripts ∞ and e denote far-field values and values extrapolated from the interior cells adjacent to the boundary, respectively, and let V_n and a be the velocity component normal to boundary and to the speed of sound (a). If the flow in the far field is subsonic, fixed and extrapolated Riemann invariants are introduced as:

$$R_\infty = V_{n\infty} - \frac{2a_\infty}{\gamma - 1}$$
$$R_e = V_{ne} - \frac{2a_e}{\gamma - 1} \quad (30)$$

Which are corresponding to incoming and outgoing waves. These invariants may be added and subtracted to obtain:

$$V_n = \frac{1}{2}(R_e + R_\infty)$$
$$a = \frac{1}{4}(\gamma - 1)(R_e - R_\infty) \quad (31)$$

At an outflow boundary, we extrapolate the tangential components of velocity, which results in:

$$\vec{V} = \vec{V}_e + \left[\hat{n}.\left(\vec{V} - \vec{V}_e\right)\right]\hat{n} \quad (32)$$

And for an inflow boundary, the tangential components are those of the free stream:

$$\vec{V} = \vec{V}_\infty + \left[\hat{n}.\left(\vec{V} - \vec{V}_\infty\right)\right]\hat{n} \quad (33)$$

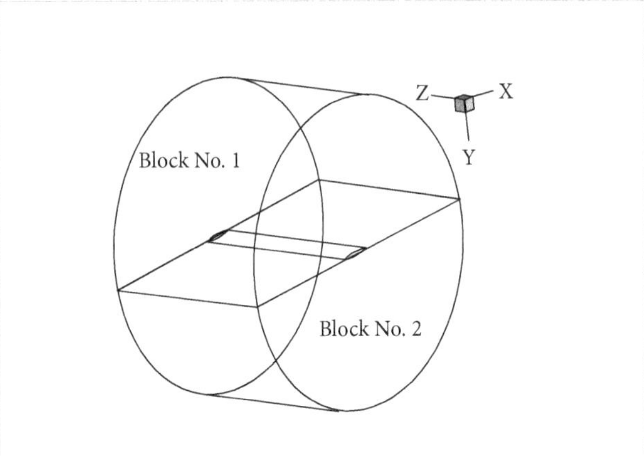

Figure 7. Computational region is covered by two blocks of structured grids.

The procedure is completed by extrapolating the entropy at an outflow boundary $S = S_e$ or setting it equal to its free stream value at inflow boundary $S = S_\infty$. The density, energy and pressure can then be calculated from a and s.

RESULTS

In this section some illustrative examples are presented as to demonstrate the capability of the proposed method for predicting the aerodynamic load and the flow field of a model helicopter rotor in hover state at various tip speeds. The rotor is a two bladed rotor with an aspect ratio of 6, which are untwisted and have rectangular planform with the NACA-0012 airfoil section. The experiments on this rotor have been presented by Caradonna and Tung (1981) at the NASA Ames Research Center. Computations are performed on a two-block mesh with 141570 nodes, on a personal computer. Figure 7 shows the computational domain which is symmetric in relation to the blade planform.

Five orders of magnitude reduction in the density residual are considered as convergence criteria. However, in most simulations for a Courant number of 2.0, the convergence criteria is not met in less than 1,600 time steps for the first iteration of the free wake model, and it is reduced to around 800 steps for the next iterations. Some of the results are discussed in the following sub-sections.

NON-LIFTING CASE

For this case, the free wake model is switched off and only the CFD calculation is performed. Pressure distribution on the blade upper and lower surfaces is shown in Fig. 8 for a tip Mach number of $M_t = 0.52$ and zero collective angle. Except near the tip, the agreement between the Euler calculation and the experimental data is obvious. The discrepancy near the tip can be caused by boundary conditions. However, the difference between numerical and excremental results is not significant.

LIFTING CASES

As stated before, for this case, the rotor downwash is first calculated by modeling the blade vortices with the BCVE vortex elements. Then, the effect of the wake and tip vortices is fed into the CFD solver and a new bound circulation is calculated from the sectional lift. Finally, a new iteration is started by correcting the shape of the wake and its strength according to the new bound circulation. The procedure is continued until the induced downwash velocity converges to a steady form. For the lifting cases, calculations are performed for five blade tip Mach numbers and in four different collective angles. For these simulations, the bound vortex is modeled by 32 vortex rings. In the near field part, the tip vortex is modeled by 3 turns of a helical vortex filament, each turn composed by six BCVE elements. The pitch of helix is initially set to $\sqrt{C_T/2}$ in this region, where the initial C_T is estimated from a collective angle by using the rotor momentum theory. The sheet vortices are also modeled by 4 filaments, each helical filament has 2 turns and each turn has six basic vortex elements. The mid part of the wake consists of 8 turns of helical filaments with the same axial pitch as the near wake filaments, i.e., $\sqrt{C_T/2}$. Also the far part of the wake consists of 50 turns with an axial pitch, which is twice the value used in the mid part.

For these simulations the number of iterations required for the convergence of induced downwash velocity is less than 300 iterations for a known circulation distribution in the free wake solver, and the number of iterations for achieving a steady distribution of downwash velocity from the combined free wake-CFD solver at low collective angles is six while it is reduced to less than 4 iterations for higher angles.

The sectional lift coefficient and the circulation distribution obtained from these calculations are shown in Fig. 9. The non-dimensional circulation (experimental or numerical) is defined as:

$$\bar{\Gamma} = \frac{1}{2}\frac{c}{R}\frac{z}{R}C_l \tag{34}$$

where C_l is the section lift coefficient at z radial station. The circulation shown in these figures is non-dimensioned by$R^2\Omega$, where Ω is the rotational velocity of the rotor and R is the rotor radius. As can be seen from these figures, the agreement with the experimental data is satisfactory. The thrust and inviscid power coefficients have also been computed and presented in Table 1. Comparison reveals that coefficients have good coincidence with experimental data (Caradonna and Tung, 1981) for all collective angles.

As a typical result, non-dimensional downwash and circulation distribution at intermediate iterations between the free wake model and the CFD solver is depicted in Fig. 10. It should be pointed out that through this iterative process, the compressibility effects in the tip region of the blades are included into the free wake model. Figure 10 shows that the downwash, or circulation distribution, is converged to a steady form in a few iterations.

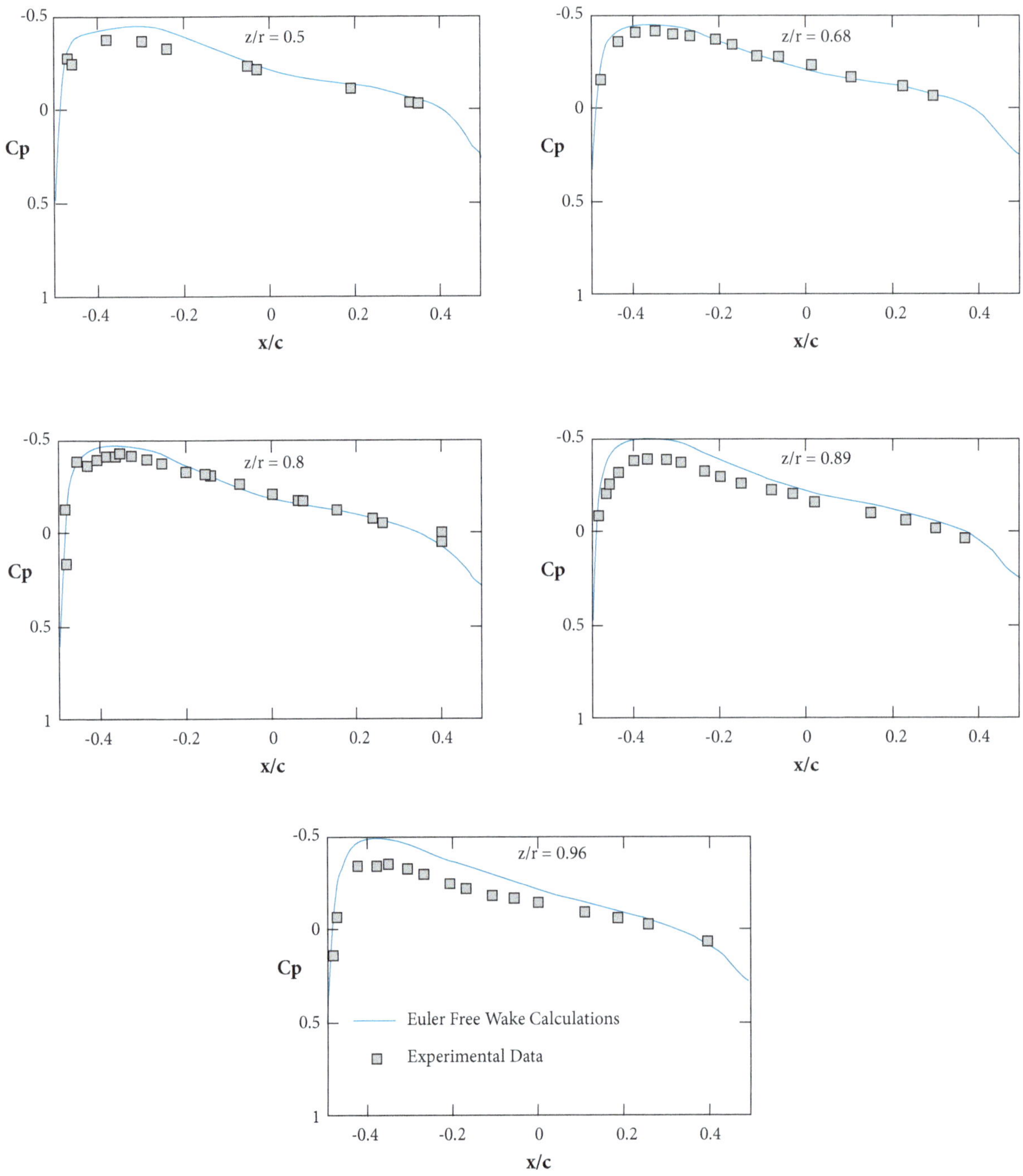

Figure 8. Pressure distribution in non-lifting case, M_t = 0.52, θ_c = 0, AR = 6.0, untwisted, untapered, NACA 0012 blade.

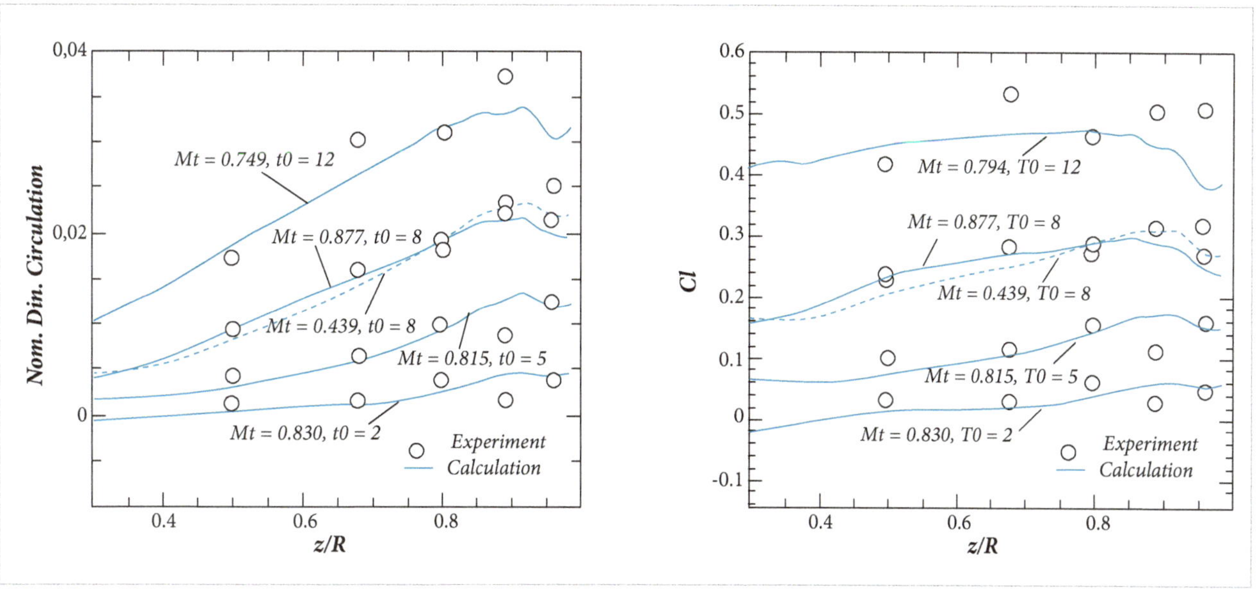

Figure 9. Lift coefficient and circulation distribution at various tip Mach numbers and collective angles.

Table 1. The computed thrust and power coefficients and comparison with experimental data.

Collective Angle (degree)	Tip Mach No.	Thrust Coefficient×10³ (Computed)	Thrust Coefficient×10³ (Experimental)	Power Coefficient×10⁴ (Computed)
2.0	0.830	0.615	0.690	0.935
5.0	0.815	2.253	2.190	1.769
8.0	0.439	4.655	4.590	3.120
8.0	0.877	4.606	4.730	4.262
12.0	0.794	7.730	7.920	7.941

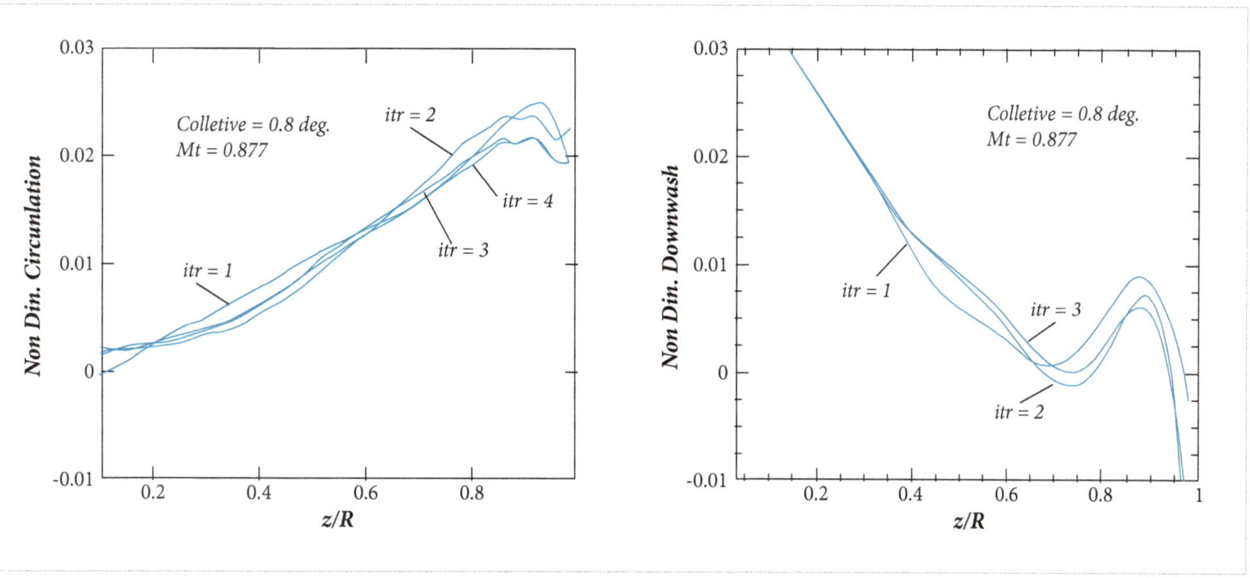

Figure 10. Downwash and circulation distribution at subsequent iterations.

Pressure distribution is shown in Figs. 11-15 for different tip Mach numbers and collective angles from 2 to 12 degrees. As shown in these figures, the agreement with experimental data is very good, especially for the position of shock wave on the upper surface of the blade.

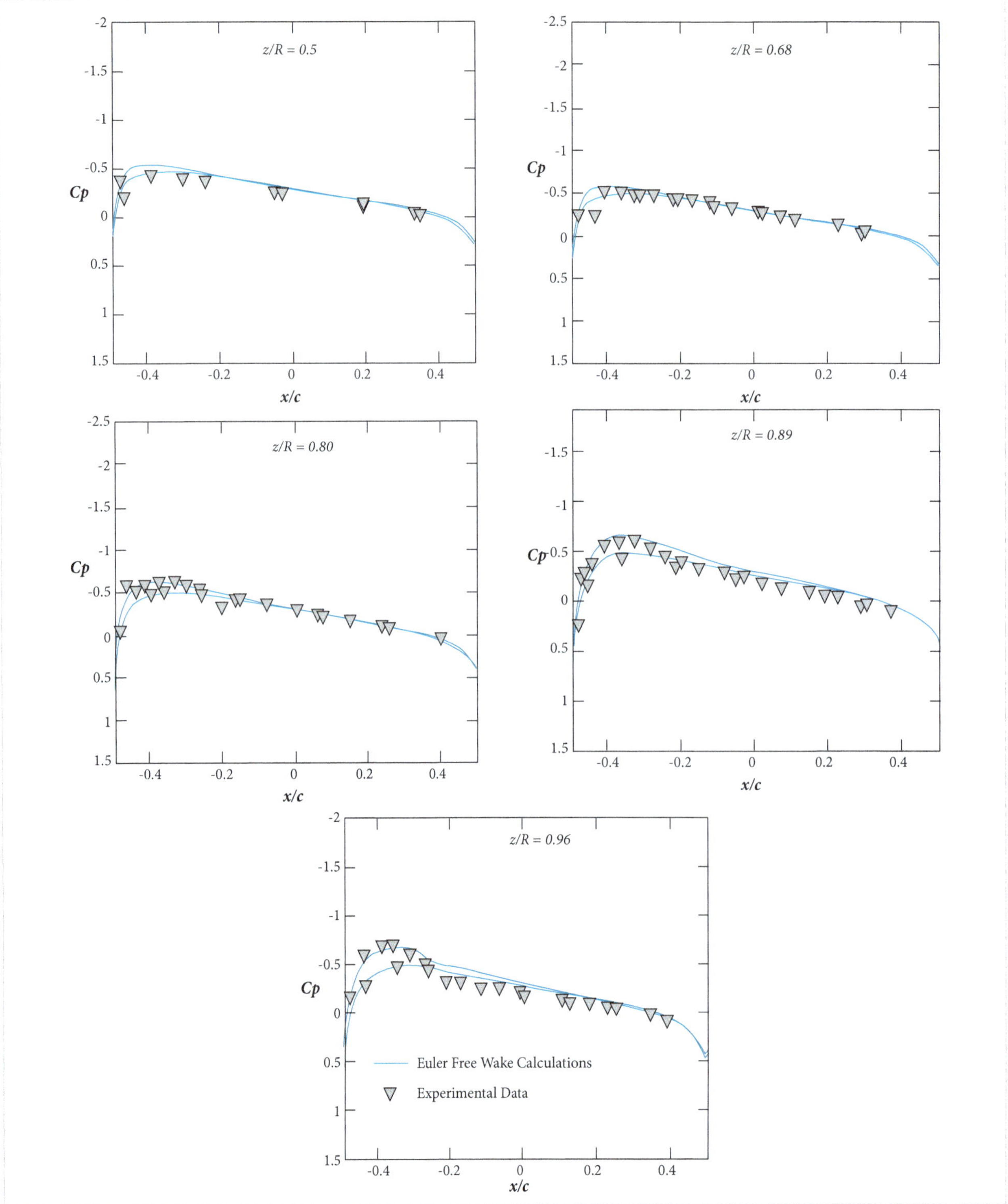

Figure 11. Pressure distribution on a lifting rotor in hover, $M_t = 0.830$, $\theta_c = 2$deg.

The flow solver can also be run in a full multi-grid (FMG) mode, which accelerates the process of convergence. Figure 16 shows the results obtained by applying the FMG approach in three levels and in a two-block cylindrical mesh. The time of computation can be decreased one order of magnitude and approximately two hours on a personal computer.

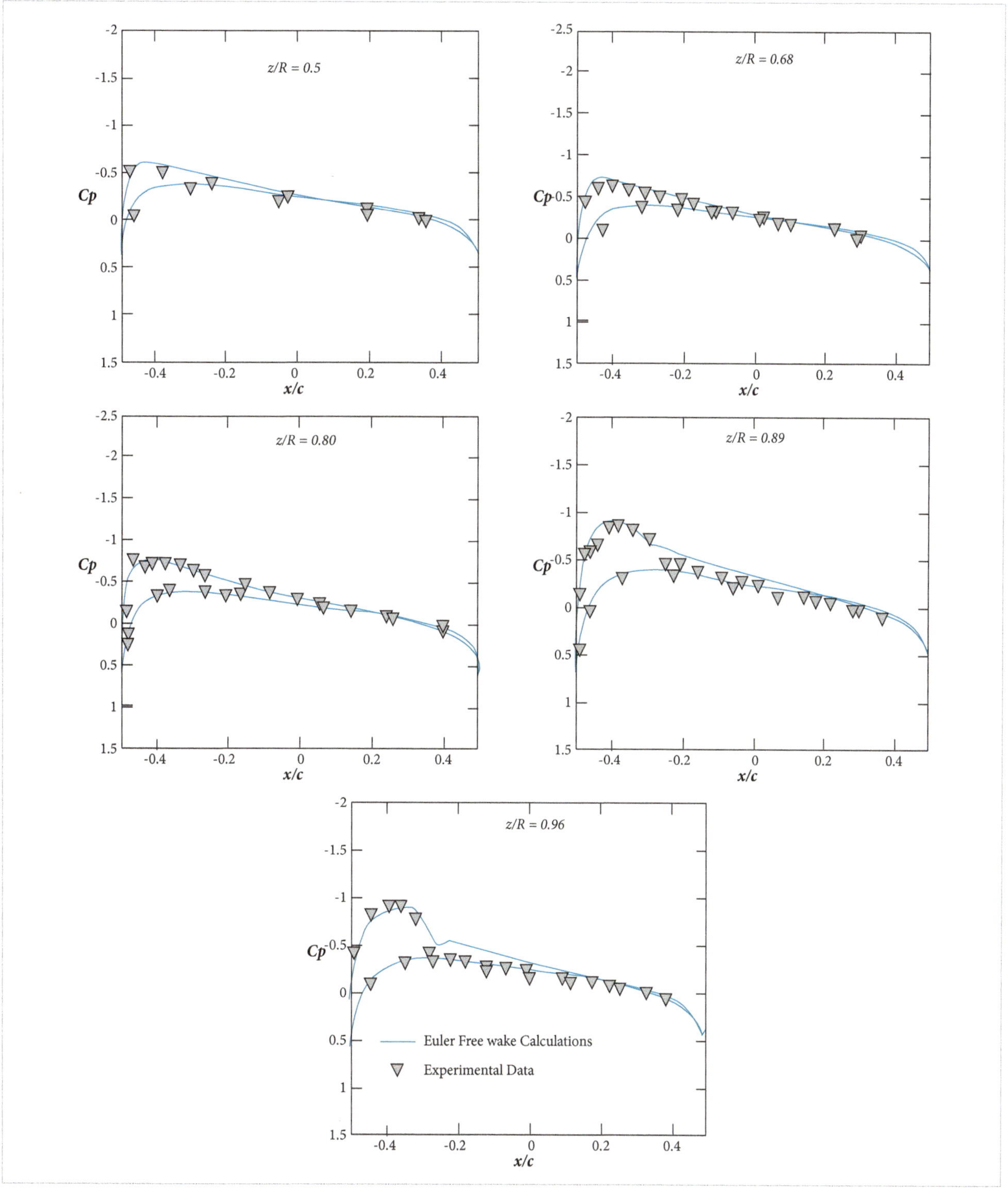

Figure 12. Pressure distribution on a lifting rotor in hover, $M_t = 0.815$, $\theta_c = 5$deg.

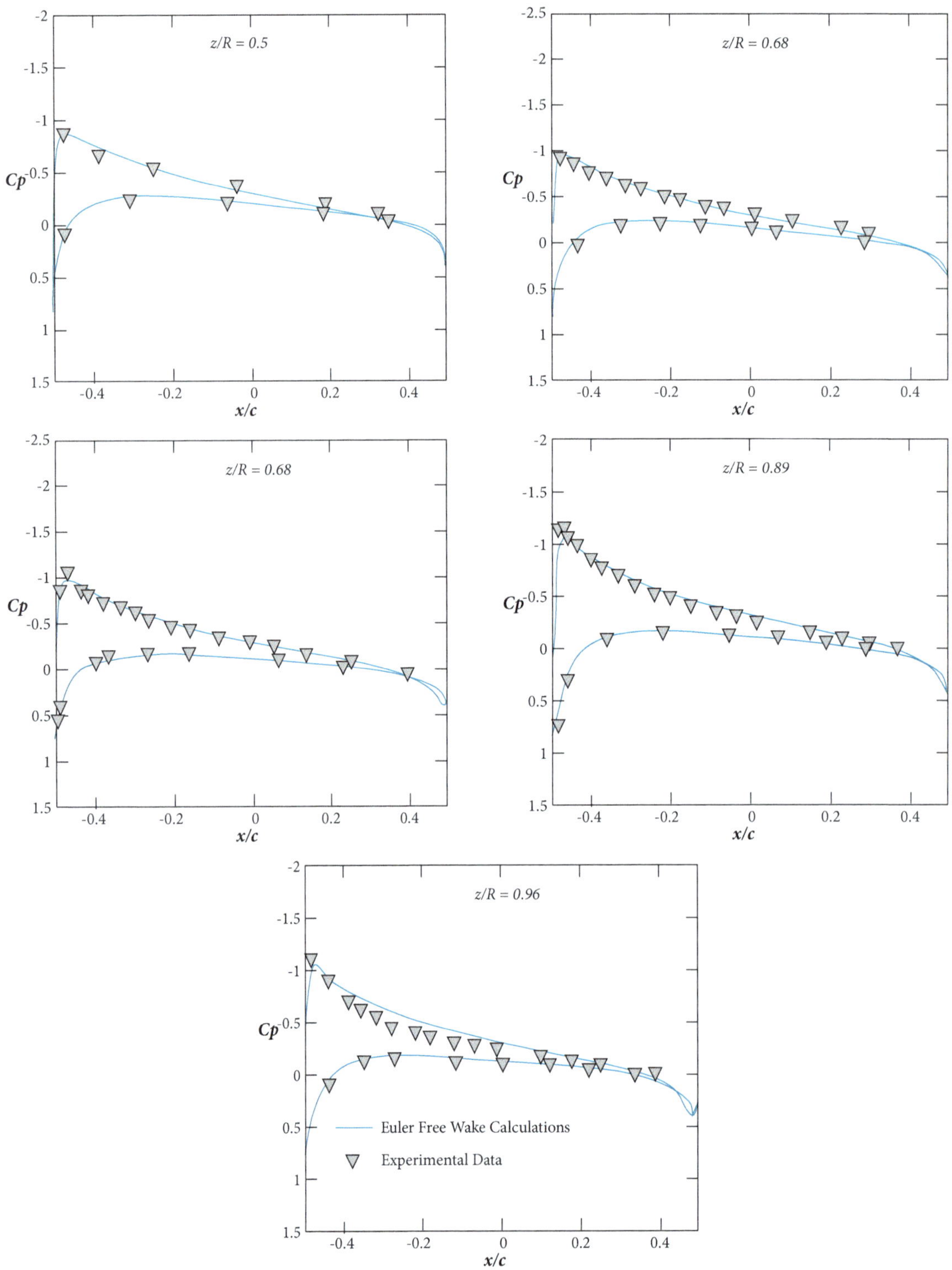

Figure 13. Pressure distribution on a lifting rotor in hover, $M_t = 0.439$, $\theta_c = 8\text{deg}$.

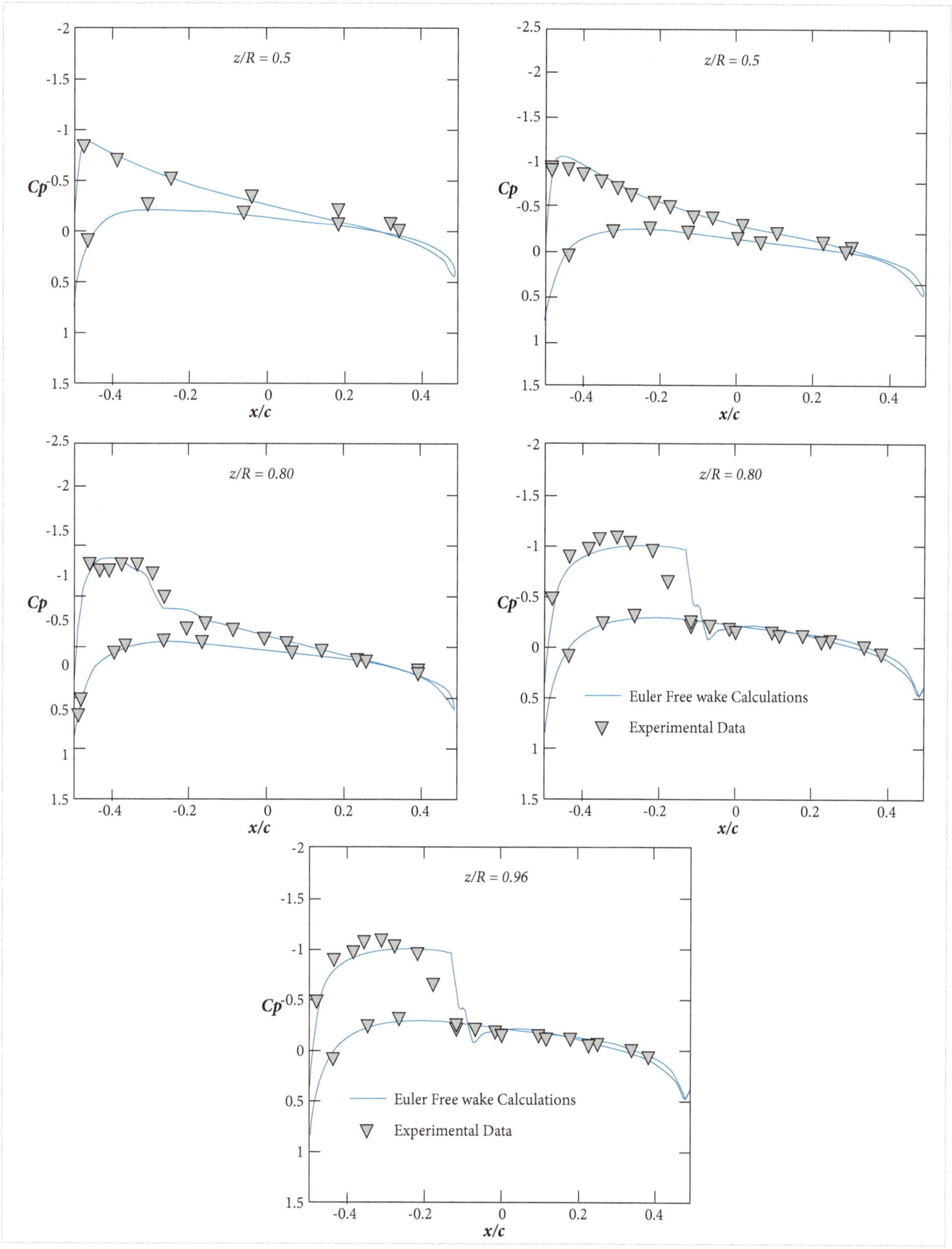

Figure 14. Pressure distribution on a lifting rotor in hover, $M_t = 0.877$, $\theta_c = 8$deg.

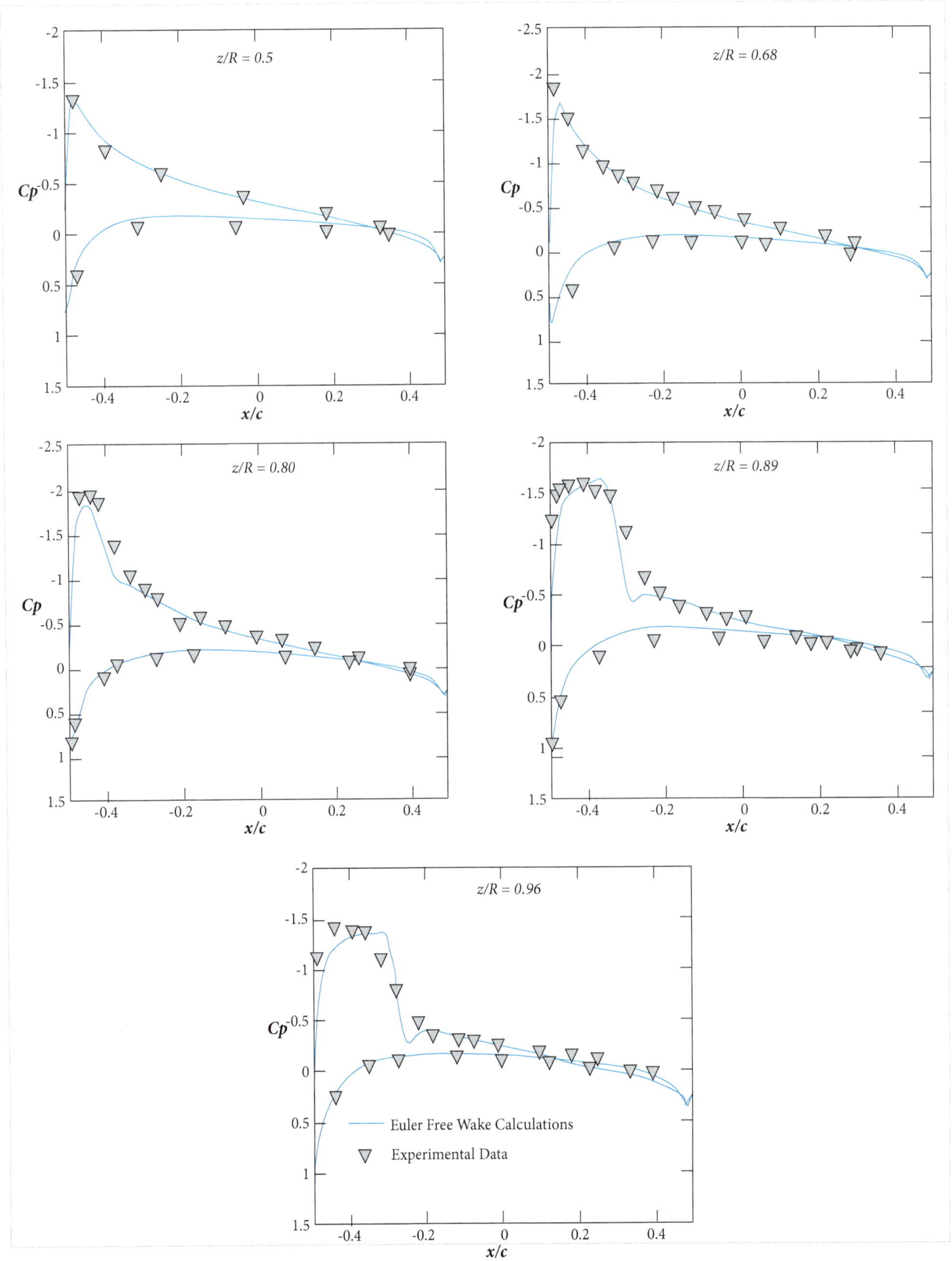

Figure 15. Pressure distribution on a lifting rotor in hover, $M_t = 0.794$, $\theta_c = 12$deg.

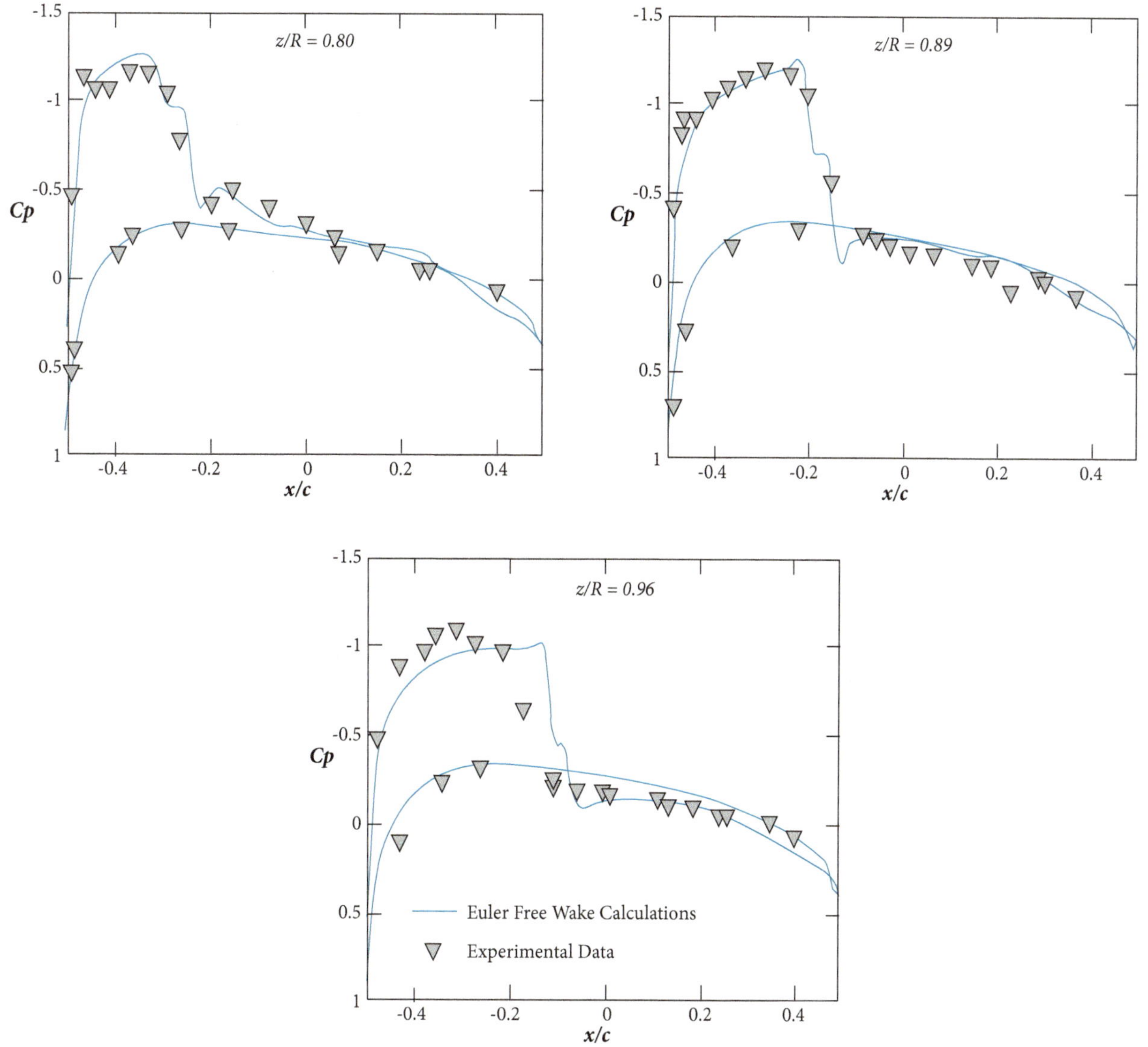

Figure 16. Pressure distribution on a lifting rotor in hover obtained from FMG approach, M_t = 0.877, θ_c = 8deg, AR = 6.0.

CONCLUSION

A new coupled free wake-CFD methodology for evaluation of aerodynamic loads on a helicopter rotor in hovering condition was introduced. In this approach, the Euler equations are solved in a rotating coordinate system and the rotor wake effects are modeled by a free wake approach and included into the CFD solver by a transpiration boundary condition at the blade surface.

The obtained results from this approach for the pressure distribution on the blade surface are promising; the performance parameters such as thrust and induced torque coefficients are particularly estimated with a rather good accuracy. The main advantage of this method is that satisfactory results can be achieved by minimum computational efforts, particularly for the grid generation problem and the required CPU time. This feature becomes important where a low fidelity accurate rotor blade flow solver is required for primarily design or optimization purposes. However, the proposed method ignores important phenomena such as the Blade-Vortex Interaction (BVI) or shock vortex interactions.

REFERENCES

Agarwal, R.K. and Deese, J.E., 1987, "Euler Calculation for Flowfield of a Helicopter Rotor in Hover" AIAA, Journal of Aircraft, Vol.24, No.4, pp. 231-238. doi: 10.2514/3.45431.

Agarwal, R.K. and Deese, J.E., 1988, "Navier-Stokes Calculations of the Flow field of a Helicopter Rotor in Hover", AIAA 26th Aerospace Sciences Meeting, Paper 88-0106. doi: 10.2514/6.1988-106.

Baskin, V.E., Vildgrube, L.S., Vozhdayev, Y.S. and Maykapar, G.I., 1967, "Theory of Lifting Airscrew", NASA TTF-823.

Batchelor, G.K., 1967, "An Introduction to Fluid Dynamics", Cambridge University Press.

Benoit, B., Dequim, A.M., Kampa K., Grünhagen, W., Basset, P.M. and Gimonent B., 2000, "HOST, a General Helicopter Simulation tool for Germany and France", In Proceeding of: American Helicopter Society, 56th Annual Forum and Technology Display, Virginia Beach, VA.

Bliss, D.B., Wachspress, D.A. and Quackenbush, T.R., 1984, "A New Methodology for Free Wake Analysis Using Curved Vortex Elements", Continuum Dynamics, Inc. Report No. 84-6.

Bliss, D.B., Wachspress, D.A. and Quackenbush, T.R., 1985, "A New Approach to the Free Wake Problem for Hovering Rotors", Proceedings of the 41st Annual Forum of the American Helicopter Society, Fort Worth, TX, pp. 463-477.

Caradonna, F.X. and Tung, C., 1981, "Experimental and Analytical Studies of a Model Helicopter Rotor in Hover", Vertica, Vol. 5, pp. 149-161.

Caradonna, F.X., Tung, C. and Desopper, A., 1984, "Finite Difference Modeling of Rotor Flows Including Wake Effects", Journal of the American Helicopter Society, Vol. 29, No. 2, pp. 26-33(8). doi: 10.4050/JAHS.29.26.

Chang, L.C. and Tung, C., 1985, "Numerical Solution of the Full Potential Equation for Rotors and Oblique Wings Using a New Wake Model", AIAA Paper 85-0268.

Chang, L.C., and Tung, C., 1987, "Euler Solution of the Transonic Flow for a Helicopter Rotor", AIAA 25th Aerospace Sciences Meeting, Paper 87-0523. doi: 10.2514/6.1987-523

Chen, C.L. and McCroskey, W.J., 1988, "Numerical Simulation of Helicopter Multi-Bladed Rotor Flow", AIAA, Aerospace Sciences Meeting, Paper 88-0046.

Chen, C.S., Velkoff, H.R. and Tung, C., 1987, "Free-Wake Analysis of a Rotor in Hover", AIAA Paper 87-1245.

Clark, D.R and Landgrebe, A.J., 1971, "Wake and Boundary Layer Effects in Helicopter Rotor Aerodynamics," AIAA 4th Fluid and Plasma Dynamics Conference, AIAA paper No. 71-581.

Clark, D.R. and Leiper, A.C., 1970, "The Free Wake Analysis: A Method for the Prediction of Helicopter Rotor Hovering Performance", Journal of the American Helicopter Society, Vol. 15, No. 1.

Crimi, P., 1965, "Theoretical Prediction of the Flow in the Wake of a Helicopter Rotor", Cornell Aeronautical Laboratory, Report CAL BB-1994-S.

Egolf, T.A. and Sparks, S.P., 1987, "A Full Potential Flow Analysis with Realistic Wake Influence for Helicopter Rotor Airload Prediction", NASA CR-4007.

Holmes, D.G. and Tong, S.S., 1985, "A Three-Dimensional Euler Solver for Turbomachinery Blade Rows", Journal of Engineering for Gas Turbines and Power, Transactions of ASME, Vol. 107, No. 2, pp. 258-264. doi: 10.1115/1.3239705.

Jameson, A. and Baker, T.J., 1983, "Solution of Euler Equations for Complex Configurations", AIAA Paper 83-1929.

Jameson, A., Schmidt, W. and Turkel, E., 1981, "Numerical Solution of the Euler Equations by Finite Volume Methods Using Runge-Kutta Time-Stepping Schemes", AIAA, 14th Fluid and Plasma Dynamics Conference, Paper 81-1259.

Jenny, D.S., Olson, J.R. and Landgrebe, A.J., 1968, "A Reassessment of Rotor Hovering Performance Prediction Methods", Journal of American Helicopter Society, Vol. 13, No. 2, pp. 1-26. doi: 10.4050/JAHS.13.1.

Johnson, W., 1980, "Helicopter Theory", Princeton University Press.

Johnson, W., 1981, "Development of a Comprehensive Analysis for Rotorcraft - I. Rotor Model and Wake Analysis", Vertica, Vol. 5, No. 2, pp. 99-129.

Kocurek, J.D. and Tangler, J.L., 1976, "A Prescribed Wake Lifting Surface Hover Performance Analysis", Journal of the American Helicopter Society, Vol. 22, No. 1, pp. 24-35. doi: 10.4050/JAHS.22.24.

Kramer, E., Hertel, J. and Wagner, S., 1988, "Computation of Subsonic and Transonic Helicopter Rotor Flow using Euler Equations", Vertica, Vol. 12, No. 3, pp. 279-291.

Landgrebe, A.J., 1969, "An Analytical Method for Predicting Rotor Wake Geometries", Journal of the American Helicopter Society, Vol. 14, No. 4, pp.20-32(13). doi: 10.4050/JAHS.14.20.

Landgrebe, A.J., 1972, "The Wake Geometry of a Hovering Helicopter Rotor and its Influence on Rotor Performance", Proceedings of the 28th Annual National Forum of the American Helicopter Society, Journal of the American Helicopter Society, Vol. 17, No. 4, pp. 3-15(13). doi: 10.4050/JAHS.17.3.

Landgrebe, A.J. and Cheney, M.C., 1972, "Rotor Wakes-key to Performance Prediction", AGARD Conference Proceedings No. 111 on Aerodynamics of Rotary Wings.

Landgrebe, A.J., Moffitt, R.C. and Clark, D.R., 1977, "Aerodynamic Technology for Advanced Rotorcraft-Part I", Journal of the American Helicopter Society, Vol. 22, No. 2, pp. 21-27. doi: 10.4050/JAHS.22.21.

Lee, S.W. and Kwon, O.J., 2006, "Aerodynamic Shape Optimization of Hovering Rotor Blades in Transonic Flow Using Unstructured Meshes", AIAA Journal, Vol. 44, No. 8, pp. 181-1825. doi: 10.2514/1.15385.

Narramore, J.C. and Vermeland, R., 1989, "Use of Navier-Stokes Code to Predict Flow Phenomena Near Stall as Measured on a 0.658-Scale V-22 Tilt-rotor Blade", AIAA Paper 89-1814.

Quackenbush, T.R., Bliss, D.B. and Wachspress, D.A., 1989, "New Free-wake Analysis of Hover Performance Using a New Influence Coefficient Method", NASA CR 4309, Journal of Aircraft, Vol. 26, No. 12, pp. 1090-1097. doi: 10.2514/3.45885.

Ramachandran, K., Tung. C. and Caradonna, F.X., 1989, "Rotor Hover Performance Prediction Using a Free-Wake Computational Fluid Dynamics Method" , Journal of Aircraft, Vol. 26, No. 12, pp. 1105-1110. doi: 10.2514/3.45887.

Roberts, T.W. and Murman, E.M., 1985 "Solution Method for a Hovering Helicopter Rotor Using the Euler Equations", AIAA Paper 85-0436.

Rosen, A. and Graber, A., 1988, "Free Wake Model of Hovering Rotors Having Straight or Curved Blades", Proceedings of the International Conference on Rotorcraft Basic Research, Journal of the American Helicopter Society, Vol. 33, No. 3, pp. 11-19(9). doi: 10.4050/JAHS.33.11.

Saberi, H.A., 1983, "Analytical Model of Rotor Aerodynamics in Ground Effect", NASA CR 166533.

Sadler, S.G., 1971, "Development and Application of a Method for Predicting Rotor Free Wake Positions and Resulting Rotor Blade Airloads", NASA CR 1911 and CR 1912.

Sankar, N.L., Wake, B.E. and Lekoudis, S.G., 1986, "Solution of the Unsteady Euler Equations for Fixed and Rotor Wing Configurations", Journal of Aircraft, Vol. 23, No. 4, pp. 283-289. doi: 10.2514/3.45301.

Srinivasan, G.R. and McCroskey, W.J., 1988, "Navier-Stokes Calculations of Hovering Rotor Flow Fields", Journal of Aircraft, Vol. 25, No. 10, pp. 865-874. doi: 10.2514/3.45673.

Strawn, R.C. and Caradonna, F.X., 1987, "Conservative Full-Potential Model for Unsteady Transonic Rotor Flows", AIAA Journal, Vol. 25, No. 2, pp. 193-198. doi: 10.2514/3.9608.

Strawn, R.C. and Tung, C., 1986, "The Prediction of Transonic Loading on Advancing Helicopter Rotors", NASA TM-88238.

Summa, J.M. and Clark, D.R., 1979, "A Lifting-Surface Method for Hover/Climb Airloads", Proceedings of the 35th Annual National Forum of the American Helicopter Society, Washington, D.C., USA.

Summa, J.M. and Maskew, B., 1981, "A Surface Singularity Method for Rotors in Hover or Climb", USAAVRADCOM TR 81-D-23.

Wake, B.E. and Sankar, L.N., 1989, "Solutions of the Navier-Stokes Equations for the Flow About a Rotor Blade", Journal of the American Helicopter Society, Vol. 34, No. 2, pp. 13-23(11).doi: 10.4050/JAHS.34.13.

Direct Numerical Simulation of the Onset of Vortex Shedding for Blunt Elongated Bodies

Marcos Aurélio Ortega[1], Roberto da Mota Girardi[1], Jorge Hugo Silvestrini[2]

ABSTRACT: This research project focuses upon the wake behind a two-dimensional blunt-trailing-edged body. Data are obtained numerically by means of a Direct Numerical Simulation code. The body has an elliptical nose followed by a straight section that ends in a blunt base. The present paper is dedicated to the analysis of the onset of the shedding process. The effort is certainly worthwhile, because, in contrast to the case of the circular cylinder, the boundary layers's separation points are defined and fixed. This allows a better assessment of the vital influence of the boundary layers upon the wake, in a controlled way. This is not the case for the circular cylinder, because, in this instance, the separation points oscillate in relation to a mean position. In the present analysis, the relationship between the onset of shedding Reynolds number, Re^h_K , and the aspect ratio, *AR*, is obtained. To this end, a wide range of aspect ratios, between 3 and 25, was investigated. The result represented by this relationship is a novelty in the literature. Values of Re^h_K are strongly influenced by the aspect ratio for the case of the short cylinders — for which *AR* is low. After *AR* about 9, the curve flattens and the influence of the aspect ratio upon the shedding Reynolds number is very mild. Besides, the paper discusses another very important aspect; the overall stability of the pre-shedding laminar bubble at the base of the body. It is important to stress that the latter study relies on the fact that the boundary layers separation points are fixed.

KEYWORDS: Wakes, Onset of shedding, Blunt-trailing-edged body, Numerical analysis.

INTRODUCTION

The aim of the present work is to study the wake behind a blunt-trailing-edged body, consisting of an elliptical nose followed by a straight section (Fig. 1). In the present paper, a study of the onset of shedding is reported, whereas in a second paper, the main topological features of the flow are discussed (Ortega *et al.*, 2012).

There is a myriad of publications in the literature related to flows about blunt bodies. It is not the aim of the present authors to present here a thorough literature review about this theme. The interested reader can find these information in the following citations. Some of the most comprehensive and authoritative historical reviews are the ones by Williamson (1996) and Dauchy *et al.* (1997). On the other hand, low Reynolds number data, relative to elliptical elongated cylinders, is rather scarce. Bearman (1965) has conducted, during the 1960's, a thorough experimental analysis of this geometrical form. Two-dimensional results were presented for a model with and without splitter plates, and the Reynolds number, based on the model chord, was made to vary in the range of 1.4×10^5 to 2.56×10^5 (Bearman, 1965). Several splitter plate lengths were investigated, and studies of base pressure, shedding frequency, hot-wire traverse distributions of mean flow velocity, and r.m.s. of velocity fluctuations, were done. The main focus of the investigation was the understanding of how those factors would influence the formation length, and, at the same time, to assess the possibility of drag reduction. In a sequence, Bearman (1967), following basically the same line of study, investigated the effect of base bleed behind the same model. Now, the Reynolds number, based on the body base height, was made to vary between 1.3×10^4 and 4.1×10^4.

1. Instituto Tecnológico de Aeronáutica – São José dos Campos/SP – Brazil 2. Pontifícia Universidade Católica do Rio Grande do Sul - Porto Alegre/RS – Brazil

Author for correspondence: Marcos Aurélio Ortega | Department of Aerodynamics – Division of Aeronautical Engineering – ITA, CEP: 12.228-001 - São José dos Campos/ SP | Brazil | Email: marcos.ts.ortega@gmail.com

The model investigated had a span of 71.1 cm, a chord of 15.24 cm, and a base height, *h*, of 2.54 cm. The aspect ratio, $AR = c/h$, was equal to 6. The frontal half ellipse had semi-major and -minor axes equal to, respectively, 12.70 cm and 1.27 cm. In the words of the author: "The rear 2.54 cm of the model was parallel sided in order that the flow left the surface smoothly at the trailing edge corners". As pointed out before, this was the main geometric characteristic that attracted our attention. Recently, Park *et al.* (2006) have investigated a new passive device for drag reduction in a flow about a bluff body with a cross-section like the one in Fig. 1. The aspect ratio was equal to 6.33 and the study involved both experimental and numerical treatments. Another work that has focused on the elongated cylinder shape is the one by Ryan *et al.* (2005). In this paper, the authors conducted a numerical Floquet analysis of the three-dimensional transition to turbulence in the wake of the body.

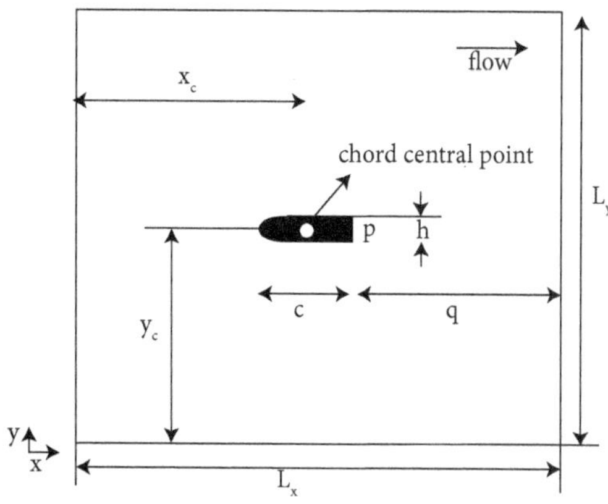

x_c, y_c = position of the body in the calculation domain, h = body base height, c = body chord length.

Figure 1. Cross-sectional view of the elongated cylinder and dimensions of computational domain.

For the case of the elongated cylinder, the onset-of-shedding process that happens at low values of the Reynolds number, the well known Hopf bifurcation, responsible for the appearance of the von Kármán vortex street, has never been investigated before. This is the main focus of this article. Values of the onset of shedding Reynolds number are numerically obtained by varying the value of the aspect ratio. For low values of the aspect ratio, the influence upon the Reynolds number in strong, but after about $AR = 9$, the curve flattens and an asymptotic behavior is observed. After that, an exploration on the influence of the boundary layers upon the shedding process was attained. For this purpose, we have "installed" small bumps on the upper and lower surfaces of the body, and have studied how those perturbations affect the conditions of the vortexes emission.

This paper is organized as follows. In the section "Computational strategy", the main aspects of the mathematical model are presented and discussed. A short review of the Hopf bifurcation is given, with the purpose of establishing the physical scenario to be tackled ahead in the study. After that, some data about validation and convergence of the code are discussed. Convergence characteristics are extremely important here because the physics is essentially time dependent. In the section "In search of the onset of shedding Reynolds number", the core of the article is divided into three sub-sections. In the first one, a strategy of investigation is designed. We have arrived at the rake of grids idea. This idea was tested, first, by dealing with the circular cylinder, taking into account the vast number of literature data for this geometry. Because the circular cylinder results were consistent, we then applied it to the elongated body. Finally, we made an effort to assess the influence of the boundary layers upon the formation and shedding of the structures at the base of the body. Some conclusions are then presented.

COMPUTATIONAL STRATEGY

Firstly, solutions for flow past a circular cylinder from time-dependent simulations of the Navier-Stokes equations on two-dimensional domains are obtained. The immediate objective is to perform code convergence tests. Secondly, the code is applied to the studies of the flow about the elongated cylinder. The code is called Incompact3d and it was originally developed by Lamballais and Silvestrini (2002), and Lamballais *et al.* (2008); it is a software for incompressible flows of the immersed boundaries type and it uses compact differences for discretization purposes.

TIME DEPENDENT SIMULATIONS

We consider the flow of an incompressible Newtonian fluid. The physical behavior is mathematically modelled by the Navier-Stokes equations, written here as:

$$\vec{\nabla}.\vec{u} = 0, \tag{1}$$

$$\frac{\partial \vec{u}}{\partial t} + \vec{u}.\vec{\nabla}\vec{u} = -\frac{1}{\rho}\vec{\nabla}P + \frac{1}{Re}\nabla^2\vec{u} + \vec{f}, \tag{2}$$

where;

ρ is the density, $P(\vec{x},t)$ is the pressure field, and $\vec{u}(\vec{x},t)$ is the velocity field.

The introduction of an external force field $\vec{f}(\vec{x},t)$ is necessary, in order to simulate the presence of the body. The equations above, if written in rotational form, are stable to aliasing errors (Kravchenko and Moin, 1997); therefore, in the numerical algorithm, the following formulation is preferred:

$$\frac{\partial \vec{u}}{\partial t} = -\frac{1}{\rho}\vec{\nabla}P^* - \vec{\omega}\times\vec{u} + \frac{1}{Re}\nabla^2\vec{u} + \vec{f}, \tag{3}$$

where;

$P^*(\vec{x},t)$ is the modified pressure field (equal to $(P+\rho\vec{u}^2/2)$), and $\vec{\omega}(\vec{x},t)$ is the vorticity field (equal to $\vec{\nabla}\times\vec{u}$). The groups of equations above are written in non-dimensional form: lengths are scaled by a typical length, l_t, — the diameter for the circular cylinder, *d*, and the base height for the elongated cylinder, *h*, — and a typical velocity, which happens to be, in both instances, the magnitude of the free-stream velocity, U_∞. The Reynolds number that results from the non-dimensionalisation is given by $Re = U_\infty . l_t / \upsilon_\infty$, where υ_∞, is the free-stream kinematic viscosity.

Time-dependent simulations based on these equations in two dimensions ($\omega \equiv 0$; $\partial/\partial_z \equiv 0$) are carried out using a sixth-order, compact-finite-difference scheme (Lele, 1992). Meshes are Cartesian and the presence of the body is simulated by virtue of an immersed-boundary technique (Goldstein *et al.*, 1993). To integrate Eq. (3), a third-order low-storage Runge-Kutta strategy was used (Lamballais, 1996). The core of the algorithm solves a Poisson equation for the pressure, which, in the sequel, acts as a projector of the velocity field onto a divergence-free space (Lamballais, 1996). Part of the forcing terms in the immersed boundary sub-routine is advanced implicitly, instead, and the Crank-Nicolson scheme is used (Lamballais, 1996). This feature improves the efficiency of the forcing mechanism (Fadlun *et al.*, 2000; Lamballais and Silvestrini, 2002). All calculations presented here were performed in 64-bits precision. Further details of the code, and many verification and validation studies, can be assessed in a series of former applications, both two- and three-dimensional (Ribeiro, 2002; Vitola, 2006; Lamballais *et al.*, 2008; Silvestrini and Lamballais, 2004; Lardeau *et al.*, 2002; Laizet *et al.*, 2009). The two-dimensional approach that is used to investigate the problem is appropriate due to the fact that, when the Reynolds number is low, the flow is everywhere laminar but the main characteristics of the great structures are mostly all present (Barkley and Henderson, 1996).

Boundary conditions are required at the limits of the computational domain. At the entrance plane — left vertical boundary in Fig. 1 —, velocity components are specified according to: $u_x = U_\infty = 1$ and $u_y = 0$. A white noise might be added, in order to speed up a process of transition that is eventually being studied. At the upper and lower boundaries, the uniform flow is enforced, i.e., $u_x = 1$ and $u_y = 0$ (Ribeiro, 2002), while along the exit plane, conditions are established according to a simplified convection equation:

$$\frac{\partial u_i}{\partial t} + U_{conv}\frac{\partial u_i}{\partial x} = 0, \tag{4}$$

where U_{conv} is made equal to the main structures mean-convection velocity at the end of each iteration. The very mild degradation of the flow, imposed by this condition, is confined to a narrow region close to the boundary, according to Akselvoll and Moin (1996). The last frontier is composed of the body surface. To cope with it, following Goldstein *et al.* (1993) and Saiki and Biringen (1996), a feedback forcing term was introduced in the momentum equation, in order to represent the presence of the solid body. This procedure presents an outstanding advantage because a Cartesian grid can always be used, independently of the body's external geometry.

HOPF STABILITY ANALYSIS

For sufficiently low Reynolds numbers, the topology in the near wake of a two-dimensional bluff body is that of the laminar twin vortices. After that, and as the Reynolds number grows, an instability corresponding to a Hopf bifurcation leads to a changing of regimes. For the case of the circular cylinder, and for a Reynolds number, somewhere between 40 and 50, the steady flow field transitions to a laminar two-dimensional shedding wake, the von Kármán vortex street — Williamson (1996) reports a Re_K^d equal to 49; Dušek *et al.* (1994) inform a slightly different range, 46 - 47, while Sumer and Fredsoe (1997) give $Re_K^d = 40$. The subscript *K* in the Reynolds number symbol is an indicative that the value corresponds to the onset of shedding, while a superscript indicates the definition reference length. To be more specific, one should remind that, at this low range, there are three limiting values of the Reynolds number, which we shall indicate by Re_c, Re_A, and Re_K. The subscripts correspond to "critical state", "absolute instability" and "onset of shedding", respectively. For the case of the circular cylinder and for a Reynolds number based on the diameter, Monkewitz (1988) reports the following: $Re_c \approx 5$, $Re_A \approx 25$, and $Re_K^d \approx 47$. The critical Reynolds number, Re_c, divides the ranges of complete stability ($Re < Re_c$) and convective instability ($Re_c < Re < Re_A$). Re_A marks the appearance of the first local absolute instability, while Re_K^d is indicative of the onset of vortexes shedding. See, for example, Huerre and Monkewitz (1985), or Bers (1983), or Drazin and Reid (2004), for an explanation of those stability concepts. In his studies Monkewitz investigated the stability of a family of incompressible bluff-body wakes by means of a linear parallel, that is, local, approach. One of the main results of this investigation is the confirmation that Re_K is really larger than Re_A, a result first established by Chomaz *et al.* (1988). In other words, this means that the Karman vortex street is a consequence of self-excited oscillations of the wake, and that shedding is not triggered by the first appearance of an absolute instability in the near wake. This is a necessary condition, but shedding is the "end product" of a saturated state, that is, global instability sets in after a whole region of the near wake has reached the absolute instability state (Monkewitz, 1988; Chomaz *et al.*, 1988; Yang and Zebib, 1989). Monkewitz (1988) suggests that for forms other than circular, the sequence $Re_c < Re_A < Re_K$ is maintained, but with different numerical values.

The majority of the studies which deal with the Hopf bifurcation is either experimental or analytical (based on some variation of the linearized Landau equation). There is also a hybrid approach, where the author significantly simplifies the physical scenario: by considering the basic flow as parallel, for example, and completing the study by numerical means (Triantafyllou *et al.*, 1986).

The approach to be followed in this work is the DNS, Direct Numerical Simulation, a numerical tool that solves the complete Navier-Stokes equations. The great advantage is to solve the equations in their original form, and, in doing so, to take into account the essential non-linearities of the physical problem. The disadvantage, which is also considerable, corresponds to the computer costs. The main drawback is the following. For a coarse grid, the elapsed time to reach a stable oscillating state is rather short (taking into account that the Reynolds number is already inside the shedding range). This time length is measured from the very beginning — iteration 1 —, when the field of flow corresponds to the initial state (usually, uniform distributions of parameters along the domain of calculation). The destabilizing interval of time is short, in this instance, because the numerical error is relatively large in view of the coarseness of the grid. The numerical error is the leading factor in the onset of shedding triggering process (this point will be explored further in this paper). On the other hand, if the grid is coarse, the overall accuracy will most probably be poor. Eventually then, the user might be tempted to refine the grid in order to improve accuracy. As a result of this initiative, the destabilizing interval of time will grow accordingly, because a refined grid will "offer" a smaller destabilizing effect. Because the process is truly an assymptotic one, refining the grid further will correspond to an ever increasing computational time (a similar effect can be observed in Figs. 2 and 4 in relation to compiler precision). The way we have adopted to avoid this physical/numerical difficulty is to search for pre-established ranges of Reynolds numbers, which will, at least under certain controllable circumstances, contain the true value of Re_K. That is, we will be looking for data in the form of $Re_K \pm \Delta Re$. In the sequel, the details of this strategy will be given and discussed.

CONVERGENCE TESTS

The main concern in this section is to check for approximation errors. For the cyclic flow computations, the results were verified to be mesh independent to a high degree of accuracy. These error checking studies were conducted considering the flow about a circular cylinder.

The cyclic flow calculation was submitted to a complete battery of tests. In former works, members of our team (Ribeiro, 2002; Vitola, 2006) have already investigated the "collection" of optimized code parameters — related both to domain geometry and mathematical algorithms. Notwithstanding this, and due to the nature of the present paper, we have decided to further investigate the convergence characteristics of code Incompact3d, by means of a grid-refinement study— following Barkley and Henderson (1996). To this end, we have used a $30d \times 24d$ grid and several resolutions, as shown in Table 1. The Reynolds number was fixed in 300. Here, an enlarged grid is preferred because errors resulting, for example, from small distance of the body to the domain entrance plane and blockage effects, can be better controlled. Results are given in Table 1 showing convergence to three digits for $\Delta/d \leq 0.0167$.

Table 1. Grid refinement study of code Incompact3d.

Δ/d	$< C_d >$	C'_d	C'_l
0.0250	1.4267	0.0589	0.6776
0.0222	1.4147	0.0584	0.6703
0.0200	1.4178	0.0585	0.6718
0.0185	1.4018	0.0578	0.6615
0.0167	1.3927	0.0562	0.6483
0.0156	1.3924	0.0562	0.6482

IN SEARCH OF THE ONSET OF SHEDDING REYNOLDS NUMBER

THE STRATEGY FOR THE CIRCULAR CYLINDER

The numerical investigation in the range $Re \approx Re_K$, that is, for values of the Reynolds number in the onset of shedding region, must be done with very much care. As soon as one approaches the exact value of Re_K, the final solution gets extremely sensitive to the various parameters that have some influence in the numerical calculation. Basically, every factor has its influence in the final shedding conditions: (i) Grid resolution (probably, the most influential one); (ii) Type of numerical algorithm; (iii) Domain overall sizes (L_x and L_y); (iv) Accuracy — if single or double precision; (v) Time step; (vi) Type of machine: if 32 or 64 bits; (vii) For codes of the immersed boundary type — as the one here employed is —, how smoothly is the perimeter of the body represented. These items bring to the stage the question of the computational error, a factor that helps to trigger the shedding process when it grows up to a certain point. The problem of the blockage comes embedded in item (iii), and is especially related to L_y. Item (iii) also includes the effect of the position of the body in relation to the domain boundaries, especially the entrance plane. The time step has an indirect effect, because as the grid is refined, most probably one will be required to lower it, in order to guarantee convergence. The Reynolds number is evidently a prime influence. Besides all those aspects, there is still to be remembered the fact that the closing-in process to Re_K is, in a way, always asymptotic, and, most certainly, the investigation will require running the code for long intervals of time. In the sequence, we will return to some of these points and discuss them further.

After some work, we devised what was judged as a good approach to the problem. The core of our strategy corresponded to the creation of a sort of a "grid rake". This means a collection of grids with different resolutions. The testing of the idea was done considering the flow about a circular cylinder. For this specific case, five meshes were used in the study, corresponding to $[n_x/d] = [n_y/d] = 12, 18, 24, 30$ and 36, where d stands for the circular cylinder diameter. The domain of calculation dimensions were adopted as $L_x = 19d$, $L_y = 12d$, and $x_c = 8d$ (these values are the results of optimizing studies by Ribeiro (2002); for a Reynolds number equal to 300, these studies indicate $n/d = 24$ as a best value for the number of grid nodes per diameter).

The fact is that, the numerical investigation of the onset of shedding neighborhood is difficult, time consuming, and very elusive. Therefore, we concentrated mainly on the grid resolution, and probed other influences more superficially. The main aim of this particular effort is to try to develop a viable and reliable criterion that could be applied to the Re_K region, especially now to the elongated cylinder studies.

We studied the flow about the circular cylinder in the range of Reynolds number from 40 to 50. Table 2 shows the wake flow state for the various cases investigated. The first column of the table gives the value of the Reynolds number, while the first line indicates the grid resolution in terms of grid points per cylinder diameter. We began by running the case $Re = 40$ and ended with $Re = 50$, and the intermediate values are shown in Table 2. Each of these values of the Reynolds number was investigated using the five grid resolutions, as shown in Table 2, $[n_x/d] = [n_y/d] = 12, 18, 24, 30, 36$. All cases were run in single precision in a 64-bits machine, and at least from $t = 0$ up to $t = 1000$ units of non-dimensional time. The dimensionless time step was always the same and equal to 0.004838, which corresponded to a safe value that would guarantee the convergence at the finer grids. For $Re = 40, 41$ and 42, there was no shedding, and the signal of the anemometers indicated damping (some "numerical" anemometers were strategically placed in certain positions along the grid.) When the Reynolds number was raised to 43 there was shedding for resolution 12/d. The first Reynolds number value for which we obtained vortexes shedding in all grids was $Re = 45$. One would be then tempted to define 45 as the value of Re_K^d for the cylinder, but there is a crucial point here. We do believe that it is basically not possible to define an absolute value of the onset of shedding Reynolds number,

Table 2. Condition of the circular cylinder wake in the onset of shedding range.

Re^d	12/*d*	18/*d*	24/*d*	30/*d*	36/*d*	36/*d* (0.01%)	36/*d* (0.1%)
50	s	s	s	s	s	–	–
47	s	s	s	s	s	–	–
45	s	s	s	s	s	–	–
44	s	s	s	d	d	s	s
43	s	d	s	d	d	d	s
42	d	d	d	d	d	d	d
41	d	d	d	d	d	–	–
40	d	d	d	d	d	–	–

s: vortexes shedding; d: vortexes damping (Calculations performed in single precision).

at least when one is using an experimental or numerical tool to probe the process. The most one can do is to search for a range of values inside which one can probably guarantee, under certain controlled conditions, that the value of Re^d_K resides.

To clarify this point, let us focus our attention upon the case "43/18d", in Table 2. In this instance, there was no shedding until $t = 1000$. On the other hand, how can one be sure that the shedding will not set in for a certain instant of time after that limit? As we have stressed before, the final stability of the laminar bubble system in the wakeside of the cylinder is a characteristic of the system *perse*. In other words, given the state of the final saturation, the onset of shedding settles in. Therefore, what might be happenning here is that, for $t = 1000$, we have not yet reached the final saturated state. If this is the case, running for a longer stretch of time would almost inevitably result in shedding due to the mounting of the overall instability, which is ultimately fed by the numerical errors, being introduced in every iteration. Putting it differently, the result in Table 2 for the 43/18 case is conditioned to the length of the pre-defined time interval, equal, in this case, to 1,000 units. To be really sure, one would have to run the case for a much longer time stretch. If the setting 43/18 is physically not a shedding case, the feeding of perturbation at each iteration would never saturate the laminar bubble system, and the machine would run indefinitely without disturbing the damping state. This is why we argue that, for numerical, as well as for experimental investigations, it is not possible to fix an absolute value of the shedding Reynolds number. In this work, we will concentrate on the effort of determining the ranges of Re_K.

A careful observation of Table 2 reveals an unexpected result relative to $Re = 43$. Because the arrangement 43/18 corresponds to a damped flow, with more reason, the combination 43/24 should also correspond to a damped case. In normal conditions, the amplitude of the numerical error distribution along the grid is the main drive for the shed triggering. All combinations in Table 2 were run in a 64-bits machine (Dell Precision 690) with single precision. In any case, the amplitude of the error diminishes as the grid is refined. After some investigation we found out that the 24 points per diameter grid does not match as best as possible, as it should, the cylinder perimeter. That is, for this resolution, some nodes get slightly mispositioned along the perimeter. This "mispositioning" is recognized considering the details of the immersed boundary strategy. In practice what happens is that the flow in the numerical calculation "sees", in this case, the cylinder as a body with an augmented rugosity in its surface. Therefore, the perturbation increased, since 43 is located in a very critical position in relation to the interval that is being investigated ($Re = 40$ - 45), i.e., exactly at the centre of the stretch, the case 43/24 came out of the calculation as a shedding case and not as a damped one.

Because the points previously stressed are too important, we would like to illustrate them further. A clear evidence of the fact that the error amplitude is decisive is given in Fig. 2. We rerun several 12 points per diameter cases, but now using double precision, and a much longer calculation that reached 10 141.26 units of time. The running of the code, for such long stretches of time, can be looked at as a severe test for the damping characteristic of the flow. Besides, one has to keep

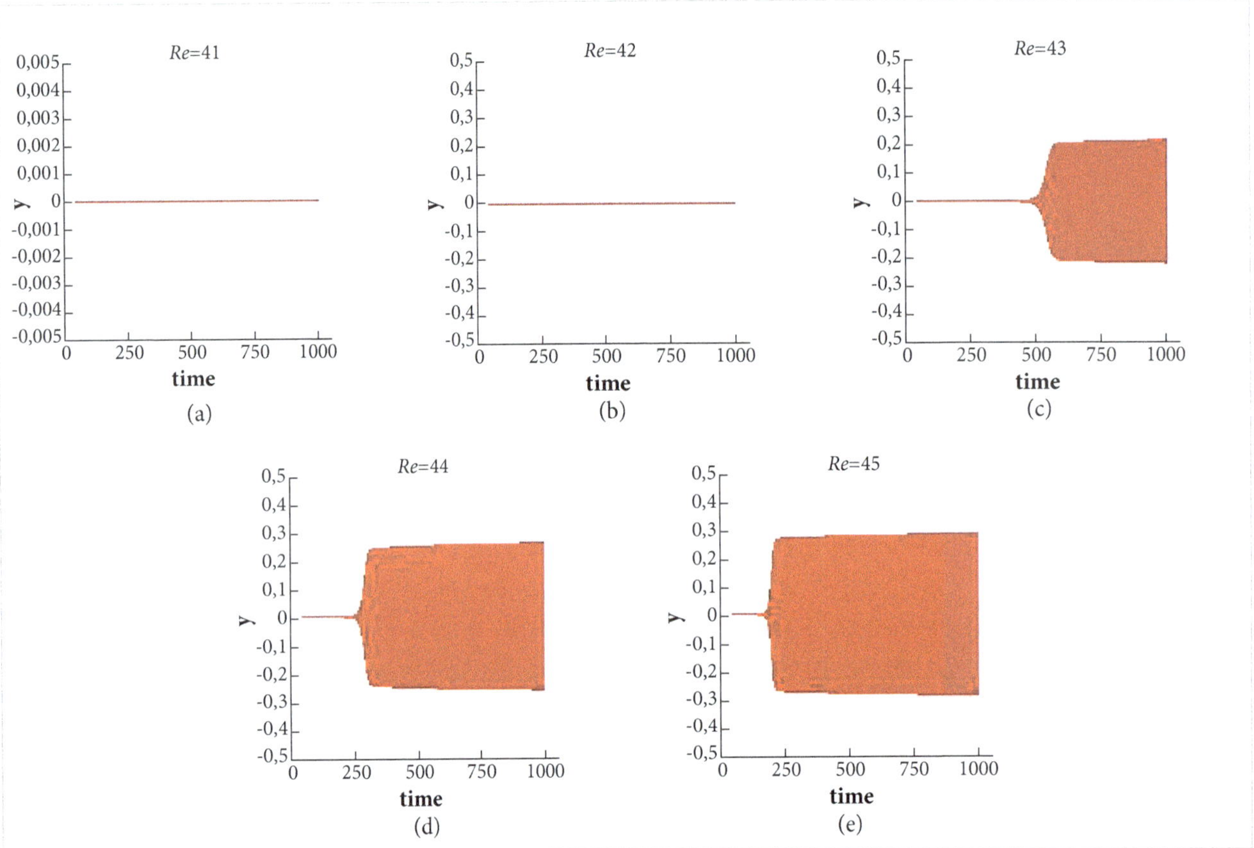

Figure 2. History of the crosswise velocity component, υ, for the circular cylinder and several values of *Re*. Grid resolution=12/*d*. Processing in double precision

in mind that we are using here the coarser grid, the one that rises the largest amount of error. The reader can perceive that shedding was damped for $Re = 41$ and $Re = 42$, and very much delayed for the other cases. The reason is evidently the fact that the truncation error has decreased when the compiling precision was increased. As a final test following this line, we introduced a white noise with amplitude of 0.0001, with the basic aim of speeding the onset of shedding. The results are shown on Fig. 3, and when compared with Fig. 2, they confirm the expected speeding of the process. When submitted to this level of excitation, the $Re = 42$ case did not "resist" and the shedding was initiated, only after about 7,500 units of time. Besides, and again from Fig. 3, one can appreciate the fact that the introduction of this level of noise corresponds basically to running the cases in single precision — compare it to Fig. 4. In fact, and the reader should keep this in mind, what is happening here is that for this grid resolution and double precision compiling, the beginning of vortexes emission was delayed, but not eliminated. The introduction of the white noise simply "pulled" the shedding process back to an earlier time, what evidently spares a lot of computing time. On the other hand this is a clear example of the difficulties associated to the study of transitional physical problems. The last word about a certain case, i.e., if the flow has finally reached the final global response — Kármán vortex shedding, Monkewitz (1988) —, or not, depends upon many factors, and much care must be exercised.

Attention was also paid to the other side of Table 2. Flows with Reynolds numbers 42, 43, and 44 were simulated again, but now subjected to higher levels of excitation. This was accomplished by raising the white noise amplitude. Columns "$36/d(0.01\%)$" and "$36/d(0.1\%)$" contain results for the 36-nodes-per-diameter grid and levels of perturbation of 0.01% and 0.1%, respectively. Again we observe that shedding was obtained for some Reynolds numbers that, otherwise, when excitation was smaller, corresponded to damped cases, at least during the length of time of numerical simulation — 1,000 units. Another clear evidence from Table 2 is the very

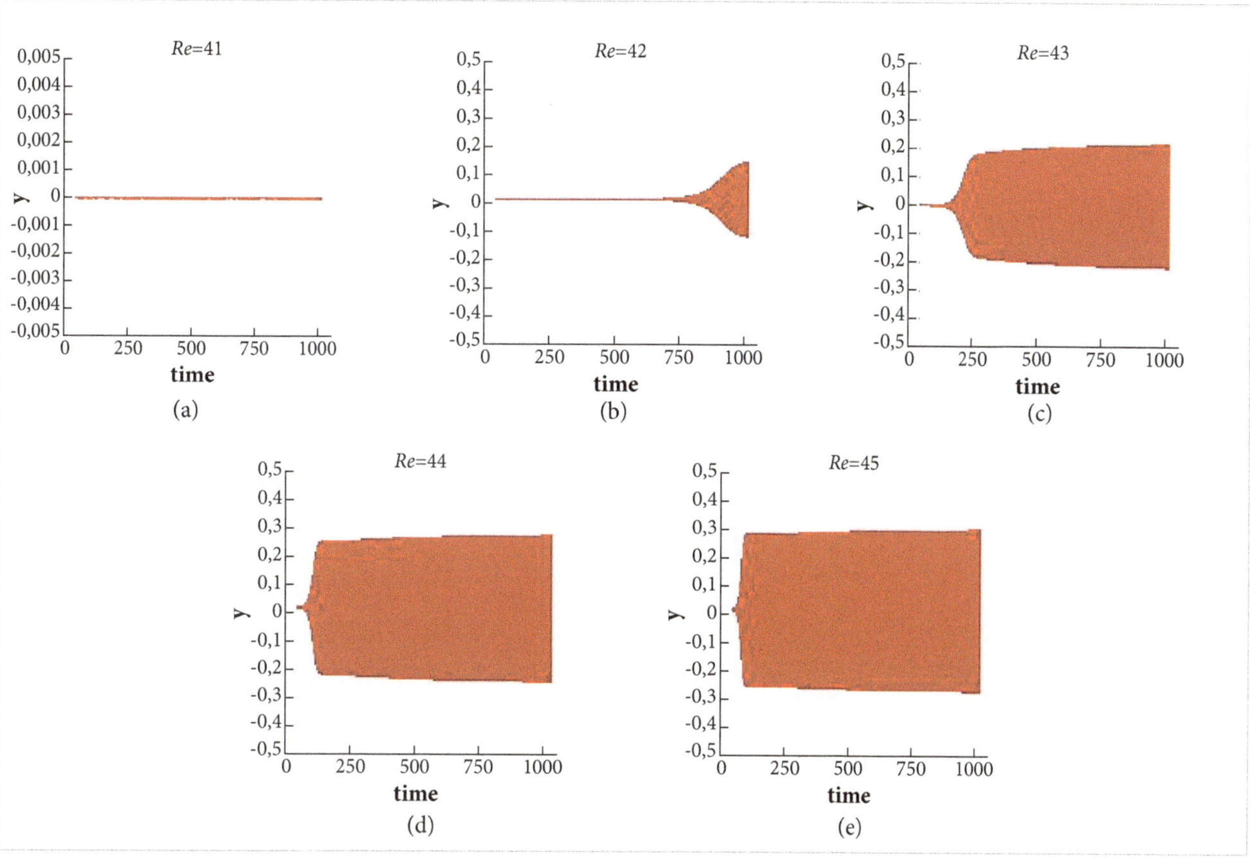

Figure 3. History of the crosswise velocity component, v, for the circular cylinder and several values of *Re*. Grid resolution=12/*d*. Processing in double precision and a 0.0001 amplitude white noise.

important fact that studies of onset of shedding, as well as every investigation on transitional phenomena, being it experimental or numerical, must bring clearly the information relative to the levels of turbulence/perturbation under which the experience was carried out. Otherwise, an isolated figure, in this specific case the value of the Reynolds number, carries the risk of getting almost meaningless. Maybe the only thing that can be assured is that the provided number furnished falls inside a very wide interval generally well established in the literature.

If one compares all those results carefully, and takes into account the arguments previously presented, it is possible to state that the "scanning" of the Reynolds number interval by a rake of grids in single precision and a relatively "small" time stretch (about 1,000 units) is a guarantee that the interval is well located, at least, say, to "first order". The error of the code in single precision (that corresponds, for the present code, to a perturbation level of 0.01% — compare Figs. 3 and 4) is capable of triggering the shedding process according to the right trends. That is, if we use, for example, double precision, what happens is that the shedding, in most cases, will be delayed, but, most certainly, not eliminated. We believe that, under those circumstances, and restrained by the conditions of the numerical simulation — among which the most influential is the grid refinement —, it is possible to state, after Table 2, that the onset of the shedding Reynolds number is located in the interval $Re_K^d = 43.5 \pm 1.5$. If one decides to reach a second order level of accuracy in the determination of Re_K, then the computational cost will be amplified by many folds. For example, to run a 36-points-per-diameter case in double precision, and reach 10 ,000 units of time, one would need about 2,100,000 iterations, what, in the case of our Dell Station 690, took about 40 (forty) days of calculation. This is because the largest time step for guaranteeing convergence is equal to 0.00483786, due to the grid spacing. For reasons of our convenience, the investigation

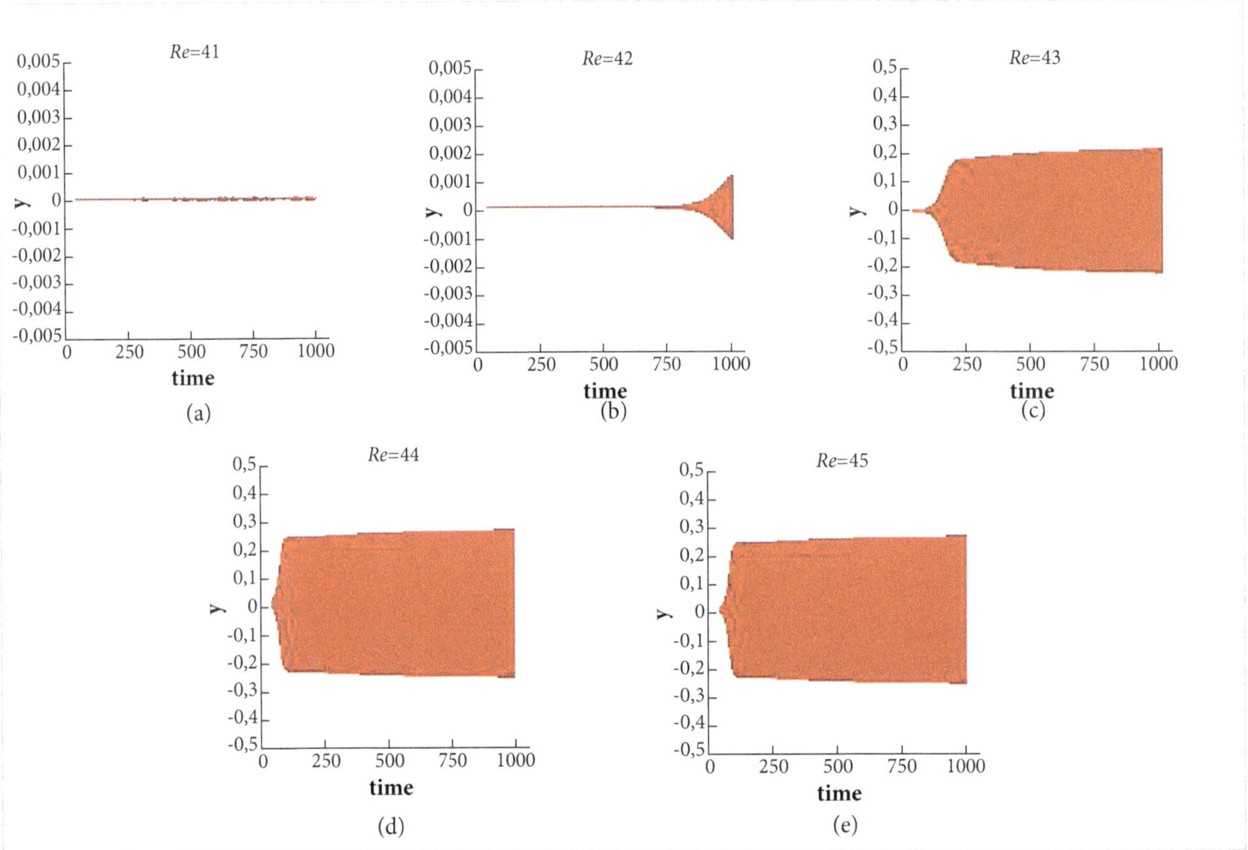

Figure 4. History of the crosswise velocity component, υ, for the circular cylinder and several values of *Re*. Grid resolution=12/d. Processing in single precision.

that is being reported here was performed using the serial version of code Incompact3d.

THE ELONGATED CYLINDER ONSET OF SHEDDING STUDY

The objective of this section is to apply to the elongated cylinder the technique that was established for the case of the circular cylinder. Before going into the details, it is important to describe how the different aspect ratios were investigated. The immediate idea would be to maintain the body base height fixed and to adapt the chord length to the desired aspect ratio. It is obvious that, in doing so, the domain of calculation total length, L_x, would increase according to the increase of the aspect ratio. Therefore, having in mind the goal of sparing grid nodes, we decided differently: to keep the chord fixed and to diminish the base height. In both cases, however, either keeping the chord length or the base height fixed, one has to care about the similarity of both the overall size of the domain of calculation and the grid resolution. This is essential in order to obtain results that can be compared on the same basis. Observe Fig. 1. The following rules were then established: (i) The ratio $L_y=h$ was kept constant; this would assure that any blockage effect, whatever its extension, would, in principle, be the same for every configuration and for every code run. (ii) The ratio $q=h$ was also kept constant; therefore, the influence of the exit plane would be felt in the same terms in every running of the code. By keeping constant the ratios $L_y=h$ and $q=h$, the same similar evolution domain was offered to the wake, irrespective of the actual investigation. (iii) The ratio of grid nodes per base height, n^h_y/h, where n^h_y is the number of crosswise grid points, allocated at the body base height, was kept the same from case to case. This figure ultimately defines n_y the total number of points along L_y.

The rake of grids for this study was composed of three choices whose number of points were: $[n^h_y/h]$ = 12, 20 and 30. It is more difficult to select the grids and the resolution in the case of the elongated body when compared to the circular cylinder. The difficulties come from the geometric constraints, to guarantee similarity, plus the fact that the total number of grid nodes for each direction has to be a product of the form $[2^a 3^b 5^c]$, in order to meet the "needs" of the fast Fourier subroutines. Table 3 shows the main grid characteristics of some of the cases.

The result of the investigation is plotted in Fig. 5. It is important to draw attention to the fact that each point in this figure is the result of a rather long process. Many Reynolds number values had to be tried, and for each of them, the extension of the time of calculation was about 1,000 units of non-dimensional time, which demanded a large number of iterations (generally, on the order of six digits). Only after that, one can be sure that the onset of shedding for that particular case was attained. A minimum number of at least 100 steady cycles of emission was made to run, before guaranteeing that the case corresponded to a real "onset of shedding" case. The calculations were performed in a 64-bits machine (Dell Precision 690), and the Fortran compiler was the Intel ifort version running in single precision. In Fig. 5, the vertical

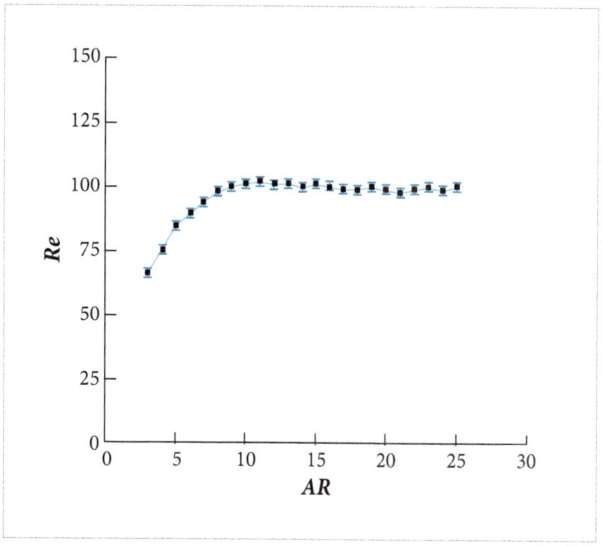

Figure 5. The dependence of Re^h_K on the elongated cylinder aspect ratio, *AR*.

Table 3. Data relative to the principal domains and grids that were used in the study of the elongated body onset of shedding. Observe that lengths are given in base heigths.

AR	n^h_y/h	$L_x=h$	$L_y=h$	n_x	n_y	x_c/h	q/h
	12	30.0	15.0	451	181	13.5	15.0
3	20	30.0	15.0	601	301	13.5	15.0
	30	30.0	15.0	751	451	13.5	15.0
	12	45.0	15.0	481	181	26.5	15.0
7	20	45.0	15.0	751	301	26.5	15.0
	30	45.0	15.0	901	451	26.5	15.0
	12	45.0	15.0	541	181	24.0	15.0
12	20	45.0	15.0	601	301	24.0	15.0
	30	45.0	15.0	901	451	24.0	15.0
	12	60.0	15.0	541	181	36.0	15.0
18	20	60.0	15.0	601	301	36.0	15.0
	30	60.0	15.0	901	451	36.0	15.0
	12	75.0	15.0	601	181	50.0	15.0
25	20	75.0	15.0	601	301	50.0	15.0
	30	75.0	15.0	901	451	50.0	15.0

bar for each point is an indication of the spreading extension of the Reynolds number as function of the grid resolution. For example, for AR = 14, the result is Re^h_K = 100 ± 2.0. Therefore, taking into account the envelope of restrictions of the numerical prediction — that we have tried to point out during the discussion so far —, it is guaranteed that the true value of Re^h_K falls inside the spreading interval defined by the vertical lines. On the other hand there is a lot of other factors that might influence this spreading range, as we have discussed previously in the case of the cylinder. However, the grid resolution is indeed the main factor. We do believe that these data (especially, Fig. 5) are important, and have not as yet been published in the literature.

Two important results stem immediately as one observes Fig. 5. Firstly, the Reynolds number grows almost linearly for small values of the aspect ratio, and after an AR equal to 9 it behaves in a plateau-like fashion, leveling at about 100. Ryan *et al.* (2005), in the context of a three-dimensional investigation, have already called attention to the increase in the onset-of-shedding Reynolds number with aspect ratio. Secondly, because the functional relationship of Re and AR is practically linear in the range 3 to 6, an extrapolation of this line to values of AR less than 3 would meet a vertical line through AR = 1 in a range of Reynolds numbers between 45 to 50. It is exactly in this region that the circular cylinder critical Reynolds number is located. It is important to remind that the circular cylinder has an aspect ratio equal to 1.

In the present two-dimensional study we learn that, for small values of AR, the Reynolds number really increases, but afterwards it levels off to an almost constant value (at least, up to AR = 25). The main reason for this behavior is suggested by Fig. 6. Instant vorticity fields are shown where values are made dimensionless regarding U_∞ and h. Following the order (a), (b), (c), and (d), for $[n^h_y/h]$ = 12, one may appreciate, respectively, snapshots of cases AR = 3, Re^h_K = 66.5, t = 1930.31 (units of non-dimensional time); AR = 6, Re^h_K = 89.5, t = 758.33; AR = 9, Re^h_K = 100.0, t = 1004.09 and AR = 12, Re^h_K = 101.0, t = 1011.41. The flow about the body and especially the boundary layers asymptote to a common pattern, what Ryan *et al.* (2005) called a "near-universal" boundary layer. Putting it differently, for low values of AR, the boundary layers are still carrying the history of their initial formation at the ogival front part. As the aspect ratio increases, the flow, especially that part near the tip of the body base, just before separation, gradually looses the influence of the ogive and behaves more like a flow along a flat plate. This is apparent for cases (c) and (d), AR = 9 and 12, where the physical scenario of the streams along the walls of the bodies is exactly alike.

THE INFLUENCE OF THE BOUNDARY LAYERS

In this section, we will investigate the possible influences of the boundary layers "running" above and below the body upon the physical scenario at the body base and near wake. The main interest at this point is to investigate some specific influences of the boundary layers upon the onset of shedding condition. One should pay special attention to this point. The present study is made possible due to the special geometry of the elongated cylinder. The presence of the two flat surfaces that constitute the body, plus the fact that the separation points are fixed, are instrumental. It is not possible to repeat the same numerical experiments with the circular cylinder, because of the everlasting excursions of the separation points during the shedding state.

The first aspect to stress is that the boundary layers at the tips of the body base, just before separation and at the onset of shedding state, are always laminar. The reader can find, in Fig. 7, data about the layers thicknesses and shape factor as functions of the aspect ratios. For each point in the figure, the Reynolds number corresponds to Re^h_K, the onset of shedding value. By the value of H, the shape factor, one can see that the boundary layers are laminar. One should remind that, for the laminar boundary layer along a flat plate and without pressure gradient H is equal to 2.59 (Schlichting, 1979). At this point of this research project, we are focused upon the low Reynolds number range, for which the boundary layers and, most of the time also the shear layers, are laminar. One can find in the literature some studies related to the influence of the turbulent boundary layers upon the characteristics of the wake — Sieverding and Heinemann (1990); Rowe *et al.* (2001) —, but in general the Reynolds numbers are much larger (the works are mostly experimental) and the emphasis lies on the analysis of "macro" parameters, for example, the influence of the boundary layer shape factor upon the shedding frequency, and studies alike. In Fig. 8, for AR = 7 and Re^h = 95, we have plotted the boundary layers profiles at the upper and lower tips of the body base, for eleven instants of time of a complete cycle of emission. Precisely in the case of Fig. 8, the cycle starts at t = 288.88 and ends at t = 290.12. Plotted in the figure, there are eleven profiles for the upper tip and

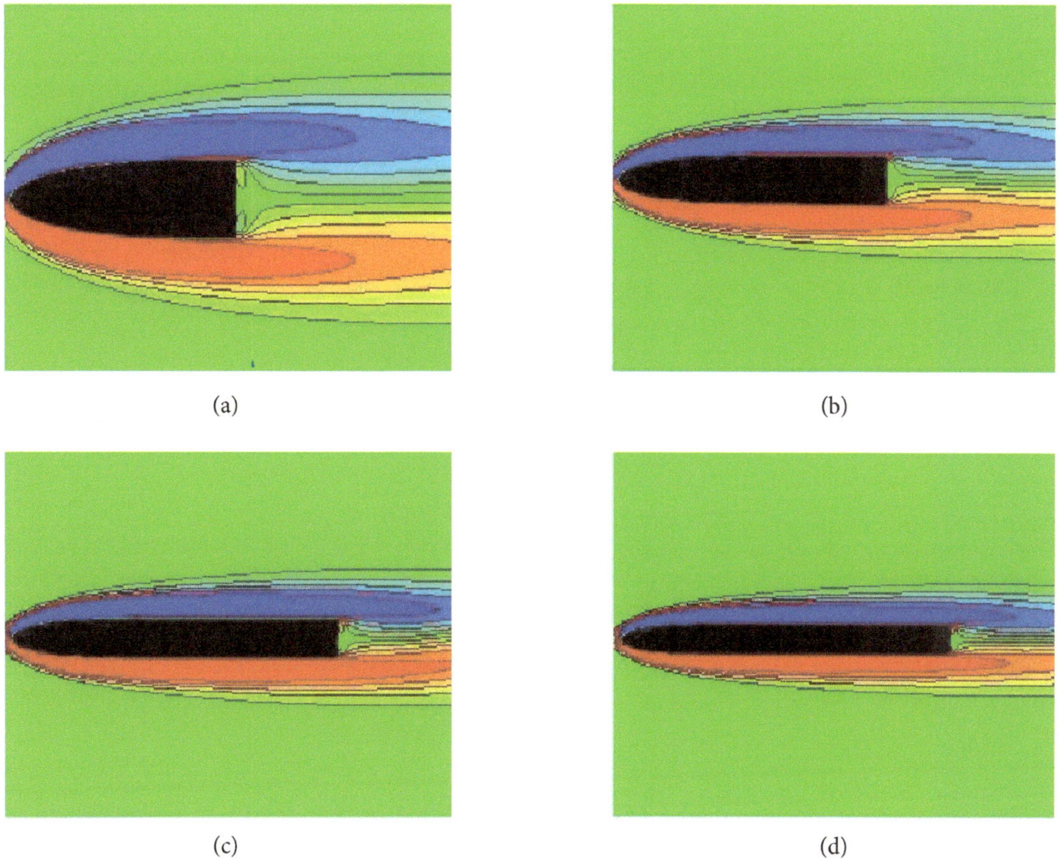

Figure 6. Instant vorticity fields at the onset of shedding *Re*. (a) $AR = 3$, $Re^h_K = 66.5$, $t = 1930.31$, $-1.33 \le \omega_z \le 1.33$, $\Delta\omega_z = 0.27$; (b) $AR = 6$, $Re^h_K = 89.5$, $t = 758.33$, $-1.17 \le \omega_z \le 1.17$, $\Delta\omega_z = 0.23$; (c) $AR = 9$, $Re^h_K = 100.0$, $t = 1004.09$, $-1.0 \le \omega_z \le 1.0$, $\Delta\omega_z = 0.2$; (d) $AR = 12$, $Re^h_K = 101.0$, $t = 1011.41$, $-1.0 \le \omega_z \le 1.0$, $\Delta\omega_z = 0.2$.

eleven for the lower one (the cycle was divided in ten equal time intervals). The case is that, the eleven profiles are basically coincident in the scale of the figure, and it seems that there is only one plotted distribution of the horizontal velocity component. What one can learn from this figure is that, in a situation where there is already vortex shedding — let us remind that for $AR = 7$ the onset of shedding Reynolds number is 94.0 —, there is virtually no variation in the boundary layers profiles during a whole cycle of emission. If we invert the argument, one may verify that, if the profiles do not change with time during a whole cycle, most probably they do not have any influence upon the mechanism of vortex liberation in the formation region. This is another indication that the global instability with the consequent vortexes evolution and shedding is a specific characteristic of the wake itself, considered as a dynamical system.

Because this is a point of paramount importance, we have focused onto it and introduced new numerical experiments. Using the geometries corresponding to $AR = 6$, 14 and 22, the following extra cases were investigated. The body shapes were slightly changed by introducing two small bumps, one at the upper wall, and another at the lower one. The bumps were made triangular, with heights equal to $1.0dy$, $0.5dy$ and $0.25dy$; the bases of the triangles were always equal to $2dx$. The position of both bumps were defined at half-base height, one base height and two base heights, upstream of the upper and lower tips of the trailing edge. The $12/h$ grid and single precision were used again in those experiences. In this part of the investigation, for every situation, we mostly used the

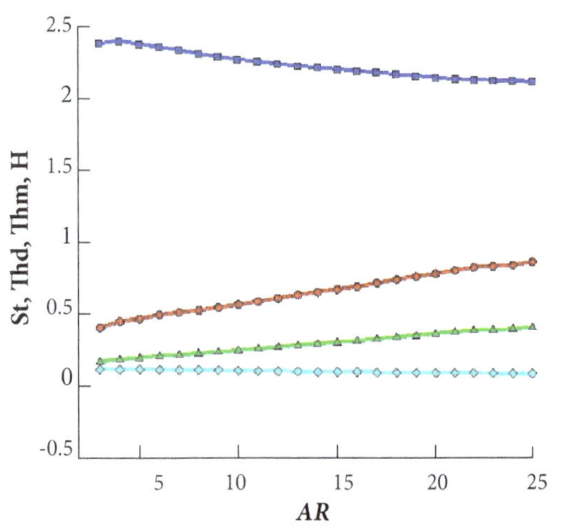

Figure 7. Strouhal number, tip boundary layers thicknesses and shape factor at the onset of shedding *Re* as function of the aspect ratio. The reference length is always the base height, *h*. *St*: Strouhal number (diamonds); T_{hd}: displacement thickness (circles); T_{hm}: moment thickness (deltas); *H*: shape factor (squares).

value of $Re^h = (Re_K^{h,12} - 1)$, that is, the onset of shedding Reynolds number for that configuration and the grid resolution equal to $12/h$, minus one unity. The main final results are presented in Table 4 and Fig. 10. In Table 4, the crossing of a row and a column corresponds to one case studied; for example, "14" and "$h/0.5dy$" indicate an AR equal to 14, and both bumps, the upper and the lower ones, positioned one base height upstream of the base tips, being the height of the bumps equal to $0.5dy$.

Details of the top tip boundary layers are depicted in Fig. 10. In pictures (a), (b), and (c), the reader can observe four lines marked with symbols: squares, circles, deltas and right triangles, corresponding, respectively, to the plain body, bumps at $0.5h/0.5dy$, bumps at $1h/1dy$, and bumps at $2h/1dy$. There are some important aspects about those profiles. The bumps introduce a distortion in the boundary layer profile in the form of a retracting defect, situated approximately between $0.1h$ and $0.6h$ from the wall. Figure 10 depicts details of those regions for a better understanding of the point. When plotted together in a normal scale, it is practically impossible to detect those differences in the value of the horizontal velocity component. Even in the extended scale of Fig. 10, it is not possible to differentiate between the plain body and the $(0.5h/0.5dy)$-bumps cases. The square and circle symbols, and their corresponding lines, are coincident. For the other instances, i.e., bumps at $1h/1dy$ and $2h/1dy$, the defects in the values of the velocity in relation to the plain body distribution, and for a distance of the wall of $y/h = 0.25$, are 0.035/0.02 ($AR = 6$), 0.02/0.01 ($AR = 14$) and 0.015/0.005 ($AR = 22$), respectively. Hence, the influence of the bumps upon the tips' boundary layers diminishes as the aspect ratio grows. By "tip" we mean the point that belongs at the same time to the flat plate and to the base of the body. By observing Fig. 9, one can grasp why this is so. As the aspect ratio grows the thickness of the boundary layers along the walls of the body also grows. Therefore, for larger values of AR, the perturbation due to the bumps has to evolve in a region of the boundary layer relatively closer to the wall, and where shear stresses are larger. This is an indication that the perturbation signal will be more dissipated until reaching the base tip. And the dissipation will be larger for larger values of the aspect ratios. This is probably the reason why there was no shedding for $AR = 22$ (Table 4).

The fact that stems from Fig. 10 and Table 4 is that we only had shedding for a value of the Reynolds number equal to $(Re_K^{h,12} - 1)$ when the perturbation amplitude grew beyond a certain level. Based on these results, one might appreciate a very important aspect of a bluff body wake. Given that the level of disturbances is sufficiently low (in the "linear" range, say), the "command" of the onset of shedding state is dictated by the wake itself. Only after sufficiently raising the level of the amplitude of the perturbation (the "non-linear" range, say), is that it is possible to disturb, from upstream, the overall state of the twin bubbles at the body's base region. This agrees with many previous studies, e.g., to name a few, Monkewitz (1988), and Yang and Zebib (1989), for the case of the circular cylinder, and Jackson (1987) for an assortment of geometrical forms. See also Mathis *et al.* (1984) and Provansal *et al.* (1987), who report important experimental studies about the Karman instability for the circular cylinder. The reader should appreciate how important this result is. In the case of the elongated cylinder, because the points of separation are fixed, one may conclude, apparently with no possible doubt, that the wake is unstable by itself; for the circular cylinder, even if the boundary layer profiles did not vary during a whole cycle, the oscillation of the separation points — which would prevent a definitive affirmation — is

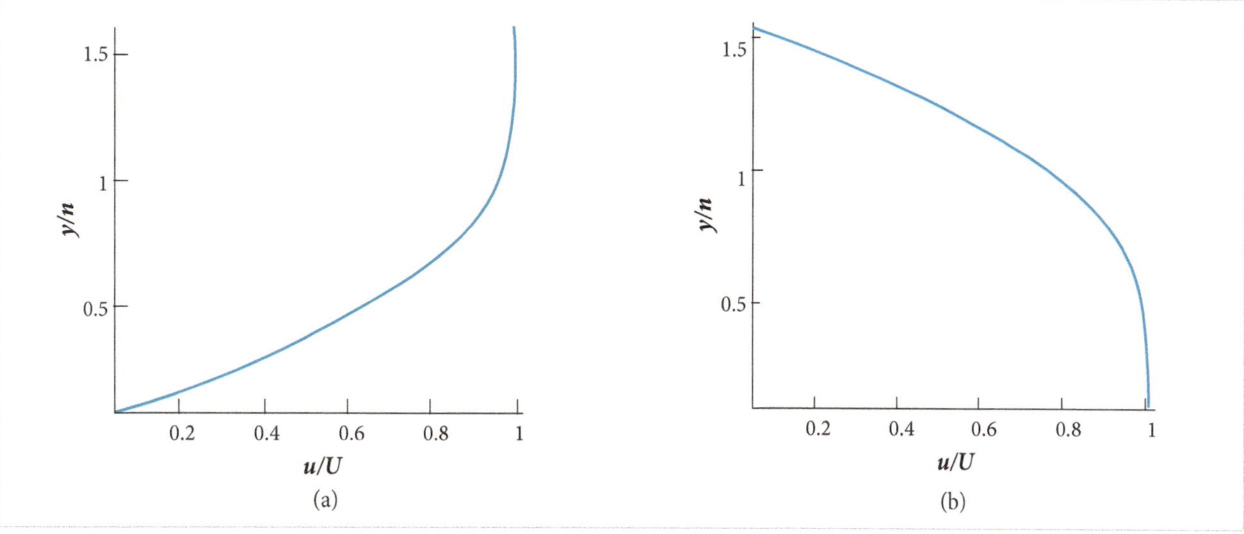

$Re^h = 95$; $AR = 7$.

Figure 8. Comparison of boundary layer profiles along a shedding cycle: (a) upper tip; (b) lower tip.

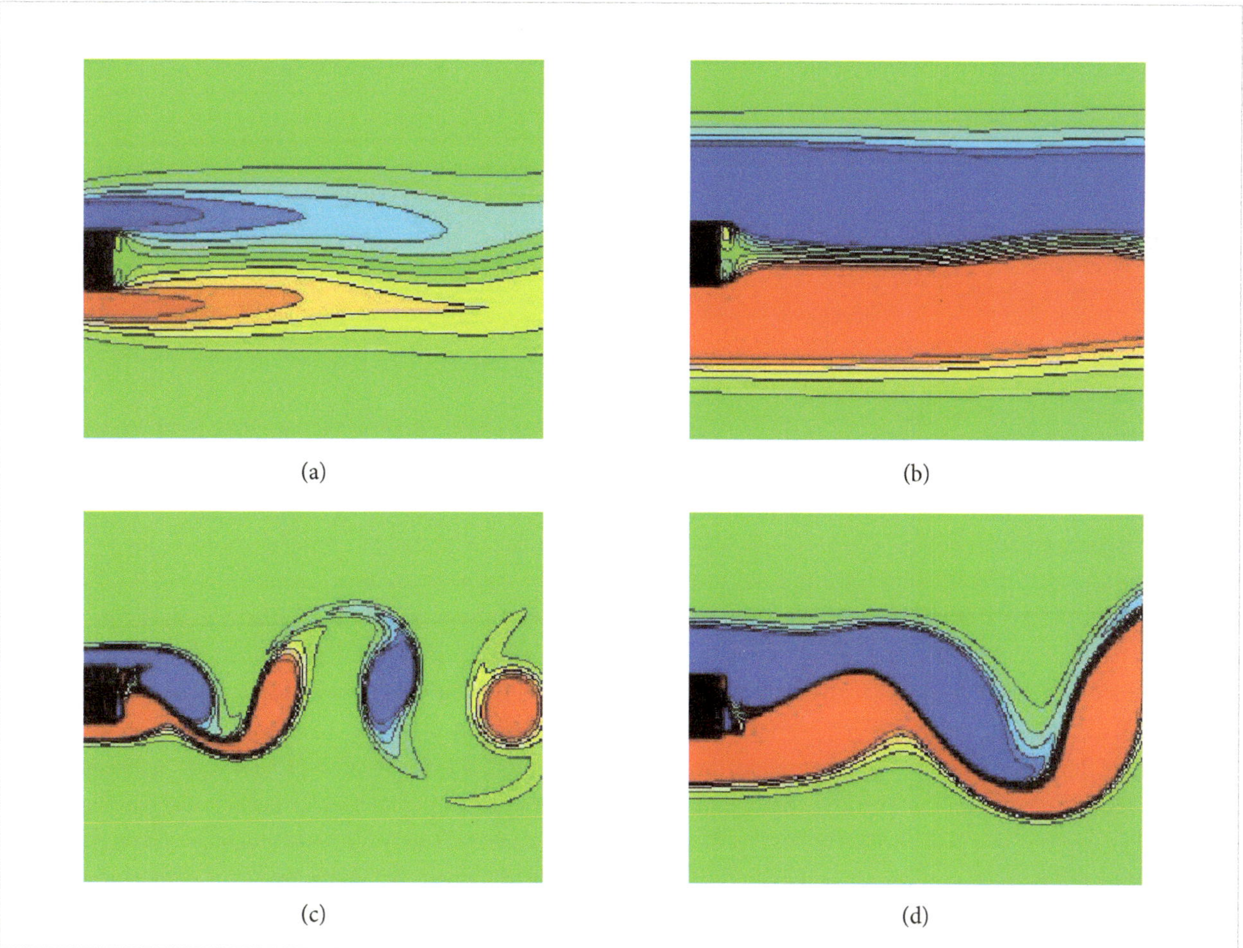

Figure 9. Instant vorticity fields at different aspect ratios: (a) $AR = 3$, $Re_K^h = 66.5$, $t = 1930.31$; (b) $AR = 3$, $Re^h = 300$, $t = 112.28$; (c) $AR = 12$, $Re_K^h = 101.0$, $t = 1008.39$; (d) $AR = 12$, $Re^h = 300$, $t = 35.81$.

Table 4. Investigation of upstream influences upon the flow about some representative aspect-ratio geometries. The Reynolds number is one unity less than the onset of shedding Reynolds number for each geometry (and a grid resolution equal to $12/h$).

AR	Re^h	$_{Re}h,12$	0.5h/0.5*dy*	*h*/*dy*	*h*/0.5*dy*	2*h*/*dy*
6	87	88	ns	s	ns	ns
14	97	98	ns	s	ns	ns
22	96	97	ns	ns	ns	ns

s: shedding state; ns: no shedding.

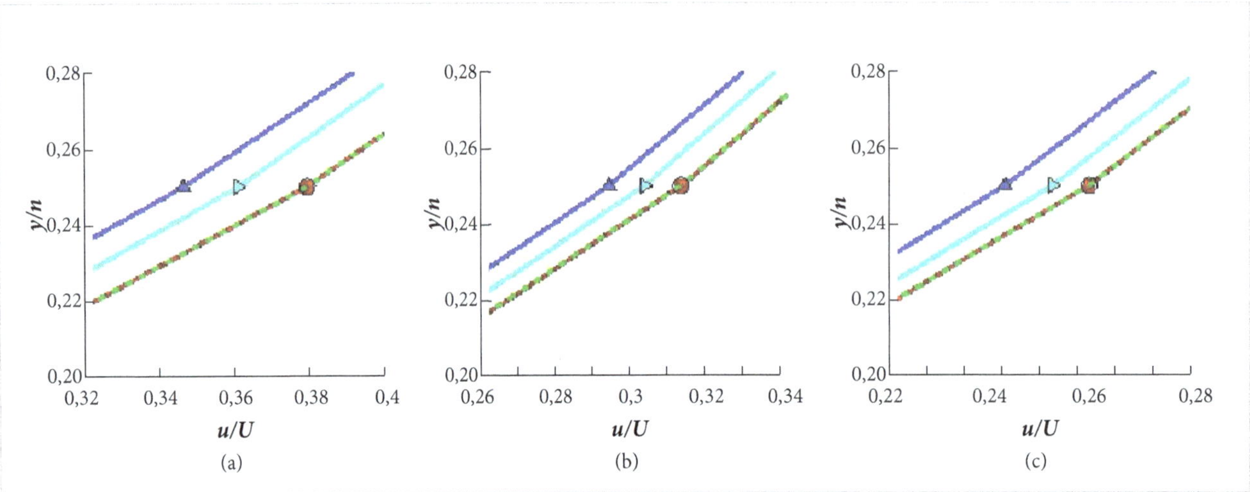

Figure 10. Details of tip boundary layer profiles. (a) AR = 6 and Re^h = 87; (b) AR = 14 and Re^h = 97; (c) AR = 22 and Re^h = 96. The symbols indicate: Squares: no bumps; Circles: bumps at 0.5h, height 0.5h; Deltas: bumps at 1h, height 1.0 d_y; Right triangles: bumps at 2h, height 1.0 d_y.

still to be considered. The reader should also be aware that we are not trying to state that for the case of the circular cylinder the wake is not unstable by itself, but that it may depend on the oscillation of the separation points. There are sufficient arguments in the literature to sustain the point that, for the circular cylinder, the wake is also self unstable. What we are trying to convey is the fact that the case of the elongated cylinder, for which the separation points are fixed, helps to better clarify and fix the point.

For the range of Reynolds number we have focused on, the boundary layers work as "conveyors" of vorticity, feeding the bubbles at the base region. This may be observed in Fig. 11. The instant bubbles are shown for the AR = 14 configuration and for several Reynolds numbers. For each frame we call attention to the value of the vorticity at certain specific points. The points at the separation stations correspond to the maxima of the vorticity at the tip boundary layer transversal section. The points located inside the bubble correspond to the approximate bubble centre, and the values on top of the upper bubble are located one base height from the vertical wall. As the Reynolds number grows towards the onset of shedding state the overall level of vorticity grows at the base region. It is evident that, with the raising of the vorticity, the length of the bubbles grows and the state of instability saturation is finally reached.

The numerical experiments represented in Fig. 5 and the data collected seems to indicate that the boundary layers feed vorticity, but the transition of the wake to another state of flow — the crossing by the point of bifurcation —, is determined by the dynamics of the bubble itself. On the other hand it is evident that much more work is needed, especially stability studies, in order to better support these conclusions.

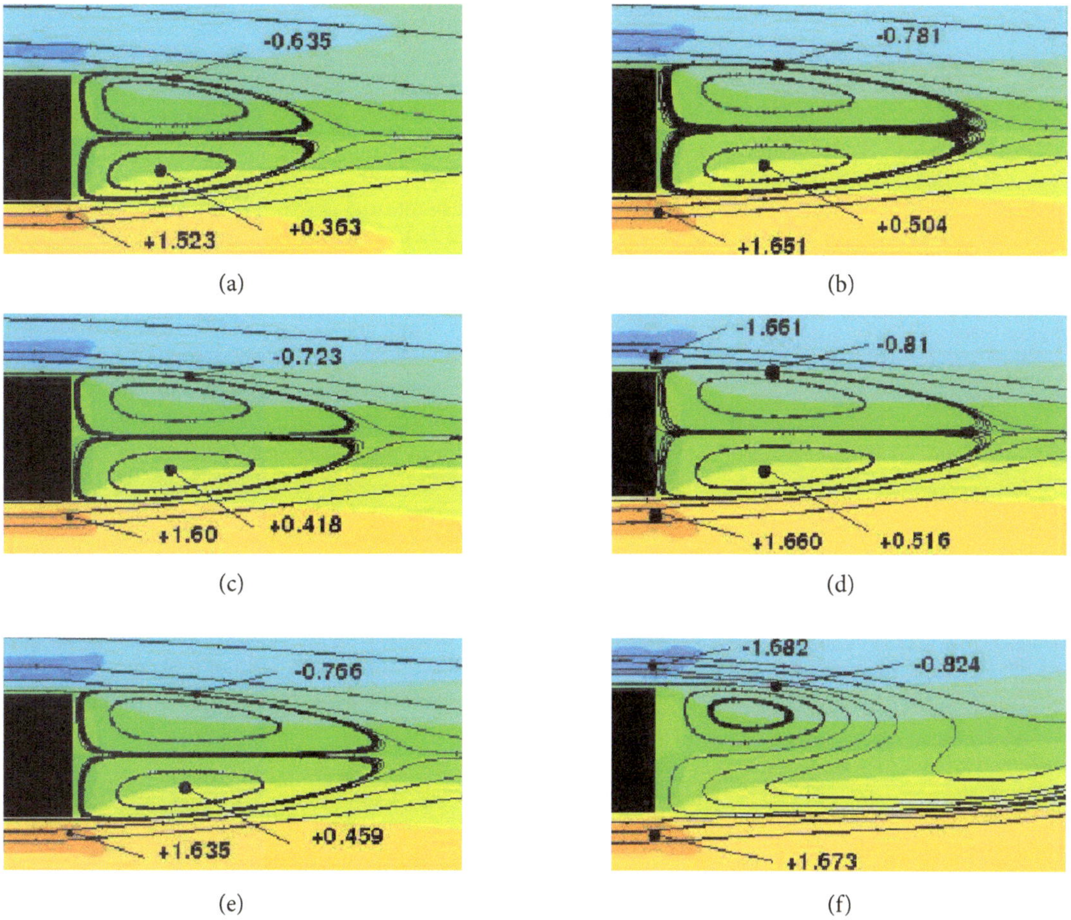

Figure 11. Evolution of the vorticity at the base region of the AR = 14 body, with varying Re. The numbers in each plot correspond to the dimensionless value of the vorticity. (a) Re^h = 93; (b) Re^h = 95; (c) Re^h = 96; (d) Re^h = 97; (e) Re^h = 98; (f) Re^h_K = 100.

CONCLUSIONS

The main goal of this research effort is to study the onset of shedding of the wake behind a blunt-trailing-edged body - an elongated cylinder with an elliptical front part. What motivated this study was the objective of obtaining, for this geometry, much of the same data (at least in quality) as there are for the circular cylinder. In this latter case, the reasons for the great interest of the engineering community are obvious, considering the myriad of applications of this geometrical form. On the other hand, the elongated body has a paramount advantage, in terms of easiness of analysis, because the points of separation are fixed. Therefore, with the elimination of the separation points excursions, some investigations and their consequent analyses are facilitated. We do believe that we have succeeded in our endeavor, and hope that the collection of results presented here will be of help for those individuals who are pursuing studies in this field. For example, the information contained in Fig. 5, and the studies carried out as a result of the boundary layers influence upon the stability of the great structures in the near wake of the body.

ACKNOWLEDGMENTS

The authors would like to acknowledge the support provided by the CNPq — *Conselho Nacional de Desenvolvimento Científico e Tecnológico* — (grant 303184/2007-8), and FAPESP — *Fundação de Amparo à Pesquisa do Estado de São Paulo* — (grant 2007/00305-5).

REFERENCES

Akselvoll, K. and Moin, P., 1996, "Large-eddy simulation of turbulent confined coannular jets", Journal of Fluid Mechanics, Vol. 315, pp. 387-411. doi:10.1017/S0022112096002479.

Barkley, D. and Henderson, R.D., 1996, "Three-Dimensional Floquet stability analysis of the wake of a circular cylinder" Journal of Fluid Mechanics, Vol. 322, pp. 215-241. doi:10.1017/S0022112096002777.

Bearman, P.W., 1965, "Investigation of the flow behind a two-dimensional model with a blunttrailing edge and fitted with splitter plates", Journal of Fluid Mechanics, Vol. 21, No 2, pp. 241-255. doi:10.1017/S0022112065000162.

Bearman, P.W., 1967, "The effect of base bleed on the flow behind a two-dimensional model with a blunt-trailing edge", Aero Quarterly, Vol. 18, pp. 207-214.

Bers, A., 1983, "Basic Plasma Physics I", In: Handbook of Plasma Physics (ed. M. N. Rosenbluth & R. Z. Sagdeev), Vol. 1, Chap. 3.2, North Holland.

Braza, M., Chassaing, P. and Ha Minh, H., 1986, "Numerical study and physical analysis of the pressure and velocity fields in the near wake of a circular cylinder", Journal of Fluid Mechanics, Vol. 165, pp. 79-130. doi: 10.1017/S0022112086003014.

Chomaz, J.M., Huerre, P. and Redekopp, L.G., 1988, "Bifurcations to local and global modes in spatially developing flows", Phys Rev Lett, Vol. 60, No 1, pp. 25-28. doi:10.1103/PhysRevLett.60.25.

Dauchy, C., Dussëk, J. and Fraunié, P., 1997, "Primary and secondary instabilities in the wake of a cylinder with free ends", Journal of Fluid Mechanics, Vol. 332, pp. 295-339.

Drazin, P.G. and Reid, W.H., 2004, "Hydrodynamic stability", 2nd Edition, Cambridge University Press, Cambridge, UK.

Dusëk, J., Fraunié, P. and Le Gal, P., 1994, "Local analysis of the onset of instability in shear flows", Phys Fluids, Vol. 6, No 1, pp. 172-186.doi: 10.1063/1.868080.

Fadlun, E.A., Verzicco, R., Orlandi, P. and Mohd-Yusof, J., 2000, "Combined immersed-boundary finite-difference methods for three-dimensional complex flow simulations", Journal of Computational Physics, Vol. 161, No 1, pp. 35-60. doi: 10.1006/jcph.2000.6484.

Goldstein, G., Handler, R. and Sirovich, L., 1993, "Modeling a no-slip flow boundary with an external force field", Journal of Computational Physics, Vol. 105, pp. 354-366. doi: 10.1006/jcph.1993.1081.

Huerre, P. and Monkewitz, P.A., 1985 "Absolute and convective instabilities in free shear layers" Journal of Fluid Mechanics, Vol. 159, pp. 151-168. doi:10.1017/S0022112085003147.

Jackson, C.P., 1987, "A finite-element study of the onset of vortex shedding in flow past variously shaped bodies", Journal of Fluid Mechanics, Vol. 182, pp. 23-45. doi: 10.1017/S0022112087002234.

Kravchenko, A.G. and Moin, P., 1997, "On the effect of numerical errors in large eddy simulations of turbulent flows", Journal of Computational Physics, Vol. 131, No 2, pp. 310-322. doi: 10.1006/jcph.1996.5597.

Laizet, S., Lamballais, E. and Vassilicos, J.C., 2009, "A numerical strategy to combine high-order schemes, complex geometry and parallel computing for high resolution DNS of fractal generated turbulence", Journal of Computer & Fluids. doi:10.1016/j.compfluid.2009.09.018.

Lardeau, S., Lamballais, E. and Bonnet, J.P., 2002, "Direct Numerical Simulation of a Jet Controlled by Fluid Injection", Journal of Turbulence, Vol. 3, No 1, pp. 002-011. doi: 10.1088/1468-5248/3/1/002.

Lamballais, E., 1996, « Simulation numériques de la turbulence dans un canal plan tournant », PhD Dissertation, Laboratoire Descoulements Gophysiques Et Industrielles, Institut National Polytechnique de Grenoble.

Lamballais, E. and Silvestrini, J.H., 2002, "Direct numerical simulation of interactions between a mixing layer and a wake around a cylinder", Journal of Turbulence, Vol. 3, pp. 028-036. doi:10.1088/1468-5248/3/1/028.

Lamballais, E., Silvestrini, J. and Laizet, S., 2008, "Direct Numerical Simulation of a Separation Bubble on a Rounded Finite-Width Leading Edge", International Journal of Heat and Fluid Flow, Vol. 29, No 3, pp. 612-625. doi:10.1016/j.ijheatfluidflow.2008.03.006.

Lele, S., 1992, "Compact finite difference schemes with spectral-like resolution", Journal of Computational Physics, Vol. 103, No 1, pp. 16-42. doi:10.1016/0021-9991(92)90324-R.

Mathis, C., Provansal, M. and Boyer, L., 1984, "The Bénard-von Kármán instability: an experimental study near the threshold", Journal de Physique – Lettres, Vol. 45, No 10, pp. 483-491. doi: 10.1051/jphyslet:019840045010048300.

Monkewitz, P., 1988, "The absolute and convective nature of instability in two-dimensional wakes at low Reynolds numbers" Physics Fluids, Vol. 31, No 5, pp. 999. doi: 10.1063/1.866720.

Ortega, M.A., Girardi, R.M. and Silvestrini, J.H., 2012, "A numerical study of the wake behind a blunt-trailing-edged body. Part 2: the topology of the flow", International Journal for Numerical Methods in Fluids, Vol. 69, No 1, pp. 29-56. doi: 10.1002/fld.2533.

Park, H., Lee, D., Jeon, W.-P., Hahn, S., Kim, J., Kim, J., Choi, J. and Choi, H., 2006, "Drag reduction in flow over a two-dimensional bluff body with a blunt trailing edge using a new passive device" Journal of Fluid Mechanics, Vol. 563, pp. 389-414. doi:10.1017/S0022112006001364.

Provansal, M., Mathis, C. and Boyer, L., 1987, "Bénard-von Kármán instability: transient and forced regimes", Journal of Fluid Mechanics, Vol. 182, pp. 1-22. doi: 10.1017/S0022112087002222.

Ribeiro, P.A.R., 2002, "The control of the vortexes shedding mechanism in the wake of circular cylinders by means of numerical simulation", Master in Science Thesis, Federal University of Rio Grande do Sul, Porto Alegre, RS, Brazil (in Portuguese).

Rowe, A., Fry, A.L.A. and Montallebi, F., 2001, "Influence of boundary-Layer thickness on base pressure and vortex shedding frequency", AIAA Journal, Vol. 39, No 4, pp. 754-756. doi: 10.2514/2.1377.

Ryan, K., Thompson, M.C. and Hourigan, K., 2005, "Three-dimensional transition in the wake of bluff elongated cylinders", Journal of Fluid Mechanics, Vol. 538, pp. 1-29. doi:10.1017/S0022112005005082.

Saiki, E.M. and Biringen, S., 1996, "Numerical simulation of a cylinder in uniform flow: Application of a virtual boundary method", Journal of Computational Physics, Vol. 123, No 2, pp. 450-465. doi:10.1006/jcph.1996.0036.

Silvestrini, J.H. and Lamballais, E., 2004, "Direct numerical simulation of oblique vortex shedding from a cylinder in shear flow", International Journal of Heat and Fluid Flow, Vol. 25, No 3, pp. 461-470. doi: 10.1016/j.ijheatfluidflow.2004.02.013.

Sieverding, C.H. and Heinemann, H., 1990, "The influence of boundary layer state on vortex shedding from at plates and turbine cascades", Journal of Turbo-machinery, Vol. 112, No 2, pp. 181-187. doi:10.1115/1.2927631.

Schlichting, H., 1979, Boundary-Layer Theory. 7th ed. ,McGraw-Hill Series in Mechanical Engineering. New York.

Sumer, B.M. and Fredsøe, J., 1997, "Hydrodynamics around cylindrical structures", Advanced Series on Ocean Engineering, World Scientific Pub Co Inc, London, UK.

Triantafyllou, G.S., Triantafyllou, M.S. and Chryssostomidis, C., 1986, "On the formation of vortex streets behind stationary cylinders", Journal of Fluid Mechanics, Vol. 170, pp. 461-477. doi: 10.1017/S0022112086000976.

Vitola, M.A., 2006, "Onset of shedding of a circular cylinder in the presence of flat plate", PhD Dissertation, Federal University of Rio Grande do Sul, Porto Alegre, RS, Brazil (in Portuguese).

Yang, X. and Zebib, A., 1989, "Absolute and convective instability of a cylinder wake", Physics of Fluids A, Vol. 1, No 4, pp. 689. doi: 10.1063/1.857362.

Williamson, C.H.K., 1996, "Vortex dynamics in the cylinder wake", Annual Review of Fluid Mechanics, Vol. 28, pp. 477-539. doi: 10.1146/annurev.fl.28.010196.002401.

Permissions

All chapters in this book were first published in JATM, by Departamento de Ciência e Tecnologia Aeroespacial (DCTA); hereby published with permission under the Creative Commons Attribution License or equivalent. Every chapter published in this book has been scrutinized by our experts. Their significance has been extensively debated. The topics covered herein carry significant findings which will fuel the growth of the discipline. They may even be implemented as practical applications or may be referred to as a beginning point for another development.

The contributors of this book come from diverse backgrounds, making this book a truly international effort. This book will bring forth new frontiers with its revolutionizing research information and detailed analysis of the nascent developments around the world.

We would like to thank all the contributing authors for lending their expertise to make the book truly unique. They have played a crucial role in the development of this book. Without their invaluable contributions this book wouldn't have been possible. They have made vital efforts to compile up to date information on the varied aspects of this subject to make this book a valuable addition to the collection of many professionals and students.

This book was conceptualized with the vision of imparting up-to-date information and advanced data in this field. To ensure the same, a matchless editorial board was set up. Every individual on the board went through rigorous rounds of assessment to prove their worth. After which they invested a large part of their time researching and compiling the most relevant data for our readers.

The editorial board has been involved in producing this book since its inception. They have spent rigorous hours researching and exploring the diverse topics which have resulted in the successful publishing of this book. They have passed on their knowledge of decades through this book. To expedite this challenging task, the publisher supported the team at every step. A small team of assistant editors was also appointed to further simplify the editing procedure and attain best results for the readers.

Apart from the editorial board, the designing team has also invested a significant amount of their time in understanding the subject and creating the most relevant covers. They scrutinized every image to scout for the most suitable representation of the subject and create an appropriate cover for the book.

The publishing team has been an ardent support to the editorial, designing and production team. Their endless efforts to recruit the best for this project, has resulted in the accomplishment of this book. They are a veteran in the field of academics and their pool of knowledge is as vast as their experience in printing. Their expertise and guidance has proved useful at every step. Their uncompromising quality standards have made this book an exceptional effort. Their encouragement from time to time has been an inspiration for everyone.

The publisher and the editorial board hope that this book will prove to be a valuable piece of knowledge for researchers, students, practitioners and scholars across the globe.

List of Contributors

Luciana Bassi Marinho Pires
Instituto Nacional de Pesquisas Espaciais - São José dos Campos/SP - Brazil
Avenida dos Astronautas, 1.758 - Jardim da Granja | CEP 12.227-010 São José dos Campos/SP - Brazil

Igor Braga de Paula
Pontifícia Universidade Católica do Rio de Janeiro - Rio de Janeiro/RJ - Brazil

Gilberto Fisch
Instituto de Aeronáutica e Espaço - São José dos Campos/SP - Brazil

Ralf Gielow
Pontifícia Universidade Católica do Rio de Janeiro - Rio de Janeiro/RJ - Brazil

Roberto da Mota Girardi
Instituto Tecnológico de Aeronáutica - São José dos Campos/SP - Brazil

Erick Giovani Sperandio Nascimento
Universidade Federal do Espírito Santo - Vitoria/ES - Brazil

Davidson Martins Moreira
Instituto de Aeronáutica e Espaço - São José dos Campos/SP - Brazil
Centro Integrado de Manufatura e Tecnologia - Salvador/BA - Brazil
Praça Marechal Eduardo Gomes, 50, São José dos Campos/SP | CEP: 12.228-901 | Brazil

Gilberto Fisch
Instituto de Aeronáutica e Espaço - São José dos Campos/SP - Brazil

Taciana Toledo de Almeida Albuquerque
Universidade Federal do Espírito Santo - Vitoria/ES - Brazil

Mohamed Okasha
International Islamic University Malaysia - Kuala Lumpur - Malaysia
Old Dominion University - Norfolk/VA - United States
Assistant Professor - Department of Mechanical Engineering - Faculty of Engineering-International Islamic University Malaysia
Jalan Gombak | P.O. Box 10 50728 Kuala Lumpur - Malaysia, Ph.D. Alumni from Old Dominion University

Brett Newman
Old Dominion University - Norfolk/VA - United States

Odenir de Almeida
Universidade Federal de Uberlândia - Uberlândia/MG - Brazil
Avenida João Naves de Ávila, 2.121 - Santa Mônica | CEP 38.408- 100 Uberlândia/MG - Brazil |

João Roberto Barbosa
Instituto Tecnológico de Aeronáutica - São José dos Campos/SP - Brazil

Juan Battaner Moro
University of Southampton - Southampton/United Kingdom

Rodney Harold Self
University of Southampton - Southampton/United Kingdom
Rogério Pirk
Instituto de Aeronáutica e Espaço - São José dos Campos/SP - Brazil
Instituto Tecnológico de Aeronáutica - São José dos Campos/SP - Brazil
Praça Marechal Eduardo Gomes, 50 - Vila das Acácias | CEP: 12.228-904 São José dos Campos/SP - Brazil

Carlos d'Andrade Souto
Instituto de Aeronáutica e Espaço - São José dos Campos/SP - Brazil
Instituto Tecnológico de Aeronáutica - São José dos Campos/SP - Brazil

Jan Klement
Tesat-Spacecom GmbH & Co - Backnang - Germany
KG, Gerberstraße 49 | D-71522 Backnang - Germany

Willer Gomes dos Santos
Instituto Nacional de Pesquisas Espaciais - São José dos Campos/SP - Brazil
Instituto Nacional de Pesquisas Espaciais | Avenida dos Astronautas 1758, São José dos Campos/SP | CEP: 12227-010 - Brazil

Evandro Marconi Rocco
Instituto Nacional de Pesquisas Espaciais - São José dos Campos/SP - Brazil

Valdemir Carrara
Instituto Nacional de Pesquisas Espaciais - São José dos Campos/SP - Brazil

Mauricio Vicente Donadon
Instituto Tecnológico de Aeronáutica - São José dos Campos/SP - Brazil
Praça Eduardo Gomes, 50 - Vila das Acácias | CEP: 12228-901, São José dos Campos/SP - Brazil

Lorenzo Iannucci
Imperial College London - London - United Kingdom

Xu Yan
Zhejiang University - School of Aeronautics and Astronautics - Hangzhou - China

Zheng Yao
Zhejiang University - School of Aeronautics and Astronautics - Hangzhou - China

Guan Fuling
Zhejiang University - College of Civil Engineering and Architecture - Hangzhou - China

Huang He
Zhejiang University - College of Civil Engineering and Architecture - Hangzhou - China

Xu Xian
Zhejiang University - College of Civil Engineering and Architecture - Hangzhou - China

Elizabeth Diane de Jesus Reuter
Instituto Nacional de Pesquisas Espaciais - Centro de Previsão de Tempo e Estudos Climáticos - São José dos Campos/SP - Brazil
Secretaria da Aviação Civil - Empresa Brasileira de Infraestrutura Aeroportuária - Guarulhos/SP - Brazil

Gilberto Fisch
Departamento de Ciência e Tecnologia Aeroespacial - Instituto de Aeronáutica e Espaço - Divisão de Ciências Atmosféricas - São José dos Campos/SP - Brazil
Instituto de Aeronáutica e Espaço - Divisão de Ciências Atmosféricas | Praça Marechal Eduardo Gomes, 50 - Vila das Acácias | CEP: 12.228-901 - São José dos Campos/SP - Brazil

Cleber Souza Correa
Departamento de Controle do Espaço Aéreo - Instituto de Controle de Tráfego Aéreo - Rio de Janeiro/RJ - Brazil

Aguinaldo Bezerra Batista Júnior
Centro de Lançamento da Barreira do Inferno - Natal/RN - Brazil
Rodovia RN 063 - Km 11 PO Box 54 | CEP: 59.140-970 | Parnamirim/RN - Brazil

Paulo Sérgio da Motta Pires
Universidade Federal do Rio Grande do Norte - Natal/RN - Brazil

Lucie Rouleau
Katholieke Universiteit Leuven - Department of Mechanical Engineering - Heverlee - Belgium
Cnam Paris - Structural Mechanics and Coupled Systems Laboratory | 2 rue Conté, 75003 | Paris - France

Rogério Pirk
Katholieke Universiteit Leuven - Department of Mechanical Engineering - Heverlee - Belgium
Departamento de Ciência e Tecnologia Aeroespacial - Instituto de Aeronáutica e Espaço - São José dos Campos/SP - Brazil

Bert Pluymers
Katholieke Universiteit Leuven - Department of Mechanical Engineering - Heverlee - Belgium

Wim Desmet
Katholieke Universiteit Leuven - Department of Mechanical Engineering - Heverlee - Belgium

Guilherme da Silveira
Departamento de Ciência e Tecnologia Aeroespacial - Instituto de Aeronáutica e Espaço - Divisão de Sistemas Espaciais - São José dos Campos/SP - Brazil
Instituto de Aeronáutica e Espaço - Divisão de Sistemas Espaciais | Praça Marechal Eduardo Gomes, 50 - Vila das Acácias | CEP: 12.228-901 - São José dos Campos/SP - Brazil

Valdemir Carrara
Instituto Nacional de Pesquisas Espaciais - Coordenação Geral de Engenharia e Tecnologia Espacial - Divisão de Mecânica Espacial e Controle - São José dos Campos/SP - Brazil

Antônio Carlos Foltran
Universidade Federal do Paraná - Departamento de Engenharia Mecânica - Curitiba/PR - Brazil

Diego Fernando Moro
Universidade Federal do Paraná - Departamento de Engenharia Mecânica - Curitiba/PR - Brazil

Nicholas Dicati Pereira da Silva
Universidade Federal do Paraná - Departamento de Engenharia Mecânica - Curitiba/PR - Brazil

Ana Eliza Gonçalves Ferreira
Universidade Federal do Paraná - Departamento de Engenharia Mecânica - Curitiba/PR - Brazil

Luciano Kiyoshi Araki
Universidade Federal do Paraná - Departamento de Engenharia Mecânica - Curitiba/PR - Brazil
Caixa Postal: 19.040 | CEP: 81.531-980 Curitiba/PR - Brazil

Carlos Henrique Marchi
Instituto de Aeronáutica e Espaço - São José dos Campos/SP - Brazil

Sônia Fonseca Costa e Silva
Instituto de Aeronáutica e Espaço - São José dos Campos/SP - Brazil
Praça Marechal Eduardo Gomes, 50 - Vila das Acácias | CEP: 12.228-904 São José dos Campos/SP - Brazil

Humberto Araújo Machado
Instituto de Aeronáutica e Espaço - São José dos Campos/SP - Brazil
Universidade Estadual do Rio de Janeiro - Resende/RJ - Brazil

Edison Bittencourt
Universidade Estadual de Campinas - Campinas/SP - Brazil

Liu Rong-qiang
Harbin Institute of Technology - Harbin, China

Zhang Jing
Harbin Institute of Technology - Harbin, China
School of Mechatronics Engineering | Harbin Institute of Technology | 92 West Dazhi Street | Nan Gang District | Post Code 150001 Harbin - China

Guo Hong-wei
Harbin Institute of Technology - Harbin, China

Deng Zong-quan
Harbin Institute of Technology - Harbin, China

William Walton Vaughan
University of Alabama - Huntsville/AL - USA
301 Sparkman Drive Northwest | 35899 Huntsville/AL - USA

Dale Leroy Johnson
NASA (retired) - USA

Rajkumar Sureshchandra Pant
Indian Institute of Technology Bombay - Mumbai/Maharashtra - India
Adi Shankaracharya Marg, Powai | 400076 Mumbai/Maharashtra - India

Halil Ersin Söken
Graduate University for Advanced Studies - Sagamihara/Kanagawa - Japan

Chingiz Hajiyev
Istanbul Technical University - Istanbul - Turquey
Aeronautics and Astronautics Faculty | Istanbul Technical University | Maslak, 34469 Istanbul - Turkey

Leandro Dantas Santana
Escola de Engenharia de São Carlos - São Carlos/SP - Brazil
Avenida Trabalhador São Carlense, 400 - Centro | CEP: 13566-590 São Carlos/SP - Brazil

Micael Carmo
Empresa Brasileira de Aeronáutica - São José dos Campos/SP - Brazil

Fernando Martini Catalano
Escola de Engenharia de São Carlos - São Carlos/SP - Brazil

Kuizhi Yue
Beijing University of Aeronautics and Astronautics - School of Aeronautic Science and Engineering - Beijing 100191China
Naval Aeronautical and Astronautical University - Department of Airborne Vehicle Engineering - Yantai 264001 - China

Bing Hu
Beijing University of Aeronautics and Astronautics - School of Aeronautic Science and Engineering - Beijing 100191China

Weigang Guo
Beijing University of Aeronautics and Astronautics - School of Aeronautic Science and Engineering - Beijing 100191China

Yunliang Wang
Naval Aeronautical and Astronautical University - Department of Airborne Vehicle Engineering - Yantai 264001 - China

Hamid Farrokhfal
Malek-Ashtar University of Technology - Isfahan - Iran
Isfahan University of Technology - Isfahan - Iran
Department of Mechanical & Aerospace Engineering | Malek-Ashtar University of Technology - 83145-115 | Shahinshahr - Isfahan - Iran

Ahmad Reza Pishevar
Isfahan University of Technology - Isfahan - Iran

Marcos Aurelio Ortega
Instituto Tecnologico de Aeronautica - Sao Jose dos Campos/SP - Brazil
Department of Aerodynamics - Division of Aeronautical Engineering - ITA, CEP: 12.228-001 - Sao Jose dos Campos/ SP | Brazil

Roberto da Mota Girardi
Instituto Tecnologico de Aeronautica - Sao Jose dos Campos/SP - Brazil

Jorge Hugo Silvestrini
Pontificia Universidade Catolica do Rio Grande do Sul - Porto Alegre/RS - Brazil